Lecture Notes in Mathematics

Edited by A. Dold and B. Eckmann

Subseries: Department of Mathematics, University of Maryland
Adviser: J. Alexander

1077

Lie Group Representations III

Proceedings of the Special Year
held at the University of Maryland, College Park 1982–1983

Edited by R. Herb, R. Johnson, R. Lipsman and J. Rosenberg

Springer-Verlag
Berlin Heidelberg New York Tokyo 1984

Editors
Rebecca Herb
Raymond Johnson
Ronald Lipsman
Jonathan Rosenberg
Department of Mathematics, University of Maryland
College Park, Maryland 20742, USA

AMS Subject Classification (1980): 22E25, 22E45, 22E70, 35H05, 58G35

ISBN 3-540-13385-2 Springer-Verlag Berlin Heidelberg New York Tokyo
ISBN 0-387-13385-2 Springer-Verlag New York Heidelberg Berlin Tokyo

Printing and binding: Beltz Offsetdruck, Hemsbach/Bergstr.
2146/3140-543210

Dedicated to the Memory of

Harish-Chandra

1923-1983

PREFACE

The Department of Mathematics of the University of Maryland conducted a Special Year in Lie Group Represenations during the academic year 1982-1983. This volume is the last (of three) containing articles submitted by the main speakers during the Special Year. Most of the invited speakers submitted articles, and virtually all of those appearing here deal with the subject matter on which the authors lectured during their visits to Maryland.

The Special Year program at Maryland represents a thriving departmental tradition—this being the fourteenth consecutive year in which such an event has taken place. As usual, the subject matter was chosen on the basis of active current research and the interests of departmental members. The modern theory of Lie Group Representations is a vast subject. In order to keep the program within bounds, the Special Year was planned around five distinct intensive periods of activity—each one (of three weeks duration) devoted to one of the main branches of current research in the subject. During those periods (approximately) eight distinguished researchers were invited to present lecture series on areas of current interest. Each visitor spent 1-3 weeks in the department and gave 2-5 lectures. In addition, during each period approximately 8-10 other visitors received financial support in order to attend and participate in the Special Year activities. Thus each period had to some extent the flavor of a mini-conference; but the length of the periods, the fact that visitors were provided with office space and the (relatively) low number of lectures per day also left ample time for private discussion and created the atmosphere of "departmental visitor" rather than "conference participant." Furthermore, as part of the Special Year the department was fortunate to have in residence D. Barbasch, J. Berstein and J-L. Brylinski for the Fall 1982 semester, and B. Blank for the Spring 1983 semester. These visitors ran semester-long seminars in Group

Representations. All of the activities of the Special Year were enthusiastically supported by the department, its faculty and graduate students.

Although most of the cost of the Special Year was borne by the department, the NSF did provide a generous amount of supplementary support. In particular, the contributions to the additional visitors were entirely funded by NSF. The Mathematics Department is grateful to the Foundation for its support of the Special Year. The Organizing Committee would also like to express its gratitude to the Department for its support. In particular the splendid efforts of Professors W. Kirwan, J. Osborn, G. Lehner, as well as of N. Lindley, D. Kennedy, D. Forbes, M. Keimig and J. Cooper were vital to the success of the Special Year. The outstanding job of preparation of manuscripts by June Slack, Anne Eberly, Linda Fiori and Vicki Hendricks was of immense help in producing this volume so quickly. Also we are grateful to Springer-Verlag for its cooperation. Finally we are very pleased that so many of our participants provided us with high quality manuscripts, neatly prepared and submitted on time. It is our conviction that the theory of Group Representations has profited greatly from the efforts of all the above people towards the Special Year.

The Editors
February 1984

INTRODUCTION

We have made a serious attempt to group the papers (within the three volumes) according to the Periods in which they were presented and according to subject matter. However we were also influenced by the time at which manuscripts became available, and by a desire to equalize the size of the volumes. This (third) volume contains papers from Periods III, IV and V of the Special Year. The programs for these periods were as follows:

PERIOD III. Analytic Aspects of Semisimple Theory—Invariant Eigendistributions, L^p-Analysis, Schwartz Space, Irreducibility Criteria, Inversion Theorems, Semisimple Symmetric Spaces, Geometric Realization of Unitary Representations

M. Flensted-Jensen	--Harmonic analysis on semisimple symmetric spaces—a method of duality
S. Helgason	--Wave equations on homogeneous spaces
A. Knapp	--Unitary representations and basic cases
P. Sally	--Tempered spectrum of SL(n) over a p-adic field
V.S. Varadarajan	--Eigenvalues and eigenfunctions on homogeneous spaces
G. Warner	--Toward the trace formula
G. Zuckerman	--Quantum physics and semisimple symmetric spaces

PERIOD IV. Orbit Method—Non-semisimple Groups, Orbital Description of Ingredients of Harmonic Analysis: Parameterization and Construction of Irreducible Representations, Characters and Plancherel Measure; Work of Duflo, Non-type I Groups

L. Corwin	--Primary projections on nilmanifolds
M. Cowling	--Some explicit intertwining operator calculations
M. Duflo	--Plancherel formula for connected type I Lie groups
R. Howe	--Symbols and orbits
H. Moscovici	--Elliptic systems and Hecke operators
R. Penney	--Applications of Kirillov theory to problems in PDE and geometry
L. Pukanszky	--Generalized symplectic geometry and unitary representations
W. Rossman	--Characters as contour integrals
M. Vergne	--Poisson-Plancherel formulas: Equivariant index and Kirillov's character formula

PERIOD V. Applications—Solvability and Hypoellipticity Criteria for Invariant Differential Operators on Lie Groups and Homogeneous Spaces, Use of Nilpotent Groups in Classical Analysis, Use of Compact Homogeneous Spaces as Testing Grounds for Problems in Differential Geometry

L. Corwin	--Solvability of left invariant operators on nilpotent Lie groups
B. Helffer	--Maximal hypoellipticity for operators constructed from vector fields
S. Helgason	--Radon transforms and differential equations
R. Howe	--Symbols and orbits
A. Koranyi	--Geometric analysis on Heisenberg type groups
H. Moscovici	--Non-vanishing results for middle L^2-cohomology of arithmetic groups
R. Penney	--Non-hypoelliptic boundary Laplacians on domains in C^n
L. Rothschild	--Analyticity of solutions of partial differential equations on nilpotent Lie groups

The additional participants during these periods of the Special Year were:

III. D. Barbasch
D. Collingwood
J. Kolk
R. Kunze
B. Orsted
R. Stanton
E. van den Ban
J. Wolf

IV. M. Andler
R. Boyer
P. Dourmashkin
H. Fujiwara
E. Gootman
K. Kumahara
L. Mantini
N. Pedersen
R. Urwin

V. W. Chang
J. Cygan
J. Faraut
D. Geller
P. Godin
R. Goodman
F. Greenleaf
K. Johnson
W. Lichtenstein
D. Mueller
M. Picardello
A. Valette

SPECIAL YEAR DATA

A. The five periods of activity of the Special Year were as follows:

I. Algebraic Aspects of Semisimple Theory -- Sept 7, 1982 - Oct. 1, 1982
II. The Langlands Program -- Nov. 1, 1982 - Nov. 19, 1982
III. Analytic Aspects of Semisimple Theory -- Jan. 24, 1983 - Feb. 11, 1983
IV. The Orbit Method -- Feb. 28, 1983 - March 18, 1983
V. Applications -- April 18, 1983 - May 6, 1983

B. The speakers and the dates of their visits were:

Period I

Thomas Enright, UCSD (9/9 - 9/22)
Anthony Joseph, Weizmann Institute (9/21 - 9/25)
Bertram Kostant, MIT (9/7 - 9/14)
George Lusztig, MIT (9/7 - 9/11)
Wilfried Schmid, Harvard (9/13 - 9/18)
David Vogan, MIT (9/27 - 10/1)
Nolan Wallach, Rutgers (9/20 - 10/1)

Period II

James Arthur, Toronto (11/1 - 11/19)
William Casselman, British Columbia (11/3 - 11/12)
Stephen Gelbart, Cornell (11/1 - 11/12)
Roger Howe, Yale (11/8 - 11/12)
Hervé Jacquet, Columbia (11/1 - 11/12)
David Kazhdan, Harvard (11/1 - 11/12)
Robert Langlands, IAS (11/1 - 11/12)
Ilya Piatetski-Shapiro, Yale (11/1 - 11/12)

Period III

Mogens Flensted-Jensen, Copenhagen (1/24 - 2/11)
Sigurdur Helgason, MIT (1/24 - 1/28)
Anthony Knapp, Cornell (2/2 - 2/4)
Paul Sally, Chicago (1/24 - 2/11)
V.S. Varadarajan, UCLA (1/24 - 2/11)
Garth Warner, Washington (2/7 - 2/8)
Gregg Zuckerman, Yale (1/24 - 2/4)

Period IV

Lawrence Corwin, Rutgers (3/7 - 3/11)
Michael Cowling, Genova (3/2 - 3/4)
Michel Duflo, Paris (2/28 - 3/11)
Roger Howe, Yale (3/7 - 3/11)
Henri Moscovici, Ohio State (3/7 - 3/18)
Richard Penney, Purdue (3/7 - 3/11)
Lajos Pukanszky, Penn (3/7, 3/11 - 3/18)
Wulf Rossman, Ottawa (2/28 - 3/4)
Michèle Vergne, MIT (3/3 - 3/15)

Period V

Lawrence Corwin, Rutgers (4/18 - 4/29)
Bernard Helffer, Nantes (4/18 - 5/6)
Sigurdur Helgason, MIT (4/18 - 4/22)
Roger Howe, Yale (4/18 - 4/22)
Adam Koranyi, Washington Univ. (4.18 - 5/6)
Henri Moscovici, Ohio State (4/25 - 4/30)
Richard Penney, Purdue (4/25 - 5/6)
Linda Rothschild, Wisconsin (4/18 - 4/22)

C. The Orgainizing Committee for the 1982-1983 Special Year in Lie Group Represenations is

Rebecca Herb
Raymond Johnson
Stephen Kudla
Ronald Lipsman (Chairman)
Jonathan Rosenberg

TABLE OF CONTENTS

Matrix Coefficients of Nilpotent Lie Groups

Lawrence Corwin

Let G be a unimodular locally compact group, and let π be an irreducible unitary representation of G on a Hilbert space $\mathcal{H}_\pi = \mathcal{H}$. A <u>matrix coefficient</u> of π is a function of the form

$$f_{v,w}(x) = \langle\pi(x)v,w\rangle,$$

where $v,w \in \mathcal{H}$. (In what follows, we adopt the convention that v and w are never 0 unless the case $v = 0$ (or $w = 0$) is trivially needed to make a result nontrivially true.) The function $f_{v,w}$ is obviously constant on cosets of $G_\pi = \operatorname{Ker}\pi$.

We say that π has L^p matrix coefficients $(1\leq p<\infty)$ if there are (nonzero!) vectors v,w such that $f_{v,w} \in L^p(G/G_\pi)$. Of course, all matrix coefficients are in L^∞. It is easy to check that if $f_{v,w} \in L^p(G/G_\pi)$, then so are $f_{\pi(x)v,w}$ and $f_{v,\pi(x)w}$ for every $x \in G$; moreover, if v is fixed, then $\{w: f_{v,w} \in L^p(G/G_\pi)\}$ and $\{w: f_{w,v} \in L^p(G/G_\pi)\}$ are subspaces of $\mathcal{H}$. Hence if π has L^p matrix coefficients, then there are dense subspaces V,W of $\mathcal{H}$ such that $f_{v,w} \in L^p(G/G_\pi)$ for all $v \in V$ and all $w \in W$. For $p = 2$, more is true: if one matrix coefficient is in L^2 and if G/G_π is unimodular, then all matrix coefficients are L^2. (See, e.g., [3], p. 278.)

Now suppose that G is a nilpotent, connected, simply connected Lie group with Lie algebra $\mathfrak{g}$. There seem to be two main papers in the literature dealing with matrix coefficients of G: Moore-Wolf ([6]) and Howe-Moore ([5]). It is not surprising, therefore, that the work I am going to describe is the result of a collaboration with Calvin Moore. (The work is still in progress; this is a preliminary report.)

The paper [6] is devoted to a study of square integrable representations (the case $p = 2$). The basic theorem is:

Theorem A. Let $\pi \in G^{\wedge}$, and let $\mathcal{O}_\pi$ be the Kirillov orbit in $\mathfrak{g}^*$ corresponding to π. Then π has L^2 matrix coefficients iff $\mathcal{O}_\pi$ is flat (a coset of a subspace of $\mathfrak{g}^*$). Equivalently: π has L^2 matrix coefficients iff for any $\ell_0 \in \mathcal{O}_\pi$, $\mathcal{O}_\pi = \ell_0 + R_{\ell_0}^{\perp}$. Here, R_{ℓ_0} = radical of $\ell_0 = \{X \in \mathfrak{g}: \ell_0([X,Y]) = 0$ for all $Y \in \mathfrak{g}\}$.)

One obvious question is whether these representations have matrix coefficients that are better than L^2. We have the following answer:

Theorem 1. Suppose that π has L^2 matrix coefficients (here and below, π is an irreducible unitary representation of the connected, simply connected nilpotent Lie group G). Let π act on H, and let v,w be C^∞ vectors for π. Then $f_{v,w}$ is a Schwartz class function on G/G_π.

Theorem 1 might be interpreted as saying that the matrix coefficients of π are in L^ε for all $\varepsilon > 0$. Incidentally, there is a similar theorem for p-adic nilpotent Lie groups; see [7].

Now suppose that π is not square integrable. Let $\tilde{G}_\pi$ be the projective kernel of π (i.e., $x \in \tilde{G}_\pi \Leftrightarrow \pi(x)$ is a multiple of I); then $|f_{v,w}|$ is constant on cosets of $\tilde{G}_\pi$, and we can ask about the behavior of $|f_{v,w}|$ on $G/\tilde{G}_\pi$. (In the groups we care about, $\tilde{G}_\pi/G_\pi \cong T^1$, so that the distinction between $\tilde{G}_\pi$ and G_π is not too important.) In [4], the following result was proved:

Theorem B. For all $v,w \in H$, $|f_{v,w}|$ vanishes at ∞ on $G/\tilde{G}_\pi$.

We can improve on this somewhat. The group $\tilde{G}_\pi$ is connected; let its Lie algebra be $\tilde{\mathfrak{g}}_\pi$. Let $\overline{G}_\pi = G/\tilde{G}_\pi$, $\overline{\mathfrak{g}}_\pi = \mathfrak{g}/\tilde{\mathfrak{g}}_\pi$. Choose a Euclidean norm, $|\ |$, on $\overline{\mathfrak{g}}_\pi$, and lift it back to $\overline{G}_\pi$ via exp.

Theorem 2. There are vectors v,w, and constants C, $\gamma > 0$ such that on $\overline{G}_\pi$,

$$|f_{v,w}(x)| \le C(1+|x|)^{-\gamma}.$$

To see why Theorem 1 is true, look at the 3-dimensional Heisenberg group H. Its infinite-dimensional representations are all the same (up to an automorphism of H_1). Write a typical element of H_1 as (x,y,z); the elements $(0,0,z)$ are central, and π can be realized on $L^2(\mathbb{R})$ by

$$(\pi(x,y,z)\phi)(t) = e^{i(z+ty)}\phi(t+x).$$

Write (x,y) for $(x,y,0)$; we really need to examine

$$F_{\phi,\psi}(x,y) = \int_{\mathbb{R}} e^{ity}\phi(t+x)\psi(x)\,dt$$

as a function on R^2. Now $(x,t) \longrightarrow \phi(t+x)\psi(t)$ is obviously in $S(\mathbb{R}^2)$ if ϕ and ψ are, and $f_{\phi,\psi}$ is obtained by a partial Fourier transform. Thus $f_{\phi,\psi} \in S(\mathbb{R}^2)$.

This example is slightly simpler than the general case, but not much. We actually prove a slightly stronger statement than Theorem 1:

<u>Theorem 1'</u>: Let H_π^∞ be the space of C^∞ vectors of π (with the standard Frechet space topology). Then the map

$$(v,w) \to f_{v,w}$$

is a C^∞ map of $H_\pi^\infty \times H_\pi^\infty$ into $S(G/G_\pi)$.

The proof is by induction (mathematical and representation - theoretic). We may assume that G_π is discrete, so that G has 1-dimensional center. Then (unless $G \cong \mathbb{R}$) G has a subgroup G_o of codimension 1 such that:

(1) π is induced from a square integrable representation π_o of G_o;

(2) G_o has a 2-dimensional center.

Assume the result for π_o. One can write down the action of π in terms of that of π_o, and this makes it fairly easy to check that $f_{v,w}$ is in $S(G/G_\pi)$. The continuity of the map $(v,w) \to f_{v,w}$ is then an application of the Uniform Boundedness and Closed Graph Theorems. (The full details of this and the other proofs will appear in a future paper.)

Theorem 2 is harder. The proof runs as follows: we may realize π on $L^2(\mathbb{R}^k)$ for some k by

$$(1) \qquad \pi(x)\phi(t) = e^{iP(x,t)}(\phi \circ \Omega)(x,t),$$

where P and Q are polynomials (and G is appropriately parametrized). Now restrict x to a cross-section for $\bar{G}_\pi$; we may arrange matters so that $P(x,0) = 0$. Write

$$(2) \qquad P(x,t) = \sum_\alpha p_\alpha(x) t^\alpha,$$

where α is a multi- index; similarly, write

$$(3) \qquad Q(x,t) = (Q_1(x,t),\ldots,Q_k(x,t)), Q_j(x,t) = \sum_\alpha q_{j,\alpha}(x) t^\alpha$$

and define

$$\|P(x)\|^2 = \sum_\alpha p_\alpha(x)^2, \quad \|Q(x)\|^2 = \sum_{j,\alpha} q_{j,\alpha}(x)^2.$$

There are three main steps in the argument, with similar statements but quite dissimilar proofs: for appropriate ϕ,ψ, there are positive constants C_j, γ_j $(1 < j < 3)$ such that

$$\begin{cases} \text{(a)} & |f_{\phi,\psi}(x)| \le C_1(1+\|P(x)\|)^{-\gamma_1}; \\ \text{(b)} & |f_{\phi,\psi}(x)| \le C_2(1+\|Q(x)\|)^{-\gamma_2}; \\ \text{(c)} & 1 + \|P(x)\|^2 + \|Q(x)\|^2 \ge C_3(1+|x|)^{\gamma_3}. \end{cases}$$

Together, these statements give the theorem.

For (a), begin by assuming that $k = 1$. (The general case can be reduced to this one by a lemma to be described below.) Since $P(0) = 0$, we can replace $\|P(x)\|$ by $\|P'(x)\|$ (where $P'(x,t) = \frac{\partial P}{\partial t}(x,t)$) in the estimate. A little work then shows that what we really need is an estimate of the following sort: if $F(t) = \sum_{j=0}^n a_j t^{j+1}/(j+1)$, then

$$\left|\int_{-1}^1 e^{iF(t)}dt\right| < C_1(1+\|a\|)^{-\gamma_1}, \quad \|a\|^2 = \sum_{j=0}^n a_j^2.$$

To get an idea of why this sort of estimate ought to be true, consider the estimate from stationary phase. Fix F, and suppose that F' has a root of multiplicity h in $[-1,1]$, but no root of higher multiplicity. Then

$$\left|\int_{-1}^{1} e^{iyF(t)}dt\right| = O(|y|^{-\frac{1}{h+1}}) \quad \text{as} \quad y \to \infty.$$

This sort of result doesn't do us any good, because the coefficients of F are all varying, but a similar sort of estimate does apply. Let $\alpha_1,\ldots,\alpha_n$ be the roots of F'; suppose that $\alpha_1,\ldots,\alpha_h$ are "small" (near $[-1,1]$) and that the rest are "large." Let R be the product of the absolute values of the large roots. Then

$$R \approx |{}^{a}h/a_n|.$$

Why? Recall that $|a_h/a_n|$ is the sum of all products of $(n-h)$ roots of F; but every term in the sum except R has a small root as a factor. A similar sort of argument shows that

$$R \approx \|a\|/|a_n|.$$

Now choose ε in a way to be described below, and snip some intervals out of $[-1,1]$ containing all points within ε of a root. We can do this so that the total length is $\leq 2h\varepsilon$; the integral over these intervals has absolute value $\leq 2h\varepsilon$. The remaining piece of $[-1,1]$ is a union of $\leq n+1$ intervals on which F' doesn't change sign. A simple integration by parts shows that if $[b_1,b_2]$ is such an interval, then

$$\left|\int_{b_1}^{b_2} e^{iF(t)}dt\right| < 2|F'(b_1)^{-1}| + 2|F'(b_2)|^{-1}.$$

It's easy to estimate $|F'(b_1)| = |a_n|\prod_{j=1}^{n}|b_1-\alpha_j|$: the first h terms are $\geq \varepsilon$, and $|b_1-\alpha_j| \approx |\alpha_j|$ for $j > h$, (since $|b_1| < 1$ and α_j

is big). So

$$|F'(b_1)|^{-1} \le (\varepsilon^h|a_n|R) \approx (\varepsilon^h\|a\|)^{-1}.$$

Thus

$$\left|\int_{-1}^{1} e^{iF(t)}dt\right| < C_1'(\varepsilon^h\|a\|)^{-1}.$$

Set $\varepsilon = \|a\|^{-\frac{1}{(h+1)}}$ and note that $h \le n$:

$$\left|\int_{-1}^{1} e^{iF(t)}dt\right| < C_1\|a\|^{-\frac{1}{(n+1)}}.$$

That's (a).

For (b), again assume $k = 1$; assume also that ϕ, ψ are bounded in absolute value by 1 and have support in $[-1,1]$. We're looking at

$$\int_{-1}^{1} e^{iP(t)}\phi(Q(x,t))\overline{\psi(t)}dt.$$

The integrand is ≤ 1 in absolute value and is 0 unless $|t| \le 1$ and $|Q(x,t)| \le 1$. This suggests an approach: for $F(t) = \sum_{j=0}^{n} a_j t^j$, let

$$E_F = \{t\colon |t| \le 1, |F(t)| \le 1\}, \quad \lambda_F = \text{measure of } E_F.$$

If we can show that $\exists c_n, \varepsilon_n > 0$:

(4) $$\lambda_F \le c_n(1+\|a\|)^{-\varepsilon_n},$$

then (b) holds.

We prove (4) by induction; in fact, we can let $\varepsilon_n = \frac{1}{n+1}$. For $n = 0$, this is trivial: $E_F = \phi$ if $\|a\| > 1$. In general, we need only worry about large values of $\|a\|$. Assume the result for degree $n - 1$, and let F be as above. If $|a_0| > 1 + \sum_{j=1}^{n}|a_j|$, then $E_F = \phi$ and we are done. Otherwise, we have $F'(t) = \sum_{j=0}^{n-1} b_j t^j$, and $\|b\|$ and $\|a\|$

are comparable in size. The set E_F is composed of at most n disjoint closed intervals (since F has an extremum between any two such). Choose $K > 0$, and look at one of the intervals in E_F. It is composed of at most $(n-1)$ disjoint closed intervals on which $|F'(t)| \le K$, plus at most n disjoint open intervals where $|F'(t)| > K$. These latter each have length $< 2/K$ (by the Mean Value Theorem), and we can use the inductive hypothesis to bound the length of the other intervals. It turns out that

$$\lambda_n < n\left(2n/K + c_{n-1}K^{\frac{1}{n}}(K+\|b\|)^{-\frac{1}{n}}\right).$$

Choose $K = (1+\|b\|)^{\frac{1}{n+1}}$ to get the result.

That leaves (c). A theorem (proved, e.g., in [4]) says that (c) holds once $\|P(x)\| + \|Q(x)\| \to \infty$ as $\|x\| \to \infty$. But if we can choose a sequence $x_n \to \infty$ with $\|P(x_n)\|$ and $\|Q(x_n)\|$ bounded, then (by passing to a subsequence) we can make $\pi(x_n)\phi$ converge to a nonzero vector ϕ_o. Then

$$<\pi(x_n)\phi,\phi_o> \not\to 0,$$

contradicting Theorem B. That almost finishes the proof.

We still need to remove the hypothesis that $k = 1$. Write $t = (t_1,t')$, with $t' \in \mathbb{R}^{k-1}$, and write $P(x,t) = \sum_j P_j(x,t')t_1^j$, $Q_i(x,t) = \sum_j Q_{i_j}(x,t')t_1^j$. Suppose that $\exists j$:

$$\inf_{|t'|_\infty \le 1} |P_j(x,t')| > C^{-1}\|P(x)\|$$

for some constant C depending only on k and deg P. Then we could prove the results for P just as in (a) above, using t_1 for t. Unfortunately, j need not exist. However, it turns out that we can reduce to this special case by an adroit change of variables. The situation for Q is essentially the same. The idea behind this reduction is not too abstruse, but the current proof is a notational nightmare.

Theorem 2 is seriously unsatisfactory in one respect: the constant γ is not specified, and we have therefore not produced a definite p for which some matrix coefficient is in L^p. The problem is in step (c) of the proof. The proof in [4] actually shows that for given P and Q, the constant γ_3 is effectively computable. But it does not let us know, for instance, whether for a given G there exists a single p such that every irreducible unitary representation has L^p coefficients.

To get this better result with the above techniques, one needs to add more information about the group G. The following result gives the sort of information needed.

<u>Proposition 1.</u> Let $\bar{G}_\pi$ have dimension n. Then one can coordinatize $\bar{G}_\pi$ so that the realization of π given in formula (1) has the following property: write P,Q as in formulas (2) and (3). Then for each j, $1 \le j \le n$, there is an α such that one of $p_\alpha(x)$, $q_{1,\alpha}(x),\ldots,$ $q_{n,\alpha}(x)$ has the form

$$x_j + \text{polynomial in } x_{j+1},\ldots,x_n.$$

The point of Proposition 1 is that it gives an easy estimate of γ_3. Suppose that $\|P(x)\|^2 + \|Q(x)\|^2 \le K^2$, and that all coefficients are of degree $\le r$. Then $|x_n| \le K$ (apply the proposition for $j = n$). Similarly

$$|x_{n-1} + g(x_n)| \le K,$$

where g is some polynomial of degree $\le r$; thus (up to a constant factor independent of K) $|x_{n-1}| \le K^r$. By a simple induction, we get $\gamma_3 \ge r^{-n}$. It follows easily that one can take $p \ge r^n(n+1)$. Since $r \le n^2$, and $\dim \tilde{G}_\pi > 1$, we get

<u>Theorem 3.</u> If $\bar{G}_\pi$ has dimension n, then π has L^p matrix coeffi-

cients for $p > n^{2n}(n+1)$.

Proposition 1 is proved by induction on n. We may assume that $G_\pi^\sim$ has dimension 1. The representation π is induced from a representation σ on a subgroup H of codimension 1, and (as in the proof of Theorem 1) we can write down π fairly explicitly in terms of σ. Once one tracks down how π acts on H_σ, the theorem emerges. As with many other proofs here, the notation is one of the major difficulties.

Once we have Theorem 3, we can ask how good an estimate it gives for the "best possible" p. The clue that is most easily available suggests that the estimate is poor. In [5], Howe and Moore prove that the matrix coefficients of π vanish at ∞ by showing that a kth tensor power of π is a subrepresentation of a cyclic representation with L^2 matrix coefficients. This suggests that π has matrix coefficients in L^{2k} (but does not prove it, since the larger representation is reducible). Howe and Moore also point out that one can take $k \le 2(n+1)^2$. (They give a better bound, which depends on some constructions in their paper.) So Theorem 3 can probably be improved.

Theorem 3 can also be improved if we look at special classes of nilpotent Lie groups. The most obvious example of this is Theorem 1; here is another.

<u>Theorem 4</u>. Suppose π is induced from a 1-dimensional representation of a <u>normal</u> subgroup M. Then π has L^p matrix coefficients whenever $p > (\dim G/M)^2 (\dim M/G_\pi^\sim)$.

The proof is a greatly simplified version of that of Theorem 2. Write a typical element of $\overline{G}$ as (x,y), $x \in M/G_\pi^\sim$ and y in a cross-section of G/M. Then one checks that:

(a) There is a constant C such that if $|y| > C$, then some coordinate of $Q(x,y,t)$ is > 1 in absolute value whenever $|t| \le 1$. That is, $|f_{\phi,\psi}(x,y)| = 0$ when $|y| > C$.

(b) $P(x,y,t)$ is linear in x, and when $|y| < C$, then $\|P(x,y)\| \asymp |x|$ uniformly (i.e., each is bounded by a constant times the other).

Now it's easy to estimate γ in Theorem 2: $\gamma = 1/\deg P$, and crude estimates give $\deg P \leq \dim(G/M)^2$. (Of course, deg P is usually smaller, so that the proof gives a better estimate.)

How good is Theorem 4? Here are two examples (for which we can actually compute some $f_{\phi,\psi}$'s).

1. Let $\mathfrak{g}$ be spanned by $X, Y_1, \ldots, Y_n$, with

$$[X,Y_1] = Y_2, \quad [X,Y_2] = Y_3, \ldots, [X,Y_{n-1}] = Y_n.$$

Assume that π is nontrivial on Y_n. Then Theorem 4 (or the better estimate involving deg P) implies that π has L^p matrix coefficients for $p > (n-1)^2$. The representation π can be realized on $L^2(\mathbb{R})$, and for $n = 3$, one can compute $f_{\phi,\phi}$ explicitly for $\phi(t) = e^{-t^2}$. It turns out that $f_{\phi,\phi} \in L^{4+\varepsilon}$. So if you believe that ϕ is the generic good vector in $L^2(\mathbb{R})$, then the estimate is sharp. (For $n > 3$, explicit calculations soon mire one in special functions.)

2. Let $\mathfrak{g}$ = algebra of all 4×4 strictly upper triangular matrices. If ℓ is in general position (nontrivial on the center), then one can realize π on $L^2(\mathbb{R}^2)$ so that Theorem 4 applies; the polynomial P is of degree 2 in t. Thus Theorem 3 says that π has matrix coefficients in $L^{6+\varepsilon}$. If $\phi(t) = e^{-|t|^2}$, then $f_{\phi,\phi} \in L^{3+\varepsilon}$ (but not L^3); one can squeeze $L^{3+\varepsilon}$ out of the proofs of Theorems 2 and 3 with some extra work.

What about other groups? The situation is (by now) well understood for semisimple groups; see [1] and [2]. The results in [5] are for algebraic groups over local fields of characteristic 0; Moore and I plan to look at such groups once we get nilpotent groups in more comprehensible shape.

Let me close with one other question suggested by our examples. Suppose that π has matrix coefficients in $L^{2+\varepsilon}$ for all $\varepsilon > 0$. Is π then square integrable? I suspect so, but I certainly don't have a proof.

References

1. Borel, A., and Wallach, N., Continuous Cohomology, Discrete Subgroups, and Representations of Reductive Groups. Annals of Mathematics Studies #94, Princeton, Princeton University Press, 1980.

2. Casselman, W., and Miličić, D., Asymptotic Behavior of Matrix Coefficients of Admissible Representations, preprint.

3. Dixmier, J., Les C*-Algebres et Leurs Représentations. Paris, Gauthier-Villars, 1964.

4. Gorin, E., Asymptotic Properties of Polynomials and Algebraic Functions of Several Variables, Uspehi Mat. Nauk 16 (1961), pp. 93-119 (English translation in Russian Mathematical Surveys 16 (1961), pp. 95-119).

5. Howe, R., and Moore, C. C., Asymptotic Properties of Unitary Representations, J. Funct. Anal. 32 (1979), pp. 72-96.

6. Moore, C. C., and Wolf, J., Square Integrable Representations of Nilpotent Lie Groups, Trans. A. M. S. 185 (1973), pp. 445-462.

7. Van Dijk, G., Square-Integrable Representations mod Z of Unipotent Groups, Compositio Math. 29 (1974), pp. 141-150.

Lawrence Corwin
Rutgers University
New Brunswick, N.J.

Primary Projections on Nilmanifolds

Lawrence Corwin

The complex exponential functions $e^{2\pi inx}$ span the irreducible (1-dimensional) subspaces of $L^2(\mathbb{Z}\backslash\mathbb{R})$ which are stable under right translation by elements of $\mathbb{R}$, and Fourier series may be viewed as the decomposition of an L^2 function by the orthogonal projections onto these subspaces. There is a natural analogue of this decomposition for other Lie groups. If G is a Lie group with a discrete, cocompact subgroup Γ, then $\Gamma\backslash G$ has a unique invariant Borel probability measure m, and G acts on $L^2(\Gamma\backslash G) = L^2(\Gamma\backslash G,m)$ unitarily by right translation. Let σ be this unitary representation. It is known [5] that σ is a discrete direct sum of irreducible unitary representations, each with finite multiplicity. Write $\sigma = \bigoplus_{\pi\in G^\wedge} n_\pi \pi$ (here, $G^\wedge$ is the set of [equivalence classes of] irreducible unitary representations of G), and let $[G:\Gamma]^\wedge = \{\pi\in G^\wedge : n_\pi > 0\}$. There is a unique closed subspace H_π of $L^2(\Gamma\backslash G)$ which is stable under π and satisfies $\sigma|_{H_\pi} \cong n_\pi\pi$. Let P_π be the orthogonal projection on H_π. Given $f \in L^2(\Gamma\backslash G,m)$, the functions $P_\pi f$, $\pi \in [G:\Gamma]^\wedge$, may be regarded as the Fourier components of f. Since P_π is usually a projection onto an infinite-dimensional space, these Fourier components generally will not be so well-behaved as the complex exponentials. One natural problem in non-Abelian harmonic analysis is to determine their properties.

In this general setting, not too much is known; the main results can be found in [1]. The C^∞ vectors for σ (those vectors on which the enveloping algebra $U(\mathfrak{g})$ of G acts) are simply the elements of $C^\infty(\Gamma\backslash G)$, and P_π maps $C^\infty(\Gamma\backslash G)$ into $C^\infty(\Gamma\backslash G)$. If $f \in C^\infty(\Gamma\backslash G)$, let $f^y(x) = f(xy)$; it is easy to show that

$$(P_\pi f)(y) = (P_\pi f^y)(e).$$

Thus P_π is determined by D_π, where D_π is the distribution given by

$$(D_\pi, f) = (P_\pi f)(e).$$

We can even estimate the order $\sigma(D_\pi)$ of D_π:

(1) $\sigma(D_\pi) \leq 1 + [\dim G/2]$, $[\ \ ]$ = greatest integer function.

Here is a sketch of the proof: P_π commutes with σ, and hence with $\sigma(X)$, when X is in the Lie algebra $\mathfrak{g}$ of G. Thus if $f \in C^k(\Gamma\backslash G)$, then $P_\pi f$ has k derivatives in $L^2(\Gamma\backslash G)$. Sobolev theory says that $P_\pi f$ is continuous if $k > \frac{n}{2}$, where $n = \dim G$.

For further information about P_π of D_π, we need more information about G; from now on, we assume that G is (connected and) nilpotent. This seems to be the most tractable case. Moore [10] made an excellent start on determining $[G:\Gamma]^\wedge$; later, Howe [6] and Richardson [15] simultaneously and independently determined $[G:\Gamma]^\wedge$ and gave a formula for the n_π. Thus we begin with some useful information.

When G is Abelian, $P_\pi f$ is always an analytic function. In the nilpotent case, people seem to have expected that $P_\pi f$ would be continuous whenever f was — i.e., that D_π would be of order 0 (a measure). Then in [16], Richardson gave an example where D_π was not a measure. Later, Penney [12] showed that D_π is a measure iff π is square integrable mod its kernel, or, equivalently, iff the Kirillov orbit $\mathcal{O}_\pi$ of π is flat. (See [9] or [13] for Kirillov theory; square integrable representations are described in [11].) Beyond that, we have estimates of $\sigma(D_\pi)$ (sharper than (1)) only in special cases.

The standard procedure for estimating $\sigma(D_\pi)$ is to find a formula for D_π and then to determine when it converges. There are a number of different formulas; unfortunately, they give essentially the same estimates in essentially the same cases. Here they are.

1. There is a "canonical formula" for D_π given in [4]. To describe it, one needs first the "canonical objects" associated with π, first defined in [13]. Kirillov theory associates π with an $\mathrm{Ad}^*(G)$-orbit of $\mathfrak{g}$. Let $\mathcal{O}_\pi$ be this orbit, and let $\ell \in \mathcal{O}_\pi$. Define

$\mathfrak{r} = \mathfrak{r}(\ell)$ = radical of $\ell = \{X\in\mathfrak{g}: \ell([X,Y]) = 0$ for all $y \in \mathfrak{g}\}$;

$\mathfrak{h}_1 = \mathfrak{h}_1(\ell)$ = smallest ideal of $\mathfrak{g}$ containing $\mathfrak{r}(\ell)$;

and define $\mathfrak{h}_2, \mathfrak{h}_3, \ldots$ inductively: given $\mathfrak{h}_k$, let

$$\ell_k = \ell|_{\mathfrak{h}_k}, \quad \mathfrak{r}_k = \text{radical of } \ell_k,$$

$$\mathfrak{h}_{k+1} = \text{smallest ideal of } \mathfrak{h}_k \text{ containing } \mathfrak{r}_k.$$

Then $\mathfrak{g} \supseteq \mathfrak{h}_1 \supseteq \mathfrak{h}_2 \supseteq \cdots$; eventually $\mathfrak{h}_k = \mathfrak{h}_{k+1}$, and then $\mathfrak{h}_k = \mathfrak{h}_m$ for all $m \geq k$. Write $\mathfrak{h}_\infty = \mathfrak{h}_\infty(\ell)$ for $\mathfrak{h}_k$, and let $\mathfrak{k}_\infty = \mathfrak{k}_\infty(\ell) = \{X\in\mathfrak{g}: \ell([X,Y]) = 0$ for all $Y \in \mathfrak{h}_\infty(\ell)\}$. Let H_∞, K_∞ be the Lie subgroups of G corresponding to $\mathfrak{h}_\infty$, $\mathfrak{k}_\infty$ respectively. Then

$$\mathfrak{h}_\infty \subseteq \mathfrak{k}_\infty, \text{ and } \mathfrak{h}_\infty \text{ is an ideal of } \mathfrak{k}_\infty;$$

$$\mathfrak{h}_\infty = \mathfrak{r}(\ell|_{\mathfrak{k}_\infty});$$

if σ_ℓ is the representation of K_∞ corresponding to $\ell|_{\mathfrak{k}_\infty}$, then σ_ℓ is scalar on H_∞ and square integrable mod H_∞;

$$\pi_\ell = \mathrm{Ind}_{K_\infty\to G}\sigma_\ell .$$

These objects are "canonical" in that if $\ell' = \mathrm{Ad}^*(x)\ell$, then $\mathfrak{h}_\infty(\ell') = (\mathrm{Ad}\,x)\mathfrak{h}_\infty(\ell)$ and $\mathfrak{k}_\infty(\ell') = (\mathrm{Ad}\,x)\mathfrak{k}_\infty(\ell)$.

Now suppose that $\pi \in [G:\Gamma]^\wedge$, and let $\Lambda = \log\Gamma \subset \mathfrak{g}$. Then one can find $\ell \in \mathcal{O}_\pi$ such that $\ell(\Lambda) \subset \mathbb{Q}$ and $\ell(\Lambda\cap\mathfrak{h}_\infty(\ell)) \subseteq \mathbb{Z}$. Define λ on $H_\infty(\ell)$ by

$$\lambda(\exp Y) = e^{2\pi i \ell(Y)}$$

and define λ^x on $xH_\infty(\ell)x^{-1} = H_\infty^x$ by $\lambda^x(h) = \lambda(x^{-1}hx)$. Define x to be integral if λ^x is trivial on $H_\infty^x \cap \Gamma$. Note that λ^x depends only on the coset $xK_\infty(\ell)$; thus we may talk about integral cosets. The "canonical formula" is

$$(2) \quad (D_\pi, f) = \sum_{x \in G/K_\infty(\ell),\ x \text{ integral}} \int_{H_\infty^x \cap \Gamma \backslash H_\infty^x} \overline{\lambda^x(h)} f(h)\, d\dot{h}.$$

This converges for all $f \in C^\infty(\Gamma\backslash G)$, and the proof of this fact presumably gives an estimate of $\sigma(D_\pi)$; I have never tried to compute this estimate, and I suspect that it is poor. When $\mathfrak{h}_1(\ell) = \mathfrak{h}_\infty(\ell)$, however, one can estimate $\sigma(D_\pi)$ fairly easily:

$$(3) \quad \sigma(D_\pi) \leq [(\mathrm{Dim}\ \mathfrak{h}_1(\ell) - \mathrm{Dim}\ \mathfrak{r}(\ell))/2] + 1 \quad \text{if} \quad \mathfrak{h}_1(\ell) = \mathfrak{h}_\infty(\ell).$$

The idea (also used in [3]) is that the sum in (2) can be transformed into a sum of Fourier coefficients for a function on a torus. Then it is not hard to transform statements about the differentiability of f into statements about the decay rate of the coefficients.

2. There is a different formula, proved in [8], for D_π (or P_π) in the case where $\mathcal{O}_\pi$ has maximal dimension. Let $F \in C^\infty(\mathfrak{g}^*)$ satisfy the following conditions:

(a) All derivatives of F grow at most polynomially;

(b) F is constant on $\mathrm{Ad}^*(G)$-orbits;

(c) F is identically 1 on $\mathcal{O}_\pi$, and is 0 on $\mathcal{O}_\rho$ if $\rho \neq \pi$ and $\rho \in [G:\Gamma]^\wedge$.

Now define a tempered distribution D_F on G as follows:

$$(D_F, \varphi) = \int_{\mathfrak{g}^*} F(\ell)(\varphi \circ \exp)^\wedge(\ell)\, d\ell,$$

where $^\wedge$ is the Fourier transform. It is shown in [7] that convolution by D_F takes $\mathcal{S}(G)$ to $\mathcal{S}(G)$. Given $f \in C^\infty(\Gamma\backslash G)$, regard f as a function on G (by lifting it); then f is also a tempered distribution. Thus there is a distribution $D^{(F)}f$ defined by

$$(D^{(F)}f,\varphi) = (f,D_F*\varphi).$$

One checks that $D^{(F)}f \in C^\infty(G)$ and is constant on the cosets Γ_x; thus $D^{(F)}f$ may be regarded as a function in $C^\infty(\Gamma\backslash G)$. It turns out that (with appropriate normalizations)

(4) $$D^{(F)}f = P_\pi f.$$

This formula yields an order estimate for P_π when $\mathfrak{h}_1(\ell) = \mathfrak{h}_\infty(\ell)$ for $\ell \in \mathcal{O}_\pi$: the estimate is the same as that in (3). If $\mathcal{O}_\pi$ does not have maximal dimension among all $\mathrm{Ad}^*(G)$-orbits in $\mathfrak{g}^*$, then one cannot find F satisfying (a)-(c) above; this procedure, therefore, gives slightly less general results than the first one.

(3) In [17], a third formula for P_π was given; it holds only in certain situations. Suppose that $\mathfrak{h}$ is an ideal which is maximal subordinate (= polarizing) for some $\ell \in \mathcal{O}_\pi$ (and hence every $\ell \in \mathcal{O}_\pi$). We now select a basis $\{X_1,\ldots,X_n\}$ for $\mathfrak{g}$ and an element $\ell_0 \in \mathcal{O}_\pi$ such that

(a) $\Gamma = \exp \mathbb{Z}X_1 \cdots \exp \mathbb{Z}X_n$;

(b) $\mathrm{Span}\{X_1,\ldots,X_j\} = \mathfrak{g}_j$ is an ideal in $\mathfrak{g}$ for each j;

(c) $\mathfrak{g}_k = \mathfrak{h}$ if $k = \dim \mathfrak{h}$;

(d) for each $j \geq k$, there exists $Y \in \mathfrak{h} \cap \Lambda$ such that

$$\ell_0([X_j,Y_j]) \neq 0 \quad \text{and} \quad \ell_0([\mathfrak{g}_{j-1},Y_j]) = 0;$$

(e) $\ell_0(X_j) \in \mathbb{Z}, \quad \forall_j$.

Let $\ell_1,\ldots,\ell_n$ be the basis dual to $X_1,\ldots,X_n$; set

$$T_\pi = \{\ell\in\mathfrak{g}^*: \ell|_{\mathfrak{h}^*} = \ell'|_{\mathfrak{h}^*} \text{ for some } \ell' \in \mathcal{O}_\pi, \text{ and } \ell(X_j) \in \mathbb{Z}, \ \forall_j\}.$$

Now define $W: \mathfrak{g} \to G$ by

$$W\left(\sum_{j=1}^{n} x_j X_j\right) = \exp x_1 X_1 \cdots \exp x_n X_n.$$

Then W is a diffeomorphism, and we may use W to identify G and $\mathfrak{g}$. Given $f \in C^\infty(\Gamma\backslash G)$, choose $f_0 \in S(G)$ such that

$$f(\Gamma x) = \sum_{\gamma\in\Gamma} f_0(x), \quad \forall x \in G.$$

Then

$$(5) \qquad (D_\pi, f) = \sum_{\ell\in\Gamma_\pi} \hat{f_0}(\ell)$$

(where f_0 is regarded as a function on $\mathfrak{g}$ via W). Now an argument similar to the one used for formula (2) yields (3). This estimate, of course, applies only to the π with a polarizing ideal; for all of these π, $\mathfrak{h}_1(\ell) = \mathfrak{h}_\infty(\ell)$. One might be able to extend the estimate in (5) to all π such that $\mathfrak{h}_1(\ell) = \mathfrak{h}_\infty(\ell)$, but it seems doubtful the formula is true in greater generality than that.

There is one other special case where $\sigma(D_\pi)$ can be estimated, and that case is the subject of the rest of this report. Describing it requires more notation and further constructions. Given $\ell \in \mathfrak{g}^*$, we constructed $\mathfrak{h}_1(\ell),\ldots,\mathfrak{h}_\infty(\ell)$. Now let

$\mathfrak{i}_1 = \mathfrak{i}_1(\ell) =$ largest ideal of $\mathfrak{g}$ in $\mathfrak{h}_\infty(\ell)$;

$\mathfrak{g}_1 = \mathfrak{g}_1^{(\ell)} =$ annihilator of $\mathfrak{i}_1$ with respect to ℓ

$= \{X\in\mathfrak{g}: \ell([X,Y]) = 0, \forall Y \in \mathfrak{i}_1\}, \quad \ell_1 = \ell|_{\mathfrak{g}_1},$

and iterate: we get

$\mathfrak{h}_\infty(\ell_1), \mathfrak{i}_2 = \mathfrak{i}_1(\ell_1) =$ largest ideal of $\mathfrak{g}_1$ in $\mathfrak{h}_\infty(\ell_1)$,

$\mathfrak{g}_2 =$ annihilator of $\mathfrak{i}_2$ with respect to ℓ_1,

$\ell_2 = \ell|_{\mathfrak{g}_2},$

and so on. It is easy to check that $\mathfrak{i}_1 \subseteq \mathfrak{i}_2 \subseteq \cdots$ and that $\mathfrak{g}_1 \supseteq \mathfrak{g}_2 \supseteq \cdots$. Furthermore,

$$\mathfrak{i}_i = \mathfrak{i}_{i+1} \Rightarrow \mathfrak{g}_i = \mathfrak{g}_{i+k} \text{ and } \mathfrak{i}_i = \mathfrak{i}_{i+k} \text{ for all } k > 0;$$

$$\mathfrak{g}_i = \mathfrak{g}_{i+1} \Rightarrow \mathfrak{i}_{i+1} = \mathfrak{i}_{i+2}. \tag{6}$$

Thus we may speak of $\mathfrak{i}_\infty = \mathfrak{i}_\infty(\ell)$ and $\mathfrak{g}_\infty = \mathfrak{g}_\infty(\ell)$. All of these objects are "canonical", in that, e.g., $\mathfrak{i}_i(\mathrm{Ad}^* x \circ \ell) = (\mathrm{Ad}\, x)(\mathfrak{i}_i(\ell))$.

One result that makes this construction interesting is the following:

<u>Lemma</u>. If $\mathfrak{r}(\ell)$ is not an ideal, then $\mathfrak{i}_1(\ell) \nsubseteq \mathfrak{r}_0(\ell)$ and $\mathfrak{g}_1(\ell) \neq \mathfrak{g}$.

A corollary is that $\mathfrak{i}_\infty(\ell) = \mathfrak{r}(\ell|_{\mathfrak{g}_\infty(\ell)})$. Thus the construction of [12] applies to $\mathfrak{i}_\infty$, $\mathfrak{g}_\infty$: if J_∞, G_∞ are the corresponding groups, and if τ_ℓ is the representation of G_∞ corresponding to $\ell|_{\mathfrak{g}_\infty}$, then τ_ℓ is scalar on J_∞ and square integrable mod J_∞, and

$$\pi_\ell = \mathrm{Ind}_{G_\infty \to G}\, \sigma_\ell .$$

Moreover, the $\mathfrak{i}_j(\ell)$ and $\mathfrak{g}_j(\ell)$ are rational if ℓ is rational. Now the reasoning of [4] gives a new "canonical formula" for D_π:

$$(D_\pi, f) = \sum_{x \in G/G_\infty(\ell),\, x \text{ integral}} \int_{\Gamma \cap J_\infty^x(\ell) \backslash J_\infty^x(\ell)} \overline{\lambda^x(h)}\, f(h)\, d\dot{h}, \tag{7}$$

where $J_\infty^x(\ell)$, λ^*, and integral x are defined essentially as they were for (2).

Formula (7) has one advantage over (2): in the case where $\mathfrak{g}_1(\ell) = \mathfrak{g}_\infty(\ell)$, it gives an estimate for $\sigma(D_\pi)$. In view of (6), $\mathfrak{g}_1(\ell) = \mathfrak{g}_\infty(\ell) \Rightarrow \mathfrak{i}_2(\ell) = \mathfrak{i}_\infty(\ell)$. If in fact $\mathfrak{i}_1(\ell) = \mathfrak{i}_\infty(\ell)$, then it is easy to check that $\mathfrak{i}_1(\ell) = \mathfrak{h}_\infty(\ell) = \mathfrak{h}_1(\ell)$; thus we are in the case where estimate (3) applies. But if $\mathfrak{i}_1(\ell) \neq \mathfrak{i}_\infty(\ell)$, then the estimate is new.

Theorem. If $\mathfrak{g}_1(\ell) = \mathfrak{g}_\infty(\ell)$, but $\mathfrak{i}_1(\ell) \neq \mathfrak{i}_\infty(\ell)$, then

$$\sigma(D_\pi) \leq 1 + \dim(\mathfrak{g}/\mathfrak{g}_1(\ell)). \tag{8}$$

Here is a sketch of the proof; details will appear in [2]. For simplicity, assume that Λ is a subgroup of $\mathfrak{g}$. Choose a "strong Malcev basis for $\mathfrak{g}$, Λ through $\mathfrak{i}_1(\ell)$"; i.e., choose a basis $\{X_1,\dots,X_n\}$ of $\mathfrak{g}$ such that:

(a) $\mathbb{Z}$-span $(X_1,\dots,X_n) = \Lambda$;

(b) $\mathfrak{h}_j = \text{span}\,(X_1,\dots,X_j)$ is an ideal of $\mathfrak{g}$ for $1 \leq j \leq n$;

(c) $\mathfrak{h}_k = \mathfrak{i}_1$ if $k = \dim \mathfrak{i}_1$.

Let $\{\ell_1,\dots,\ell_n\}$ be the dual basis of $\mathfrak{g}$; give $\mathfrak{g},\mathfrak{g}^*$ Euclidean norms by declaring the X_j and ℓ_j respectively to be orthonormal bases.

Now consider a typical term on the right side of formula (7),

$$\int_{\Gamma\cap J_\infty^x\backslash J_\infty^x} \overline{\lambda^x(h)} f(h)\,\dot{d}h.$$

This integral amounts to an integral over a torus. (If J_∞^x is Abelian, it _is_ an integral over a torus; otherwise one first integrates over cosets of the commutator subgroup.) To make notation simpler, assume that J_∞^x is Abelian. If X is a left invariant vector field on J_∞^x, then integration by parts gives

$$\begin{aligned}\int_{\Gamma\cap J_\infty^x\backslash J_\infty^x} \lambda^x(h)(X^p f)(h) &= (-1)^p \int_{\Gamma\cap J_\infty^x\backslash J_\infty^x} \overline{X^p(\lambda^x)(h)} f(h)\,dh \\ &= (2\pi i \ell^x(X))^p \int_{\Gamma\cap J_\infty^x\backslash J_\infty^x} \overline{\lambda^x(h)} f(h)\,\dot{d}h.\end{aligned}$$

It is easy to show that

$$\max_{\|X\|=1} |\ell^x(X)| = \text{distance from the origin to } (\mathrm{Ad}^x J_\infty^x)\ell^x = d(x), \text{ say.}$$

Thus if f is p times differentiable, the sum in (7) is bounded by

$$(9) \qquad \sum_{x \in G/G_\infty(\ell),\, x \text{ integral}} \|f\|_p \, d(x)^{-p},$$

where $\|f\|_p$ is an appropriate p^{th} Sobolev norm. All we need to do now is to estimate $d(x)$.

To do that, we use a standard theorem on parametrizing $\mathrm{Ad}^*(G)$-orbits (see, e.g., [14], pp. 50 ff). Suppose that $m = \dim \mathcal{O}_\ell$. Then there is a polynomial map $P\colon \mathbb{R}^m \to \mathfrak{g}^*$, $P(t_1,\ldots,t_m) = P(t) = \sum_{j=1}^{n} P_j(t)\ell_j$, such that:

(a) $\mathcal{O}_\ell = \text{Image } P$;

(b) there are indices $j_1,\ldots,j_m$ such that $P_{j_i}(t) = \alpha_i t_i +$ a polynomial in $t_1,\ldots,t_{i-1}$, and α_i is an integer;

(c) if $j_i < j < j_{i+1}$, then P_j depends only on $t_1,\ldots,t_i$.

Furthermore, because $\mathfrak{i}_1$ is one of the $\mathfrak{h}_j$, one can check that for any $x \in G$, there are numbers $u_1,\ldots,u_{i_0}$ $(i_0 = \dim(G/G_\infty(\ell)))$ such that

$$(\mathrm{Ad}^*G_1^x)\ell^x = (\mathrm{Ad}^*x)(\mathrm{Ad}^*G_1)\ell = \{P(u_1,\ldots,u_{i_0},t_{i_0+1},\ldots,t_m):$$
$$t_{i_0+1},\ldots,t_m \in \mathbb{R}\}.$$

This makes it easy to estimate $d(x)$. Let $u(x) = (u_1,\ldots,u_{i_0})$. Then

$$d(x) \geq C_0(1 + |u(x)|),$$

where $|\ |$ is the standard norm and C_0 is some constant. If x is integral, then $\alpha u(x)$ is integral (where $\alpha = (\alpha_1,\ldots,\alpha_{i_0})$). It is now easy to see that (9) converges for $p > \dim G/G_\infty(\ell)$, and the theorem follows.

The obvious question about the theorem is: is it vacuous? The answer is "no". In fact, there have been two standard "test examples" where the previous results did not apply:

<u>Example 1.</u> $\mathfrak{g}$ is the (8-dimensional) Lie algebra of all strictly upper

triangular 5×5 matrices (a_{ij}) with $a_{23} = a_{34}$ and $a_{24} = 0$. Then all $\ell \in \mathfrak{g}^*$ in "general position" have $\mathfrak{h}_1(\ell) \neq \mathfrak{h}_2(\ell) = \mathfrak{h}_\infty(\ell)$.

Example 2. $\mathfrak{g}_n$ is the Lie algebra of all $n \times n$ strictly upper triangular $n \times n$ matrices, and ℓ is defined by

$$\ell(a_{ij}) = a_{1n}.$$

Then ℓ is not in general position (for $n > 4$). Here,

$$\mathfrak{h}_\infty(\ell) = \mathfrak{h}_j(\ell) \neq \mathfrak{h}_{j-1}(\ell), \quad j = [\tfrac{n}{2}].$$

In these examples, however,

$$\mathfrak{g}_1(\ell) = \mathfrak{g}_\infty(\ell),$$

and the above theorem applies. For that matter, $\mathfrak{i}_1(\ell) = \mathfrak{h}_\infty(\ell)$ in these examples. I do not know of an example where that is not the case, though I have no reason to believe that it always holds.

While I do not know of a representation $\pi \in [G:\Gamma]^\wedge$ where the order estimates are not covered by (3) or (8), I certainly believe that such examples exist. I intend to search for some in the near future.

Bibliography

1. Auslander, L., and Brezin, J., Uniform Distribution on Solvmanifolds, Advances in Math. 71 (1971), pp. 111-144.

2. Corwin, L., "Order Estimates for Irreducible Projections in L^2 of a Nilmanifold", preprint.

3. Corwin, L., and Greenleaf, F.P., "Integral Formulas with Distribution Kernels for Irreducible Projections in L^2 of a Nilmanifold", J. Funct. Anal. 23 (1976), pp. 255-284.

4. Corwin, L., Greenleaf, F.P., and Penney, R., "A Canonical Formula for the Distribution Kernels of Primary Projections in L^2 of a Nilmanifold", Comm. Pure. Appl. Math. 30 (1977), pp. 355-372.

5. Gelfand, I.M., Graev, M.I., and Piatetskii-Shapiro, I.I., Representation Theory and Automorphic Functions, Philadelphia, W.B. Saunders, 1969.

6. Howe, R., "On Frobenius Duality for Unipotent Algebraic Groups over Q", Am. J. Math. 93 (1971), pp. 163-172.

7. Howe, R., "On a Connection between Nilpotent Groups and Oscillatory Integrals Associated to Singularities", Pac. J. Math. 73 (1977), pp. 329-364.

8. Jenkins, J., "Primary Projections on L^2 of a Nilmanifold", J. Funct. Anal. 32 (1979), pp. 131-138.

9. Kirillov, A.A., "Unitary Representations of Nilpotent Lie Groups", Uspechi Mat. Nauk 17 (1962), pp. 57-110.

10. Moore, C.C., "Decomposition of Unitary Representations Defined by Discrete Subgroups of Nilpotent Groups", Ann. of Math. 82 (1965), pp. 146-182.

11. Moore, C.C., and Wolf, J., "Square Integrable Representations of Nilpotent Lie Groups", Trans. A.M.S. 185 (1973), pp. 445-462.

12. Penney, R., "Central Idempotent Measures on a Nilmanifold", J. Funct. Anal. 36 (1980), pp. 255-271.

13. Penney, R., "Canonical Objects in Kirillov Theory on Nilpotent Lie Groups", Proc. A.M.S. 66 (1977), pp. 175-178.

14. Pukanszky, L., Leçons sur les Représentations des Groupes, Paris, Dunod, 1967.

15. Richardson, L., "Decomposition of the L^2 space of a General Compact Nilmanifold", Amer. J. Math. 93 (1971), pp. 173-190.

16. Richardson, L., "A Class of Idempotent Measures on Compact Nilmanifolds", Acta. Math. 135 (1975), pp. 129-154.

17. Richardson, L., "Poisson Summation on Kirillov Orbits", Math. Annalen, 239 (1979), pp. 229-240.

Lawrence Corwin
Rutgers University
New Brunswick, N.J.

SOLVABILITY OF LEFT INVARIANT DIFFERENTIAL OPERATORS ON NILPOTENT LIE GROUPS

by

Lawrence Corwin
Rutgers University

What follows is a survey of various conditions (necessary, sufficient, or both) for solvability of left invariant operators on nilpotent Lie groups. It is incomplete for a variety of reasons. First of all, I have dealt almost exclusively with results and techniques that rely on the left invariance and the Lie group structure; many operators can be shown to be solvable because, for example, they are elliptic or strongly hyperbolic, but these considerations are usually ignored below. I have also undoubtedly omitted some aspects of the theory because of carelessness or ignorance, for which I apologize. Finally, some results have been slighted to keep the paper manageable.

I have tried to give correct references to the main results given here. However, the lack of a reference does not necessarily mean that the result is new.

I have benefited especially greatly in writing this account from conversations with two friends, colleagues, and collaborators: Fred Greenleaf and Linda Rothschild.

1. Generalities

For simplicity, G will be a connected unimodular group in what follows; its Lie algebra will be $\mathfrak{G}$.

We let L be a left invariant differential operator on G. As is standard, we say that:

(a) L is <u>locally solvable</u> at $x \in G$ if there is an open neighborhood U of x such that for every $f \in C_c^\infty(U)$ there is a function $u \in C^\infty(U)$ with $Lu = f$ on U;

(b) L is <u>semiglobally solvable</u> on G if for every $U \subseteq G$ with compact support and every $f \in C_c^\infty(U)$ there is a function $u \in C^\infty(U)$ with $Lu = f$ on U;

(c) L is <u>globally solvable on</u> G if $L(C^\infty(G)) = C^\infty(G)$;

(d) L is <u>globally solvable for</u> $\mathcal{D}'(G)$ if $L(\mathcal{D}'(G)) = \mathcal{D}'(G)$;

(e) L <u>has a fundamental solution</u> if there is a distribution ξ such that $L\xi = \delta$ (= point mass at the identity).

If in addition there is a notion of Schwartz class functions on G, so that $\mathcal{S}(G)$ and $\mathcal{S}'(G)$ make sense, then we say that

(f) L has a <u>tempered fundamental</u> solution if there is a distribution $\xi \in \mathcal{S}'(G)$ with $L\xi = \delta$;

(g) L is <u>globally solvable for</u> $\mathcal{S}'(G)$ if $L(\mathcal{S}'(G)) = \mathcal{S}'(G)$.

(Most of these notions can be extended to more general operators on manifolds, but we shall not bother.) There are a number of formal relations among these notions, which we shall now explore.

The left invariance of L can be exploited in two very simple ways. First of all, local solvability of L at x is equivalent to local solvability at e, and we therefore work near e. Second of all, we can use convolutions. Let λ and ρ denote respectively the left and right regular representations of G:

$$(\lambda(x)f)(y) = f(x^{-1}y), \quad (\rho(x)f)(y) = f(yx),$$

and define $f'(x) = f(x^{-1})$. For $f, g \in L^1(G)$, we have

$$(f*g)(x) = \int_G f(xy)g(y^{-1})dy = \int_G f(xy^{-1})g(y)dy \ (= \int_G (\lambda(x^{-1})f)'(y)g(y)dy)$$

$$= \int_G f(y)g(y^{-1}x)dy \ (= \int_G f(y)(\rho(x)g)'(y)dy).$$

Thus if $\xi \in \mathcal{D}'(G)$ and $\varphi \in C_c^\infty(G) = \mathcal{D}(G)$, we define $\xi*\varphi$, $\varphi*\xi$ by

$$(\xi*\varphi)(x) = (\xi,(\rho(x)\varphi)'), \quad (\varphi*\xi)(x) = (\xi,(\lambda(x^{-1})\varphi)').$$

We define convolution similarly on other appropriate spaces ($\xi \in \mathcal{S}'(G)$ and $\varphi \in \mathcal{S}(G)$, for instance). Note, incidentally, that

$$(\rho(x)\varphi)' = \lambda(x)(\varphi'), \quad (\lambda(x)\varphi)' = \rho(x)(\varphi').$$

If L is a left invariant differential operator, then $L \in \mathfrak{U}(\mathfrak{G})$, the universal enveloping algebra of $\mathfrak{G}$. Define the transpose map $L \to L^t$ on $\mathfrak{U}(\mathfrak{G})$ as the unique anti-automorphism with

$$L^t = -L \quad \text{if} \quad L \in \mathfrak{G}.$$

Then a routine calculation (given, e.g., in [52]) shows that

$$L\varphi = \varphi*L^t .$$

For $\xi \in \mathcal{D}'(G)$, define $\xi' \in \mathcal{D}'(G)$ by

$$(\xi',\varphi) = (\xi,\varphi').$$

For $\xi \in \mathcal{D}'(G)$ and $\varphi, \psi \in C_c^\infty(G)$, we have

$$(\xi,\varphi*\psi) = (\xi*\psi',\varphi) = (\varphi'*\xi,\psi).$$

Thus if $\eta \in \mathcal{E}'(G)$, $\mathcal{E}(G) = C^\infty(G)$, we define $\xi*\eta$, $\eta*\xi$ by

$$(\xi*\eta,\varphi) = (\xi,\varphi*\eta'), \quad (\eta*\xi,\varphi) = (\xi,\eta'*\varphi).$$

These definitions, too, extend to other appropriate spaces. Recall ([58], p. 158) that convolution of distributions is associative if at most one of the distributions is not in $\mathcal{E}'(G)$. Thus for $L \in \mathfrak{U}(\mathfrak{G})$, we have

$$L(\xi * \eta) = (\xi * \eta) * L^t = \xi * (\eta * L^t) = \xi * L\eta$$

when either ξ or η is in $\mathcal{E}'(G)$ (and in particular when $\eta \in \mathcal{D}(G)$).

Proposition 1. The following are equivalent conditions on the operator L:

(a) L is locally solvable.

(a') L has a local fundamental solution (i.e., there is an open neighborhood U of e and there is a distribution ξ on U with $L\xi = \delta$ on U).

(a") There is an open neighborhood V of e such that if $\eta \in \mathcal{E}'(V)$, then there is a distribution $u \in \mathcal{D}'(V)$ such that $Lu = \eta$ on V. (That is, L is locally solvable for $\mathcal{E}'$.)

(a'") There is an integer $k > 0$ and there is an open neighborhood V of e such that if $f \in C_c^m(V)$ and $m \geq k$, then there is a function $u \in C^{m-k}(V)$ with $Lu = f$ on V.

(a^{iv}) There is an open neighborhood U of e such that if $f \in C_c^\infty(U)$, then there is a distribution $u \in \mathcal{D}'(U)$ with $Lu = f$ on U.

Proof. Obviously (a"') $\Rightarrow$ (a^{iv}) and (a") $\Rightarrow$ (a').

(a')$\Rightarrow$(a"). Let V be a symmetric neighborhood of e ($V = V^{-1}$) with $V^2 \subseteq U$. If $\eta \in \mathcal{E}'(V)$, let $u = \eta * \xi$. We have $Lu = \eta * L\xi$. If $\varphi \in C_c^\infty(V)$, then $\operatorname{supp}(\eta' * \varphi) \subseteq U$; hence

$$(Lu, \varphi) = (\eta * L\xi, \varphi) = (L\xi, \eta' * \varphi) = (\delta, \eta' * \varphi) = (\eta, \varphi),$$

or $Lu = \eta$ on V.

(a')⇒(a"'). Let U_0 be a neighborhood of e whose closure is in U, and let k be the order of $\xi|U_0$. Choose V to be a symmetric neighborhood of e with $V^2 \subseteq U_0$. Given f, let $u = f*\xi$. Then $u \in C^{m-k}(V)$, and the above calculation shows that $Lu = f$ on V.

(a^{iv})⇒(a). This is essentially the content of [30], Theorem 6.1.3. Alternatively, we may appeal to a theorem in [16]. Let V be a symmetric neighborhood of e with $V^2 \subsetneq U$. Given $f \in C_c^\infty(V)$, we may write $f = \sum_{j=1}^{n} g_j * h_j$, with the $g_j, h_j \in C_c^\infty(V)$. Let $v_j \in \mathcal{D}'(U)$ satisfy $Lv_j = h_j$ on U, and set $u = \sum_{j=1}^{n} g_j * h_j$. Then $Lu = f$ on V.

(a)⇒(a'). This is proved in [53]; we repeat the proof here. It suffices to show that for some $m > 0$, we can always solve $Lu = f$ on U when $f \in C_c^m(U)$ (here, u may be in $\mathcal{D}'(U)$). For let Δ be a right invariant elliptic operator on G; for instance, one might let $\Delta\varphi = (X_1^2 + \ldots + X_n^2)*\varphi$, where $\{X_1, \ldots, X_n\}$ is a basis for $\mathfrak{G}$. Then Sobolev theory and the theory of elliptic operators (see, e.g., [1]) imply that for large enough k, and for any relatively compact open set W, Δ^k has a fundamental solution on W which is a C^m function. (In fact, k can be chosen to depend only on m and $n = \dim G$.) By using a cutoff function, we can find an open neighborhood U_0 of e and a function $f \in C_c^m(U)$ with $\Delta^k f = \delta$ on U_0. Now solve $Lu = f$. Since L and Δ^k commute, $L(\Delta^k u) = \Delta^k f = \delta$ on U_0.

To show that L is "locally solvable in C_c^m, we use an argument that goes back at least to [29]. Let A_j be a basis for the elements of $U(\mathfrak{G})$ of order $\leq j$. The usual topology of $C_c^\infty(U^-)$ is defined by the seminorms

$$\|\varphi\|_{\infty;j} = \sum_{D \in A_j} \|D\varphi\|_\infty .$$

Define a topology on $C_c^\infty(U)$ by the seminorms $\|L^t\varphi\|_{\infty;j}$, and consider the bilinear form on $C_c^\infty(U^-) \times C_c^\infty(U)$ defined by

$$(\varphi,\psi) = \int_U \varphi(x)\psi(x)dx.$$

This is obviously continuous in φ for fixed ψ. If φ is fixed, choose $u \in C^\infty(U)$ with $Lu = \varphi$. Then integration by parts gives

$$(\varphi,\psi) = \int_U (Lu)(x)\psi(x)dx = \int_U u(x)(L^t\psi)(x)dx ,$$

which is continuous in ψ. It follows from the Banach-Steinhaus Theorem (see [59], p. 354) that this bilinear form is jointly continuous. Hence there are integers m, m', plus a constant C, with

$$|(\varphi,\psi)| \leq C\|\varphi\|_{\infty;m}\|L^t\psi\|_{\infty;m'} .$$

This inequality now extends to all $\varphi \in C_c^m(U^-)$. That means that if $f \in C_c^m(U^-)$, then the map of $L^t(C_c^\infty(U))$ to C given by $L^t\psi \mapsto \int f(x)\psi(x)dx$ is continuous. By the Hahn-Banach Theorem, it extends to a continuous map u on $C_c^\infty(U)$. Then

$$(u,L^t\psi) = (f,\psi), \quad \text{all} \quad \psi \in C_c^\infty(U);$$

that is, $Lu = f$. This completes the proof.

A similar result holds for semiglobal solvability.

<u>Proposition</u> 2. The following conditions on L are equivalent:

(b) L is semiglobally solvable.

(b') For every open neighborhood U of e with compact closure, there is a distribution ξ on U with $L\xi = \delta$ on U.

(b") For every open set U with compact closure and every $\eta \in \mathcal{E}'(U)$, there is a distribution $u \in \mathcal{D}'(U)$ with $Lu = \eta$ on U.

(b'") For every open set U with compact closure there is an integer $k = k(U)$ such that if $f \in C_c^\infty(U)$ and $m \geq k$, then there is a function $u \in C^{m-k}(U)$ with $Lu = f$ on U.

(b^{iv}) For every open set U with compact closure and every $f \in C_c^\infty(U)$, there is a distribution u with $Lu = f$ on U.

The proof is essentially the same as that in Proposition 1.

Condition (b"') suggests a further definition. We say that

(h) L is <u>uniformly semiglobally solvable</u> on G if there is an integer k such that for every open neighborhood U of e with compact closure there is a distribution ξ on U of order $\leq k$ with $L\xi = \delta$ on U.

We might, in fact, say that L is uniformly semiglobally solvable of order $\leq k$. (See [13].) Now easy modifications of the previous proofs give:

<u>Proposition 3</u>. The following conditions on L are equivalent:

(h) L is uniformly semiglobally solvable.

(h') There is an integer k such that for every open U with compact closure and every $\eta \in \mathcal{E}'(U)$, there is a distribution $u \in \mathcal{D}'(U)$ with $Lu = \eta$ on U and order $(u) \leq k +$ order (η).

(h") There is an integer k such that for every open U with compact closure and every $f \in C_c^m(U)$ with $m \geq k$, there exists $u \in C^{m-k}(U)$ with $Lu = f$ on U.

Before getting to the main theorem, we list five (!) more conditions on L:

(j) L is <u>globally solvable on</u> $C^m(G)$ <u>of finite order</u> if there is an integer k such that whenever $m \geq k$ and $f \in C^m(G)$, there exists $u \in C^{m-k}(G)$ with $Lu = f$.

(k) L is <u>globally solvable</u> on $C^m(G)$ if for every $f \in C^m(G)$ there exists $u \in \mathcal{D}'(G)$ with $Lu = f$.

(ℓ) L is <u>globally solvable on distributions of finite order</u> if for every $\xi \in \mathcal{D}'$ of finite order, one can always find $u \in \mathcal{D}'(G)$ with $Lu = \xi$.

(m) L is <u>globally solvable of finite order on distributions of finite order</u> if there exists k such that for every $\xi \in \mathcal{D}'(G)$ of finite order, there exists $u \in \mathcal{D}'(G)$ of finite order with $Lu = \xi$ and order $(u) \leq k +$ order (ξ).

(n) L <u>has a fundamental solution of finite order</u> if there is a distribution $\mathcal{D}'(G)$ of finite order with $Lu = \delta$.

<u>Theorem</u> 1. Let G be a connected, simply connected nilpotent Lie group. The above conditions on L are related as follows:

$$\begin{array}{ccc}
(g) & & \\
\Downarrow & & \\
(f) & & (d) \\
\Downarrow & & \\
(h)\Leftrightarrow(j)\Leftrightarrow(m)\Leftrightarrow(n) & & \Downarrow \\
\Downarrow & & \\
(k) & \Rightarrow & (\ell) \\
 & & \Downarrow \\
 & & (e) \\
 & & \Downarrow \\
 & & (b)\Leftrightarrow(c) \\
 & & \Downarrow \\
 & & (a)
\end{array}$$

(This means that (h), etc. imply (k) for some m and that (k) for some m implies (ℓ).)

Proof. The definitions imply trivially that (g)⇒(f)⇒(n)⇒(e), that (d)⇒(ℓ)⇒(e)⇒(b')⇒(a'), that (m)⇒(n), that (c)⇒(b), that (j)⇒(k), that (j)⇒(h"), and that (n)⇒(h). So it suffices to show that (b)⇒(c), that (h)⇒(m), that (h)⇒(j), and that (k)⇒(ℓ).

(b)⇒(c): it is known that this implication holds whenever G is L-convex; i.e., if for every compact $K \subset G$, there is a compact $K' \subset G$ such that if $u, f \in \mathcal{E}'(G)$, $Lu = f$, and supp $f \subseteq K$, then supp $u \subseteq K'$.

(There is an abstract proof that (b) and L-convexity are equivalent to (c); see [59], pp. 392-394. The direct proof (given in a special case, but in a way that lets the proof generalize) is given as Theorem 3.5.5 of [30].) That G is L-convex for any left invariant L is proved in [19] and [6].

We note one related fact which will be useful below. Let the sets A_j be as before, and define the Sobolev space $H^s(G)$ $(s = 0, 1, \ldots)$ to be the closure of $C_c^\infty(G)$ under the norm

$$\|\varphi\|_{2;s}^2 = \sum_{D \in A_s} \|D\varphi\|_2^2 .$$

We may regard $H^s(G)$ as a (dense) subspace of $L^2(G)$. Now define $H^s_{loc}(G)$ by saying that $f \in H^s_{loc}(G)$ if $\varphi f \in H^s(G)$ for every $\varphi \in C_c^\infty(G)$. One can define seminorms on H^s_{loc} as in chapter 31 of [59]; thus H^s_{loc} is a Frechet space. Now let P be an elliptic left invariant operator of order k. Then standard elliptic theory and the above argument (in [59] or [30]) show that $P(H^s_{loc}) = H^{s-k}_{loc}$ for all $s \geq k$. (One can extend these definitions to all $s \in \mathbb{Z}$, and then the above result holds for all s.)

(h)⇒(m). It is proved in [13] that (h)⇒(n), and the proof is easily adapted to show that (h)⇒(m). Here is a fairly detailed sketch. Let $\{\Omega_m\}$ be an increasing sequence of open sets (each containing the closure of the previous one) with compact closure whose union is G. For $\varphi \in C_c^\infty(\bar{\Omega}_m)$, define

$$|\varphi|_{j;m} = \sup_{D \in A_j, x \in \Omega_m} |D\varphi(x)|;$$

let $\|\ \|_{j;m}$ be the dual norm for distributions of order $\leq j$ on $\bar{\Omega}_m$. (This notation clashes with earlier notation, and will be used only in part of the proof.) Note that there is an integer s such that if ξ has order ℓ, then we can approximate ξ arbitrarily closely in the $\|\ \|_{\ell+s;m}$ norm by C^∞ functions. (Write $\xi = \Delta\xi_0$, where Δ is elliptic and ξ_0 is a function; approximate ξ_0 by $\varphi \in C^\infty$ in the sup norm, and apply Δ to each side.) Let $\ell_0 = \ell+s$, let k be as in (h), and let $k_0 = k+\ell_0$. It is not hard to show, using (h), that for fixed ℓ_0, there is a sequence of numbers $\{K_m\}$ such that for all $f \in C_c^\infty(\Omega_m^-)$ we can find $u \in C_c^\infty(\Omega_{m+1})$ with

(1) $Lu = f$ on Ω_m;

(2) $\|u\|_{k_0;m} \le K_m \|f\|_{\ell_0;m}$.

Now let $\xi \in \mathcal{D}'(G)$ have order ℓ, and let $\{\varepsilon_m\}$ be a decreasing sequence of positive numbers such that $\sum_{m=1}^{\infty} \varepsilon_m(1+K_m)$ converges. We find functions $u_m, f_m \in C_c^\infty(\Omega_{m+1})$, $m = 1,2,\ldots,$ with

(1') $Lu_m = f_m$;

(2') $\|f_m - \xi\|_{\ell_0;m} < \varepsilon_m$;

(3') $\|u_{m+1} - u_m\|_{k_0;m} < 2^{-m} + 2K_m\varepsilon_m$.

It is then easy to see that the u_m converge to a distribution u of order $\le k + \ell_0$ such that $Lu = \xi$.

The u_m, f_m are defined inductively. For $m = 1$, they are easy to define; see [13] for details. The rest of the induction also follows the lines given in [13]. Given u_m and f_m, we choose $g_m \in C_c^\infty(\overline{\Omega}_m)$ with

$$\|\xi - f_m - g_m\|_{\ell_0;m} < \varepsilon_{m+1}/2.$$

Some easy estimates show that $\|g_m\|_{\ell_0;m} < 2\varepsilon_m$. Let $v_m \in C_c^\infty(\Omega_{m+1})$ satisfy $Lv_m = g_m$ on Ω_m and $\|v_m\|_{k_0;m} \le K_m\|g_m\|_{\ell_0;m}$. Then

$$\|v_m\|_{k_0;m} \le 2K_m\varepsilon_m, \quad L(u_m + v_m) = f_m + G_m ,$$

where $G_m = L(v_m)$ satisfies $G_m|_{\Omega_m} = g_m$. So

$$\|\xi - f_m - G_m\|_{\ell;m} < \varepsilon_{m+1}/2 .$$

Choose $F_{m+1} \in C_c^\infty(\Omega_{m+1})$ so that $F_{m+1}|_{\Omega_m} = f_m + g_m$ and $\|F_{m+1} - \xi\|_{k_0;m+1} < \varepsilon_{m+1}$, and set $h_{m+1} = F_{m+1} - f_m - G_m$. Pick $w_{m+1} \in C_c^\infty(\Omega_{m+2})$ with $Lw_{m+1} = h_{m+1}$ on Ω_{m+1}. Now $h_{m+1}|_{\Omega_m} = 0$, G is L-convex, and L is semiglobally solvable; thus we can find $W_{m+1} \in C_c^\infty(\Omega_{m+2})$ with $LW_{m+1}|_{\Omega_{m+1}} = 0$ and $\|W_{m+1} - w_{m+1}\|_{k_0;m} < 2^{-m}$; see pp. 393-394 of [59] for a proof. Finally, set $u_{m+1} = u_m + v_m + w_{m+1}$, $f_{m+1} = Lu_{m+1}$. Then (1') holds, and on Ω_{m+1} we have

$$f_{m+1} = L(u_m+v_m)+L(w_{m+1})-L(W_{m+1}) = f_m+G_m+h_{m+1} = F_{m+1} ,$$

so that (2') holds. Moreover,

$$\|u_{m+1}-u_m\|_{\ell_0;m} \le \|v_m\|_{\ell_0;m}+\|W_{m+1}-w_{m+1}\|_{\ell_0;m} < 2K_m\varepsilon_m+2^{-m} ,$$

which verifies (3'). This completes the proof.

(h)⇒(j). A direct proof seems to be difficult. Instead, we reason as follows: let Δ be a right invariant elliptic operator of order m. The spaces H^s_{loc} are the same whether left or right invariant operators are considered, and Δ maps $H^m_{loc}(G)$ onto H^0_{loc}, as mentioned earlier. Suppose that $f \in C^m(G)$. Then $f \in H^m_{loc}(G)$, and $\Delta f \in H^0_{loc}$. As (m) holds we can find a distribution of order $\le k_0$ with $L\xi = \Delta f$. Now choose u_0 with $\Delta u_0 = \xi$. Sobolev theory says that if m is sufficiently large, then $u_0 \in C^{m-k_1}(G)$ for some k_1 depending only on k_0 and dim G. Set $v_0 = Lu_0$. then

$$\Delta(v_0 - f) = \Delta(Lu_0 - f) = L\xi - \Delta f = 0.$$

As Δ is elliptic, $v_0 - f$ is smooth. Since $(h)\Rightarrow(n)\Rightarrow(e)\Rightarrow(c)$, we can solve $Lu_1 = v_0$ with $u_1 \in C^\infty(G)$. Set $u = u_0 - u_1$; $u \in C^{m-k_1}(G)$, and $Lu = f$.

$(k)\Rightarrow(\ell)$. Let Δ again be a right invariant elliptic operator. Then, as noted before, Δ satifies (h) and hence (m). If Δ is of high enough order, then we can find $f \in C^m(G)$ with $\Delta f = \xi$. Let $Lu = f$; then $L(\Delta u) = \xi$, and we are done.

What about the other implications? The state of our knowledge appears to be the following:

1. If $G = \mathbb{R}^n$ (so that L is a constant coefficient operator), then L satisfies (g) and (d), and hence all the other properties. (For (d), see [20], [41] or section 3.6 of [30]; for (g), see [28], [40], [2], [4], or [3].)

2. For a connected, simply connected nilpotent Lie group, no example is known of a left invariant L which is locally solvable, but does not satisfy all of (b) through (m). On the other hand, virtually the only such operators for which we can verify all of (a) through (m) are those that become constant coefficient operators with respect to some global coordinate system. (That is, we can verify all the properties only in "trivial" cases. The reason for the word "virtually" is that a few special cases may possibly have been checked.) There are, incidentally, a number of examples of biinvariant operators on nilpotent Lie groups which become constant coefficient operators in appropriate coordinates; Roger Howe pointed out some interesting examples in one of his talks at this conference.

3. There are examples of non-nilpotent Lie groups G (still connected and simply connected) and left invariant operators L for which (c) holds, but not (d). In fact, if G is a real semisimple group of (real) rank ≥ 2, then one can take L to be the Casimir operator, which is biinvariant; see [57]. Duflo ([18]) has given a similar example of a biinvariant operator on a solvable Lie group satisfying (c), but not (d).

4. There are examples of left invariant operators L on nilpotent Lie groups G which are not locally solvable. Indeed, Hans Lewy's original example (in [38]) of an unsolvable operator is a left invariant operator on the 3-dimensional Heisenberg group. Moreover, if G is nilpotent and every left invariant complex vector field on G is locally solvable, then G is Abelian; see [5] for a proof.

5. Properties (d) (global solvability in $\mathcal{D}'$) and (g) (global solvability in $\mathcal{S}'$) have not been given much attention in the literature, probably because there are few general approaches. (For a discussion of (d) for general differential operators, see [31].) We shall not consider them further here.

2. Sufficient Conditions for Solvability

One basic method for proving that left invariant operators on G are locally solvable uses the Plancherel Theorem for G. Here is a summary of what will be needed. G acts on $\mathfrak{G}$ by Ad, and on $\mathfrak{G}^*$ by the contragredient, Ad*; Kirillov theory ([34] or [45]) gives a bijective map between Ad*-orbits of $\mathfrak{G}^*$ and irreducible representations of G. Identify G and $\mathfrak{G}$ via exp, and let $\wedge$ be the Fourier transform on $\mathfrak{G}$ (so that if $f \in S(G)$, say, then $f^\wedge \in S(\mathfrak{G}^*)$. Let $\mathcal{O}_\pi \subsetneq \mathfrak{G}^*$ be the orbit corresponding to $\pi \in G^\wedge$--the dual space of (equivalence classes of) irreducible unitary representations. There is a unique Ad*-invariant measure $d_\pi(\ell_0)$ on $\mathcal{O}_\pi$ so that for all $f \in S(G) = S(\mathfrak{G})$,

$$\operatorname{Tr}\pi(f) = \int_{\mathcal{O}_\pi} f^\wedge(\ell_0)\, d_\pi(\ell_0) .$$

In particular, $\pi(f)$ is trace class for $f \in S(G)$. There are explicit descriptions of $d_\pi(\ell_0)$; see [47]. (A related description is found in [9].) Moreover, one can find a subspace V of $\mathfrak{G}^*$, a Zariski-open subspace U of $\mathfrak{G}^*$, and a Zariski-open subset U_0 of V such that:

(1) $\quad \mathrm{Ad}^*(G)U_0 = U.$

(2) $\quad$ If $\mathcal{O}_\pi \subset U_1$ then $\mathcal{O}_\pi \cap U_0$ consists of one element (ℓ_π, say).

(3) $\quad$ Each orbit $\mathcal{O}_\pi \subset U$ is the graph of a polynomial map $P_\pi\colon V \to V'$, where V' is a fixed complementary subspace of V; $P_\pi(0) = \ell_\pi$, and the P_π vary rationally with ℓ_π.

The set U_0 parametrizes the "generic" representations of G; for $\ell \in U_0$, we let π_ℓ satisfy $\ell \in \mathcal{O}(\pi_\ell)$, and write d_ℓ for d_{π_ℓ}, etc. The Plancherel formula can now be stated: there is an Ad*-invariant polynomial $\mathrm{Pf}\colon \mathfrak{G}^* \to \mathbb{R}$ such that for $f, g \in S(G)$,

(2.1) $f(e) = \int_{U_0} \mathrm{Tr}(\pi_\ell(f))\,|Pf(\ell)|\,d\ell$ ($d\ell$ = Lebesgue measure on V);

(2.2) $\langle f,g\rangle = \int_{U_0} \mathrm{Tr}(\pi_\ell(g)^*\pi_\ell(f))\,|Pf(\ell)|\,d\ell$ ($\langle,\rangle$ is the L^2 inner product).

Strictly speaking, (2.1) is Fourier inversion, while (2.2) is the Plancherel formula. There are explicit formulas for Pf; see [47] or [7]. Note that $\int_{\mathfrak{G}^*} f d\ell = \int_{U_0} \int_{O_\ell} |Pf(\ell)|\, f(\ell_0) d_\ell(\ell_0) d\ell$.

Perhaps the first general result proved about solvability of left invariant operators on nilpotent Lie groups was the following:

<u>Theorem 2</u> ([45]). Let L be a bi-invariant differential operator on the (connected, simply connected) nilpotent Lie group G (so that $L \in \mathfrak{Z}(\mathfrak{U}(\mathfrak{G}))$). Then L has a central tempered fundamental solution ξ_0 (thus $\xi_0 * \varphi = \varphi * \xi_0$, $\forall \varphi \in S(G)$).

The proof runs as follows: we may assume (replacing L by LL^* if necessary) that L is a positive operator. For each $\pi \in G^\wedge$, $\pi(L)$ is a nonnegative multiple of the identity (one proof of this uses Schur's lemma). Define $L^\wedge$ on $\mathfrak{G}^*$ by $\pi(L^t) = L^\wedge(\ell)I$ for $\ell \in O_\pi$. Then $L^\wedge$ is an Ad*-invariant polynomial. The map $s \to (L^\wedge)^s$ is a holomorphic function into the space of tempered distributions on $\mathfrak{G}^\wedge$ when $\mathrm{Re}(s) > 0$, and it extends to a meromorphic function into $S'(\mathfrak{G}^*)$; see [2], [4], or [3]. Moreover, $(L^\wedge)^s$ has a Laurent series expansion at each point in $\mathbb{C}$, and all the coefficients can be shown to be Ad*(G)-invariant elements of $S'(\mathfrak{G}^*)$. Define L^s to be the inverse Fourier transform of $(L^\wedge)^s$. Then $s \to L^s$ is a meromorphic map into $S'(\mathfrak{G}) = S'(G)$, and the coefficients of the Laurent series of L^s about any point are Ad(G)-invariant (= conjugation-invariant). Then expand L^s in a Laurent series about $s = -1$:

$$L^s = \sum_{j=-n}^{\infty} (s+1)^j \xi_j .$$

The key fact to check is that $L^s * L = L^{s+1}$ (where the convolution is in G). Assume this for now. About $s = 0$, the Laurent series is $L^s = \delta + \sum_{j=1}^{\infty} s^j \eta_j$, say. Since Laurent series are unique, we see (by convolving the first series with L and writing s for $(s+1)$) that $\xi_0 * L = \delta$; also ξ_0 is conjugation-invariant (= central) on G.

To prove that $L^s * L = L^{s+1}$, we may assume $\mathrm{Re}(s) > 0$ (both sides are meromorphic). Then

$$(L^s * L, \varphi) = (L^s, \varphi * L^t) = \int_{\mathfrak{G}^*} L^\wedge(\ell)^s (\varphi * L^t)^\wedge(\ell)\, d\ell \qquad (L^t = L^{\text{transpose}}!)$$

$$= \int_{U_0} |\mathrm{Pf}(\ell)|\, L^\wedge(\ell)^s \int_{\mathcal{O}_\ell} (\varphi * L^t)^\wedge(\ell_0)\, d_\ell(\ell_0)\, d\ell$$

$$= \int_{U_0} |\mathrm{Pf}(\ell)| L^\wedge(\ell)^s \mathrm{Tr}\pi_\ell(\varphi * L^t)\, d\ell = \int_{U_0} |\mathrm{Pf}(\ell)|\, L^\wedge(\ell)^s L^\wedge(\ell) \mathrm{Tr}\pi_\ell(\varphi)\, d\ell$$

$$= \int_{U_0} |\mathrm{Pf}(\ell)|\, L^\wedge(\ell)^{s+1} \mathrm{Tr}\pi_\ell(\varphi)\, d\ell = (L^{s+1}, \varphi),$$

as one sees by retracing steps. We're done.

Note. Duflo [17] has shown that if G is any Lie group and $L \in \mathfrak{Z}(\mathfrak{U}(\mathfrak{G}))$, then L is locally solvable. When G is simply connected and completely solvable, elements in $\mathfrak{Z}(\mathfrak{U}(\mathfrak{G}))$ are globally solvable ([56], plus [19] or [6]); when G is simply connected, semisimple, and noncompact, the Casimir operator is globally solvable ([49]), but not all bi-invariant operators are ([61]).

The basic technique for proving local (and, indeed, semiglobal) solvability of left invariant operators on nilpotent Lie groups is given in [53]; it stems from an idea in [52], and has been expanded in various papers, among them [7], [39], and [8]. The following account is essentially the one in [7].

The generic representations π_ℓ described earlier are also called the representations "in general position." According to Kirillov theory, these representations can be realized on $L^2(\mathbb{R}^k)$, $k = \dim V$, in a manner that varies measurably with ℓ, in that the map $(x,\ell) \mapsto \langle\pi_\ell(x)f,g\rangle$ is measurable for all $f, g \in L^2(\mathbb{R}^k)$. (In fact, one can even make this map C^∞ in x and ℓ near any given $\ell_0 \in U_0$ when f and g are in $S(\mathbb{R}^k)$ = space of C^∞ vectors for the π_ℓ.) Thus it makes sense to talk of a measurable map $\ell \mapsto T_\ell \in B(L^2(\mathbb{R}^k))$, where $B(H)$ = algebra of bounded linear operators on H.

Define $L \mapsto L^*$ to be the unique conjugate linear automorphism of $U(\mathfrak{G})$ given on $\mathfrak{G}$ by $X^* = -X$. A straightforward induction shows that for any unitary representation π, $\pi(L^*) = \pi(L)^*$. The relationship between $\pi(L)$ and representations of L^t is more complicated. Recall that the contragredient representation π^t of π is defined on H', the dual space of H (not H, which is its own <u>conjugate</u> dual) by

$$\pi^t(x) = (\pi(x^{-1}))^t ,$$

where $A \mapsto A^t$ is the map $B(H) \to B(H')$ defined by $(Av,v') = (v,A^t v')$ for all $v \in H$, $v' \in H'$. Then one can check that $\pi(f)^t = \pi^t(f')$ (remember: $f'(x) = f(x^{-1})$) and that

$$(\pi(L))^t = \pi^t(L^t).$$

(One proof is given in [7].) When $H = L^2(\mathbb{R}^k)$, one can identify H' with $L^2(\mathbb{R}^k)$ via the bilinear form $(f,g) = \int_{\mathbb{R}^k} fg\,dx$. If π is irreducible, then so is π^t, and $O_{\pi^t} = -O_\pi$; i.e., $\ell \in O_\pi$ iff $-\ell \in O_{\pi^t}$. That is, $\pi_\ell^t = \pi_{-\ell}$ for $\ell \in U_0$.

We also need the formula (used earlier) that $\pi(L\varphi) = \pi(\varphi)\pi(L^t)$.

<u>Theorem</u> 3. Let $L \in U(\mathfrak{G})$ satisfy:

(1) For each $\ell \in U_0$, $\pi_\ell(L)$ has a bounded right inverse, T_ℓ.

(2) The T_ℓ vary measurably with ℓ.

(3) There are nonzero $Ad^*(G)$-invariant polynomials P_1, P_2 on $\mathfrak{G}^*$ such that $|P_1(\ell)|\|T_\ell\| \le |P_2(\ell)|$ for all $\ell \in U_0$.

Then L is semiglobally solvable.

Here is a sketch of the proof. It follows from Theorem 4.8.12 of [15] that there are elements Z_1, Z_2 of $\mathfrak{Z}(\mathfrak{U}(\mathfrak{G}))$ uniquely defined by

$$\pi_\ell(Z_1) = P_1(\ell)I, \ \pi_\ell(Z_2) = P_2(\ell)I.$$

Let $A_\ell = (T^*_{-\ell})^t$. Then $\pi_\ell((L^*)^t)A_\ell = I$ for each ℓ.

Now let $\varphi \in S(G)$. We define $u = F(\varphi)$ by

$$<\psi,u> = \int_{U_0} |Pf(\ell)|\overline{P_1(-\ell)}Tr(A_\ell\pi_\ell(\varphi)*\pi_\ell(\psi))d\ell \quad \text{if} \quad \psi \in S(G).$$

Then (formally) we have

$$<\psi,Lu> = <L^*\psi,u> = <\psi*(L^*)^t,u> = \int_{U_0} |Pf(\ell)|\overline{P_1(-\ell)}Tr(A_\ell\pi_\ell(\varphi)*\pi_\ell(\psi)\pi_\ell((L^*)^t))d\ell.$$

But

$$Tr(A_\ell\pi_\ell(\varphi)*\pi_\ell(\psi)\pi_\ell((L^*)^t)) = Tr(\pi_\ell(\varphi)*\pi_\ell(\psi)\pi_\ell((L^*)^t)A_\ell) = Tr(\pi_\ell(\varphi)*\pi_\ell(\psi));$$

Thus (since $\pi_\ell((Z_1^t)^*) = \overline{P_1(-\ell)}I$)

$$<\psi,Lu> = \int_{U_0} |Pf(\ell)|Tr(\pi_\ell((Z_1^t)^*)\pi_\ell(\varphi)*\pi_\ell(\psi))d\ell = \int_{U_0} |Pf(\ell)|Tr(\pi_\ell(Z_1\varphi)*\pi_\ell(\psi))d\ell$$

$$= <\psi,Z_1\varphi>,$$

from formula (2.2). Now suppose that $f \in C_c^\infty(G)$ and that U is an open set with compact closure. Then (Theorem 2) we can find $\varphi \in C_c^\infty$ with $Z_1\varphi = f$ on U, and $u = F(\varphi)$ solves $Lu = f$ on U.

We need to justify the formal equality. Let $\|\ \|_2$ be the Hilbert-Schmidt norm. The integrand (with $|Pf(\ell)|$ omitted) is bounded by

$$|P_1(-\ell)|\,\|A_\ell\|\,\|\pi_\ell(\varphi)^*\|_2\|\pi_\ell(\psi)\|_2$$

$$\leq |P_2(-\ell)|\,\|\pi_\ell(\varphi)^*\|_2\|\pi_\ell(\psi)\|_2 = \|\pi_\ell(Z_2\varphi)^*\|_2\|\pi_\ell(\psi)\|_2.$$

Now the Cauchy-Schwarz inequality and the Plancherel theorem together show that $u \in L^2(G)$ (and that $\|u\|_2 \leq \|Z_2\varphi\|_2$; here the norms are L^2 norms). Since $D*u = F(D^* * \varphi)$ for $D \in U(\mathfrak{G})$, all derivatives of u are in L^2; thus u is C^∞, and we are done.

One can squeeze a certain amount more out of this general method. For simplicity, assume that G is stratified; this means that $\mathfrak{G} = \bigoplus_{j=1}^{k} \mathfrak{G}_j$ as a vector space, with $[\mathfrak{G}_i, \mathfrak{G}_j] \subseteq \mathfrak{G}_{i+j}$ (and $\mathfrak{G}_j = (0)$ for $j > k$). Then the map $\alpha_t = \mathfrak{G} \to \mathfrak{G}$, defined when $t > 0$ by $\alpha_t|_{\mathfrak{G}_i} = t^i I$, is an automorphism. Say that $L \in \mathfrak{U}(\mathfrak{G})$ is homogeneous of degree d if $\alpha_t(L) = t^d L$ for all $t > 0$. Let B_d be a base for the homogeneous elements of degree d. Given a representation π of G on $H(\pi)$, the space $H^\infty(\pi)$ of C^∞ vectors of π is the intersection of the domains of all the $\pi(L)$, $L \in U(\mathfrak{G})$. Define the dth Sobolev norm on these vectors by

$$\|v\|_d^2 = \sum_{d' \leq d} \sum_{D \in B_{d'}} \|\pi(D)v\|^2,$$

and complete to get the space $H^d(\pi)$; define H^{-d} by duality. One can then weaken hypothesis (1) to require only that there are integers d_1, d_2 such that for all $\ell \in U_0$, $\pi_\ell(L)$, regarded as a map from $H^{d_1}(\pi_\ell)$ to $H^{d_2}(\pi_\ell)$, has a bounded right inverse T_ℓ (and that the inverses satisfy (2) and (3)). For details, see [8].

Two problems arise almost immediately when one tries to apply Theorem 3:

1. The hypotheses are usually hard to check. Even in the comparatively easy cases when $L = L^*$, Theorem 3 "reduces" the local solvability of L to a question of spectral analysis of a family of (self-adjoint) partial differential operators.

2. In many cases, the hypotheses are not met: on some subset of U_0, $\pi_\ell(L)$ is not invertible. Then L may still be semiglobally solvable, but one must modify the proof of Theorem 3 to show this fact.

As a consequence, there have been a number of papers written to prove that certain classes of operators are locally solvable. Here is a list (incomplete) of examples of what can be done.

For most of these examples, we deal with 2-step nilpotent Lie groups G. Then G is stratified, with $\mathfrak{G}_2 = [\mathfrak{G},\mathfrak{G}]$; $\mathfrak{G}_1$ can be any complement to $\mathfrak{G}_2$. If π is any irreducible representation of G trivial on $G_2 = \operatorname{Exp} \mathfrak{G}_2$, then π is 1-dimensional. Let $L = \sum_{j=0}^{m} L_j$, where $L_m \neq 0$ and L_j is homogeneous of degree j. We say that L is <u>elliptic in the generating directions</u> if $\pi(L_m) \neq 0$ whenever π is trivial on G_2 and π is not the trivial representation.

If $\eta \in \mathfrak{G}_2^*$, there is a bilinear form B_η on $\mathfrak{G}_1 \times \mathfrak{G}_1$ defined by

$$B_\eta(X,Y) = \eta([X,Y]).$$

We say that G is a group of type (H), or an (H)-group ([42]) if B_η is nondegenerate whenever $\eta \neq 0$. At the other extreme, we can consider groups for which B_η always has nonzero radical. (Note that B_η is antisymmetric; thus the radical is nonzero if $\dim\mathfrak{G}_1$ is odd.)

<u>Example</u> 1. For an H-group, the irreducible unitary representations of G are of two sorts. If $\ell \in \mathfrak{G}^*$ satisfies $\ell|_{\mathfrak{G}_2} = 0$, then the $\mathrm{Ad}^*(G)$ orbit of ℓ is $\ell + \mathfrak{G}_2^\perp$, and the representation corresponding

to ℓ is determined by $\eta = \ell|_{\mathfrak{G}_2}$; we denote it by π_η. If $\ell|_{\mathfrak{G}_2} = 0$, then (as was mentioned) the corresponding representation is 1-dimensional.

<u>Theorem</u> 4. If L is elliptic in the generating directions and there is no (nonempty) open set of elements $\eta \in \mathfrak{G}_2$ such that $\pi_\eta(L)$ has kernel, then L is semiglobally solvable.

This theorem was proved in [54] for the special case when L is homogeneous. The idea of the proof is this: we may assume (replacing L by LL^*) that L is self-adjoint. Put a Euclidean norm, $|\ |$, on $\mathfrak{G}_2^*$; the homogeneity of L means that we need only concern ourselves with the representations π_η, $|\eta| = 1$. We know enough about L and the π_η to conclude that the operators $\pi_\eta(L)$ have compact resolvent and vary analytically with η (in the sense of [33]). Hence the eigenvalues and eigenspaces of $\pi_\eta(L)$ vary analytically. If 0 is never an eigenvalue, Theorem 3 applies. Otherwise, 0 is an eigenvalue of finite multiplicity on a closed set of codimension ≥ 1, and we can invert $\pi_\eta(L)$ on a subspace of $\mathcal{H}(\pi_\eta)$ of finite codimension. The method of Theorem 3 then lets us reduce the problem to the following one: let $A: \mathbb{R}^k \to M_n(\mathbb{R})$ be a nonzero analytic matrix-valued function. Then we need a distribution ξ which inverts A_1 in the sense that if $\varphi: \mathbb{R}^k \to M_n(\mathbb{R})$ has compact support, then

$$(A\varphi, \xi) = \int \mathrm{Tr}\,\varphi(x)\,dx.$$

(Here, ξ is in the dual of the space of C_c^∞ functions from $\mathbb{R}^k$ to $M_n(\mathbb{R})$.) The existence of ξ follows from the theorem (proved first in [40]) on division by analytic functions.

The theorem as stated above follows from similar considerations, plus a result of Melin ([42]); see [13] for details.

<u>Example</u> 2. Let L be a homogeneous second order operator on a 2-step nilpotent Lie group; assume again that L is elliptic in the generating directions. By choosing appropriate bases $\{X_1, \ldots, X_p\}$

of $\mathfrak{G}_1$ and $\{T_1,\ldots,T_q\}$ of $\mathfrak{G}_2$, one can write L as

$$L = \sum_{j=1}^{p} X_j^2 + \sum_{j=1}^{q} c_j T_j \ , \ c_j \in \mathbb{C}.$$

Theorem ([35]): Suppose that $\mathrm{Re}(c_j) \neq 0$ for some j. Then L is semiglobally solvable.

One proof (essentially along the lines of that in [35]) is this: one can determine the operators $\pi_\ell(L)$ quite explicitly. In appropriate models for π_ℓ, they are essentially Hermite operators (i.e., operators for the harmonic oscillator). Thus one can show that Theorem 3 does apply.

Example 3. Let L be as in Example 2.

Theorem 5. If the form B_η (defined before Example 1) is always degenerate, then L is semiglobally solvable.

There is a proof in [53] involving considerations like those in Example 1. Another proof, given in [36], runs as follows: for simplicity, assume that G is the product of a Heisenberg group H with $\mathbb{R}$. Then a representation π of G can be written as $\pi_1 \otimes \pi_2$, where π_1 is a representation of H and π_2 is a representation of $\mathbb{R}$. There are difficulties in complexifying representations of H, but complexifying representations of $\mathbb{R}$ is easy: a typical one takes x to $\exp 2\pi i\zeta x$, $\zeta \in \mathbb{C}$. To prove the theorem, one uses the method of Theorem 3, but one first applies the Cauchy Integral Formula to transform the integral into one involving complex representations π of G where it is easy to see that the inverses for $\pi(L)$ exist and are appropriately bounded in norm.

This technique has other applications; in [37], Lévy-Bruhl applies it to operators on certain 3-step nilpotent Lie groups.

We turn now to another technique for proving local solvability. Suppose that G has a discrete subgroup Γ such that $\Gamma\backslash G$ is compact. Then G acts in a natural way on $\Gamma\backslash G$, and this leads to a representation ρ of G on $L^2(\Gamma\backslash G)$:

$$(\rho(x)f)(\Gamma y) = f(\Gamma yx) .$$

It is known ([22], p. 23) that ρ decomposes as a discrete direct sum of irreducibles, each with finite multiplicity. Moreover, the irreducibles appearing in the sum can be described; see, e.g., [32], [50], or [9]. Let $(G:\Gamma)^\wedge$ be the set of (equivalence classes of) irreducibles occurring in ρ; for $\pi \in [G:\Gamma]^\wedge$, let P_π be the projection onto the π-primary subspace of $L^2(\Gamma\backslash G)$. Now let L be the operator on G. Suppose that $\pi(L)$ is right invertible for all $\pi \in [G:\Gamma]^\wedge$; let A_π be the inverse. Given $\varphi \in C^\infty(\Gamma\backslash G)$, we could try to solve $Lu = \varphi$ by setting $P_\pi(u) = A_\pi^t P_\pi(\varphi)$; under reasonable conditions on the norms $\|A_\pi\|$, this does indeed define an element u of $L^2(\Gamma\backslash G)$. Since we can map a small neighborhood of $e \in G$ diffeomorphically into $\Gamma\backslash G$, this proves local solvability.

One obvious problem with this procedure is that $\pi(L)$ will generally not be invertible for all $\pi \in [G:\Gamma]^\wedge$. (For instance, π could be the trivial representation.) But this problem can often be avoided by solving $Lu = Z\varphi$, where $Z \in \mathfrak{Z}(\mathfrak{U}(\mathfrak{G}))$ is chosen to be zero on the troublesome representations. Then one applies Theorem 2 to prove that L is locally solvable.

Here is one theorem that can be proved by this method, followed by an example:

<u>Theorem</u> 6 ([11]). Let G,Γ,L be as above; let $[G:\Gamma]' = \{\pi \in [G:\Gamma]^\wedge : \pi$ is in general position$\}$. Put a norm, $|\ |$, on the subspace V_0 of $\mathfrak{G}^*$ used to parametrize the representations in general position. If:

(1) for all $\pi_\ell \in [G:\Gamma]'$, $\pi_\ell(L)$ has a bounded right inverse A_ℓ;

(2) there is a polynomial Q on V_0 with $\|A_\ell\| \le |Q(\ell)|$ for all ℓ with $\pi_\ell \in [G:\Gamma]'$;

(3) $\mathfrak{G}$ has a rational ideal $\mathfrak{H}$ which is maximal subordinate (in the sense of Kirillov theory) for all $\ell \in U_0$;

then L is locally solvable.

Example 4. Let G be the 3-dimensional Heisenberg group; $\mathfrak{G}$ is spanned by X, Y, and Z, with $[X,Y] = Z$. Let $L = X^2 + Y^2 + p(Z)$, where p is a polynomial. Then L is semiglobally solvable unless $p(Z) = (2n+1)iZ$ for some $n \in \mathbb{Z}$. To prove this, let π_ℓ be the representation of G with $\pi_\ell(Z) = i\ell I$, $\ell \neq 0$. Then π_ℓ can be realized on $L^2(\mathbb{R})$ so that

$$\pi_\ell(L) = |\ell|\left(\frac{d^2}{dt^2} - t^2\right) + p(i\ell).$$

The eigenvalues of $\pi_\ell(L)$ are $p(i\ell) + |\ell|(2n+1)$, $n = 0,1,2,\ldots$; if 0 is not an eigenvalue, then $\pi_\ell(L)$ has an inverse A_ℓ, and $\|A_\ell\| = |\lambda_\ell|^{-1}$, where λ_ℓ is the smallest eigenvalue (in absolute value) of $\pi_\ell(L)$. Given $r > 0$, there is a discrete cocompact subgroup Γ_r of G such that $[G:\Gamma_r]' = \{\pi_{mr}: m \in \mathbb{Z}, m \neq 0\}$. Thus Theorem 6 applies (to prove local solvability) if we can choose r, $C > 0$, and a polynomial Q such that for all integers $m \neq 0$ and $n \ge 0$,

$$|p(imr) + |mr|(2n+1)|\,|Q(mr)| \ge C.$$

This amounts to a statement about approximating irrationals by rationals. To see why, consider $p(Z) = -Z^2$. Then we want Q with

$$|mr|(2n+1 - |mr|)\,|Q(mr)| \ge C.$$

We can ignore the factor of $|mr|$, which can only help the estimate. Let $\Omega(t) = t^k$; we need

$$\left|\frac{2n+1}{m} - r\right| \geq \frac{C'}{|m^{k+1}|}, \quad C' = C/r^k .$$

But this is true (for $k = 2$) if r is, say, a quadratic irrational number; see, e.g., Chapter 10 of [27]. Thus L is locally solvable. In fact, we can show that for every open set $U \subset G$ with compact closure, there exists Γ_r such that the above inequality holds and U maps diffeomorphically into $\Gamma_r\backslash G$. Thus one can prove semiglobal solvability (and, in fact, uniform semiglobal solvability).

The proof for general p is similar.

The above technique raises other questions about global solvability on $\Gamma\backslash G$. This is a tricky matter; for example, d/dx is not globally solvable on $\mathbb{R}/\mathbb{Z}$ (consider $du/dx = 1$), but $d/dx + 1$ is. See [51] and [14] for some results.

Finally, there is the matter of fundamental solutions. The above techniques often show that L is uniformly semiglobally solvable; then of course, L has a fundamental solution of finite order. In [21], Folland showed how to construct tempered fundamental solutions for some homogeneous operators. Another result along these lines will be described in the next section.

3. Necessary Conditions for Solvability

There is a considerable literature on the question of local solvability of partial differential equations in $\mathbb{R}^n$ (with variable coefficients); necessary conditions are given in, e.g., [29] and [45] (for some classes of operators). These results often apply to left invariant operators. Here, as earlier, we consider only theorems that apply specifically to nilpotent Lie groups.

Recall that if W is a sufficiently small open set in U_0 (the parametrizing space for the representations in general position), then the representations π_ℓ, $\ell \in W$, can be realized on the same Hilbert space, $L^2(\mathbb{R}^k)$ $(k = \dim V)$, so that if $\varphi, \psi \in \mathcal{S}(\mathbb{R}^k)$, then the functions $(\ell,x) \to \langle\pi_\ell(x)\varphi,\psi\rangle$ varies smoothly in x and ℓ. Moreover, these representations have the property that $H^\infty(\pi_\ell) = \mathcal{S}(\mathbb{R}^k)$. On W, therefore, we can define a smoothly varying map $\ell \to \varphi_\ell$, $\varphi_\ell \in H^\infty(\pi_\ell)$; this means that the map $(\ell,x) \to \langle\pi_\ell(x)\varphi_\ell,\psi\rangle$ is smooth in ℓ and x. We can extend this notion to all of U_0 (by partitions of unity, for instance).

<u>Theorem</u> 7 ([12]). Let G be a stratified nilpotent Lie group, and let L be a homogeneous left invariant operator on G. Suppose that there is a smoothly varying map $\ell \to \varphi_\ell \in H^\infty(\pi_\ell)$ on U_0, not identically 0, with $\varphi_\ell \in \operatorname{Ker} \pi_\ell(L^t)$ for all ℓ. Then L is not locally solvable.

In other words, if $\pi_\ell(L^t)$ has nontrivial kernel in $H^\infty(\pi_\ell)$ for all ℓ in an open (nonvoid) set in U_0, and if one can make elements of the kernel vary smoothly, then L is not locally solvable.

The proof divides into two parts:

1. Suppose that we can find $\psi \in \mathcal{S}(G)$, $\psi \neq 0$, with $L^t\psi = 0$. Then L is not locally solvable. For if L is locally solvable, then (Proposition 1) L has a local fundamental solution. We can use this solution and the homogeneity of L to get the following: let $|\ |$

be a homogeneous norm on $\mathfrak{G}$ (see [21]), and lift $|\ |$ to G via exp. Then for every $\chi \in C_c^\infty(G)$ and every integer m, we can find functions $u_m \in C_c^\infty(G)$ such that

$$Lu_m = \chi \text{ on the "ball" } |x| \leq m;$$

$$u_m(x) = 0 \text{ if } |x| \geq m+1;$$

$$\|Lu_m\|_\infty \text{ grows at most polynomially with } m.$$

Now

$$(\chi,\psi) = (Lu_m,\psi) + (\chi - Lu_m,\psi)$$

$$= (u_m, L^t\psi) + (\chi - Lu_m,\psi) = (\chi - Lu_m,\psi),$$

since $L^t\psi = 0$; as $\chi - Lu_m$ is polynomially bounded in m and vanishes for $|x| \leq m$ and as ψ decreases more rapidly than any polynomial, the last term goes to 0. Hence $(\chi,\psi) = 0$ for all χ, which is impossible.

2. Now we need to show that if the φ_ℓ exist as in the hypotheses, then there exists $\psi \in S(G)$ with $L^t\psi = 0$. One straightforward approach is this: let P_ℓ be the projection on span φ_ℓ, and let $\alpha \in C_c^\infty(U_0)$ have compact support contained in the interior of supp $(\ell \mapsto \varphi_\ell)$. Now define ψ by

$$\pi_\ell(\psi') = \alpha(\ell)P_\ell, \ \ell \in U_0 \quad (\psi'(x) = \psi(x^{-1});$$

$$\pi(\psi') = 0 \text{ for all } \pi \in G^\wedge \text{ not in general position.}$$

Then $L^t\psi$ satisfies

$$(\pi_\ell^t(L^t\psi))^t = (\pi_\ell^t(\psi)\pi_\ell^t(L))^t = \pi_\ell(L^t)\pi_\ell(\psi') = \alpha_\ell\pi_\ell(L^t)P_\ell = 0.$$

Thus we are done if $\psi \in S(G)$. That is, we now need a sort of Paley-Wiener theorem.

The problem is that it is difficult to give a proof that $\psi \in S(G)$.

(For one characterization of the Fourier transforms of elements of $\mathcal{S}(G)$, see [26].) Instead, one needs a trick. Suppose that L (on G) satisfies the hypotheses of Theorem 7, and suppose that G is of codimension 1 in G_1. Regard L as an element of $\mathfrak{U}(\mathfrak{G}_1)$, where $\mathfrak{G}_1$ is the Lie algebra of G_1. One can then show that the hypotheses of Theorem 5 are satisfied for representations of G_1. By induction, therefore, we may replace G with any nilpotent Lie group H containing G. We can always imbed G in the group N_n of unipotent upper triangular $n \times n$ matrices (for large n), and then it is known (see [10]) that the function ψ_n, constructed as above, but on N_n, is in $\mathcal{S}(N_n)$. Moreover, $L^t \psi_n = 0$; restrict ψ_n to an appropriate G-coset to get ψ.

As one example, the theorem shows that the Lewy operator $L = X + iY$ on the 3-dimensional Heisenberg group is unsolvable. One can realize the representations π_ℓ (described in Example 4) on $L^2(\mathbb{R})$ so that

$$\pi_\ell(L) = |\ell|^{1/2}\left(\frac{d}{dt} + t\right) \quad \text{if} \quad \ell < 0.$$

Then $e^{-t^2} \in \operatorname{Ker} \pi_\ell$ for all $\ell < 0$. Similarly, the operators $X^2 + Y^2 + inZ$, $n \in \mathbb{Z}$, are unsolvable.

One can also use Theorem 7 to show that for homogeneous operators of the sort described in Theorem 4, the sufficient condition given there for semiglobal solvability is also necessary for local solvability. The hypotheses of Theorem 7 can be checked by using the fact that the π_ℓ form an analytic family of operators. (More recently, in [60], Lévy-Bruhl has shown that if G is stratified ($\mathfrak{G} = \bigoplus_{j=1}^{k} \mathfrak{G}_j$), if L is homogeneous, and if $\pi(L)$ is invertible for all nontrivial $\pi \in G^\wedge$ with $\pi|_{\exp \mathfrak{G}_k}$ trivial, then L is unsolvable whenever $\operatorname{Ker}\pi_\ell(L) \cap H^\infty(\pi_\ell) \neq (0)$ for an open set of $\ell \in U_0$.)

The sufficient condition of Theorem 4 for semiglobal solvability is also necessary for local solvability of operators that are elliptic in the generating directions on 2-step groups (even when the operators are inhomogeneous and the group is not an H-group). But the proof is different.

Theorem 8 ([13]). Let G be a 2-step nilpotent Lie group, and let L be a left invariant operator on G, elliptic in the generating directions. Suppose that there is a nonzero function $f \in L^2(G)$ with $L^t f = 0$. (The function f exists iff $\pi_\ell(L^t)$ has nonzero kernel on a nonempty open set of elements $\ell \in U_0$; $\operatorname{Ker} \pi_\ell(L^t)$ is necessarily in $H^\infty(\pi_\ell)$.) Then L is not locally solvable.

The idea of the proof is a modification of the argument in [24]. The first step is to show that if $L^2(G) \cap \operatorname{Ker} L^t \neq 0$, then there is a map $P: L^2(G) \to \operatorname{Ker} L^t$ with the following properties:

1. $P = P^t$ (P^t is defined by $\int_G (Pf)g\,dx = \int f(P^t g)\,dx$);

2. P is pseudolocal (i.e., if $f \in L^2(G)$ is C^∞ near x, then Pf is C^∞ near x);

3. P is not infinitely smoothing (i.e., there exists $f \in L^2(G)$ such that Pf is not C^∞.

For simplicity in describing the rest of the argument, assume that for each k there is a function $f_k \in L^2(G) \cap C^k(G)$ such that Pf_k is not infinitely smooth. We may assume (using the pseudolocality of P) that f_k has compact support. From Proposition 1, we may find $u_k \in L^2(G)$ with $Lu_k = f_k$ near e when k is sufficiently large. Pick such a k. Then $f_k - Lu_k \equiv 0$ near e, and so $P(f_k - Lu_k)$ is C^∞ near e. But

$$PL = (L^t P^t)^t = 0.$$

Therefore $P(f_k - Lu_k) = Pf_k$, which is not C^∞ near e. This gives a contradiction, and the theorem follows. (The argument in [13] is similar in outline, but proceeds by microlocalization.)

As noted, Theorems 8 and 4 show that for left invariant operators L on H-groups that are elliptic in the generating directions, L is locally solvable iff L is globally solvable iff $(\mathrm{Ker}\ L^t) \cap L^2(G) = \{0\}$. This result suggests the question of determining Range L when L is unsolvable. For the Lewy operator, this question was answered in [24]; see [23] for further information on Range L when G is a Heisenberg group.

The last result to be considered here gives information about the existence of tempered fundamental solutions. We begin with an easy lemma.

Proposition 5. Let G_1 be a connected subgroup of G; let $L \in \mathfrak{U}(\mathfrak{G}_1) \subseteq \mathfrak{U}(\mathfrak{G})$. Then:

(1) L is locally solvable on G iff L is locally solvable on G_1;

(2) L has a fundamental solution on G iff L has a fundamental solution on G_1;

(3) L has a tempered fundamental solution on G iff L has a tempered fundamental solution on G_1.

Here is a proof of (2) which is easily adapted to the other parts (since (1) means that L has a local fundamental solution). By induction, we may assume that G_1 has codimension 1 in G. If $L\xi_1 = \delta_{G_1}$, then define $\xi \in \mathcal{D}'(G)$ by $(\xi,\varphi) = (\xi_1, \varphi|_{G_1})$; then $L\xi = \delta_G$. If $L\xi = \delta_G$, then write G as a semidirect product of $\mathbb{R}$ with G_1, so that $G = \mathbb{R} \times G_1$ as a manifold. Choose $\psi \in C_c^\infty(\mathbb{R})$ with $\psi(0) = 1$, and define $\psi_1 \in \mathcal{D}'(G)$ by $(\xi_1, \varphi_1) = (\xi, \psi \otimes \varphi_1)$. Then $L\xi_1 = \delta_{G_1}$.

Proposition 5 means that in questions involving fundamental solutions, one can always replace G with any larger group. (Of course, one pays a price; analysis on the larger group may be harder.) Now call the nilpotent Lie group G <u>special</u> if

G is stratified;

G has 1-dimensional center;

G has square-integrable representations (i.e., for all $\ell \in \mathfrak{G}^*$ such that $\ell \neq 0$ on the center $\mathfrak{Z}$ of $\mathfrak{G}$, $B_\ell(X,Y) = \ell([X,Y])$ is nondegenerate on $\mathfrak{G}/_{\mathfrak{Z}}$; see [44]);

$\mathfrak{G}$ has an ideal $\mathfrak{H}$ which is polarizing for all $\ell \in \mathfrak{G}^*$ such that $\ell|_{\mathfrak{H}} \neq 0$. (Note: $\mathfrak{H}$ is necessarily Abelian.)

It is proved in [8] that every stratified Lie group can be imbedded in a special Lie group so that the stratification is preserved. (In fact, every Lie group can be imbedded in a special Lie group.) Thus it suffices, when considering homogeneous left invariant operators, to deal with special Lie groups.

<u>Theorem</u> 9 ([8]). Let G be a special Lie group; let $L \in \mathfrak{U}(\mathfrak{G})$ be homogeneous. For a representation π of G on $H(\pi)$, define the Sobolev spaces $H^d(\pi)$ as in the discussion after Theorem 3. If L has a tempered fundamental solution, then there are integers d_1, d_2 such that for all $\ell \in U_0$, the operator $\pi_\ell(L)$ (regarded as a map from $H^{d_1}(\pi_\ell)$ to $H^{d_2}(\pi_\ell)$ has a bounded right inverse. If G is a Heisenberg group, the converse holds as well.

The first half of this theorem is not hard to prove. Because G is square integrable, the parametrizing subspace V of $\mathfrak{G}^*$ can be taken to be any complement to $\mathfrak{Z}^\perp$, and hence 1-dimensional. Then $U_0 \cong \mathbb{R} - \{0\}$, and the homogeneity means that one needs inverses only on π_1 and π_{-1}. The existence of a fundamental solution implies easily that on the weighted Sobolev spaces of G, one has an inequality like

$$(3.1) \qquad \exists s,s': \left|\int_G f(x)g(x)dx\right| \leq C\|f\|_s\|L^t g\|_{s'}, \; f,\ g \in S(G);$$

if one takes a partial Fourier transform in the variables of $\mathfrak{Z}$, the action of L decomposes into an action on representation spaces, and one can easily read off an inequality like (3.1) involving the Sobolev spaces $H^d(\pi_1)$ (and one for $H^d(\pi_{-1})$. The result on inverses now follows.

Theorem 3 says that if the $\pi_\ell(L)$ have right inverses on appropriate Sobolev spaces, then L is semiglobally solvable. The proof of the converse for Heisenberg groups G uses the group structure heavily. We can regard the group as $\mathbb{R}^{2m} \times \mathbb{R}$, with $\{0\} \times \mathbb{R}$ central; a typical element can be written as (x,z), with $x \in \mathbb{R}^{2m}$. Let Z span the center of $\mathfrak{G}$, and for simplicity assume that Z is not a factor of L. Theorem 3 implies that we can find a tempered distribution u and an integer $k \geq 0$ such that

$$Lu = Z^k\delta.$$

(In this case, Z generates $\mathfrak{Z}(\mathfrak{U}(\mathfrak{G}))$.) Since $Z = \frac{\partial}{\partial z}$, the (constant coefficient) theorem on solvability in $S'(\mathbb{R}^n)$ says that we can find $\xi_0 \in S'(G)$ with

$$Z^k\xi_0 = u.$$

Let $L\xi_0 = \alpha_0$; then

$$Z^k(\alpha_0 - \delta) = 0.$$

Set $\alpha = \alpha_0 - \delta$. If we can find a tempered distribution ξ with $L\xi = \alpha$, we are done; then $L(\xi_0 - \xi) = \delta$.

Suppose for definiteness that $Z\alpha = 0$; the general case $(Z^k\alpha = 0)$

is done by a similar argument and an induction on k. We can write L in coordinates as

$$L = L_0(D_x) + \sum_{j=1}^{m} L_j(x,D_x)Z^j,$$

where L_0 is a polynomial in $D_x = (\frac{\partial}{\partial x_1}, \ldots, \frac{\partial}{\partial x_n})$, and $L_0(D_x) \neq 0$. It follows from an example in Chapter IV of [58] that $\alpha = \beta(x) \otimes 1$, since $Z\alpha = 0$. Now we use the facts that $\beta \in S(\mathbb{R}^{2m})$ and L_0 is a constant coefficient operator to find $\xi_1 \in S'(\mathbb{R}^{2m})$ with $L_0(D_x)\xi_1 = \beta$. But then

$$L(\xi_1 \otimes 1) = \beta \otimes 1,$$

and we are done. (This argument does not work on other special groups; there, the sufficiency of the condition is an open question.)

This report has necessarily concentrated on known results, but it should be apparent how many questions are still open. Here are two among many:

1. Is there a left invariant L which has a fundamental solution, but no tempered fundamental solution? (I personally doubt it.)

2. Theorem 3 and its extensions give sufficient conditions for local solvability of homogeneous L; Theorem 6 gives a necessary condition. An example in [12] shows that the existence of right inverses to $\pi_\ell(L)$ for all $\ell \in U_0$ is not sufficient for local solvability. What are good necessary and sufficient conditions?

Bibliography

[1] Agmon, S., Lectures on Elliptic Boundary Value Problems. Princeton, Van Nostrand, 1965.

[2] Atiyah, M., Resolution of Singularities and Division of Distributions, Comm. Pure and Appl. Math. 23(1970), 145-150.

[3] Bernstein, I.N., The Analytic Continuation of Generalized Functions with respect to a Parameter, Functional Analysis and its Applications 6(1972), 273-285.

[4] Bernstein, I.N., and Gelfand, S.I., Meromorphy of the Function P^λ, Functional Analysis and its Applications 3(1969), 68-69.

[5] Cerézo, A., and Rouvière, F., Résolubilité locale d'un opérateur différentiel invariant du prémier ordre, Ann. Sci. Éc. Norm. Sup. 4(1971), 21-30.

[6] Chang, W., Invariant Differential Operators and P-Convexity of Solvable Lie Groups, Advances in Math. 46(1982), 284-304.

[7] Corwin, L., A Representation - Theoretic Criterion for Local Solvability of Left Invariant Operators on Nilpotent Lie Groups, Transactions Am. Math. Soc. 264(1981), 113-120.

[8] Corwin, L., Criteria for Solvability of Left Invariant Operators on Nilpotent Lie Groups, Transactions Am. Math. Soc., to appear.

[9] Corwin, L., and Greenleaf, F.P., Character Formulas and Spectra of Compact Nilmanifolds, J. Funct. Anal. 21(1976), 123-154.

[10] Corwin, L., and Greenleaf, F.P., Fourier Transforms of Smooth Functions on Certain Nilpotent Lie groups, J. Funct. Anal. 37(1980), 203-217.

[11] Corwin, L., and Greenleaf, F.P., Solvability of Certain Left Invariant Differential Operators by Nilmanifold Theory, Comm. Pure and Appl. Math., to appear.

[12] Corwin, L., and Rothschild, L.P., Necessary Conditions for Local Solvability of Homogeneous Left Invariant Differential Operators on Nilpotent Lie Groups, Acta. Math. 147(1981), 265-288.

[13] Corwin, L., and Rothschild, L.P., Solvability of Transversally Elliptic Differential Operators on Nilpotent Lie Groups, preprint.

[14] Cygan, J., and Richardson, L., Global Solvability on 2-Step Compact Nilmanifolds, preprint.

[15] Dixmier, J., Enveloping Algebras. Amsterdam, North-Holland, 1977.

[16] Dixmier, J., and Malliavin, P., Factorisations de fonctions et de vecteurs indéfiniment différentiables, Bull. Sc. Math. 102(1978), 305-330.

[17] Duflo, M. Opérateurs différentiels bi-invariants sur un groupe de Lie, Ann. Sci. Éc. Norm. Sup. 10(1977), 265-288.

[18] Duflo, M., Séminaire Goulaouic-Schwartz, 1977-1978.

[19] Duflo, M., and Wigner, D., Convexité pour les opérateurs différentiels invariants sur les groupes de Lie, Math. Zeit. 61(1979), 61-80.

[20] Ehrenpreis, L., Solution of some Problems of Division III, Am. J. Math. 78(1956), 685-715.

[21] Folland, G.B., Subelliptic Estimates and Function Spaces on Nilpotent Lie Groups, Ark. før Math. 13(1979), 161-207.

[22] Gelfand, I.M., Graev, M.I., and Piatetskii-Shapiro, I.I., Representation Theory and Automorphic Functions. Philadelphia, Saunders, 1969.

[23] Geller, D., Local Solvability and Homogeneous Distributions on the Heisenberg Group, Comm. in Partial Diff. Eqs. 5(1980), 475-560.

[24] Greiner, P., Kohn, J.J., and Stein, E., Necessary and Sufficient Conditions for Solvability of the Levy Equation, Proc. Nat. Acad. Sci. (U.S.A.) 72(1975), 3787-3789.

[25] Grigis, A., and Rothschild, L.P., A Criterion for Analytic Hypoellipticity on a Class of Differential Operators with Polynomial Coefficients, Annals of Mathematics, to appear.

[26] Hai, Nghiem Xuan, La transformation de Fourier Plancherel analytique des groupes de Lie, II: les groupes nilpotents. Prepublications, Université de Paris-Sud (Orsay), 81T23.

[27] Hardy, G.H., and Wright, E.M., An Introduction to the Theory of Numbers, 4th Ed. Oxford, the Clarendon Press, 1950.

[28] Hörmander, L., On the Division of Distributions by Polynomials, Ark. før Math. 3(1958), 555-568.

[29] Hörmander, L., Differential Operators of Principal Type, Math. Ann. 140(1960), 124-146.

[30] Hörmander, L., Linear Partial Differential Operators. New York, Springer, 1969.

[31] Hörmander, L., On the Existence and Regularity of Solutions of Linear Pseudo-Differential Equations, L'Ens. Math. 17(1971), 99-163.

[32] Howe, R., Frobenius Reciprocity for Unipotent Algebraic Groups over Q, Am. J. Math. 93(1971), 163-172.

[33] Kato, T., Perturbation Theory for Linear Operators. New York, Springer, 1976.

[34] Kirillov, A.A., Unitary Representations of Nilpotent Lie Groups, Uspehi Mat. Nauk. 17(1962), 57-110.

[35] Lévy-Bruhl, P., Resolubilité locale de certains operateurs invariants du second ordre sur des groupes de Lie nilpotents, Bull. Sc. Math. 104(1980), 369-391.

[36] Lévy-Bruhl, P., Application de la formule de Plancherel à la résolubilité d'opérateurs invariants à gauche sur des groupes nilpotents d'ordre deux, Bull. Sc. Math. 106(1982), 171-191.

[37] Lévy-Bruhl, P., Resolubilité locale d'opérateurs homogénes invariants à gauche sur certains groupes de Lie nilpotents de rang trois, preprint.

[38] Lewy, H., An Example of a Smooth Linear Partial Differential Equation Without Solution, Ann. Math. 66(1957), 155-158.

[39] Lion, G., Hypoellipticité et résolubilité d'opérateurs differéntiels sur des groupes nilpotents de rang 2, Comptes Rendus Acad. Sc. (Paris) 290(1980), 271-274.

[40] Lojasiewicz, S., Sur le probleme de division, Studia Math. 18(1959), 87-136.

[41] Malgrange, B., Sur la propagation de la regularité de solutions des équations constants, Bull. Math. Soc. Sci. Math. Phys. R.P. Roumanie 3(53)(1959), 433-440.

[42] Melin, A., Parametrix Constructions for some Classes of Right Invariant Differential Operators on the Heisenberg Group, preprint.

[43] Metivier, G., Hypoellipticité analytique sur des groupes nilpotents de rang 2, Duke Math. J. 47(1980), 195-222.

[44] Moore, C.C., and Wolf, J., Square Integrable Representations of Nilpotent Groups, Transactions A.M.S. 185(1973), 445-462.

[45] Nirenberg, L., and Treves, J.F., On Local Solvability of Linear Partial Differential Equations, Part I: Necessary Conditions. Comm. Pure and Appl. Math. 23(1970), 1-38.

[46] Pukanszky, L., Leçons sur les Representations des Groupes. Paris, Dunod, 1967.

[47] Pukanszky, L., On the Characters and the Plancherel Formula of Nilpotent Groups, J. Funct. Anal. 1(1967), 255-280.

[48] Raïs, M., Solutions élémentaires des operateurs differentiels bi-invariants sur un groupe de Lie nilpotent, Comptes Rendus Acad. Sci. (Paris) 273(1971), 495-498.

[49] Rauch, J., and Wigner, D., Global Solvability of the Casimir Operator, Ann. of Math. 103(1976), 229-236.

[50] Richardson, L., Decomposition of the L^2 Space of a General Compact Nilmanifold, Am. J. Math. 93(1971), 173-190.

[51] Richardson, L., Global Solvability on Compact Heisenberg Manifolds, Transactions A.M.S. 273(1982), 309-318.

[52] Rockland, C., Hypoellipticity of the Heisenberg Group — Representation-Theoretic Criteria, Transactions A.M.S. 240 (1978), 1-52.

[53] Rothschild, L.P., Local Solvability of Left-Invariant Operators on the Heisenberg Group, Proc. Am. Math. Soc. 74(1979), 383-388.

[54] Rothschild, L.P., Local Solvability of Second Order Operators on Nilpotent Lie Groups, Ark. før Math. 19(1981), 145-175.

[55] Rothschild, L.P., and Tartakoff, D., Inversion of Analytic Matrices and Local Solvability of some Invariant Differential Operators on Nilpotent Lie Groups, Comm. Partial Diff. Eq. 6(1981), 625-650.

[56] Rouvière, F., Sur la resolubilité locale des opérateurs bi-invariants, Annali Scuola Normale Superiore (Pisa) 3(1976), 231-244.

[57] Rouvière, F., Solutions distributions de l'opérateurs de Casimir, Comptes Rendus Acad. Sci. (Paris) 282(1976), 853-856.

[58] Schwartz, L., Théorie des Distributions. Paris, Hermann, 1966.

[59] Treves, J.F., Topological Vector Spaces, Distributions, and Kernels. New York, Academic Press, 1967.

[60] Lévy-Bruhl, P., Condition nécessaire de résolubilité locale d'opérateurs invariants sur les groupes nilpotents, preprint.

[61] Rouvière, F., Invariant Differential Equations on Certain Semi-simple Lie Groups, Transactions Am. Math. Soc. 243(1978), 97-114.

HARMONIC ANALYSIS ON HEISENBERG TYPE GROUPS FROM A GEOMETRIC VIEWPOINT

Michael Cowling and Adam Korányi

Introduction

This paper is a progress report on our attempt to understand harmonic analysis on the Heisenberg group and on its generalizations from the viewpoint of dilations and rotations. The groups to which this point of view can be most successfully applied are the nilpotent parts in the Iwasawa decomposition of simple Lie groups of real rank one. R^n occurs in this way in $SO(n,1)$ and the Heisenberg group H_n in $SU(n+1,1)$; it is known from classification theory (which we will not use) that there exist two more types of such groups. These nilpotent groups will be the main objects of our study. They admit a one-parameter group of dilations and a large group of rotations; in addition, there is on them a natural analogue of inversion with respect to a sphere. This is a geometric structure which can be well exploited for the study of harmonic analysis of the group itself and of the ambient simple group.

Much of what we do applies to a somewhat larger class, the class of H-type groups introduced by A. Kaplan [11]. These groups are very convenient to work with, so whatever we can we will do in this more general setting. A considerable part of the results we will report on are not new. Our methods, however, are largely new, and we shall attempt to give an easily accessible unified treatment of the subject.

In §1 we recall the definition of H-type groups and give simple proofs of some of their main properties. We also show how the nilpotent Iwasawa components of the rank one simple groups fit into this class. In §2 we first discuss the analytic continuation of certain families of distributions which generalise the classical Riemann-Liouville operators. Then we proceed to determine the Fourier transforms (in the sense of the nilpotent group) of these families. Part of these results were obtained in [1] and [2]; it is mainly for the special case of the Heisenberg groups that we obtain here a considerably more general version. This latter result was also found earlier by D. Geller [7]; we will give here our own proof which is different and, at least from our point of view, simpler than Geller's.

In §3 we study the (non-unitary) principal series representations of the simple Lie groups of real rank one. The families of distributions from §2 arise here as intertwining operators; they can be used to describe the complementary series of unitary representations. For the spherical principal series a study of this type was made in [1] and [2] where the results of Kostant [15] were reproved with the aid of the Fourier transform of the nilpotent group. The novelty in the present paper is that in the case of $SU(n,1)$ and of its universal covering group we can also handle the principal series induced from non-trivial characters of the parabolic group. In this way we get a new simple proof of some results of Flensted-Jensen [5]. We note that in the case of SU(2,1) these results amount to a complete description of the entire complementary series.

In §4 we describe some applications. It was shown in [2] that the fundamental solution of the invariant sublaplacian of every H-type group can immediately be obtained from the analytic continuation of our families of homogeneous kernels. Here we note that the same is true for a more general class of operators studied by Folland and Stein in [6]. Then we will describe how group representation theory can be used to define a generalised Kelvin transformation on the nilpotent parts of the simple groups of real rank one. For the Heisenberg group such a generalisation was given in [13]; here it is done, in the general case, in a more conceptual way as an application of the results of §3. This approach to the Kelvin transform was also known to E.M. Stein to whom we express our thanks for some useful discussions. In conclusion, we will make some remarks about the use of the Kelvin transform for the construction of harmonic polynomials.

Chapter 1

Groups of Type H

A group of type H is a connected, simply connected Lie group whose Lie algebra is of type H.

A Lie algebra of type H is defined as follows [10]. Let $\mathfrak{n} = \mathfrak{v} \oplus \mathfrak{z}$ be a direct sum of real Euclidean spaces with a Lie algebra structure such that $\mathfrak{z}$ is the center and for all $X \in \mathfrak{v}$ with $|X| = 1$ the map $\mathrm{ad}(X)$ is a surjective isometry from the orthogonal complement $\mathfrak{v} - \ker \mathrm{ad}(X)$ onto $\mathfrak{z}$.

The flexibility of this notion is due to the fact that there is the following equivalent definition.

$\mathfrak{n} = \mathfrak{v} \oplus \mathfrak{z}$ is a direct sum of real Euclidean spaces; there is a linear map $j : \mathfrak{z} \to \mathrm{End}\,\mathfrak{v}$ such that, for all $X \in \mathfrak{v}$, $Y \in \mathfrak{z}$,

$$|j(Y)X| = |Y|\,|X| \tag{1.1}$$

$$j(Y)^2 = -|Y|^2 I \tag{1.2}$$

and the Lie algebra structure on $\mathfrak{n}$ is given by the condition that $\mathfrak{z}$ is the center and

$$\langle Y, [X,X']\rangle = \langle j(Y)X, X'\rangle \tag{1.3}$$

for all $Y \in \mathfrak{z}$; $X, X' \in \mathfrak{v}$. (Observe that, given the first two properties, it follows by polarizing (1.1) and using (1.2) that $j(Y)$ is skew-symmetric; therefore (1.3) does indeed define a Lie bracket.)

We give the proof of the equivalence, slightly simplifying [10].

Suppose $\mathfrak{n}$ is of type H (in the sense of the first definition). Then the equation (1.3) <u>defines</u> a linear map j, and we only have to prove (1.1) and (1.2).

Let $0 \neq X \in \mathfrak{v}$ and $Y \in \mathfrak{z}$ be given. Then, by the isometry property of $|X|^{-1}\mathrm{ad}(X)$ there is a unique X_1 in $\mathfrak{v} - \ker \mathrm{ad}(X)$ such that

$$|X|^{-1}[X, X_1] = Y$$

and $|X_1| = |Y|$. To determine X_1 we note that for all $X' \in \mathfrak{v} - \ker \mathrm{ad}(X)$ we have (by the isometry property and by (1.3))

$$\langle X_1, X'\rangle = |X|^{-2}\langle [X, X_1], [X, X']\rangle = |X|^{-1}\langle Y, [X, X']\rangle$$

$$= |X|^{-1}\langle j(Y)X, X'\rangle .$$

Hence

$$X_1 = |X|^{-1} j(Y)X .$$

The relation $|X_1| = |Y|$ now gives (1.1). To prove (1.2) we note that $j(Y)$ is skew-symmetric by (1.3); hence, using (1.1),

$$-\langle j(Y)^2 X, X\rangle = \langle j(Y)X, j(Y)X\rangle = |Y|^2 \langle X,X\rangle .$$

Polarizing this identity gives

$$-\langle j(Y)^2 X , X'\rangle = |Y|^2 \langle X,X'\rangle$$

for all $X' \in \mathfrak{v}$. Hence we have (1.2).

To prove the converse, we assume (1.1), (1.2), (1.3) and we show that for any given $X \in \mathfrak{v}$ with $|X| = 1$ and $Y \in \mathfrak{z}$ we can find $X_o \in \mathfrak{v}$ such that $X_o \perp \ker \mathrm{ad}(X)$, $[X,X_o] = Y$ and $|X_o| = |Y|$. We claim that $X_o = j(Y)X$ has these properties. In fact, by (1.3) it is clearly orthogonal to $\ker \mathrm{ad}(X)$ and by (1.1), $|X_o| = |Y|$. Also, using (1.3) and then polarizing (1.1) we get, for all $Y' \in \mathfrak{z}$,

$$\langle Y', [X, j(Y)X]\rangle = \langle j(Y')X , j(Y)X\rangle = \langle Y',Y\rangle$$

which gives $[X, j(Y)X] = Y$, finishing the proof.

(We may note the general relation

$$[X, j(Y)X] = |X|^2 Y \tag{1.4}$$

which played a role in both parts of the proof.)

On a group of type H there is a natural gauge [10], [3] defined by

$$N(g) = (|X|^4 + |Y|^2)^{1/4} \tag{1.5}$$

for $g = \exp(X + Y/4)$, $X \in \mathfrak{v}$, $Y \in \mathfrak{z}$. Given any other element $g' = \exp(X' + Y'/4)$ we have

$$N(gg') \leqq N(g) + N(g') .$$

<u>Proof.</u> (Simplification of [3].) By the Campbell-Hausdorff formula we have

$$gg' = \exp(X + X' + 1/4(Y + Y' + 2[X,X'])) .$$

We may write

$$N(gg')^2 = \left| |X+X'|^2 + i\,|Y+Y'+2[X,X']| \right|$$

(where the outer bars denote the absolute value of a complex number). It follows that

$$N(gg')^2 \leqq \left| |X|^2 + i|Y| \right| + \left| |X'|^2 + i|Y'| \right|$$

$$+ 2\left| \langle X,X' \rangle + i|[X,X']| \right| .$$

The two first terms on the right are $N(g)^2$ and $N(g')^2$, so it suffices to prove that the third term is majorized by $2N(g)N(g')$. This will follow if we prove that

$$\langle X,X' \rangle^2 + |[X,X']|^2 \leqq |X|^2 |X'|^2 .$$

To prove this, we write $X' = X_1 + X_2$ with X_1 in $\ker \operatorname{ad}(X)$ and X_2 in its orthogonal complement. The left hand side reduces to

$$\langle X,X_1 \rangle^2 + |[X,X_2]|^2$$

and the right hand side to

$$|X|^2 |X_1|^2 + |X|^2 |X_2|^2 .$$

The inequality becomes trivial, since $|[X,X_2]| = |X|\,|X_2|$ by the definition of type H.

The main importance of H type groups for us is that this class includes the nilpotent part in the Iwasawa decomposition of a simple Lie group of real rank one. This can be checked using the classification, or proved from general semisimple theory [14]. Here we give another general proof (essentially that of [17]) which is very short but is based on some relatively deep results.

Let G be a non-compact simple group of real rank one, $\mathfrak{g}$ its Lie algebra. As usual, we consider a Cartan decomposition $\mathfrak{g} = \mathfrak{k} + \mathfrak{p}$ and a maximal Abelian subspace $\mathfrak{a}$ in $\mathfrak{p}$. We denote the positive root spaces with respect to $\mathfrak{a}$ by $\mathfrak{g}_\alpha$ and $\mathfrak{g}_{2\alpha}$; the nilpotent Iwasawa algebra is then $\mathfrak{n} = \mathfrak{g}_\alpha + \mathfrak{g}_{2\alpha}$. It can happen that one of the root spaces is zero; this happens exactly when $G = SO(n,1)$. In this case $\mathfrak{n}$ is clearly Abelian, and we have a degenerate special case of a group of type H. In order not to complicate our statements in the following we exclude this trivial case and assume that $\mathfrak{g}_\alpha$ and $\mathfrak{g}_{2\alpha}$ are non-zero.

As usual, we denote by M the centralizer of $\mathfrak{a}$ in K, the subgroup corresponding to $\mathfrak{k}$. M then operates on $\mathfrak{g}_\alpha$ and on $\mathfrak{g}_{2\alpha}$ by the adjoint representation and preserves the usual Euclidean structure $\langle U,V\rangle_o = -B(U,\theta V)$ where B is the Killing form and θ the Cartan involution. Kostant's double transitivity theorem (see e.g. [20]) states that M acts transitively on $S_1 \times S_2$, the product of the unit spheres in $\mathfrak{g}_\alpha$ and $\mathfrak{g}_{2\alpha}$.

Note that the Euclidean structure on an H-type algebra $\mathfrak{n} = \mathfrak{v} + \mathfrak{z}$ is not unique: Clearly, we may change the norm on $\mathfrak{v}$ by a factor p and on $\mathfrak{z}$ by p^2.

Now we fix any Euclidean structure of the form $\langle,\rangle = c_1\langle,\rangle_o$ on $\mathfrak{v}$. We take $X \in \mathfrak{g}_\alpha$ with $|X| = 1$ and $ad(X) \neq 0$. The stabilizer M_X of X in M is then transitive on S_2. Since $ad(X)m \cdot X' = m \cdot ad(X)X'$ for $m \in M_X$, we have that $ad(X)$ is surjective. It also follows that we can fix a Euclidean structure $\langle,\rangle = c_2\langle,\rangle_o$ on $\mathfrak{z}$ so that $ad(X)$ is an isometry of $\ker(ad\,X)^\perp$ onto $\mathfrak{g}_{2\alpha}$. This gives us an algebra of type H: In fact, any element in $\mathfrak{v}$ of length 1 is of the form $m \cdot X$ with some $m \in M$. Since m is an automorphism and an isometry, it follows that $ad(m \cdot X)$ has the same properties as $ad(X)$.

Our statement is proved except for finding the value of c_2. This can be determined using the formula $[Y,\theta Y] = B(Y,\theta Y)H_{2\alpha}$ from [9, p. 54], similarly as in [14]. In fact, choosing $c_1 = (p+4q)^{-1}$ where p,q are the dimensions of $\mathfrak{g}_\alpha + \mathfrak{g}_{2\alpha}$, we get $c_2 = c_1$. We also have the formula

$$j(Y)X = ad(Y)\theta X \tag{1.6}$$

which follows immediately from (1.3) and from the invariance of the Killing form.

Chapter 2

Homogeneous Kernels on H-type Groups

In this section, we analyse in detail a family of homogeneous kernels on Heisenberg groups, and sketch a few generalisations to arbitrary H-type groups. Some of our results are not new. In particular, our Theorem 2.6 has also been proved by D. Geller [7], who has priority. But we feel our methods are simpler, so that our proof is not without merit.

For ease of notation, we consider a group N of type H, with Lie algebra $\mathfrak{n} = \mathfrak{v} + \mathfrak{z}$, and set

$$(v,z) = \exp(v + z/4) \qquad v \in \mathfrak{v} \ , \ z \in \mathfrak{z} \ .$$

We concentrate on the Heisenberg group H_n: here $\mathfrak{v} = \mathbf{C}^n$, $\mathfrak{z} = \mathbf{R}$, and the multiplication is given by the formula

$$(v',t')(v,t) = (v' + v \ , \ t' + t + 2\,\mathrm{Im}(v' v^*))$$

($v', v \in \mathbf{C}^n$, $t', t \in \mathbf{R}$; $v' v^*$ is the Hermitean inner product of v' and v). We define functions A and $\bar{A} : H_n \to \mathbf{C}$ by the rule

$$A(v,t) = |v|^2 + it \qquad \bar{A}(v,t) = |v|^2 - it$$

so that $A\bar{A} = N^4$. Let p be a real-homogeneous harmonic polynomial of degree m on $\mathbf{C}^n$, and denote by $K_{\sigma,\tau}$ the following kernel:

$$K_{\sigma,\tau}(v,t) = p(v) A^{\sigma}(v,t)\, \bar{A}^{\tau}(v,t) \ .$$

It is clear that $K_{\sigma,\tau}$ is locally integrable if

$$m + 2\mathrm{Re}(\sigma + \tau) + 2n + 2 > 0 \ ,$$

so $K_{\sigma,\tau}$ defines a distribution (obviously tempered) by integration. We shall show that $K_{\sigma,\tau}$ may be continued meromorphically into $\mathbf{C}^2$ as a distribution-valued function, with simple poles when

$$-(m + \sigma + \tau + n + 1) \in \mathbf{N} \ .$$

The residues at these poles are differential operators supported in $(0,0)$. We calculate, reasonably explicitly, the Fourier transform of $K_{\sigma,\tau}$ (Theorem 2.6).

The main tools in our study of $K_{\sigma,\tau}$ are the theory of functions and a mysterious differential equation satisfied by $K_{\sigma,\tau}$. We denote by V_j, $\bar{V}_j$ and T the following distributions on H_n, with support $(0,0)$:

$$\left.\begin{aligned} V_j(u) &= (\partial/\partial v_j)u(0,0) \\ \bar{V}_j(u) &= (\partial/\partial \bar{v}_j)u(0,0) \\ T(u) &= (\partial/\partial t)u(0,0) \end{aligned}\right\} \quad u \in \mathcal{D}(H_n) \qquad ,$$

and set $L_\gamma = (-1/2)\Sigma_{j=1}^{n} (V_j * \bar{V}_j + \bar{V}_j * V_j) + i\gamma T$. Then we find that (Theorem 2.3)

$$L_\gamma * K_{\sigma,\tau} = c\, K_{\sigma-1,\tau} + c'\, K_{\sigma,\tau-1}$$

(c and c' depend on γ, σ and τ). The calculation of the Fourier transform of $K_{\sigma,\tau}$ is a long but elementary consequence of this relation.

At the end of this section we treat groups of type H. We consider kernels of the following form:

$$K_\zeta(v,z) = p(v)\, |v|^{2s}\, N^\zeta(v,z)$$

(where $s \in R^+$), describe their meromorphic continuations, and for $p = 1$ and $s = 0$, calculate their Fourier transform.

Before we begin our study of homogeneous kernels, we shall remind the reader of a few well known facts about spherical harmonics in R^n, and recall the definition of the irreducible unitary representations of H-type groups.

We define a symmetric bilinear form on polynomials:

$$\langle\langle p,q \rangle\rangle = [p(\partial)q](0) \ ,$$

(where $p(\partial) = \Sigma_\alpha a_\alpha (\partial/\partial x_1)^{\alpha_1} \ldots (\partial/\partial x_n)^{\alpha_n}$ if $p(x) = \Sigma_\alpha a_\alpha x_1^{\alpha_1} \ldots x_n^{\alpha_n}$). Setting $((p,q)) = \langle\langle p,\bar{q} \rangle\rangle$, we obtain an inner product. This may be used to show that, if p is any polynomial of degree m, there is a unique "decomposition":

$$p(x) = \Sigma_{j=0}^{[m/2]} |x|^{2j}\, p_j(x) \ ,$$

with p_j harmonic (and of degree at most $m-2j$). Here [ ] is the integral part function.

Lemma 2.1. *If* p *and* q *are harmonic polynomials on* R^n *and* q *is homogeneous of degree* m , *then*

$$\langle\langle p,q\rangle\rangle = 2^{m-1}\pi^{-n/2}\Gamma(n/2+m)\int_{S^{n-1}} dx' p(x')q(x') ,$$

where dx' *is surface measure on* S^{n-1} .

Proof. It is well known that the action of $SO(n)$ on the space of homogeneous harmonic polynomials of degree m is irreducible. By Schur's Lemma it follows that $\langle\langle\, ,\,\rangle\rangle$ is proportional to the inner product in $L^2(S^{n-1})$. Calculating the norm of $(x_1+ix_2)^m$ in both, one finds that the constant of proportionality is $2^{m-1}\pi^{-n/2}\Gamma(n/2+m)$. It is also well known (and easy to see) that $H_m \perp H_k$ for $m \neq k$, with respect to both inner products.

■

On C^n , the space of (real-) homogeneous harmonic polynomials of degree m decomposes into $m+1$ irreducible $SU(n)$ modules, containing the polynomials of type (k,ℓ) , with $k,\ell \in N$, $k+\ell=m$ (see e.g. [18, §12.2]). A polynomial p is of type (k,ℓ) if

$$p(\lambda z) = \lambda^k \bar{\lambda}^{\ell} p(z) .$$

For such polynomials, the following Euler identities hold:

$$\Sigma_{j=1}^n z_j \partial/\partial z_j\, p = kp \; ; \; \Sigma_{j=1}^n \bar{z}_j \partial/\partial \bar{z}_j\, p = \ell p .$$

We now recall the irreducible representations of H_n . Apart from the characters $(v,t)\to\chi(v)$, with $\chi \in \hat{C}^n$, these are parametrised by $\pm\lambda$, $\lambda \in R^+$. In the so-called Fock model, the Hilbert space $\mathcal{H}_\lambda$ of the representation π_λ is the set of all holomorphic functions ξ on C^n for which

$$\|\xi\|^2 = \int_{C^n} dv |\xi(v)|^2 e^{-2\lambda|v|^2} < \infty$$

(here dv is the usual Lebesgue measure on C^n) and

$$\pi_\lambda(v,t)\xi(w) = \xi(w+v)\exp(-\lambda[\,|v|^2 + 2w\cdot v^* + it]) .$$

It is well known that π_λ is irreducible and that a basis for $\mathcal{H}_\lambda$ is given by the monomials, suitably normalized. These are C^∞ vectors for π_λ. The representation $\pi_{-\lambda}$ is the contragredient of π_λ. It may be realised on the space of antiholomorphic functions on C^n, square integrable relative to the measure $e^{-2\lambda|v|^2}dv$, by the formula

$$\pi_{-\lambda}(v,t)\xi(w) = \xi(w+v)\exp(-\lambda[\,|v|^2 + 2v\cdot w^* - it]) .$$

It is straightforward to check that, if ξ is a polynomial of degree d in $\mathcal{H}_{\pm\lambda}$, then

$$\pi_{\pm\lambda}(L_\gamma)\xi = (n+2d\pm\gamma)\lambda\xi .$$

The irreducible infinite dimensional unitary representations of a general H-type group may be parametrised by $S^{q-1}\times R^+$, where S^{q-1} is the unit sphere in $\mathfrak{z}$. For $\epsilon\in S^{q-1}$, we define $\pi_{\epsilon\lambda}$ on the space of all functions ξ on $\mathfrak{v}$, holomorphic relative to the complex structure $j(\epsilon)$, for which

$$\|\xi\|^2 = \int_{\mathfrak{v}} dv|\xi(v)|^2 e^{-2\lambda|v|^2} < \infty$$

(dv is Lebesgue measure on $\mathfrak{v}$), by the formula

$$[\pi_{\epsilon\lambda}(v,z)\xi](w) = \xi(v+w)\exp(-\lambda[\,|v|^2 + 2\{\langle w,v\rangle + i\langle\epsilon,[w,v]\rangle\} + i\langle\epsilon,z\rangle]) .$$

(Note that the term in braces is the Hermitean inner product on $\mathfrak{v}$ relative to the complex structure $j(\epsilon)$. On any H-type group, we let $L^{(\mathfrak{z})}$ be the following distribution, supported in $(0,0)$:

$$L^{(\mathfrak{z})}(u) = \Delta^{(\mathfrak{z})}u(0,0) \qquad u\in\mathcal{D}(N) ,$$

where $\Delta^{(\mathfrak{z})}$ is the usual Laplacean on $\mathfrak{z}$ (with sign to make it a positive operator). Then, obviously,

$$\pi_{\epsilon\lambda}(L^{(\mathfrak{z})}) = \lambda^2 .$$

On the Heisenberg group H_n, we consider $K_{\sigma,\tau}$:

$$K_{\sigma,\tau}(v,t) = p(v)A^\sigma(v,t)\bar{A}^\tau(v,t) ,$$

where p is a homogeneous harmonic polynomial on $\mathbb{C}^n$ of degree m. As long as $\mathrm{Re}(\sigma+\tau)$ is not too negative, $K_{\sigma,\tau}$ is locally integrable and so defines a distribution. We shall now prove that $K_{\sigma,\tau}$ may be continued meromorphically into $\mathbb{C}^2$.

We denote by ψ the characteristic function of the "unit ball" $\{(v,t) : |v|^4+|t|^2<1\}$, and by $\tilde{p}(\partial)_o$ the distribution on H_n, supported in $(0,0)$, defined by

$$\tilde{p}(\partial)_o(u) = [p(\partial)(u|_{\mathbb{C}^n})](0), \quad u\in\mathcal{D}(H_n) .$$

<u>Theorem 2.2.</u> <u>The distribution-valued function</u> $(\sigma,\tau)\to K_{\sigma,\tau}$ <u>extends meromorphically into</u> $\mathbb{C}^2$ <u>from</u> $\{(\sigma,\tau) : m+2\mathrm{Re}(\sigma+\tau)+2n+2>0\}$, <u>with simple poles when</u> $m+\sigma+\tau+n+1$ <u>is a non-positive integer, and no other singularities.</u> <u>Further,</u>

$$\lim_{m+\sigma+\tau+n+1\to 0}(m+\sigma+\tau+n+1)K_{\sigma,\tau}$$

$$= 2^{1-n-2m}\pi^{n+1}[\Gamma(-\sigma)\Gamma(-\tau)]^{-1}\tilde{p}(\partial)_o .$$

<u>Finally, if</u> $-n-1-m/2-J/2<a\leqq b<-n-1-m/2$, <u>the following estimate holds whenever</u> $\mathrm{Re}(\sigma+\tau)\in[a,b]$ <u>and</u> $|\mathrm{Im}(\sigma+\tau)|\geqq 1$:

$$|K_{\sigma,\tau}(u)|\leqq C\|u\|_{C^{J+1}}\exp(\pi|\mathrm{Im}(\sigma-\tau)|/2), \quad u\in\mathcal{D}(H_n) .$$

<u>Proof.</u> Obviously $K_{\sigma,\tau}=\psi K_{\sigma,\tau}+(1-\psi)K_{\sigma,\tau}$, and $(\sigma,\tau)\to(1-\psi)K_{\sigma,\tau}$ admits an analytic extension into $\mathbb{C}^2$. If $m+2\mathrm{Re}(\sigma+\tau)+2n+2<0$, then $(1-\psi)K_{\sigma,\tau}$ is integrable, and so, if $\mathrm{Re}(\sigma+\tau)\in[a,b]$,

$$\text{(2.1)} \qquad |(1-\psi)K_{\sigma,\tau}(u)|$$

$$\leqq \|u\|_{C^o}\|(1-\psi)K_{\sigma,\tau}\|_1$$

$$\leqq C\|u\|_{C^o}\iint_{|v|^4+|t|^2\geqq 1}dvdt\,|v|^m|A^{\sigma}(v,t)|\,|\bar{A}^{\tau}(v,t)|$$

$$\leqq C\|u\|_{C^o}\iint_{|v|^4+|t|^2\geqq 1}dvdt\,|v|^m N^{2\mathrm{Re}(\sigma+\tau)}(v,t)\exp(\pi|\mathrm{Im}(\sigma-\tau)|/2)$$

$$\leq C\|u\|_{C^0} \exp(\pi|\operatorname{Im}(\sigma-\tau)|/2) .$$

It remains to discuss $\psi K_{\sigma,\tau}$.

Fix an arbitrarily big positive integer J . We shall describe the continuation into $\{(\sigma,\tau) : m+2\operatorname{Re}(\sigma+\tau)+2n+2+J>0\}$ of the distribution-valued function $\psi K_{\sigma,\tau}$. This, of course, suffices.

Take u in $\mathcal{S}(H_n)$, and group together terms of the same homogeneity in its Taylor expansion in $(0,0)$. Then

$$u(v,t) = \Sigma^J_{h,j=0} u_j(v)t^h + u_J(v,t) ,$$

where $u_j(v)$ is homogeneous of degree j and $u_J(v,t) = o((|v|^2+|t|^2)^{J/2})$ as $(v,t)\to(0,0)$. Clearly, if $m+2\operatorname{Re}(\sigma+\tau)+2n+2>0$,

$$\iint dvdt\, \psi(v,t)K_{\sigma,\tau}(v,t)\, u(v,y)$$

$$= \Sigma^J_{h,j=0} \iint dvdt\, \psi(v,t)K_{\sigma,\tau}(v,t)u_j(v)t^h$$

$$+ \iint dvdt\, \psi(v,t)K_{\sigma,\tau}(v,t)u_J(v,t) .$$

The last term of this sum continues analytically into $\{(\sigma,\tau):m+2\operatorname{Re}(\sigma+\tau)+2n+2+J>0\}$, and furthermore, if $-(n+1+J/2+m/2)< a$ and $\operatorname{Re}(\sigma+\tau)\in[a,b]$, then

(2.2) $$|\iint dvdt\, \psi(v,t)K_{\sigma,\tau}(v,t)\, u_J(v,t)|$$

$$\leq [\iint dvdt\, \psi(v,t)|K_{\sigma,\tau}(v,t)|\,(|v|^2+|t|^2)^{J/2}]$$

$$\sup\{(|v|^2+|t|^2)^{-J/2}\, u_J(v,t) : |v|^4+|t|^2\leq 1\}$$

$$\leq C\|u\|_{C^{J+1}} \exp(\pi|\operatorname{Im}(\sigma-\tau)|/2) .$$

by an argument like that used to prove (2.1) and an estimate for the remainder term in the Taylor series.

It is now our aim to show that the expression

$$\iint dvdt\, \psi(v,t)K_{\sigma,\tau}(v,t)u_j(v)t^h$$

continues meromorphically, to examine its poles, and to estimate its growth.

Straightforward integration $(v = w^{1/2} v'$, where $w = |v|^2$) gives

(2.3)
$$\iint dvdt\, \psi(v,t) K_{\sigma,\tau}(v,t) u_j(v) t^h$$

$$= (1/2)\{ \iint_{\substack{w \geq 0 \\ w^2+t^2 \leq 1}} dwdt\, w^{(m+j+2n-2)/2} (w+it)^{\sigma} (w-it)^{\tau} t^h \} .$$

$$\{ \int_{S^{2n-1}} dv'\, p(v') u_j(v') \}$$

$$= (1/2)\{ \int_0^1 r\,dr\, r^{(m+j)/2+\sigma+\tau+h+n-1} \}\{ \int_{-\pi/2}^{\pi/2} d\theta (\cos\theta)^{(m+j+2n-2)/2} (\sin\theta)^h e^{i(\sigma-\tau)\theta} \}$$

$$\{ \int_{S^{2n-1}} dv'\, p(v') u_j(v') \}$$

$$= [m+j+2\sigma+2\tau+2h+2n+2]^{-1} F((m+j+2n-2)/2 ; h ; \sigma-\tau)$$

$$\{ \int_{S^{2n-1}} dv'\, p(v') u_j(v') \}$$

where, for $a,b \in \mathbb{N}$, $F(a;b;c)$ is defined by the absolutely convergent integral

$$F(a;b;c) = \int_{-\pi/2}^{\pi/2} d\theta\, (\cos\theta)^a (\sin\theta)^b e^{ic\theta} .$$

From the theory of spherical harmonics, the integral over S^{2n-1} vanishes unless $j = m+2g$, for some $g \in \mathbb{N}$. Then

$$\Sigma_{h,j=0}^{J} \iint dvdt\, \psi(v,t) K_{\sigma,\tau}(v,t) u_j(v) t^h$$

$$= \Sigma_{h=0}^{J} \Sigma_{g=0}^{[(J-m)/2]} (1/2)(m+\sigma+\tau+n+1+g+h)^{-1}$$

$$F(m+g+n-1 ; h ; \sigma-\tau)\{ \int_{S^{2n-1}} dv'\, p(v')\, u_{m+2g}(v') \} .$$

Clearly the function

$$(\sigma,\tau) \to \Gamma(m+\sigma+\tau+n+1)^{-1} \iint dvdt\, \psi(v,t) K_{\sigma,\tau}(v,t) u(v,t)$$

admits an analytic continuation into C^2, or equivalently, $\psi K_{\sigma,\tau}$ admits a meromorphic continuation whose only singularities are simple poles when $-(m+\sigma+\tau+n+1) \in N$.

The residue of $K_{\sigma,\tau}$ at a pole is a distribution supported in $(0,0)$. We shall explicitly calculate the residue in the first pole.

$$\lim_{(m+\sigma+\tau+n+1)\to 0} (m+\sigma+\tau+n+1) \iint dvdt\, K_{\sigma,\tau}(v,t) u(v,t)$$

$$= (1/2) F(m+n-1\,;\,0\,;\,\sigma-\tau) \{ \int_{S^{2n-1}} dv'\, p(v') u_m(v') \}$$

$$= 2^{1-n-2m} \pi^{n+1} [\Gamma(-\sigma)\Gamma(-\tau)]^{-1} \tilde{p}(\partial)_o(u) \,,$$

from Lemma 2.1, and the explicit formula (see, e.g., H. B. Dwight [4], 851.502)

$$F(a;0;c) = 2^{-a} \pi \Gamma(a+1) [\Gamma \frac{(a-c+2)}{2} \Gamma \frac{(a+c+2)}{2}]^{-1} .$$

We conclude the demonstration by estimating $F(a;b;c)$. By definition,

$$F(a;b;c) = \int_{-\pi/2}^{\pi/2} d\theta (\cos\theta)^a (\sin\theta)^b e^{ic\theta}$$

so, for $a,b \in N$,

$$|F(a;b;c)| \leqq \int_{-\pi/2}^{\pi/2} d\theta |e^{ic\theta}|$$

$$\leqq \pi \exp(\pi |\mathrm{Im}(c)|/2) .$$

From (2.3), if $|\mathrm{Im}(\sigma+\tau)| \geqq 1$, then

$$|\iint dvdt\, \psi(v,t) K_{\sigma,\tau}(v,t) u_j(v) t^h|$$

$$\leqq C \|u\|_{C^J} \exp(\pi |\mathrm{Im}(\sigma-\tau)|/2) .$$

This estimate, together with (1) and (2), shows that if $\mathrm{Re}(\sigma+\tau) \in [a,b]$

$$|K_{\sigma,\tau}(u)| \leqq C\|u\|_{C^{J+1}} \exp(\pi|\operatorname{Im}(\sigma-\tau)|/2)$$

if $|\operatorname{Im}(\sigma+\tau)| \geqq 1$, and concludes the demonstration. ∎

We now study $L_\gamma * K_{\sigma,\tau}$. First of all, we note that

$$V_j * f = \partial f/\partial v_j - i\bar{v}_j\, \partial f/\partial t \qquad \bar{V}_j * f = \partial f/\partial \bar{v}_j + i v_j\, \partial f/\partial t$$

$$T * f = \partial f/\partial t.$$

It follows immediately (cf. A. Korányi [13], pp 178-180) that

$$V_j * A = 2\bar{v}_j \qquad \bar{V}_j * \bar{A} = 2v_j$$

$$V_j * \bar{A} = 0 \qquad \bar{V}_j * A = 0$$

$$T * A = i \qquad T * \bar{A} = -i.$$

Theorem 2.3. *Suppose that* p *is of type* (k,ℓ) *(and is still harmonic). Then*

$$L_\gamma * K_{\sigma,\tau} = -\sigma(n+\gamma+2\ell+2\tau)K_{\sigma-1,\tau} - \tau(n-\gamma+2k+2\sigma)K_{\sigma,\tau-1}$$

and

$$K_{\sigma,\tau} * L_\gamma = -\sigma(n+\gamma+2k+2\tau)K_{\sigma-1,\tau} - \tau(n-\gamma+2\ell+2\sigma)K_{\sigma,\tau-1}.$$

Proof. By meromorphic continuation, it is enough to prove these when $\operatorname{Re}(\sigma)$ and $\operatorname{Re}(\tau)$ are large and positive, so that $K_{\sigma,\tau}$ is differentiable in the usual sense. Next, if we prove the first identity, the second will follow from the fact that $(f*g)^+ = g^+ * f^+$, where $f^+(v,t) = f(\bar{v},t)$ (clearly $(L_\gamma)^+ = L_\gamma$).

To prove the first identity we use direct calculation. Since p is harmonic, $(1/2)\,\Sigma_{j=1}^n \partial^2/\partial v_j \partial \bar{v}_j\, p = 0$, so

$$-L_o * K_{\sigma,\tau}$$

$$= (1/2)\,\Sigma_{j=1}^n [V_j * \bar{V}_j + \bar{V}_j * V_j] * [pA^\sigma \bar{A}^\tau]$$

$$= (1/2)\ \Sigma_{j=1}^{n} [p(\overline{V}_j * V_j * A^{\sigma})\overline{A}^{\tau} + pA^{\sigma}(V_j * \overline{V}_j * \overline{A}^{\tau})$$

$$+ 2(V_j * p)A^{\sigma}(\overline{V}_j * \overline{A}^{\tau}) + 2(\overline{V}_j * p)(V_j * A^{\sigma})\overline{A}^{\tau} + 2p(V_j * A^{\sigma})(\overline{V}_j * \overline{A}^{\tau})]$$

$$= (1/2)\ \Sigma_{j=1}^{n} [2\sigma p(\overline{V}_j * \overline{v}_j A^{\sigma-1})\overline{A}^{\tau} + 2\tau p A^{\sigma}(V_j * v_j \overline{A}^{\tau-1})$$

$$+ 4\tau v_j (V_j * p)A^{\sigma}\overline{A}^{\tau-1} + 4\sigma \overline{v}_j(\overline{V}_j * p)A^{\sigma-1}\overline{A}^{\tau} + 8\sigma\tau |v_j|^2 pA^{\sigma-1}\overline{A}^{\tau-1}]$$

$$= \sigma n p A^{\sigma-1}\overline{A}^{\tau} + \tau n p A^{\sigma}\overline{A}^{\tau-1}$$

$$+ 2\tau k p A^{\sigma}\overline{A}^{\tau-1} + 2\sigma\ell p A^{\sigma}\overline{A}^{\tau-1} + 2\sigma\tau [A + \overline{A}] p A^{\sigma-1}\overline{A}^{\tau-1}$$

$$= \sigma(n + 2\ell + 2\tau) p A^{\sigma-1}\overline{A}^{\tau} + \tau(n + 2k + 2\sigma) p A^{\sigma}\overline{A}^{\tau-1}\ .$$

Furthermore,

$$i\gamma T * K_{\sigma,\tau}$$

$$= -\gamma\sigma p A^{\sigma-1}\overline{A}^{\tau} + \gamma\tau p A^{\sigma}\overline{A}^{\tau-1}\ ,$$

whence the desired result.

We now come to the function-theoretic part of our approach.

Lemma 2.4. Suppose that $g : C \to C$ is entire and that $g(z+1) = g(z)$. If there exist constants C and K, $K < 2\pi$, such that

$$|g(\sigma + it)| \leq C \exp(K|t|), \quad \sigma, t \in R\ ,$$

then g is constant.

Proof. Let $h : C\backslash\{0\} \to C$ be defined by the formula

$$h(z) = g(\log(z)/2\pi i)\ , \quad z \in C \setminus \{0\}\ .$$

Clearly h is well defined and analytic. If $|z| \leq 1$, then

$$|h(z)| \leq C \exp(K|\mathrm{Im}(\log(z)/2\pi i)|)$$

$$= C|z|^{-K/2\pi}\ ,$$

so the singularity of h is 0 is removable; i.e. h extends to an entire function, and

$$|h(z)| \leqq C , \qquad |z| \leqq 1$$

so

$$|g(\sigma - it)| \leqq C, \quad \sigma \in R , \ t \in R^+ .$$

Analogously

$$|g(\sigma + it)| \leqq C, \quad \sigma \in R , \ t \in R^+$$

so by Liouville's theorem g is constant. ■

Now fix d, k, ℓ and n in N, λ in R^+ and ϵ in $\{\pm 1\}$. Let H be the domain

$$\{(\sigma,\tau) \in C^2 : 2\mathrm{Re}(\sigma+\tau) + k + \ell + 2n + 2 < 0\} .$$

Theorem 2.5. Suppose that $g : H \to C$ is meromorphic, with simple poles when $-(\sigma+\tau+k+\ell+n+1) \in N$ and no other singularities. Suppose also that if $a \leqq \mathrm{Re}(\sigma+\tau) \leqq b < -(n+1+k/2+\ell/2)$, then, whenever $|\mathrm{Im}(\sigma+\tau)| \geqq 1$,

$$|g(\sigma,\tau)| \leqq C \exp(\pi |\mathrm{Im}(\sigma-\tau)|/2) ,$$

and also that whenever $(\sigma,\tau) \in H$ and $\gamma \in C$,

(2.4)
$$(n + 2d + \epsilon\gamma)\lambda\, g(\sigma,\tau)$$
$$= -\sigma(n+\gamma+2\ell+2\tau) g(\sigma-1,\tau) - \tau(n-\gamma+2k+2\sigma) g(\sigma,\tau-1) .$$

Then

$$g(\sigma,\tau) = C\lambda^{-\sigma-\tau} \frac{\Gamma(\sigma+\tau+k+\ell+n+1)}{\Gamma(-\sigma)\Gamma(-\tau)} h_\epsilon(\sigma,\tau) ,$$

where

$$h_{+1}(\sigma,\tau) = \Gamma(d-\ell-\tau)/\Gamma(d+\sigma+k+n+1)$$

and

$$h_{-1}(\sigma,\tau) = \Gamma(d-k-\sigma)/\Gamma(d+\tau+\ell+n+1) .$$

If $\epsilon = +1$ and $d < \ell$ or if $\epsilon = -1$ and $d < k$, then $g = 0$.

Proof. We consider the case $\epsilon = +1$. Fix σ in $C \backslash Z$ and let $\gamma = -n - 2\ell - 2\tau$. Then, from (2.4),

(2.5) $$2\lambda(d-\ell-\tau)g(\sigma,\tau) = -2\tau(n+k+\ell+\sigma+\tau)g(\sigma,\tau-1) .$$

Let $h(\tau) = \lambda^{\tau}\Gamma(-\tau)g(\sigma,\tau)[\Gamma(\sigma+\tau+k+\ell+n+1)\Gamma(d-\ell-\tau)]^{-1}$. It is straightforward to check that $h(\tau) = h(\tau+1)$, using the functional equation of the Γ-function. We may use (2.5) to show inductively that $g(\sigma,\tau) = 0$ if $\tau = 0,1,2,\ldots$ if $d-\ell \notin N$, or that $g(\sigma,\tau) = 0$ if $\tau = 0,1,2,\ldots,d-\ell-1$ if $d-\ell \in N$. The poles of g are annihilated by the factor $\Gamma(\sigma+\tau+k+\ell+n+1)^{-1}$. Therefore h is analytic. Finally,

$$|h(\tau)| \leqq C\exp(\pi|\mathrm{Im}(\tau)|)$$

(it is enough to check this in a strip of width 1!) so by Lemma 2.4, h is constant. Therefore there exists $C(\sigma)$ such that

$$g(\sigma,\tau) = C(\sigma)\lambda^{-\tau}\Gamma(\sigma+\tau+k+\ell+n+1)\Gamma(d-\ell-\tau)\Gamma(-\tau)^{-1} .$$

If $d-\ell \notin N$, the fact that the only poles of g are when $-(\sigma+\tau+k+\ell+n+1) \in N$ imply that $C(\sigma) = 0$, and so $g = 0$. If $d \geqq \ell$, then we proceed by letting σ vary.

Put $\gamma = n+2k+2\sigma$, then

$$2\lambda(n+d+k+\sigma)g(\sigma,\tau) = -2\sigma(n+\sigma+\tau+k+\ell)g(\sigma-1,\tau) ,$$

and so

(2.6) $$\lambda(d+n+k+\sigma)C(\sigma) = -\sigma C(\sigma-1) .$$

Since $C(\sigma) = g(\sigma,\tau)\lambda^{\tau}\Gamma(-\tau)[\Gamma(\sigma+\tau+k+\ell+n+1)\Gamma(d-\ell-\tau)]^{-1}$, $C(\sigma)$ may be extended to an entire function. From the functional equation (2.6) we may now deduce that $C(\sigma) = 0$ if $\sigma \in N$ or if $-(\sigma+d+n+k+1) \in N$ (by induction). Consequently, the function

$$\tilde{h}:\sigma \to \lambda^{\sigma}\Gamma(-\sigma)\Gamma(d+\sigma+k+n+1)C(\sigma)$$

is entire, $\tilde{h}(\sigma) = \tilde{h}(\sigma+1)$, and $\tilde{h}$ is of "slow growth at infinity". Lemma 2.4 implies that $\tilde{h}$ is constant, and so

$$g(\sigma,\tau) = C\lambda^{-\sigma-\tau}\,\frac{\Gamma(\sigma+\tau+k+\ell+n+1)}{\Gamma(-\sigma)\Gamma(-\tau)}\,\frac{\Gamma(d-\ell-\tau)}{\Gamma(d+\sigma+k+n+1)}$$

as claimed.

To treat the case $\epsilon = -1$, note that interchanging σ with τ and k with ℓ effectively changes γ to $-\gamma$ on the right hand side of (2.4).

We may now describe the Fourier transform of $K_{\sigma,\tau}$ (where p is of type (k,ℓ)).

Theorem 2.6. *Fix monomials* ξ *and* η *in* $\mathcal{K}_{\epsilon\lambda}$, $(\epsilon=\pm 1, \lambda\in\mathbb{R}^+)$, *of degrees* c *and* d *respectively, and let* $g_{\epsilon\lambda}$ *be the meromorphic function* $(\sigma,\tau)\to \langle\pi_{\epsilon\lambda}(K_{\sigma,\tau})\xi,\eta\rangle$. *Then*

$$g_{\epsilon\lambda}(\sigma,\tau) =$$

$$2^{1-n-2k-2\ell}\pi^{n+1}\lambda^{-\sigma-\tau-k-\ell-n-1}\langle\pi_{\epsilon\lambda}(\tilde{p}(\partial)_0)\xi,\eta\rangle \frac{\Gamma(\sigma+\tau+k+\ell+n+1)}{\Gamma(-\sigma)\Gamma(-\tau)}\, h_\epsilon(\sigma,\tau) .$$

where

$$h_{+1}(\sigma,\tau)=\Gamma(d-\ell-\tau)/\Gamma(d+\sigma+k+n+1)$$

and

$$h_{-1}(\sigma,\tau)=\Gamma(d-k-\sigma)/\Gamma(d+\tau+\ell+n+1) .$$

In particular, $g_{\epsilon\lambda}=0$ *unless* $d-\ell=c-k\in\mathbb{N}$ *if* $\epsilon=+1$, *and* $d-k=c-\ell\in\mathbb{N}$ *if* $\epsilon=-1$.

Proof. By Theorem 2.2, $g_{\epsilon\lambda}$ is defined in $\{(\sigma,\tau)\in\mathbb{C}^2 : 2\mathrm{Re}(\sigma+\tau)+k+\ell+2n+2<0\}$, since $K_{\sigma,\tau}$ is the sum of a distribution of finite order and of compact support, and of an integrable function. Further, g has simple poles when $-(\sigma+\tau+k+\ell+n+1)\in\mathbb{N}$, and no other singularities. Moreover, if $\mathrm{Re}(\sigma+\tau)$ is bounded and $|\mathrm{Im}(\sigma+\tau)|\geq 1$, then

$$|g_{\epsilon\lambda}(\sigma,\tau)|\leq C\exp(\pi|\mathrm{Im}(\sigma-\tau)|/2) .$$

We observe that, since

$$L_\gamma * K_{\sigma,\tau} = -\sigma(n+\gamma+2\ell+2\tau)K_{\sigma-1,\tau} - \tau(n-\gamma+2k+2\sigma)K_{\sigma,\tau-1} ,$$

the following recursion relation holds:

$$\begin{aligned}&(n+2d+\epsilon\gamma)\lambda\, g_{\epsilon\lambda}(\sigma,\tau)\\ &= \langle\pi_{\epsilon\lambda}(K_{\sigma,\tau})\xi , \pi_{\epsilon\lambda}(L_\gamma)\eta\rangle\\ &= \langle\pi_{\epsilon\lambda}(L_\gamma * K_{\sigma,\tau})\xi,\eta\rangle\end{aligned}$$

$$= -\sigma(n+\gamma+2\ell+2\tau)g_{\epsilon\lambda}(\sigma-1,\tau) - \tau(n-\gamma+2k+2\sigma)g_{\epsilon\lambda}(\sigma,\tau-1) .$$

By Theorem 2.5,

$$g_{\epsilon\lambda}(\sigma,\tau) = C_{\epsilon\lambda}\lambda^{-\sigma-\tau}\frac{\Gamma(\sigma+\tau+k+\ell+n+1)}{\Gamma(-\sigma)\Gamma(-\tau)}h_\epsilon(\sigma,\tau) ,$$

with h_ϵ as above; $C_{\epsilon\lambda}=0$ if $\epsilon=+1$ and $d<\ell$, or if $\epsilon=-1$ and $d<k$. However, it is also true that

$$K_{\sigma,\tau} * L_\gamma = -\sigma(n+\gamma+2k+2\tau)K_{\sigma-1,\tau} - \tau(n-\gamma+2\ell+2\sigma)K_{\sigma,\tau-1} ;$$

analogously we may deduce that

$$g_{\epsilon\lambda}(\sigma,\tau) = \tilde{C}_{\epsilon\lambda}\lambda^{-\sigma-\tau}\frac{\Gamma(\sigma+\tau+k+\ell+n+1)}{\Gamma(-\sigma)\Gamma(-\tau)}\tilde{h}_\epsilon(\sigma,\tau) ,$$

where

$$\tilde{h}_{+1}(\sigma,\tau) = \Gamma(c-k-\tau)/\Gamma(d+\sigma+\ell+n+1)$$

and

$$\tilde{h}_{-1}(\sigma,\tau) = \Gamma(c-\ell-\sigma)/\Gamma(d+\tau+k+n+1) .$$

These formulae, combined with those preceding show that $C_{\epsilon\lambda}=0$ unless $c-k=d-\ell \in \mathbb{N}$ if $\epsilon=+1$, or $c-\ell=d-k\in\mathbb{N}$ if $\epsilon=-1$.

In order to evaluate $C_{\epsilon\lambda}$, we consider the first pole of $K_{\sigma,\tau}$. From Theorem 2.2, we obtain that

$$\lim_{(\sigma+\tau+k+\ell+n+1)\to 0}(\sigma+\tau+k+\ell+n+1)g_{\epsilon\lambda}(\sigma,\tau)$$

$$= 2^{1-n-2k-2\ell}\pi^{n+1}[\Gamma(-\sigma)\Gamma(-\tau)]^{-1}\langle\pi_{\epsilon\lambda}(\tilde{p}(\partial)_o)\xi,\eta\rangle .$$

On the other hand, from what we have just proved it follows that

$$\lim_{(\sigma+\tau+k+\ell+n+1)\to 0}(\sigma+\tau+k+\ell+n+1)g_{\epsilon\lambda}(\sigma,\tau)$$

$$= C_{\epsilon\lambda}\lambda^{k+\ell+n+1}[\Gamma(-\sigma)\Gamma(-\tau)]^{-1}\lim_{(\sigma+\tau+k+\ell+n+1)\to 0}h_\epsilon(\sigma,\tau)$$

$$= C_{\epsilon\lambda}\, \lambda^{k+\ell+n+1} [\Gamma(-\sigma)\Gamma(-\tau)]^{-1}.$$

We conclude that

$$\langle \pi_{\epsilon\lambda}(K_{\sigma,\tau})\xi,\eta\rangle = 2^{1-n-2k-2\ell}\pi^{n+1}\lambda^{-\sigma-\tau-k-\ell-n-1}\langle \pi_{\epsilon\lambda}(\tilde{p}(\partial)_o\xi,\eta\rangle$$

$$\frac{\Gamma(\sigma+\tau+k+\ell+n+1)}{\Gamma(-\sigma)\Gamma(-\tau)}\, h_\epsilon(\sigma,\tau) ,$$

as required.

Note that the function theory alone showed that $C_{\epsilon\lambda} = 0$ for many values of c and d. In particular $\langle \pi_{\epsilon\lambda}(\tilde{p}(\partial)_o\xi,\eta\rangle = 0$ for these values. This can also be shown directly.

∎

It is reasonable to ask how much of the above generalises to an H-type group. The first comment which must be made is that if $\dim(\mathfrak{z}) = 1$, the complex structure on $\mathfrak{v}$ is essentially unique, while if $\dim(\mathfrak{z}) > 1$, this is no longer true. So in the general case, while it makes sense to discuss harmonic polynomials of degree n, it makes no sense to treat those of type (k,ℓ). In general, A and $\bar{A}$ do not make sense (at least as complex-valued functions) but $A+\bar{A}$ and $A\bar{A}$ do. Thus it is reasonable to expect an analogue of Theorem 2.2 dealing with kernels of the form

$$K_\zeta(v,z) = p(v)\,|v|^{2s} N^\zeta(v,z)$$

(for $s \geq 0$), but it is unreasonable to expect an analogue of Theorem 2.3 unless $p = 1$ (constants are holomorphic in any complex structure!)

We shall state without proof a result on the meromorphic continuation of the kernels K_ζ and a differential equation for the case $p = 1$, $s = 0$. We have been unable to make explicit such a result for the case $p = 1$, $s \in \mathbb{R}^+$ though for abstract reasons it should exist. We deduce that N^{2-p-2q} is, up to a constant, the fundamental solution to L_o on N, and describe the Fourier transform of N^ζ.

In the rest of this section, we use the following notation: we let $X_1, \ldots, X_p$, and $Y_1, \ldots, Y_q$ be orthonormal bases of $\mathfrak{v}$ and of $\mathfrak{z}$ and, abusively, let X_j and Y_k also denote the distributions supported in $(0,0)$ such that

$$X_j(u) = (d/dt)\big|_{t=0}\, u(\exp(tX_j))$$

$$u \in \mathcal{D}(N)\ .$$

$$Y_k(u) = (d/dt)\big|_{t=0}\, u(\exp(tY_k/4))$$

We let $L^{(\mathfrak{v})}$ and $L^{(\mathfrak{z})}$ be the distributions $-\Sigma_{j=1}^{p} X_j * X_j$ and $-\Sigma_{k=1}^{q} Y_k * Y_k$ respectively. Then, if N is the Heisenberg group

$$L_o = (1/4)L^{(\mathfrak{v})}\ .$$

We study the distribution-valued function K_ζ. It is clear that if $\mathrm{Re}(\zeta) + m + 2s + p + 2q > 0$, the kernel is locally integrable, and so defines a distribution, by integration.

<u>Theorem 2.7. The distribution-valued function</u> $\zeta \to K_\zeta$ <u>continues meromorphically from</u> $\{\zeta \in \mathbb{C} : \mathrm{Re}(\zeta) + m + 2s + p + 2q > 0\}$ <u>into</u> $\mathbb{C}$, <u>with simple poles when</u> $-(\zeta + 2n + 2s + p + 2q)/2 \in \mathbb{N}$, <u>and no other singularities. Further,</u>

$$\lim_{(\zeta + 2m + 2s + p + 2q) \to 0} (1/2)(\zeta + 2m + 2s + p + 2q)K_\zeta$$

$$= 2^{1-2m-p/2}\,\pi^{(p+q+1)/2}\,\frac{\Gamma((2m+2s+p)/4)}{\Gamma((2m+2s+p+2q)/4)\Gamma((2m+p)/4)\Gamma((2m+p+2)/4)}\cdot\tilde{p}(\partial)\ .$$

<u>Finally, if</u> $-(J + m + 2s + p + 2q) < a \leq b < -(m + 2s + p + 2q)$, <u>then the following estimate holds whenever</u> $\mathrm{Re}(\zeta) \in [a,b]$ <u>and</u> $|\mathrm{Im}(\zeta)| \geq 1$:

$$|K_\zeta(u)| \leq C\|u\|_{C^{J+1}}, \quad u \in \mathcal{D}(N)\ .$$

<u>Proof.</u> Like that of Theorem 2.2. ∎

The following substitutes for Theorem 2.3 can be proved.

<u>Theorem 2.8. The following identities hold:</u>

(i) $$L^{(\mathfrak{v})} * N^\zeta(v,z) = N^\zeta * L^{(\mathfrak{v})}(v,z) = -\zeta(\zeta + p + 2q - 2)|v|^2 N^{\zeta - 4}(v,z)$$

(ii) $$L^{(\mathfrak{v})} * L^{(\mathfrak{v})} * N^\zeta + B(\zeta)L^{(\mathfrak{z})} * N^\zeta = C(\zeta)N^{\zeta - 4}\ ,$$

<u>where</u> $B(\zeta) = -4(\zeta + p + 2q - 2)^2$ <u>and</u> $C(\zeta) = \zeta(\zeta + 2q - 2)(\zeta + p + 2q - 2)(\zeta + p + 2q - 4)$.

Proof. See M. Cowling [2]. If $q = 1$, these results may be derived from Theorem 2.3: for (i), consider $L_0 * N^\zeta = L_0 * A^{\zeta/2}\bar{A}^{\zeta/2}$, and for (ii), consider $L_\gamma * L_{-\gamma} * N^\zeta$, with $\gamma = (n + \zeta/2)$.

■

Corollary 2.9. The following identity holds:

$$L^{(0)} * N^{2-p-2q} = N^{2-p-2q} * L^{(0)} = \frac{2^{4-p/2}\pi^{(p+q+1)/2}}{\Gamma(p/4)\Gamma((p+2q-2)/4)} \cdot \delta .$$

Proof. Consider $L^{(0)} * N^{2-p-2q+\varepsilon}$ as $\varepsilon \to 0$. Theorem 2.7 shows that the residue arising is a precise multiple of the Dirac δ.

■

This result was first proved, in this generality, by A. Kaplan [11]. In order to describe the Fourier transform of N^ζ, we recall that the Fourier transform of a "polyradial" function on an H-type group is an operator which acts by scalars on the homogeneous polynomials of degree d (in the model of $\pi_{\varepsilon\lambda}$ described earlier). See F. Ricci [17] for a fuller discussion of this point.

Theorem 2.10. Let $a_{\lambda,d}(\zeta)$ be the following function:

$$a_{\lambda,d}(\zeta) = 2^{1-p/2}\pi^{(p+q+1)/2}\lambda^{-\zeta/2} b_d(\zeta), \quad \zeta \in \mathbb{C}$$

where

$$b_d(\zeta) = \frac{\Gamma((\zeta+r)/2)}{\Gamma(-\zeta/4)\Gamma((2-2q-\zeta)/4)} \frac{\Gamma(d+(2-2q-\zeta)/4)}{\Gamma(d+(2+2q+2p+\zeta)/4)} .$$

Then the Fourier transform $\pi_{\varepsilon\lambda}(N^\zeta)$ is the operator which acts by the scalar $a_{\lambda,d}(\zeta)$ on the homogeneous polynomials of degree d.

Proof. See M. Cowling [2]. Alternatively, the methods of proof of Theorem 2.6 may be applied.

Chapter 3

Homogeneous Kernels in Group Representation Theory

In this section, we describe how homogeneous kernels of the type analysed in §2 arise in group representation theory, and derive some representation-theoretic corollaries of our study. We shall discuss in some detail the representations of $SU(n+1,1)^{\sim}$ (the universal covering group of $SU(n+1,1)$, with $n \geq 1$) induced from characters of a parabolic subgroup, and then briefly describe the class one principal series of the real rank one simple Lie group.

The group $SU(n+1,1)$ acts by linear transformation on $\mathbb{C}^{n+2}$, preserving the form $\varphi(z) = (\Sigma_{j=0}^{n}|z_j|^2) - |z_{n+1}|^2$. In particular, $SU(n+1,1)$ preserves the cone $C = \{z : \varphi(z) < 0\}$ and its boundary ∂C; $SU(n+1,1)$ therefore acts on Γ, the projective cone, a variety in $\mathbb{P}^{n+1}(\mathbb{C})$, and on its boundary $\partial\Gamma$. We write $G = PU(n+1,1)$ for the group of fractional linear transformations of Γ and $\partial\Gamma$ arising in this way, and $\tilde{G}$ for its universal covering group, which is also the universal covering group of $SU(n+1,1)$.

We define the family of infinite-dimensional representations $\pi_{\mu,\lambda}$ of $\tilde{G}$ to be the representation of $\tilde{G}$ induced from the characters $\chi_{\mu,\lambda}$ of a parabolic subgroup $\tilde{P}$. Adapting the general theory of intertwining operators of G. Schiffmann [19], and A.W. Knapp and E.M. Stein [12] to the case in hand, we prove that the kernels $K_{\sigma,\tau}$ of §2 are the intertwining operators for the representations $\pi_{\mu,\lambda}$. From the results of §2 we obtain immediately Flensted-Jensen's description of the complementary series of $SU(n+1,1)^{\sim}$ coming from the representations $\pi_{\mu,\lambda}$ (this description was obtained at the time of [5], but not published; cf. also Kraljević [16]).

Finally, we discuss rapidly the complementary series of the class one principal series of the real rank one simple Lie groups. As corollary to the analysis of §2, we obtain some results of B. Kostant [15].

We shall describe $PU(n+1,1)$ very concretely. We remind the reader that Γ may be explicitly realised as D, the unit disc in $\mathbb{C}^{n+1}$ or as the Siegel domain $S = \{(z_o, z) \in \mathbb{C} \times \mathbb{C}^n : \mathrm{Re}(z_o) > |z|^2\}$, by identifying (w_o, w) in D $(w_o \in \mathbb{C}, w \in \mathbb{C}^n)$ with $\mathbb{C}^*(w_o, w, 1)$ in Γ, and (z_o, z) in S with $\mathbb{C}^*((z_o - 1)/2, z, (z_o+1)/2)$ in Γ. We implement the equivalence between these realisations by the map $\Phi : S \to D$:

$$\Phi(z_o, z) = ((z_o - 1)/(z_o + 1), 2z/(z_o + 1))$$

and

$$\Phi^{-1}(w_o, w) = ((1 + w_o)/(1 - w_o), w/(1 - w_o)) .$$

Several subgroups of G are of interest: $K, K_s, K_a, M, M_s, M_a, A, N$ and $\bar{N}$. On D, K is just the usual rotation group $U(n+1)$ (acting on the right); the action of K on S is rather messy. The semisimple part of K is denoted K_s, i.e. $K_s = SU(n+1)$, and the centre is denoted K_a. We define h_r by the formula

$$(w_o, w)h_r = (e^{ir}w_o, e^{ir}w) ,$$

for (w_o, w) in D and r in $\mathbb{R}$, and then $K_a = \{h_r : r \in [0, 2\pi)\}$. If we realise Γ as S, then we obtain the formula

$$(z_o, z)h_r = \Phi^{-1}(e^{ir}\Phi(z_o, z))$$

$$= \left(\frac{z_o \cos(r/2) - i \sin(r/2)}{\cos(r/2) - i z_o \sin(r/2)}, \frac{e^{ir/2} z}{\cos(r/2) - i z_o \sin(r/2)} \right) .$$

In particular, $(z_o, z)h_\pi = (1/z_o, -z/z_o)$; this action on S is called inversion.

The subgroup M of K is composed of those rotation which fix the w_o-axis: M is isomorphic to $U(n)$. By M_s and M_a we denote the semisimple and abelian parts of M, i.e. $M_s = SU(n)$, and M_a is the centre of M. If $m \in M = U(n)$, then

$$(w_o, w)m = (w_o, wm) \qquad (w_o, w) \in D$$

and

$$(z_o, z)m = (z_o, zm) \qquad (z_o, z) \in S$$

(we beg the reader's indulgence for the abusive notation); we define m_r to be the transformation

$$(w_o, w)m_r = (w_o, e^{ir}w), \quad (w_o, w) \in D$$

The action of the subgroups A, N, and $\bar{N}$ is most easily seen on S; set, for $(z_o, z) \in S$,

$$(z_o, z)a_s = (e^{2s}z_o, e^s z)$$

$$(z_o, z)n_{v,t} = (z_o + it + |v|^2 + 2zv^*, z+v)$$

$$\bar{n}_{v,t} = h_\pi n_{v,t} h_\pi ;$$

these A, N and $\bar{N}$ are the groups of all the transformations a_s, with s in $\mathbb{R}$, $n_{v,t}$ and $\bar{n}_{v,t}$ with v in $\mathbb{C}^n$, t in $\mathbb{R}$ respectively. It is easy to check that $h_\pi a_s h_\pi = a_{-s}$, that M and A normalise N and $\bar{N}$, that M is the centraliser of A in K, and so on. We depart slightly from the usual notation, setting $P = MA\bar{N}$; then P is the isotropy group of $(0,0)$ in ∂S or of $(-1,0)$ in ∂D. Of course, $G = KAN$ and $G = MA\bar{N}N \cup MA\bar{N}h_\pi$ (the Iwasawa and Bruhat decompositions of G), and N and h_π generate G.

Let $\tilde{K}, \tilde{A}, \tilde{\bar{N}}$ and $\tilde{N}$ be the universal covering groups of $K, A, \bar{N}$ and N. Then, topologically, $\tilde{G} = \tilde{K}\tilde{A}\tilde{N}$. Let $\pi : \tilde{G} \to G$ be the canonical projection. Then π is an isomorphism of $\tilde{A}$ onto A, $\tilde{N}$ onto N, and $\tilde{\bar{N}}$ onto $\bar{N}$, while $\tilde{K}$ may be naturally parametrised by $\mathbb{R} \times SU(n+1)$, and then $\pi(r,k) = e^{ir}k$ in $U(n+1)$. We let $\tilde{K}_a$ and $\tilde{K}_s$ be the subgroups $\{\tilde{h}_r = (r,1) : r \in \mathbb{R}\}$ and $\{(0,k), k \in SU(n+1)\}$ of $\tilde{K}$. Then $\pi : \tilde{K}_s \to K_s$ is an isomorphism, while $\pi : \tilde{K}_a \to K_a$ is the covering of the torus by the line. Let $Z = \pi^{-1}(\{e\})$. Then Z is generated by z_1:

$$z_1 = (2\pi/(n+1), e^{-2\pi i/(n+1)})$$

in $\mathbb{R} \times SU(n+1)$, and is free. Finally, let $\tilde{M}$ be $\pi^{-1}(M)$ and $\tilde{P}$ be $\pi^{-1}(P)$; clearly $\tilde{P} = \tilde{M}\tilde{A}\tilde{\bar{N}}$.

Let $\tilde{M}_s$ be $\tilde{M} \cap \tilde{K}_s$ $(\tilde{M}_s = SU(n))$ and let $\tilde{M}_a$ be the subgroup of $\tilde{M}$ of elements of the form $\tilde{M}_r = (rn/(n+1), u_r)$ in $\mathbb{R} \times SU(n+1)$,

$$(w_o, w)u_r = (e^{irn/(n+1)}w_o, e^{ir/(n+1)}w), \quad (w_o, w) \in D,$$

so that $\pi(\tilde{m}_r) = m_r$. Note that $\tilde{M}_a \cap Z$ is generated by z_1^n, and that $z_1\tilde{m}_r \in \tilde{M}_s$ when $r = -2\pi/n$. Then $\tilde{M} = \tilde{M}_a\tilde{M}_s$ and $Z \subset \tilde{M}$.

The not necessarily unitary characters of $\tilde{P}$ are parametrised by $\mathbf{C} \times \mathbf{C}$. We define $\chi_{\mu,\lambda} : \tilde{P} \to \mathbf{C}^*$ by the formula

$$\chi_{\mu,\lambda}(\tilde{m}\tilde{m}_r \tilde{a}_s \tilde{\bar{n}}) = \chi_{\mu,\lambda}(\tilde{m}_r \tilde{a}_s) = \exp(i\mu r + \lambda s) ,$$

where $\tilde{m} \in \tilde{M}_s$, $\tilde{\bar{n}} \in \tilde{\bar{N}}$, and $\tilde{a}_s \in \tilde{A}$ with $\pi(\tilde{a}_s) = a_s$. These characters project to characters of P if $\mu \in n\mathbb{Z}$. We define, in particular,

$$\rho(\tilde{m}\tilde{m}_n \tilde{a}_s \tilde{\bar{n}}) = \exp((n+1)s) .$$

We consider the representation of $\tilde{G}$ induced from the character $\chi_{\mu,\lambda}$ of $\tilde{P}$. This means that we take the space of measurable functions ξ on $\tilde{N}$ (by "measurable function" we really mean equivalence class of measurable functions defined almost everywhere, modulo the equivalence relation of coinciding almost everywhere) and extend them to $\tilde{G}$ by requiring that

$$\xi(\tilde{p}\tilde{n}) = \chi_{\mu,\lambda}(\tilde{p})\rho(\tilde{p})\xi(\tilde{n}) .$$

(Note that $\tilde{P} \cap \tilde{N} = \{\tilde{e}\}$, and that $\tilde{P}\tilde{N}$ is an open dense submanifold of $\tilde{G}$, whose complement is a closed submanifold of lower dimension, and hence of measure zero; the extended functions are therefore defined almost everywhere on $\tilde{G}$, and are determined by their restrictions to $\tilde{N}$). We may norm the space by using, for instance, an L^p-norm on $\tilde{N}$. The representation $\pi_{\mu,\lambda}$ is now the right translation representation on this space:

$$[\pi_{\mu,\lambda}(\tilde{g})\xi](\tilde{g}') = \xi(\tilde{g}'\tilde{g}) .$$

The "extra factor" $\rho(\tilde{p})$ in the extension formula guarantees that if $\chi_{\mu,\lambda}$ is unitary (i.e. if μ is real and λ is imaginary) then $\pi_{\mu,\lambda}$ acts unitarily on $L^2(\tilde{N})$.

To describe the representation $\pi_{\mu,\lambda}$ in yet more concrete terms observe that, for $\tilde{n}$, $\tilde{n}'$ in $\tilde{N}$, $\tilde{m}$ in $\tilde{M}$, $\tilde{a}$ in $\tilde{A}$

$$[\pi_{\mu,\lambda}(\tilde{n})\xi](\tilde{n}') = \xi(\tilde{n}'\tilde{n})$$

and

$$[\pi_{\mu,\lambda}(\widetilde{m}\widetilde{a})\xi](\widetilde{n}) = \xi(\widetilde{n}\,\widetilde{m}\widetilde{a})$$

$$= \xi(\widetilde{m}\widetilde{a}\,(\widetilde{m}\widetilde{a})^{-1}\widetilde{n}\,\widetilde{m}\widetilde{a})$$

$$= \chi_{\mu,\lambda}(\widetilde{m}\widetilde{a})\,\rho(\widetilde{a})\xi((\widetilde{m}\,\widetilde{a})^{-1}\widetilde{n}\widetilde{m}\widetilde{a}) .$$

Since $(\widetilde{m}\widetilde{a})^{-1}\widetilde{n}(\widetilde{m}\widetilde{a}) \in \widetilde{N}$, and $\widetilde{N}$ acts simply transitively on ∂S , we may explicitly calculate $\widetilde{n}' = (\widetilde{m}\widetilde{a})^{-1}\widetilde{n}\,(\widetilde{m}\widetilde{a})$ by observing that

$$o\cdot\widetilde{n}' = o\cdot(\widetilde{m}\widetilde{a})^{-1}\widetilde{n}(\widetilde{m}\widetilde{a}) = o\cdot\widetilde{n}(\widetilde{m}\widetilde{a}) .$$

Next, since $\widetilde{N}$ and $\widetilde{K}_a$ generate $\widetilde{G}$ (a fortiori $\widetilde{N}, \widetilde{M}, \widetilde{A}$ and $\widetilde{K}_a$ generate $\widetilde{G}$). $\pi_{\mu,\lambda}$ will be completely specified if we describe $\pi_{\mu,\lambda}(\widetilde{h}_r)$. We may write

$$[\pi_{\mu,\lambda}(\widetilde{h}_r)\xi](\widetilde{n}) = \xi(\widetilde{m}\widetilde{h}_r)$$

$$= \xi(\widetilde{p}_r\widetilde{h}_r) = \chi_{\mu,\lambda}(\widetilde{p}_r)\rho(\widetilde{p}_r)\xi(\widetilde{n}_r)$$

(say), at least for small enough r : the set $\widetilde{P}\widetilde{N}$ is open in $\widetilde{G}$. Moreover, the set is diffeomorphic to $\widetilde{P}\times\widetilde{N}$, and so $\widetilde{p}_r$ and $\widetilde{n}_r$ vary smoothly with r ; obviously $\widetilde{p}_o = \widetilde{e}$. We may calculate $\widetilde{n}_r$ using the fact that

$$o\cdot\widetilde{n}\widetilde{h}_r = o\cdot\widetilde{p}_r\widetilde{n}_r = o\cdot\widetilde{n}_r \quad ;$$

the problem remains to calculate $\widetilde{p}_r$, or rather $\chi_{\mu,\lambda}(\widetilde{p}_r)\rho(\widetilde{p}_r)$. This may be done by calculating $\chi_{n,n}(\widetilde{p}_n)$, for knowledge of this and the continuity of $\widetilde{p}_r$, together with the fact that $\widetilde{p}_o = \widetilde{e}$, determine the $\widetilde{M}_a$-and $\widetilde{A}$-components of $\widetilde{p}_r$.

Recall that the $(2n+1)$-dimensional tangent space of ∂S at 0 , $T(\partial S)_o$, has a 2n-dimensional subspace which admits a natural complex structure. Split the complexification of this subspace into its holomorphic and anti-holomorphic parts $T^{1,0}(\partial S)_o$ and $T^{0,1}(\partial S)_o$, and let $T^{n,0}(\partial S)_o$ be the exterior product $\Lambda^n T^{1,0}(\partial S)_o$. Then an elementary calculation shows that if p_* is the map of $\Lambda T^{1,0}(\partial S)_o$ induced by the differential of the map $\gamma \to \gamma_p$, then

$$vp_* = \chi_{n,n}(p)v , \quad v \in T^{n,0}(\partial S)_o .$$

We prefer a more explicit formulation of the above, which we obtain by taking coordinates. We separate out the group theoretic part, which will be useful later on.

Theorem 3.1. *Fix* $(v,t)\in C^n\times R$, *and let* $\tilde{n}$ *be* $\tilde{n}_{v,t}$. *Then for all* r *in* R *such that* $C(r)\neq 0$, *where*

$$C(r)=\cos(r/2)+t\sin(r/2)-i|v|^2\sin(r/2),$$

we may write

$$\tilde{n}\tilde{h}_r=\tilde{p}_r\tilde{n}_r,$$

with $\tilde{p}_r\in\tilde{P}$ *and* $\tilde{n}_r$ *in* $\tilde{N}$. *Moreover,* $\tilde{n}_r$ *has coordinates* $(v(r),t(r))$ *where*

$$v(r)=C(r)^{-1}e^{ir/2}v$$

and

$$t(r)=\mathrm{Im}(C(r)^{-1}[|v|^2\cos(r/2)+i(t\cos(r/2)-\sin(r/2))]),$$

and

$$\tilde{p}_r=\tilde{m}_{u(r)}\tilde{a}_{s(r)}\tilde{m}\tilde{\bar{n}},$$

where $\tilde{m}\in\tilde{M}_s$, $\tilde{\bar{n}}\in\tilde{\bar{N}}$, *and*

$$s(r)=-\log|C(r)|$$

while, if $v\neq 0$, u *is that continuous increasing function such that*

$$e^{inu(r)}=e^{inr/2}(\bar{C}/|C|)^{n+2},$$

and if $v=0$, *then*

$$u(r)=r/2+(n+2)k\pi/n,$$

$$r\in(\tau+2(k-1)\pi,\ \tau+2k\pi),$$

where $0<\tau<2\pi$ *and* $\tan(\tau/2)=-1/t$. *In particular, if* $|v|>0$, *then* $u(\pi)=\frac{n+1}{n}\pi+\frac{n+2}{n}\arg(\bar{A})$, *and* $s(\pi)=-\log(|A|)$, *where* $A=|v|^2+it$.

Proof. If $C(r)\neq 0$,

$$(0,0)\tilde{n}\tilde{h}_r = (A,v)\tilde{h}_r$$

$$= (C(r)^{-1}[\cos(r/2)A - i\sin(r/2)],\ C(r)^{-1}e^{ir/2}v)\ .$$

Certainly, then, with $\tilde{n}_r$ as above,

$$(0,0)\tilde{n}\tilde{h}_r = (0,0)\tilde{n}_r\ ,$$

and $(0,0)\tilde{n}\tilde{h}_r\tilde{n}_r^{-1} = (0,0)$.

Consequently, there exists $\tilde{p}_r$ in $\tilde{P}$ such that

$$\tilde{n}\tilde{h}_r = \tilde{p}_r\tilde{n}_r\ .$$

In order to calculate the $\tilde{M}_a\tilde{A}$ component of $\tilde{p}_r$, we look at the action of $\tilde{p}_r$ on $T^{1,0}(\partial S)_o$. Remembering that this space is $\tilde{P}$-invariant, and putting $B = |v|^2 + it + 2zv^*$ and $D = \cos(r/2) - iB\sin(r/2)$ (note that $C = \cos(r/2) - iA\sin(r/2)$), we have that

$$(0,z)\tilde{n}\tilde{h}_r\tilde{n}_r^{-1} = (B,\ v+z)\tilde{h}_r\tilde{n}_r^{-1}$$

$$= (*,\ D^{-1}e^{ir/2}(v+z))\tilde{n}_r^{-1}$$

$$= (*,\ D^{-1}e^{ir/2}(v+z) - C^{-1}e^{ir/2}v)$$

$$= (*,\ (D^{-1} - C^{-1})e^{ir/2}v + D^{-1}e^{ir/2}z)$$

$$= (*,\ D^{-1}2iz\cdot v^*\sin(r/2)C^{-1}e^{ir/2}v + D^{-1}e^{ir/2}z)$$

$$= (0,\ 2iC^{-2}\sin(r/2)e^{ir/2}z\cdot v^*v + C^{-1}e^{ir/2}z) + O(z^2)\ .$$

Consequently, on $T^{n,0}(\partial S)_o$, we find that

$$(\tilde{p}_r)_* = e^{inr/2}C^{-n-1}\bar{C}$$

i.e.

$$\chi_{n,n}(\tilde{m}_{u(r)}\tilde{a}_{s(r)}) = e^{inr/2}C^{-n-1}\bar{C}\ .$$

We see immediately that $s(r) = -\log|c|$, and that

$$e^{inu(r)} = e^{inr/2}(C/|C|)^{n+2}\ .$$

If $v \neq 0$, then C is never 0, so that $u: R \to R$ is continuous. Since

$$\tan(\arg(\overline{C})) = |v|^2/[\cotan(r/2)+t] ,$$

which is increasing in each of its intervals of definition, u is increasing.

To treat the case $u=0$, note that the difficulty arises in the defintion of $\arg(\overline{C})$. However, it is always clear that

$$\begin{aligned}\tilde{n}\tilde{h}_{2k\pi} &= \tilde{h}_{2k\pi}\tilde{n} \\ &= \tilde{m}_{2k\pi(n+1)/n}\tilde{m}\tilde{n}\end{aligned}$$

for some $\tilde{m}$ in $\tilde{M}_s$, and this is enough to be able to draw our conclusions. The case $u=\pi$, $v \neq 0$ is straightforward.

■

It is now easy to describe the action of the representation $\pi_{\mu,\lambda}$ on functions on $\tilde{N}$. To do so, we shall identify $\tilde{n}_{v,t}$ (the element of $\tilde{N}$ whose image in N is $n_{v,t}$) with (v,t).

Corollary 3.2. The induced representation $\pi_{\mu,\lambda}$ acts on functions on $\tilde{N}$ by the following formulae:

$$[\pi_{\mu,\lambda}(v',t')\xi](v,t) = \xi((v,t)(v',t'))$$

$$[\pi_{\mu,\lambda}(\tilde{m}\tilde{a}_s)\xi](v,t) = \chi_{\mu,\lambda}(\tilde{m}\tilde{a}_s)\rho(\tilde{a}_s)\xi(e^s v\tilde{m}, e^{2s}t) ,$$

$$[\pi_{\mu,\lambda}(\tilde{h}_r)\xi](v,t) = e^{i\mu u(r)}e^{-(\lambda+n+1)\log(|C|)}\xi(v(r), t(r)) .$$

We now recall the intertwining operators. It is clear that, if it makes sense, the operator $A = A(\mu,\lambda)$

$$A\xi(\tilde{g}) = \int_{\tilde{N}} d\tilde{n}\, \xi(\tilde{n}\tilde{h}_\pi\tilde{g})$$

maps the representation space of $\pi_{\mu,\lambda}$ into that of $\pi_{\mu,-\lambda}$. For instance

$$\begin{aligned}A\xi(\tilde{a}_s\tilde{g}) &= \int_{\tilde{N}} d\tilde{n}\, \xi(\tilde{n}\tilde{h}_\pi\tilde{a}_s\tilde{g}) \\ &= \int_{\tilde{N}} d\tilde{n}\, \xi(\tilde{n}\tilde{a}_{-s}\tilde{h}_\pi\tilde{g})\end{aligned}$$

$$= \int_{\tilde{N}} d\tilde{n}\, \xi(\tilde{a}_{-s}\tilde{a}_s\tilde{n}\tilde{a}_{-s}\tilde{h}_\pi\tilde{g})$$

$$= \int_{\tilde{N}} d\tilde{n}\, e^{-(\lambda+n+1)s}\, \xi(\tilde{a}_s\tilde{n}\tilde{a}_{-s}\tilde{h}_\pi\tilde{g})$$

$$= \int_{\tilde{N}} d\tilde{n}\, e^{-\lambda s}\, e^{(n+1)s}\, \xi(\tilde{n}\tilde{h}_\pi\tilde{g})$$

$$= \chi_{\mu,-\lambda}(\tilde{a}_s)\rho(\tilde{e}_s)A\xi(\tilde{g}) \quad .$$

Corollary 3.3. The intertwining operator $A(\mu,\lambda)$ is up to a constant, given by (left) convolution with the kernel $K_{\sigma,\tau}(v,t) = (|v|^2+it)^\sigma(|v|^2-it)^\tau$ where

$$2\sigma = \lambda - n - 1 - \mu' \ ; \quad 2\tau = \lambda - n - 1 + \mu'$$

$$n\mu' = (n+2)\mu \ .$$

Proof. We write $A(v,t)$ for $|v|^2+it$, and $\tilde{n}^+ = (v^+, t^+)$ for $\tilde{n}_\pi$. Then

$$v^+ = -A(v,t)^{-1}v$$

and

$$t^+ = -t\,A(v,t)^{-1}\bar{A}(v,t)^{-1} \ .$$

The map $\tilde{n}\to\tilde{n}^+$ is an involution of $\tilde{N}$ whose Jacobian determinant is given by the rule

$$d\tilde{n}^+/d\tilde{n} = [A(\tilde{n})\,\bar{A}(\tilde{n})]^{-(n+1)} \ .$$

Take (v,t) in $\tilde{N}$ with $v\neq 0$. We may write

$$\tilde{n}\tilde{h}_\pi = \tilde{m}_{u(\pi)}\tilde{a}_{s(\pi)}\tilde{m}\tilde{\bar{n}}\tilde{n}^+ \ ,$$

where $\tilde{m}\in\tilde{M}_s$ and $\tilde{\bar{n}}\in\tilde{\bar{N}}$. From Theorem 3.1,

$$nu(\pi) = (n+1)\pi + (n+2)\arg(\bar{A}(v,t)) \ ,$$

and

$$s(\pi) = -\log(|A(v,t)|) \ .$$

In order to make explicit the dependence of everything on $\tilde{n}$, we now write

$$\tilde{n}\tilde{h}_\pi = \tilde{m}_{u(\tilde{n})}\tilde{a}_{s(\tilde{n})}\tilde{\bar{m}}\tilde{n} .$$

Then

$$A\xi(\tilde{n}') = \int_{\tilde{N}} d\tilde{n}\ \xi(\tilde{n}\tilde{h}_\pi\tilde{n}')$$

$$= \int_{\tilde{N}} d\tilde{n}\ \xi(\tilde{m}_{u(\tilde{n})}\tilde{a}_{s(\tilde{n})}\tilde{\bar{m}}\tilde{n}\tilde{n}')$$

$$= \int_{\tilde{N}} d\tilde{n}\ \exp(i\mu u(\tilde{n}) + (\lambda + n + 1)s(\tilde{n}))\xi(\tilde{n}^+\tilde{n}') .$$

By changing the variable of integration to $(\tilde{n}^+)^{-1}$ and inserting the values of $u(\tilde{n})$ and $s(\tilde{n})$, we obtain that

$$A\xi(\tilde{n}') = \exp(i\mu\pi(n+1)/n)\, K_{\sigma,\tau} * \xi(\tilde{n}') ,$$

where the powers of A and $\bar{A}$ are calculated using values of the argument in $(-\pi/2, \pi/2)$, and σ and τ are as above.

It is worth pointing out that μ' is perhaps a more natural parameter than μ : in fact since $\chi_{\mu,\lambda}$ "is" a character of P if $\mu \in n\mathbf{Z}$, and $SU(n+1,1)$ is a $(n+2)$-fold covering of G, the representations of $SU(n+1,1)$ correspond to those $\pi_{\mu,\lambda}$ with $\mu' \in \mathbf{Z}$.

We already know how to interpret A as a convolution with a tempered distribution.

Consider the sesquilinear form on $\mathcal{S}(\tilde{N})$ given by the rule

$$B_{\mu,\lambda}(\xi,\eta) = \int_{\tilde{N}} d\tilde{n}\ A(\mu,\lambda)\xi(\tilde{n})\bar{\eta}(\tilde{n}).$$

It is easy to check, using the intertwining property of $A(\xi,\eta)$, that

$$B_{\mu,\lambda}(\xi,\eta) = B_{\mu,\lambda}(\pi_{\mu,\lambda}(\tilde{g})\xi,\ \pi_{\mu,\lambda}(\tilde{g})\eta)$$

for $\mu,\lambda \in \mathbf{R}$. If this form is, up to a constant, positive definite, then we have a unitary representation of $\tilde{G}$, called a "complementary series" representation. However, the positivity of $B_{\mu,\lambda}$ is an immediate consequence of the Fourier transform calculation of §2.

<u>Corollary 3.4.</u> <u>The complementary series representations of</u> $\tilde{G}$ <u>occur when</u>

$$|\lambda| + |\mu'| < n+1$$

or when $|\mu'| > n+1$ and $|\lambda/2| < \min\{|(|\mu'| - n - 1)/2 - k| : k \in \mathbb{N}\}$.

Proof. From Theorem 2.6, for homogeneous polynomials ξ and η of degree d in the representation space of $\mathcal{V}_{\epsilon\lambda}$,

$$\langle \pi_{\epsilon\lambda}(K_{\sigma,\tau})\xi, \eta\rangle = C(\sigma,\tau,\lambda) h_{\epsilon}(\sigma,\tau)\langle \xi,\eta\rangle ,$$

where

$$h_1(\sigma,\tau) = \Gamma(d-\tau)/\Gamma(d+\sigma+n+1)$$

and

$$h_{-1}(\sigma,\tau) = \Gamma(d-\sigma)/\Gamma(d+\tau+n+1) .$$

(In the above formulae, $\lambda \in \mathbb{R}^+$ parametrises the representations of $\tilde{N}$; hereafter, λ will be a parameter of $\pi_{\mu,\lambda}$). Then $B_{\mu,\lambda}$ is positive definite, up to a constant, if and only if the signs of h_1 and h_{-1} are the same , for all d in $\mathbb{N}$. This means that the signs of

$$H_{\epsilon}(d) = \Gamma\left(\frac{2d-\lambda+n+1-\epsilon\mu'}{2}\right) \Big/ \Gamma\left(\frac{2d+\lambda+n+1-\epsilon\mu'}{2}\right)$$

must be the same, for all d in $\mathbb{N}$ and ϵ in $\{\pm 1\}$.

Since the characterisation of which λ and μ' satisfy this condition is invariant if we change the signs of λ or of μ' , we shall suppose henceforth that $\lambda \geqq 0$, $\mu' \geqq 0$.

Now

$$H_1(d+1)/H_1(d) = (2d-\lambda+n+1-\mu')/(2d+\lambda+n+1-\mu')$$

and

$$H_{-1}(d+1)/H_{-1}(d) = (2d-\lambda+n+1+\mu')/(2d+\lambda+n+1+\mu') .$$

In order that both these ratios be positive, it is necessary and sufficient that

a) $2d-\lambda+n+1-\mu' > 0$

or

b) $2d+\lambda+n+1-\mu' < 0$.

The second condition cannot be satisfied for all d in $\mathbb{N}$, and so either a) holds for all d in $\mathbb{N}$, in which case $\lambda + \mu' < n + 1$, or, for some d_o in $\mathbb{N}^*$,

$$2d_o - \lambda + n + 1 - \mu' > 0$$

$$2(d_o - 1) + \lambda + n + 1 - \mu' < 0,$$

and in this case certainly $\lambda < 1$ and $\mu' > n + 1$. So we should only seek complementary series if $\lambda + \mu' < n + 1$, or if $\mu' > n + 1$ and $\lambda < 1$.

Now $H_{-1}(0) > 0$, and so for complementary series it is necessary that $H_1(0) > 0$. If $\lambda + \mu' < n + 1$, this is certainly true, and moreover a) is satisfied, so we do have complementary series in this case. We conclude by considering the case $\mu' > n+1$.

If $\mu' > n+1$, and $\lambda < 1$, then

$$H_1(0) = \Gamma\left(\frac{-\lambda + n + 1 - \mu'}{2}\right) / \Gamma\left(\frac{\lambda + n + 1 - \mu'}{2}\right).$$

Here at least one of the arguments is negative. The Γ-function (of a real argument) changes sign at the poles of the Γ-function. Since

$$\left|\left(\frac{-\lambda + n + 1 - \mu'}{2}\right) - \left(\frac{\lambda + n + 1 - \mu'}{2}\right)\right| < 1,$$

$H_1(0)$ is positive if and only if $(-\lambda + n + 1 - \mu')/2$ and $(\lambda + n + 1 - \mu')/2$ lie between the same two poles, i.e. if and only if

$$\lambda/2 < \min\{|(\mu' - n - 1)/2 - k| : k \in \mathbb{N}\}.$$

Now, if this condition is satisfied, the inequality a) holds if $d > [(\mu' - n - 1)/2]$, while b) holds if $d \leq [(\mu' - n - 1)/2]$, and so again we have complementary series in this case. ∎

We conclude this section with a discussion of the situation regarding the class-one complementary series of the real rank one groups.

The class-one complementary series arises when we take the family of representations $\pi_{1,\lambda}$ of a group G, induced from a representation of the form $ma_s\bar{n} \to e^{s\lambda}\rho(a_s)$ of a parabolic subgroup $P = MA\bar{N}$. This representation may also be realised on a space of functions on N, and it is also possible to construct an operator $A(1,\lambda)$ which intertwines $\pi_{1,\lambda}$ and $\pi_{1,-\lambda}$. This operator may be realised by convolution on N with the kernel $N^{\lambda - r}$, where N here is the norm function and $r = p + 2q$. By arguments similar to those of Corollary 3.4, we may prove the following (cf. [2]).

<u>Corollary 3.5</u>. <u>The class-one complementary series of the real-rank one simple Lie groups not locally isomorphic to</u> $SO_o(n,1)$ <u>occurs when</u> $\lambda \in (-p-2, p+2)$.

This result was first proved by B. Kostant [15].

Chapter 4

The Kelvin Transform

It was indicated in §2 how the results on analytic continuation can be used to find the fundamental solution of $L^{(b)}$ (Cor. 2.9). The same reasoning can be applied in the case of the Heisenberg group H_n to the operators L_γ, introduced in §2 ; the following result was originally proved by Folland and Stein [6] via direct computations.

Proposition 4.1. Let $\gamma \in \mathbb{R}$ and write

$$\alpha = \frac{n-\gamma}{2}, \quad \beta = \frac{n+\gamma}{2} . \tag{4.1}$$

Then

$$L_\gamma * (A^{-\alpha}\bar{A}^{-\beta}) = (A^{-\alpha}\bar{A}^{-\beta}) * L_\gamma = c_\gamma \delta_{(e)}$$

where

$$c_\gamma = 2^{2-n}\pi^{n+1}\Gamma(\alpha)^{-1}\Gamma(\beta)^{-1} .$$

Proof. We write the identity of Theorem 2.3 for $p \equiv 1$, $\tau = -\beta$,

$$L_\gamma * (A^\sigma \bar{A}^{-\beta}) = (A^\sigma \bar{A}^{-\beta}) * L_\gamma = 2\beta(\alpha+\sigma)A^\sigma \bar{A}^{-\beta-1} .$$

Now we let $\sigma \to -\alpha$. The limit of the right hand side is evaluated in Theorem 2.2. Using $\beta\Gamma(\beta) = \Gamma(\beta+1)$ we have our result.

■

Corollary 4.2. When $\gamma \in \mathbb{R}$ and $|\gamma| \neq n, n+2, \dots,$ the function

$$G_\gamma = c_\gamma^{-1} A^{-\alpha}\bar{A}^{-\beta}$$

is a fundamental solution for L_γ. In particular, L_γ is hypoelliptic.

We proceed to describe the generalised Kelvin transformation, first in the case of H_n, then more generally.

Suppose $\gamma \in \mathbb{R}$, $|\gamma| \neq n, n+2, \dots$ and retain the notation (4.1). From Proposition 4.1 and Corollary 3.3 it is then clear that L_γ is (up to constant) the inverse of the intertwining operator $A(\mu, 1)$ where μ is such that $\mu' = -\gamma$ (in the notation of Corollary 3.3). Therefore,

$$L_\gamma * (\pi_{\mu,-1}(\tilde{g})\xi) = \pi_{\mu,1}(\tilde{g})(L_\gamma * \xi) \tag{4.2}$$

for all ξ in the representation space of $\pi_{\mu,-1}$. (In what sense this equality is true, and exactly what functions ξ are in the representation space is not an entirely trivial question (cf. [19]). It is clear, however, that (4.2) holds in every possible sense when both ξ and $\pi_{\mu,-1}(\tilde{g})\xi$ are smooth and have compact support.)

We shall now write out (4.2) explicitly for the case $\tilde{g} = \tilde{h}_\pi$ (where $\tilde{h}_r$ is as defined in §3). First, using Corollary 3.2 and using the notation $n^+ = (v^+, t^+)$ introduced in the proof of Corollary 3.3 we obtain, for every λ,

$$[\pi_{\mu,\lambda}(\tilde{h}_\pi)\xi](v,t) =$$

$$k\left(\frac{A(v,t)}{|A(v,t)|}\right)^\gamma e^{-(\lambda+n+1)\log|A(v,t)|}\xi(v^+, t^+)$$

with k a constant independent of λ. Specialising this to $\lambda = \pm 1$, we can rewrite (4.2) and obtain the following result.

<u>Proposition 4.3.</u> <u>Defining the "Kelvin transform" K_γ for functions f smooth on $H_n \setminus \{e\}$ by</u>

$$K_\gamma(f)(v,t) = A(v,t)^{-\alpha}\bar{A}(v,t)^{-\beta} f(v^+, t^+) \tag{4.3}$$

<u>we have</u>

$$L_\gamma * K_\gamma(f) = |A|^{-2} K_\gamma(L_\gamma * f). \tag{4.4}$$

<u>It follows that the Kelvin transform carries harmonic functions (in the sense $L_\gamma * f = 0$) to harmonic functions.</u>

This result was originally proved in [13] by use of the Folland-Stein fundamental solution and by direct computation. Greiner and Koornwinder [8] gave a new interpretation of L_γ and proved (4.4) more conceptually by considering line bundles over complex projective space. In the present proof (4.4) appears as a fact of representation theory.

There is now no difficulty in extending this result to the nilpotent Iwasawa component of any simple group G of real rank one. Of course, we have to restrict ourselves to $L^{(\mathfrak{d})}$ since L_γ for $\gamma \neq 0$ has no general analogue.

Writing, as usual, $G = KAN$ for the Iwasawa decomposition and M for the centraliser of A in K, there exists an essentially unique element w in K normalising but not centralising A and such that $w^2 = e$. (In the case of $SU(n+1,1)$, $w = \tilde{h}_\pi$.) Then, writing $\bar{N} = wNw$ and $P = MA\bar{N}$, we have $G = PN \cup Pw$ (Bruhat decomposition), and an involution, to be denoted h, is defined on $N\setminus\{e\}$ by $Ph(n) = Pnw$ (see [9], [12], [19]). Using the structure of H-type group described here in §1, the involution h is explicitly computed in [2]. Writing $n = \exp(X + Y/4)$, $h(n) = n^+ = \exp(X^+ + Y^+/4)$ and adjusting for a slight difference in our normalisations the result is

$$X^+ = -N(X,Y)^{-4}(|X|^2 - j(Y))X$$

(4.5)

$$Y^+ = -N(X,Y)^{-4}Y .$$

From the remarks at the end of §3 and from Corollary 2.9 it follows that $L^{(\mathfrak{d})}$ intertwines the representation $\pi_{1,-1}$ of G with $\pi_{1,1}$. Writing out the intertwining relation explicitly for the element w we obtain the following result

Proposition 4.4. *Defining the Kelvin transform K by*

$$Kf = (f \circ h)N^{2-p-2q}$$

we have, for all smooth functions f on $N\setminus\{e\}$

$$L^{(\mathfrak{d})} * K(f) = N^{-4}K(L^{(\mathfrak{d})} * f) .$$

In conclusion we make some remarks about harmonic polynomials on N ("harmonic" in the sense of $L^{(\mathfrak{d})} * p = 0$). In [13] we showed that Maxwell's classical construction of the harmonic polynomials on $\mathbb{R}^n$ extends to H_n. This construction also extends to the class of groups N under consideration here. In fact, let us define for any D in the convolution algebra $\mathfrak{A}$ of distributions on N supported at $\{e\}$,

$$\mu_1(D) = K(G * D)$$

where we wrote $G = N^{2-p-2q}$ for brevity. It is clear from Propositions 4.3 and 4.1 that $\mu_1(D)$ is harmonic on $N\setminus\{e\}$. We show that $\mu_1(D)$ is a polynomial (then it is necessarily harmonic everywhere on N). This is done by induction: Since the elements of $\mathfrak{d}$ generate the Lie algebra of N and hence $\mathfrak{A}$, it suffices to prove that if $p = \mu_1(D)$ is a polynomial, then so is $\mu_1(D * X)$ for all X in $\mathfrak{d}$. The definitions of μ_1 and K give

$$\mu_1(D * X) = K(K(p) * X)$$

$$= G[(G * X) \circ h]p + [(p \circ h) * X] \circ h \; .$$

Now a simple computation with the aid of (4.5) shows that the coefficient of p in the first term of the right hand side is a polynomial, and that the second term is obtained by applying to p a linear differential operator with polynomial coefficients. This finishes the induction.

It is clear that the kernel of μ_1 contains the right ideal $L^{(\mathfrak{b})} * \mathfrak{G}$ of $\mathfrak{G}$. Therefore, denoting by $\mathcal{H}$ the linear space of harmonic polynomials on N, we have an induced map

$$\mu : \mathfrak{G}/L^{(\mathfrak{b})} * \mathfrak{G} \to \mathcal{H} \; .$$

Proposition 4.5. μ is a linear isomorphism.

Proof. The argument of [13, p. 182] applies without change. ■

REFERENCES

[1] M. Cowling. "Unitary and uniformly bounded representations of some simple Lie groups," in "Harmonic analysis and group representations". C.I.M.E., Liguori, Napoli (1982).

[2] M. Cowling. "Harmonic analysis on some nilpotent groups." To appear in "Topics in modern harmonic analysis", Istituto Nazionale di Alta Matematica, Roma.

[3] J. Cygan. "Subadditivity of homogeneous norms on certain nilpotent Lie groups". Proc. Amer. Math. Soc., 83 (1981), 69-70.

[4] H.B. Dwight. "Tables of integrals and other mathematical data". 4th ed., 4th printing, Macmillan, New York (1966).

[5] M. Flensted-Jensen. "Spherical functions on a simply connected semisimple Lie group II, the Paley-Wiener theorem for the rank one case." Math. Ann., 228 (1977), 65-92.

[6] G.B. Folland and E.M. Stein. "Estimates for the ∂_b-complex and analysis on the Heisenberg group". Comm. Pure Appl. Math., 27 (1974), 429-522.

[7] D. Geller. "Spherical harmonics, the Weyl transform and the Fourier transform on the Heisenberg group." To appear.

[8] P.C. Greiner and T.H. Koornwinder. "Variations on the Heisenberg spherical harmonics." To appear.

[9] S. Helgason. "A duality for symmetric spaces with applications to group representations." Advances in Math., 5 (1970), 1-154.

[10] A. Kaplan. Fundamental solutions for a class of hypoelliptic PDE generated by composition of quadratic forms." Trans. Amer. Math. Soc., 258 (1980), 147-153.

[11] A. Kaplan. "Riemannian nilmanifolds attached to Clifford modules." Geom. Dedicata, 11 (1981), 127-136.

[12] A.W. Knapp and E.M. Stein. "Intertwining operators for semisimple groups." Ann. of Math., 93 (1971), 489-578.

[13] A. Korányi. "Kelvin transforms and harmonic polynomials on the Heisenberg group." J. Funct. Anal., 49 (1982), 177-185.

[14] A. Korányi. "Geometric properties of Heisenberg type groups." Advances in Math. To appear.

[15] B. Kostant. "On the existence and irreducibility of certain series of representations," in "Lie groups and their representations." Ed.: I.M. Gelfand, J. Wiley, New York (1975).

[16] H. Kraljević. "Representations of the universal covering group of the group SU(n,1)." Glasnik Mat., 8 (28) (1973), 23-72.

[17] F. Ricci. "Harmonic analysis on groups of type H." To appear.

[18] W. Rudin. "Function theory on the unit ball of C^n." Springer, New York (1980).

[19] G. Schiffmann. "Intégrales d'entrelacement et fonctions de Whittaker." Bull. Soc. Math. France, 99 (1971), 3-72.

[20] N.R. Wallach. "Harmonic analysis on homogeneous spaces." Marcel Dekker, New York (1973).

A. Korányi
Washington University
St. Louis, Missouri

M. Cowling
University of Genova
Genova, Italy

ON THE PLANCHEREL FORMULA FOR ALMOST ALGEBRAIC REAL LIE GROUPS

Michel Duflo
CNRS and University of Paris VII

Introduction. In this paper I consider some problems in harmonic analysis for a class of real Lie groups which I call "almost algebraic". More precisely, an almost algebraic group is a triple $(G,\Gamma,\underline{G})$, where G is a real separable Lie group, Γ a discrete subgroup of the center of G, $\underline{G}$ a linear algebraic Lie group defined over $\mathbb{R}$, such that G/Γ is open (for the Hausdorff topology) in the group $\underline{G}(\mathbb{R})$ of real points of $\underline{G}$.(1)

There are two main reasons to consider this class of groups. First, let H be a connected, simply connected Lie group. Then its derived group G can be given a structure of an almost algebraic group and Pukanszky has shown in [Pu] that several questions in harmonic analysis can be reduced (however in a nontrivial way) to questions about G. Second, this class is very convenient: it is stable under all the constructions which occur in the Mackey inductive process, and the groups G are type I (by Harish Chandra and Dixmier [Di 1]).

In my paper [Du3] I constructed a set of unitary irreducible representations of a Lie group G which, when G has an almost algebraic structure, is sufficiently large to decompose the regular representation. Here, I present a survey of some work done in [Du 3,5], [Li 1,2,3], [Bou], [Kha 1,2] on these representations, including some improvements in the almost algebraic case. I then write a rather explicit form of the Plancherel formula for G.

Of course the general setting is the orbit method introduced by Kirillov in the case of nilpotent Lie groups (cf. [Kil,2]). See the

---

(1) Notes are at the end of the paper.

nice survey of M. Vergne [Ve].

One of the main ideas in the Plancherel formula is the notion of a strongly regular linear form on the dual $\underline{g}^*$ of the Lie algebra $\underline{g}$ of G. It was introudced in a series of lectures in a preceding special year (in 1978) at College Park (it is reproduced as an appendix to [Du3]). I am very happy to have the opportunity now to report again on the subject. I would like to thank R. Herb, R.L. Lipsman and J. Rosenberg for their invitation and their hospitality. I would also like to thank June Slack and Vicki Hendricks for the hard job of typing this manuscript.

Notation. The following notations will be used without references.

$\mathbb{N} = \{0,1,2 \ldots\}$, $\mathbb{Z}$ = integers, $\mathbb{R}$ = real numbers, $\mathbb{C}$ = complex numbers. We choose one square root i of -1.

If V is a vector space, V^* is the dual space. If V is real, $V_{\mathbb{C}}$ is its complexification, and we denote $\bar{v}$ the conjugate of an element $v \in V_{\mathbb{C}}$. If W is a subspace of V, we denote by $W^{\perp}$ its orthogonal in V^*. When we do not mention the base field of a vector space, it is either $\mathbb{R}$ or $\mathbb{C}$.

If G is a locally compact group, we denote by $\hat{G}$ its unitary dual, that is the set of equivalence classes of irreducible continuous unitary representations of G in a Hilbert space.

If V is a finite dimensional real vector space, we identify V^* and $\hat{V}$ by the mapping $f \to e^{if}$. If dv is a Lebesgue measure on V, the dual measure df on V^* is the Lebesgue measure normalized to have the inversion formula

$$\iint e^{if(v)}\phi(v)dv\, df = \phi(0).$$

If V is a vector space, x an endormorphism of V, and V_1, V_2 two subspaces of V normalized by x, such that $V_1 \supset V_2$, we denote by x_{V_1/V_2} the endomorphism of V_1/V_2 induced by x.

If a group G acts in a set X, we denote by $G(x)$ the stabilizer of a point $x \in X$. If a Lie algebra $\underline{g}$ acts in a vector space V, we denote by $\underline{g}(v)$ the annihilator of v.

If V is a vector space, we denote by $S(V)$ its symmetric algebra. If $\underline{g}$ is a Lie algebra, we denote by $U(\underline{g})$ its enveloping algebra, and $\beta : S(\underline{g}) \to U(\underline{g})$ the symmetrization.

If V is a vector space, W a subspace, and $f \in V^*$, we denote by $f|V$ the restriction of f to W.

If $\underline{g}$ is a Lie algebra and $g \in \underline{g}^*$, we denote by B_g the bilinear form $X,Y \to g([X,Y])$ on $\underline{g} \times \underline{g}$. It induces a symplectic structure, still called B_g, on $\underline{g}/\underline{g}(g)$ (cf. [Ki3]).

If G is a real Lie group with Lie algebra $\underline{g}$, if $X \in \underline{g}$ and if $\phi \in C^{\infty}(G)$, we define

$$(R(X)\phi)(x) = \frac{d}{d\varepsilon} (x \exp \varepsilon X)\Big|_{\varepsilon = 0}.$$

We extend R to $U(\underline{g}_{\mathbb{C}})$ in the usual way.

Table of Contents

I. Some facts on the metaplectic representation

I.1 The metaplectic group

Let (V,B) be a real symplectic vector space, that is to say, V is a finite dimensional real vector space, and B a non degenerate skew form on $V \times V$. We denote $Sp(B)$ (or $Sp(V)$ if B is clear from the context) the group of automorphisms of (V,B); it is the symplectic group.

Suppose $V \neq \{0\}$. The fundamental group of $Sp(V)$ is $\mathbb{Z}$ (cf. [Che]). Thus, up to equivalence, there exists a unique connected two-fold covering group of $Sp(V)$, called the metaplectic group, and denoted by $Mp(V)$. Let e be the non trivial element of the kernel of the projection mapping. There is a central extension

$$1 \to \{1,e\} \to Mp(V) \to Sp(V) \to 1 \ . \tag{1}$$

For convenience, if $V = \{0\}$, we denote by $Mp(V)$ the group of order two $\{1,e\}$, so that (1) is still valid.

I.2 The mataplectic representation

Let $\underline{n}$ be the Lie algebra with underlying space $V \oplus \mathbb{R}E$, and bracket $[v + tE, v' + t'E] = B(v,v')E$. It is the Heisenberg Lie algebra, and $Sp(V)$ acts as a group of automorphisms of $\underline{n}$ by the formula

$$x(v + tE) = x(v) + tE \quad (x \in Sp(V), \ v \in V, \ t \in \mathbb{R}). \tag{2}$$

Let N be the simply connected Lie group with Lie algebra $\underline{n}$. Exponentiating (2), we get an action of $Sp(V)$ in N.

Let T be an irreducible unitary representation(2) in a Hilbert space $\mathcal{H}$, such that

$$T(\exp(tE)) = e^{it}\mathrm{Id} \qquad (t \in \mathbb{R}). \tag{3}$$

By the Stone-Von Neumann theorem, the class of T is unique. By Shale's theorem [Sha], there exists a unique representation, denoted by S, of $Mp(V)$ in $\mathcal{H}$, such that

$$S(\hat{x})T(n)S(\hat{x})^{-1} = T(x(n)) \tag{4}$$

($n \in N$, $\hat{x} \in Mp(V)$, x its image in $Sp(V)$,

$$S(e) = -Id \quad . \tag{5}$$

The representation S is called the metaplectic representation.[3] For the proof and historical facts, see [Li-Ve].

I.3 Definition of the character $\rho_{\underline{\ell}}$

Let $\underline{\ell}$ be a lagrangian subspace of $V_{\mathbb{C}}$. We denote by $Sp(V)_{\underline{\ell}}$ the normalizer of $\underline{\ell}$ in $Sp(V)$, and by $Mp(V)_{\underline{\ell}}$ its inverse image in $Mp(V)$. We define

$$q(\underline{\ell}) = \text{number of strictly negative eigenvalues} \tag{6}$$

of the hermitian mapping representing the hermitian form $v \to iB(v,\bar{v})$ on $\underline{\ell}$.

The space $\underline{\ell}$ is a (commutative) subalgebra of $\underline{n}_{\mathbb{C}}$, and thus acts in the space $\mathcal{H}^{\infty}$ of differentiable vectors of T. We consider the vector spaces $H_j(\underline{\ell},\mathcal{H}^{\infty})$, $j \in \mathbb{N}$. By definition they are the homology groups of a complex

$$\to \Lambda^j\underline{\ell} \otimes \mathcal{H}^{\infty} \to \Lambda^{j-1}\ell \otimes \mathcal{H}^{\infty} \to \quad \cdots \quad \to \mathcal{H}^{\infty} \to 0 . \tag{7}$$

The group $Mp(V)_{\underline{\ell}}$ acts in this complex through its actions in $\underline{\ell}$ and in $\mathcal{H}^{\infty}$.

Recall that (cf. [Du 3])

$$\dim H_j(\underline{\ell},\mathcal{H}^{\infty}) = \begin{cases} 0 & \text{if } j \neq q(\underline{\ell}) \\ 1 & \text{if } j = q(\underline{\ell}) . \end{cases} \tag{8}$$

The action of $Mp(V)_{\underline{\ell}}$ in (7) gives an action in $H_{q(\underline{\ell})}(\underline{\ell},\mathcal{H}^{\infty})$. We denote by $\rho_{\underline{\ell}}$ the character of $Mp(V)_{\underline{\ell}}$ thus defined. These were introduced in [Du3], and play a fundamental role in this paper.

I.4 Some properties of the character $\rho_{\underline{\ell}}$

The two main properties are (cf. [Du 3] I.9):

$$\rho_{\underline{\ell}}(e) = -1 \tag{9}$$

$$\rho_{\underline{\ell}}(\hat{x})^2 = \det(x_{\underline{\ell}}) \quad (\hat{x} \in Mp(V)_{\underline{\ell}}, \ x \text{ its image in } Sp(V)_{\underline{\ell}}). \tag{10}$$

These formulas determine $\rho_{\underline{\ell}}$ when $Sp(V)_{\underline{\ell}}$ is connected. This is the case when $\underline{\ell}$ is totally complex(4), but not in general.

Let $\underline{\ell},\underline{\ell}'$ be two lagrangian subspaces of $V_{\mathbb{C}}$, and let $x \in Sp(V)_{\underline{\ell}} \cap Sp(V)_{\underline{\ell}'}$. Let $\hat{x}$ be an inverse image of x in $Mp(V)$. From (9), it follows that $\rho_{\underline{\ell}}(\hat{x})\rho_{\underline{\ell}'}(\hat{x})^{-1}$ does not depend on the choice of $\hat{x}$. We denote it by $\rho_{\underline{\ell}}\rho_{\underline{\ell}'}^{-1}(x)$. From (10), we obtain:

$$\rho_{\underline{\ell}}\rho_{\underline{\ell}'}^{-1}(x) = \varepsilon_{\underline{\ell},\underline{\ell}'}(x) \det(x_{\underline{\ell}/\underline{\ell}\cap\underline{\ell}'}) \tag{11}$$

where $\varepsilon_{\underline{\ell},\underline{\ell}'}$ is a character of $Sp(V)_{\underline{\ell}} \cap Sp(V)_{\underline{\ell}'}$ with values in $\{\pm 1\}$.

To describe $\varepsilon_{\underline{\ell},\underline{\ell}'}$, we need some notations. Let $W = \bigcap_{j\in\mathbb{N}} (1-x)^j(V)$. It is a symplectic subspace of V stable under x. Moreover, let $\underline{m} = W_{\mathbb{C}} \cap \underline{\ell}$, $\underline{m}' = W_{\mathbb{C}} \cap \underline{\ell}'$. Then $\underline{m}$ and $\underline{m}'$ are lagrangian in $W_{\mathbb{C}}$. We define $q(\underline{m})$ and $q(\underline{m}')$ as in (6). We define:

$n(x,\underline{\ell})$ = number (counted with multiplicities) of eigenvalues of $x_{\underline{\ell}\cap V}$ which belong to $]1,\infty[$. (12)

Then (cf. [Du-He-Ve[):

$$\varepsilon_{\underline{\ell},\underline{\ell}'}(x) = (-1)^n, \text{ with}$$

$$n = q(\underline{m}) + q(\underline{m}') + n(x,\underline{\ell}) + n(x,\underline{\ell}') + \dim \underline{m}/\underline{m} \cap \underline{m}'. \tag{13}$$

I.5 The function δ

Let $\underline{\ell}$ be a positive lagrangian subspace of $\underline{\ell}$ (that is to say, $q(\underline{\ell}) = 0$). We define a unitary character $\delta_{\underline{\ell}}$ of $Mp(V)_{\underline{\ell}}$ by the formula

$$\delta_{\underline{\ell}}(\hat{x}) = \rho_{\underline{\ell}}(\hat{x}) \left|\rho_{\underline{\ell}}(\hat{x})\right|^{-1} \quad (\hat{x} \in Mp(V)_{\underline{\ell}}). \tag{14}$$

Let $x \in Sp(V)$. There always exists a positive lagrangian $\underline{\ell}$ normalized by x. Let $\underline{\ell}$ and $\underline{\ell}'$ be two such lagrangians. It is easy to see that formulas (11) and (13) imply:

$$\delta_{\underline{\ell}}(\hat{x}) = \delta_{\underline{\ell}'}(\hat{x}) \tag{15}$$

($\hat{x}$ is an inverse image of x in $Mp(V)$). This can also be deduced from [Au-Ko]. We define

$$\delta(\hat{x}) = \delta_{\underline{\ell}}(\hat{x}) \tag{16}$$

for any $x \in Mp(V)$, $\underline{\ell}$ positive lagrangian subspace of $V_{\mathbb{C}}$ normalized by x. (The function δ was defined, with the notation $\delta^{\frac{1}{2}}$, in a different way in [Dul].)

II. Representations of Nilpotent Lie Groups

I.1. Polarizations

Let G be a Lie group, $\underline{g}$ its Lie algebra, $g \in \underline{g}^*$. A polarization is a subalgebra $\underline{b}$ of $\underline{g}_{\mathbb{C}}$ such that $\underline{b}$ contains $\underline{g}(g)$, and $\underline{b}/\underline{g}(g)_{\mathbb{C}}$ is lagrangian in $(\underline{g}/\underline{g}(g))_{\mathbb{C}}$. If g or $\underline{g}$ are not clear from the context, we say polarization at g of $\underline{g}$.[5]. We define

$$q(\underline{b}) = q(\underline{b}/\underline{g}(g)_{\mathbb{C}}) \qquad \text{(cf. I (6))}. \tag{1}$$

If g is not clear from the context, we write $q(\underline{b}) = q^g(\underline{b})$. We say that $\underline{b}$ is positive if $q(\underline{b}) = 0$.

A real polarization is a subalgebra $\underline{h}$ of $\underline{g}$ such that $\underline{h}_{\mathbb{C}}$ is a polarization.

Let D be a group of automorphisms of G which fixes g. We denote by $D^{\underline{g}}$ the subgroup of $D \times Mp(\underline{g}/\underline{g}(g))$ consisting of pairs $(d,\hat{x})$ such that $d_{\underline{g}/\underline{g}(g)} = x$, where x is the image of $\hat{x}$ in $Sp(\underline{g}/\underline{g}(g))$. We use the notation $(1,e) = e$; and usually denote by $\hat{d}$ an element of $D^{\underline{g}}$ of the form $(d,\hat{x})$.

The group $D^{\underline{g}}$ is a central extension of D:

$$1 \to \{1,e\} \to D^{\underline{g}} \to D \to 1. \tag{2}$$

If $\underline{\ell}$ is a lagrangian subspace of $(\underline{g}/\underline{g}(g))_{\mathbb{C}}$, and if $\hat{d} = (d,\hat{x})$ is an element of $D^{\underline{g}}$ such that $d\underline{\ell} \subset \underline{\ell}$, we define:

$$\rho_{\underline{\ell}}(\hat{d}) = \rho_{\underline{\ell}}(\hat{x}), \quad \delta(\hat{d}) = \delta(\hat{x}). \tag{3}$$

If g is not clear from the context, we shall write $\rho^g_{\underline{\ell}} = \rho_{\underline{\ell}}$, $\delta^g = \delta$. If $\underline{b}$ is a polarization normalized by d, we shall write

$$\rho_{\underline{b}} = \rho_{\underline{b}/\underline{g}(g)_{\mathbb{C}}}. \tag{4}$$

II.2. Representations of nilpotent groups

We suppose moreover that G is a connected and simply connected nilpotent Lie group.

Let $\underline{b}$ be a positive polarization at g such that $b + \overline{b}$ is a subalgebra. Such polarizations exist. We define $\underline{d} = \underline{b} \cap \underline{g}$, we denote by D the subgroup of G with Lie algebra $\underline{d}$. We choose a G-invariant positive Radon measure μ on G/D. We denote by $H'_{\underline{b}}$ the prehilbert space of C^{∞} functions ϕ on G such that

$$R(X)\phi = -ig(X)\phi, \quad X \in \underline{b} \tag{5}$$

$$\int_{G/D} |\phi|^2 \, d\mu < \infty. \tag{6}$$

Note that (5) implies that $|\phi^2|$ is well defined on G/D. We denote by $H_{\underline{b}}$ the completion of $H'_{\underline{b}}$. We denote by $T_g^{\underline{b}}$ the representation of G in $H_{\underline{b}}$ obtained by left translation .

The following theorem is due to Kirillov [Ki 1] for real polarizations, and Auslander and Kostant [Au-Ko] in general.

(7) <u>Theorem</u>. The representation $T_g^{\underline{b}}$ is irreducible (in particular, H_b is not zero). Its class is independent of $\underline{b}$.

This class will be denoted by T_g (or T_g^G if G is not clear from the context). If x is an automorphism of G (which then acts also in $\underline{g}^*$ and $\hat{G}$) it is clear by "transposition of structure" that ${}^xT_g = T_{x(g)}$. In particular, if g and g' are in the same G-orbit Ω, we have $T_g = T_{g'}$. We write $T_g = T_{\Omega}$. Theorem (7) defines a mapping from the set $\underline{g}^*/G$ of orbits of G in $\underline{g}^*$ to $\hat{G}$.

(8) <u>Theorem</u> (Kirillov [Ki 1]). The mapping $\Omega \to T_{\Omega}$ is a bijection from $\underline{g}^*/G$ onto $\hat{G}$.

Let $d \in D$. There exists a positive polarization $\underline{b}$ such that $\underline{b} + \overline{\underline{b}}$ is a subalgebra and which is normalized by d. ([Au-Ko], cf. [Be]). We define an operator in $H_{\underline{b}}$ by the formula:

$$(S'_{\underline{b}}(d)\phi)(x) = |\det d_{\underline{g}/\underline{d}}|^{-1/2} \phi(d^{-1}(x)). \tag{9}$$

It is unitary, and satisfies

$$S'_{\underline{b}}(d)T_g^{\underline{b}}(x)S'_{\underline{b}}(d)^{-1} \quad = \quad T_g^{\underline{b}}(d(x)). \qquad (10)$$

(11) Theorem (Auslander-Kostant [Au-Ko]). The operator $S'_{\underline{b}}(d)$ does not depend on $\underline{b}$.

Theorem (11) means the following: let $\underline{b}'$ be another positive polarization such that $\underline{b}'+\bar{\underline{b}}'$ is a subalgebra and normalized by d. Choose a unitary operator $U : H_{\underline{b}} \to H_{\underline{b}'}$ intertwining $T_g^{\underline{b}}$ and $T_g^{\underline{b}'}$. Then $S'_{\underline{b}}(d) = U^{-1} S'_{\underline{b}'}(d)U$. We will denote by $S'(d)$ the class of $S'_{\underline{b}}(d)$.

Usually S' is not a representation of D, but only a projective representation. Let $\hat{d} \in D^g$. We define

$$S(\hat{d}) \quad = \quad \delta(\hat{d})S'(d) \qquad (12)$$

(where d is the image of $\hat{d}$). If g is not clear from the context, we write $S = S_g$.

(13) Theorem (Duflo [Du 1]). S is a representation of D^g.

(14) Remark. Another construction of S, using only real polarizations, is due to Lion (cf. [Li-Ve] and [Du 3] ch. V). It has the advantage of extending to p-adic nilpotent groups, but we are not concerned with that here.

(15) Remark. Consider the Heisenberg group N of Section I.2. Let $E^* \in \underline{n}^*$ be such that $E^*(E) = 1$, $E^*(V) = \{0\}$. Then, with the notations of I.2, $T = T_{E^*}$, and the S defined by (12) is the same as the S of I.2.

We say that a unitary representation U of D^g is odd if $U(e) = -\mathrm{Id}$. The representation S is odd. We denote by $U \otimes ST_g$ the representation of the semi-direct product $D \times G$ defined in the Hilbert tensor product of the spaces of U and T_g by the formula:

$$(U \otimes ST_g)(dx) = U(\hat{d}) \otimes S(\hat{d})T_g(x) \tag{16}$$

where $d \in D$, $\hat{d} \in D^{\underline{g}}$ is an inverse image of d, $x \in G$. This does not depend on the choice of $\hat{d}$, and it is a representation because of (10).

III. First construction of the representations $T_{g,\tau}$

III.1 Good polarizations

Let $\underline{g}$ be a real Lie algebra, and $g \in \underline{g}^*$. A good polarization $\underline{b}$ is a polarization which is solvable, and satisfies the following Pukanszky's type condition: let $g' \in \underline{g}^*_{\mathbb{C}}$ be such that $g|\underline{b} = g'|\underline{b}$, then $\dim \underline{g}_{\mathbb{C}}(g') = \dim \underline{g}(g)_{\mathbb{C}}$.

It is not always true that g has a good polarization. However, this can be determined by the following steps.

(1) Lemma (cf. [Du 3]). If $\underline{g}$ is reductive, g has a good polarization $\underline{b}$ if and only if $\underline{g}(g)$ is a Cartan subalgebra of $\underline{g}$, and then $\underline{b}$ is a Borel subalgebra containing $\underline{g}(g)$.

(2) Lemma. Let $\underline{a}$ be an ideal of $\underline{g}$ contained in $\ker g$. Let $\underline{g}' = \underline{g}|\underline{a}$ and $g' \in \underline{g}'^*$ be induced from $\underline{g}$. Then g has a good polarization if and only if g' has a good polarization.

(3) Lemma. Let $\underline{n}$ be a nilpotent ideal of $\underline{g}$. Define $n = g|\underline{n}$, $\underline{g}_1 = \underline{g}(n)$, $g_1 = g|\underline{g}_1$. Then g has a good polarization if and only if g_1 has a good polarization.

III.2 Admissible linear forms

Let G be a real Lie group with Lie algebra $\underline{g}$ and let $g \in \underline{g}^*$. As in II.1, we can define the two fold covering group $G(g)^{\underline{g}}$ of $G(g)$. We define $X(g)$ (or $X_G(g)$ if necessary) to be the set of classes of unitary representations τ of $G(g)^{\underline{g}}$ such that $\tau(e) = -\mathrm{Id}$ and $\tau(\exp X) = \exp ig(X)\mathrm{Id}$ for $X \in \underline{g}(g)$. We say that g is admissible (or G-admissible) if $X(g)$ is not empty. We denote by $X^{irr}(g)$ the subset of irreducible classes.

In some cases $X(g)$ can be given a simple description. Suppose for instance that there exists a lagrangian subspace $\underline{\ell}$ of $(\underline{g}/\underline{g}(g))_{\mathbb{C}}$ which is positive and $G(g)$ invariant. Then $\delta = \delta^g$ is a unitary character of $G(g)^{\underline{g}}$ such that $\delta(e) = -1$. Let's denote also by δ the differential of δ. Then $\delta \in ig(g)^*$. Let $\tilde{X}(g)$ be the set of classes of unitary representations σ of $G(g)$ such that $\sigma(\exp X) = \exp(if(X) + \delta(X))\mathrm{Id}$ for $X \in \underline{g}(g)$. Let $\tilde{X}^{irr}(g)$ be the subset of irreducible classes. Then the mapping $\tau \mapsto \delta\tau$ is a bijection of $X(g)$ onto $\tilde{X}(g)$, sending $X^{irr}(g)$ onto $\tilde{X}^{irr}(g)$. When G is connected solvable, the condition above is satisfied. It is also satisfied when G is connected semi-simple. It explains why the consideration of the covering groups $G(g)^{\underline{g}}$ can be avoided without too many problems in these particular cases.

We denote by X (or X_G if necessary) the set of pairs (g,τ) where g is admissible and has a good polarization, and $\tau \in X(g)$. We denote by X^{irr} the subset of irreducible classes.

III.3 <u>The representations</u> $T_{g,\tau}$ <u>when</u> G <u>is reductive and connected</u>

In this paragraph G is a connected reductive group, with Lie algebra $\underline{g}$.

Let $\underline{h}$ be a Cartan subalgebra of $\underline{g}$. We write Δ (or $\Delta(\underline{g}_{\mathbb{C}},\underline{h}_{\mathbb{C}})$) for the roots of $\underline{h}_{\mathbb{C}}$ in $\underline{g}_{\mathbb{C}}$, $\underline{g}^{\alpha} \subset \underline{g}_{\mathbb{C}}$ for the root space corresponding to $\alpha \in \Delta$, $h_{\alpha} \in \underline{h}_{\mathbb{C}}$ for the coroot corresponding to α. We write

$$\underline{h} = \underline{t} \oplus \underline{a} \oplus \underline{z} \tag{4}$$

where $\underline{z}$ is the center of $\underline{g}$ and $i\underline{t} \oplus \underline{a}$ is the real linear span of coroots.

Let H be the centralizer of $\underline{h}$ in G, M the centralizer of $\underline{a}$ in G, $\underline{m}$ the Lie algebra of M, $\Delta_{\underline{m}}$ the subset $\Delta(\underline{m}_{\mathbb{C}}, \underline{h}_{\mathbb{C}})$ of Δ. The roots in $\Delta_{\underline{m}}$ are precisely the roots which are purely imaginary on $\underline{h}$. If $\sum$ is a subset of Δ, we denote by $\rho(\sum)$ the half sum of elements of $\sum$. If $\alpha \in \Delta$, we denote by ξ_{α} the character of H by which H acts in $\underline{g}^{\alpha}$.

An element $\lambda \in \underline{h}^*_{\mathbb{C}}$ is said to be regular (or $\underline{g}$-regular if there is some ambiguity) if $\lambda(h_\alpha) \neq 0$ for all $\alpha \in \Delta$. In an analogous way, we define $\underline{m}$-regular elements of $\underline{h}^*_{\mathbb{C}}$.

If $\alpha \in \Delta$, denote by $\underline{s}^\alpha_{\mathbb{C}}$ the Lie algebra $\underline{g}^\alpha \oplus \underline{g}^{-\alpha} \oplus \mathbb{C}\underline{h}_\alpha$. If $\alpha \in \Delta_{\underline{m}}$, $\underline{s}^\alpha_{\mathbb{C}}$ is the complexification of $\underline{s}^\alpha = \underline{s}^\alpha_{\mathbb{C}} \cap \underline{g}$. Recall that $\alpha \in \Delta_{\underline{m}}$ is called compact if $\underline{s}^\alpha \simeq \underline{su}(2)$, and non compact if $\underline{s}^\alpha \approx \underline{s\ell}(2,\mathbb{R})$. We write $\Delta_{\underline{m},c}$ and $\Delta_{\underline{m},n}$ for the set of compact and non compact roots respectively.

Let $\lambda \in \underline{h}^*_{\mathbb{C}}$ be an element such that

$$\lambda \text{ is } \underline{m}\text{-regular} \tag{5}$$

$$\lambda|\underline{t} \in i\underline{t}^*. \tag{6}$$

We define

$$\Delta^+_{\underline{m}}(\lambda) = \{\alpha \in \Delta^+_{\underline{m}},\ \lambda(h_\alpha) > 0\} \tag{7}$$

$$\Delta^+_{\underline{m},n}(\lambda) = \Delta^+_{\underline{m}}(\lambda) \cap \Delta_{\underline{m},n};\ \Delta^+_{\underline{m},c}(\lambda) = \Delta^+_{\underline{m}}(\lambda) \cap \Delta_{\underline{m},c} \tag{8}$$

$$\delta^\lambda = \rho(\Delta^+_{\underline{m},n}(\lambda)) - \rho(\Delta^+_{\underline{m},c}(\lambda)). \tag{9}$$

Following Vogan [Vo 2], an M-regular unitary pseudo-character is a pair (Λ,λ) where $\lambda \in i\underline{h}^*$ satisfies (5) and Λ is a unitary representation of H such that $d\Lambda = (\lambda + \delta^\lambda)\mathrm{Id}$. We shall say that (Λ,λ) is irreducible if λ is irreducible, $\underline{g}$-regular if λ is $\underline{g}$-regular.(6) We denote by $R(H)$ the set of classes of M-regular unitary pseudo-characters, $R^{irr}(H)$ the irreducible subset, $R(H,\lambda)$ the set of classes with differential $(\lambda + \delta^\lambda)\mathrm{Id}$, $R^{irr}(H,\lambda) = R(H,\lambda) \cap R^{irr}(H)$.

Let $W(G,H)$ be the Weyl group (= normalizer of H in G modulo H). It obviously acts in $R(H)$ and $R^{irr}(H)$.

Let $(\Lambda,\lambda) \in R(H)$. A class $\pi(\Lambda,\lambda)$ of unitary representations is defined (essentially by Harish Chandra). Let us recall this construction. Let $F = \{x \in H,\ x$ centralizes $\underline{m}$ and $|\xi_\alpha(x)| = 1$ for all $\alpha \in \Delta\}$. It is known that

$$H \cap M_0 = H_0 \quad \text{and} \quad H = FH_0. \tag{10}$$

Let $\pi^{M_0}(\lambda)$ be the unitary irreducible representation of M_0, square integrable modulo the center of M_0, associated by Harish-Chandra to λ.(7) It can be characterized in the following manner. Let $\underline{k}_{\underline{m}}$ be the Lie algebra $(\underline{h}_{\mathbb{C}} + \sum_{\alpha\in\Delta_{\underline{m},c}} \underline{g}^{\alpha}) \cap \underline{g}$. Let K_{M_0} be the corresponding analytic subgroup. Then the restriction of $\pi^{M_0}(\lambda)$ to K_{M_0} contains the (finite dimensional) irreducible unitary representation of K_{M_0} with dominant weight $\lambda + \delta^{\lambda}$ with respect to $\Delta^{+}_{\underline{m},c}$ as a minimal K_{M_0}-type (cf. [Vo 1]).

We define a representation of FM_0 in the Hilbert tensor product of the spaces of Λ and of $\pi^{M_0}(\lambda)$ by the formula

$$\pi^{FM_0}(\Lambda,\lambda)(yx) = \Lambda(y) \otimes \pi^{M_0}(\lambda)(x) \quad (x \in M_0,\ y \in F). \tag{11}$$

Let MN be a parabolic subgroup of G with Levi component M and unipotent radical N. Then $\pi(\Lambda,\lambda)$ is the class of the representation of G induced by the representation $\pi^{FM_0} \otimes \mathrm{Id}_N$ of the group FM_0N. This class does not depend on the choice of N.

Recall that, if Λ is irreducible and λ g-regular, the class $\pi(\Lambda,\lambda)$ is irreducible (cf. [Vo 2], lemma 4.7).

Consider now an element $g \in \underline{g}^*$, which has a good polarization and is admissible. Let $\tau \in X(g)$.

We know that $\underline{g}(g)$ is a Cartan subalgebra of $\underline{g}$. Let us denote it by $\underline{h}$, and use the notations above. Then $G(g) = H$ (this is because G is connected). Let $\lambda = ig|\underline{h}$. It is a $\underline{g}$-regular element of $\underline{h}^*$. There exist positive lagrangian subspaces $\underline{\ell}$ in $(\underline{g}/\underline{g}(g))_{\mathbb{C}}$ which are $G(g)$-invariant. For instance, let $\underline{n}$ be the Lie algebra of the group N. We can choose

$$\underline{\ell} = \sum_{\alpha\in\Delta^+_{\underline{m},n}(\lambda)} \underline{g}^{\alpha} \oplus \sum_{\alpha\in\Delta^+_{\underline{m},n}(\lambda)} g^{-\alpha} \oplus \underline{n}_C$$

(in general, $\underline{\ell}$ is not a subalgebra). The function δ^g on $G(g)^{\underline{g}}$ is a character (cf. III.2) and it is clear that its differential is δ^{λ}. We obtain a g-regular pseudo-character

$$(\tau\delta^g,\lambda) \in R(H,\lambda). \tag{12}$$

We define

$$T_{g,\tau} = \pi(\tau\delta,\lambda). \tag{13}$$

III.4 <u>The representations</u> $T_{g,\tau}$ <u>when</u> G <u>is reductive</u>

In this paragraph, G is a Lie group and $\underline{g}$ its Lie algebra. We choose a linear form $g \in \underline{g}^*$ which ahs a good polarization and is admissible, and $\tau \in X(g)$.

Recall that $\underline{g}(g)$ is a Cartan subalgebra of $\underline{g}$. We denote it by $\underline{h}$ and use the notations of III.3 with the following addition: H is the centralizer of $\underline{h}$ in G. The following inclusions are obvious:

$$G(g) \supset H \supset H \cap G_0 \supset H_0, \tag{14}$$

but they can be strict inclusions.

Define: $\Delta^+ = \{\alpha \in \Delta,\ \mathrm{Im}\,\lambda(h_\alpha) > 0, \text{ or } \mathrm{Im}\,\lambda(h_\alpha) = 0 \text{ and } \mathrm{Re}\,\lambda(h_\alpha) > 0\}$, $\underline{n}^+ = \sum_{\alpha\in\Delta^+} \underline{g}^{\alpha}$, $\underline{b}^+ = \underline{h}_C \oplus \underline{n}^+$. Obviously, $\underline{b}^+$ is a polarization at g, it is stable under G(g). Let $\underline{e}^+ = \underline{b}^+ + \overline{\underline{b}^+} \cap \underline{g}$, and $\gamma = g|\underline{a}$. It is easy to see that $\underline{e}^+$ is a parabolic subalgebra of $\underline{g}$ whose reductive part is the centralizer of γ (identified to a linear form on $\underline{g}$ null on $\underline{t} + \underline{z} + [\underline{h},\underline{g}]$) and unipotent part $\left(\sum_{\substack{\alpha\in\Delta\\ \alpha(\gamma)>0}} \underline{g}^{\alpha}\right) \cap \underline{g}$.

Let $\underline{b}$ be any polarization at g which is stable under G(g) and such that $\underline{e} = (\underline{b} + \overline{\underline{b}}) \cap \underline{g}$ is a subalgebra of $\underline{g}$. It is a parabolic subalgebra of $\underline{g}$. Let $\underline{u}$ be the unipotent radical of $\underline{e}$, and $\underline{r}$ the reductive subalgebra of $\underline{e}$ which contains $\underline{h}$ and such that $\underline{r} + \underline{u} = \underline{e}$. We let R_0 and U be the analytic subgroups of G corresponding to $\underline{r}$ and $\underline{u}$ respectively. We consider the groups $R = G(g)R_0$ and $E = G(g)R_0U$. Let $r = g|\underline{r}$. It is obvious that $R(r) = G(g)$ and thus r has a good polarization. Let $x \in G(g)$. Let $\hat{x}$ be a representative of x in $G(g)^{\underline{g}}$, and $\tilde{x}$ a representative in $R(r)^{\underline{r}}$. We define an operator $\tau'(\tilde{x})$ in the space of τ by the formula

$$\delta^g(\hat{x})\tau(\hat{x}) = \delta^r(\tilde{x})\tau'(\tilde{x}). \tag{15}$$

As δ^g and δ^r are not necessarily characters (because there does not always exist a positive lagrangian subspace $\underline{\ell}$ of $(\underline{r}/\underline{r}(r))_{\mathbb{C}}$ stable under $G(g)$) it is not absolutely obvious that τ' is a representation, but it is not difficult to prove. One can check, for instance, that τ' is equal to the representation $\tau_{\underline{e}}$ defined in [Bou], end of §2. The mapping $\tau \to \tau'$ is a bijection of $X_G(g)$ onto $X_R(r)$, preserving irreducibility.

Let us consider $\underline{c} = \underline{b} \cap \underline{r}_{\mathbb{C}}$ and $\underline{v}$, the unipotent radical of $\underline{c}$. It is a lagrangian subspace of $(\underline{r}/\underline{r}(r))_{\mathbb{C}}$, totally complex (i.e. $\overline{\underline{c}} \cap \underline{c} = \{0\}$). This implies that $\underline{h}$ is a fundamental Cartan subalgebra of $\underline{r}$, and that $R_0 \cap H = H_0$. Thus $X^{irr}_{R_0}(r)$ contains exactly one element, that is the character χ such that δ^R_χ is the character of H_0 of differential $\lambda + \delta^\lambda$. Let π^0 be the representation $T^{R_0}_{r,\chi}$ of R_0 defined in III.3. Let H^∞ be the set of C^∞ vectors of the representation π^0. Recall the definition $q^r(\underline{c})$ (II.(1)). Let $\rho(\underline{c}) = \rho(\Sigma)$ where Σ is the set of roots of $\underline{h}_c$ in $\underline{v}$. If V is an $\underline{h}$-module, and $\mu \in \underline{h}^*_{\mathbb{C}}$, we denote by V_λ the space $\{v \in V, \exists n \in \mathbb{N}, (H - \lambda(H))^n v = 0 \text{ for } H \in \underline{h}\}$. Then:

$$\dim H_j(\underline{v}, H^\infty)_{\lambda+\rho(\underline{v})} = \begin{cases} 0 & \text{if } j \neq q^r(\underline{c}) \\ 1 & \text{if } j = q^r(\underline{c}). \end{cases} \tag{16}$$

This formula is a generalization of theorems of Kostant and Schmid, and is essentially due to Vogan (cf. [Du3] III, lemma 3). In [Du3], it is proven that there exists a unique unitary representation S of $R(r)^{\underline{r}}$ in the space $\mathcal{H}$ of the representation π^0 such that

$$S(\tilde{x})\pi^0(y)S(\tilde{x})^{-1} = \pi^0(xyx^{-1}) \tag{17}$$

where $x \in G(g)$, $\tilde{x}$ is a representative in $R(r)^{\underline{r}}$, $y \in R_0$, and such that the action of $R(r)^{\underline{r}}$ in $H_{q^r(\underline{c})}(\underline{v},\mathcal{H}^\infty)_{\lambda+\rho(\underline{v})}$, induced from the action in $\underline{v}$ and $\mathcal{H}^\infty$, is the character $\rho^r_{\underline{c}}$ of $R(r)^{\underline{r}}$ defined in II.1.

We define a representation π of R in the Hilbert tensor product of the spaces of the representation τ and π^0 by the formula

$$\pi(xy) = \tau'(\tilde{x}) \otimes S(\tilde{x})\pi^0(y) \tag{18}$$

(x and y are as above).

We define a representation $T_{g,\tau,\underline{b}}$ of G by the following formula

$$T_{g,\tau,\underline{b}} = \mathrm{Ind}_E^G(\pi \otimes \mathrm{Id}_U). \tag{19}$$

We define

$$T_{g,\tau} = T_{g,\tau,\underline{b}^+}. \tag{19}$$

(21) Remark. Suppose $G = G_0$. Implicit in this definition of $T_{g,\tau}$ is the fact that it coincides with the class defined in III.3 (cf. [Du 3]).

Bouaziz [Bou] proves the following

(22) Proposition. Suppose that the center of (G_0,G_0) is finite. Then $T_{g,\tau,\underline{b}} = T_{g,\tau}$.

(22a) Remark. The hypothesis in (22) is certainly not necessary. It is there only because [Bou] uses results on intertwining operators, etc. for G_0 for which references exist only in this case.

It's proved in [Du 3] that $T_{g,\tau}$ and τ have isomorphic commuting algebras.

III.5 Definition of $T_{g,\tau}$: the general case

In this paragraph, we define the classes $T^G_{g,\tau}$ for all separable Lie groups G, $(g,\tau) \in X_G$. The definition is by induction on the dimension of $\underline{g}$. To make the construction possible, we will require the following property of $T^G_{g,\tau}$.

Let J be the subgroup of $G(g)$ acting trivially in $\underline{g}/\underline{g}(g)$. If $\hat{x} \in G(g)^{\underline{g}}$ is the inverse image of an element $x \in J$, it is obvious that $\delta^g(\hat{x}) = \pm 1$. Thus there exists a unique homomorphism $r^g: J \to G(g)^{\underline{g}}$ such that $\delta(r^g(x)) = 1$, and $r^g(x)$ is an inverse image of x. We want:

(23) Suppose that C is a subgroup of the center of G and that $\tau(r^g(x)) = \psi(x)\mathrm{Id}$, where ψ is a unitary character of C. Then $T_{g,\tau}(x) = \psi(x)\mathrm{Id}$ for $x \in C$.

Suppose first that $\dim \underline{g} = 0$. Then G is a discrete group, $g = 0$. We define

$$T_{g,\tau}(x) = \tau(r^g(x)) \quad \text{for} \quad x \in G. \tag{24}$$

Obviously (23) is satisfied.

We suppose now that $\dim \underline{g} > 0$ and $T^{G_1}_{g_1,\tau_1}$ is defined whenever $\dim \underline{g}_1 < \dim \underline{g}$, $(g_1,\tau_1) \in X_{G_1}$. We suppose that $T^{G_1}_{g_1,\tau_1}$ satisfies (23). Let $(g,\tau) \in X_G$.

Let $\underline{n}$ the nilpotent radical of $\underline{g}$ (that is the largest nilpotent ideal). We denote by N the analytic subgroup of G with Lie algebra $\underline{n}$. It is closed, invariant, nilpotent and connected. Let $n = g|\underline{n}$. The Lie algebra $\underline{q} = \ker g \cap \underline{n}(n)$ is an ideal in $\underline{g}(n)$. Let $\underline{g}_1 = \underline{g}(n)/\underline{q}$. We consider two cases.

<u>First case</u>. $\dim \underline{g}_1 = \dim \underline{g}$. In this case $\underline{g}$ is reductive. We define $T_{g,\tau}$ as in III.6. It can easily be seen that it satisfies (23).

<u>Second case</u>. $\dim \underline{g}_1 < \dim \underline{g}$.

We introduce some notations. We let $\underline{h} = \underline{g}(n)$, $h = g|\underline{h}$. The group $G(n)^{\underline{n}}$ (defined in II.1) will be denoted by H. Because N is nilpotent, $N(n)^{\underline{n}}$ is the direct procut of $(N(n)^{\underline{n}})_0$ by $\{1,e\}$, and $(N(n)^{\underline{n}})_0$ is canonically isomorphic to $N(n)$. Because g is admissible, there exists a (necessarily unique) character χ of $N(n)^{\underline{n}}$ such that $d\chi = in|\underline{n}(n)$ and $\chi(e) = -1$. Let Q be the connected component of $\ker \chi$. It is a closed invariant subgroup of H with Lie algebra $\underline{q}$. Let $G_1 = H/Q$ and $p:H \to G_1$ the canonical projection.

The group $H(h)^{\underline{h}}$ contains $(G(g)^{\underline{n}})^{\underline{h}}$. Let $x \in G(g)$, and let $\hat{x}$ be a representative in $G(g)^{\underline{g}}$, $\ddot{x}$ a representative in $G(g)^{\underline{n}}$, $\tilde{\ddot{x}}$ a representative of $\ddot{x}$ in $(G(g)^{\underline{n}})^{\underline{h}}$. We define an operator $\tau'(\tilde{\ddot{x}})$ in the space of τ by the formula

$$\tau'(\tilde{\ddot{x}})\delta^h(\tilde{\ddot{x}})\delta^n(\ddot{x}) \quad = \quad \tau(\hat{x})\delta^g(\hat{x}). \tag{25}$$

The group $H(h)^{\underline{h}}$ is equal to $(G(g)^{\underline{n}})^{\underline{h}}(N(n)^{\underline{n}})^{\underline{h}}$. As $N(n)^{\underline{n}}$ acts trivially in $\underline{h}/\underline{n}(n)$, the map r^h, from $N(n)^{\underline{n}}$ into $(N(n)^{\underline{n}})^{\underline{h}}$, is defined. It can be proved that there exists a unique element $\tau' \in X_H(h)$ such that

$$\tau'(y) \quad \text{is defined by (25) if} \quad y \in (G(g)^{\underline{n}})^{\underline{h}} \tag{26}$$

$$\tau'(r^h(y)) \; = \; \chi(y)\mathrm{Id} \quad \text{if} \quad y \in N(n)^{\underline{n}}. \tag{27}$$

It is clear that $G_1(g_1) = H(h)^{\underline{h}}/r^h(Q)$, and that there exists a unique element $\tau_1 \in X_{G_1}(g_1)$ such that $\tau' = \tau_1 \circ p$. By induction, $T^{G_1}_{g_1,\tau_1}$ is defined. We let $U = T^{G_1}_{g_1,\tau_1} \circ p$. Applying (23) to the group $C = N(n)^{\underline{n}}/Q$, it follows that

$$U(y) \quad = \quad \chi(y)\mathrm{Id} \quad \text{if} \quad y \in N(n)^{\underline{n}}. \tag{28}$$

Let $\tilde{N}$ be the simply connected covering group of N, and consider the semi-direct product $G(n) \times \tilde{N}$. There is a canonical surjection:

$$G(n) \times \tilde{N} \to G(n)N. \qquad (29)$$

In II.2, we defined a representation $U \otimes S_n T_n^N$ of $G(n) \times \tilde{N}$. Because of (28), it is easy to see that it is trivial on the kernel of (29) and defines a representation (which, by "abuse of notation", we will denote by $T^{G_1}_{g_1,\tau_1} \otimes S_n T_n^N$) of $G(n)N$. We define:

$$T^G_{g,\tau} = \mathrm{Ind}^G_{G(n)N}(T^{G_1}_{g_1,\tau_1} \otimes S_n T_n^N). \qquad (30)$$

(31) Remark. When $\underline{g}$ is reductive and $\dim \underline{g}_1 < \dim \underline{g}$, $T^G_{g,\tau}$ has been defined in two different ways (in III.4, and by (30)). It is easy to see that the two definitions agree.

III.6 Functorial properties of the representations $T_{g,\tau}$

Let G be a separable Lie group with Lie algebra $\underline{g}$.

(32) Theorem ([Du3]).

(i) Let $(g,\tau) \in X$. The commuting ring of $T_{g,\tau}$ is isomorphic to the commuting ring of τ. In particular, if $(g,\tau) \in X^{irr}$, then $T_{g,\tau} \in \hat{G}$.

(ii) Let a be an automorphism of G. It acts in X and in the space of classes of representations of G. Then $T_{a(g,\tau)} = {}^aT_{g,\tau}$. In particular, if (g,τ) and (g',τ') are in the same G orbit in X, then $T_{g,\tau} = T_{g',\tau'}$.

(iii) Let $(g,\tau) \in X$, $(g',\tau') \in X$ and assume that $g \notin g'G$. Then $T_{g,\tau}$ and $T_{g',\tau'}$ are disjoint.

(iv) Let $(g,\tau) \in X$, $(g,\tau') \in X$. The space of intertwining operators between $T_{g,\tau}$ and $T_{g,\tau'}$ is isomorphic to the set of intertwining operators between τ and τ'.

(v) See (23).

The Theorem implies that the map $(g,\tau) \mapsto T_{g,\tau}$ induces an injective map $X/G \hookrightarrow \hat{G}$.

Let $\underline{n}$ be a nilpotent ideal of $\underline{g}$. We assume that the analytic subgroup N of G is closed and G-invariant. Let $(g,\tau) \in X$. We define n, $\underline{h}$, h, H, $\underline{q}$, Q, $\underline{g}$, G_1, g_1, τ_1 as in III.5.

(33) Theorem ([Du 5] ch III)

$$T^G_{g,\tau} = \mathrm{Ind}^G_{G(n)N}(T^{G_1}_{g_1,\tau_1} \otimes S_n T^N_n).$$

(34) Remark. This theorem is true by definition when $\underline{n}$ is the unipotent radical of $\underline{g}$.

Khalgui [Kha 2] gives a generalization of theorem 33 to other invariant subgroups of G (such as the derived group (G,G) if G is connected simply connected). This generalization is fundamental when one is interested in non-almost algebraic groups, to reduce problems to the almost algebraic group (G,G). As this is not the goal of this paper, I do not pursue this matter.

III.7 Normal irreducible representations, characters

Let G be a connected Lie group with Lie algebra $\underline{g}$. Let $(g,\tau) \in X^{irr}$, so that $T_{g,\tau} \in \hat{G}$. Generalizing results of Auslander-Kostant and Pukanszky, Khalgui proves [Kha 2]

(35) Theorem. The representation $T_{g,\tau}$ is normal (or G.C.R., or postliminaire in other terminology) if and only if the two conditions below are satisfied:

$$\dim \tau < \infty \tag{36}$$

$$Gg \text{ is locally closed in } \underline{g}^*. \tag{37}$$

Recall that $T_{g,\tau}$ is called traceable if for any C^∞ density α on G with compact support, the operator $T_{g,\tau}(\alpha) = \int T_{g,\tau}(x)\,d\alpha(x)$ is traceable. Then it has a character, which is the generalized function $\mathrm{tr}\, T_{g,\tau}(x)$ such that

$$\int \operatorname{tr} T_{g,\tau}(x)\,d\alpha(x) = \operatorname{tr} T_{g,\tau}(\alpha).$$

If $X \in \underline{g}$, define

$$j(X)^{\frac{1}{2}} = \left|\det\left(\frac{e^{\frac{adx}{2}} - e^{-\frac{adx}{2}}}{adx}\right)\right|. \tag{38}$$

Let $\Omega = Gg$. Recall that it has a canonical two form σ such that

$$\sigma_g(X_g, Y_g) = g([X,Y]). \tag{39}$$

Let $2d = \dim \Omega$, and β_Ω the measure:

$$\beta_\Omega = (2\pi)^{-d}(d!)^{-1}|\sigma^d|. \tag{40}$$

We consider β_Ω as a positive Borel measure on $\underline{g}^*$ concentrated on Ω. Generalizing results of Kirillov, Rossmann and others, Khalgui proves [Kha 1], [Kha 3]

(39) Theorem. Suppose $T_{g,\tau}$ normal. Suppose moreover that $\underline{g}(g)$ is nilpotent. The representation $T_{g,\tau}$ is traceable if and only if β_Ω is tempered. In this case, there is a neighborhood of 0 in $\underline{g}$ in which the following identity of generalized functions holds:

$$j(X)^{\frac{1}{2}}\operatorname{tr} T_{g,\tau}(\exp X) = \dim \tau \int_{\underline{g}^*} e^{if(X)}\,d\beta_\Omega(f).$$

Corresponding to the character formula of theorem (39), there is an infinitesimal character formula. For simplicity assume that G is connected. Let $Z(\underline{g}_{\mathbb{C}})$ be the center of $U(\underline{g}_{\mathbb{C}})$; $I(\underline{g}_{\mathbb{C}})$ the subalgebra of G-invariant elements of $S(\underline{g}_{\mathbb{C}})$, $a : Z(\underline{g}_{\mathbb{C}}) \to I(\underline{g}_{\mathbb{C}})$ the algebra isomorphism defined in [Du 2]. Let $u \in Z(\underline{g}_{\mathbb{C}})$. Then $T_{g,\tau}(u)$ (acting for instance in the space of C^∞ vectors) is scalar. I proved in [Du 5], th. IV.19:

(40) Theorem. Let $u \in Z(\underline{g}_{\mathbb{C}})$. Then

$$T_{g,\tau}(u) = a(u)(ig)\,\mathrm{Id}.$$

IV. The representations $T_{g,\tau}$ for almost algebraic groups

IV.1 Conventions

In this chapter, we consider an almost algebraic group $(G,\Gamma,\underline{\underline{G}})$. A subalgebra $\underline{h}$ of $\underline{g}$ is called algebraic if it is the Lie algebra of an algebraic subgroup of $\underline{\underline{G}}$, unipotent if it the Lie algebra of a unipotent subgroup of $\underline{\underline{G}}$. An algebraic subalgebra $\underline{h}$ of $\underline{g}$ has a unipotent radical: it is the largest unipotent ideal of $\underline{h}$ and will be denoted by ${}^u\underline{h}$. An algebraic subalgebra $\underline{h}$ said to be reductive if ${}^u\underline{h} = \{0\}$.(9)

An algebraic group $\underline{\underline{H}}$ is said to be reductive if its Lie algebra $\underline{h}$ is reductive (all algebraic groups we consider are defined over $\mathbb{R}$). By a theorem of Mostow, $\underline{\underline{G}}$ has a maximal reductive subgroup $\underline{\underline{R}}$, and two such maximal reductive subgroups are conjugate by $\underline{\underline{U}}(\mathbb{R})$, where $\underline{\underline{U}}$ is the unipotent radical of $\underline{\underline{G}}$, $\underline{\underline{G}}$ is a semidirect product $\underline{\underline{R}}\,\underline{\underline{U}}$, and $\underline{\underline{G}}(\mathbb{R})$ a semidirect product $\underline{\underline{R}}(\mathbb{R})\,\underline{\underline{U}}(\mathbb{R})$. We call $\underline{\underline{R}}$ (or its Lie algebra $\underline{r}$) a reductive factor of $\underline{\underline{G}}$ (or of $\underline{g}$). Let R be the inverse image of $\underline{\underline{R}}(\mathbb{R})$ in G. Let U be the analytic subgroup of G corresponding to $\underline{u}$. Then $(R,\Gamma,\underline{\underline{R}})$ is an almost algebraic group, and G is a semidirect product $R\,U$. The group R will also be called a reductive factor of G. Two such reductive factors are conjugate by U.

IV.2 Coisotropic subalgebras

Let $g \in \underline{g}^*$. A subalgebra $\underline{p}$ of $\underline{g}$ is said to be coisotropic if the subspace $\underline{p}^g$ of $\underline{g}$ defined by

$$\underline{p}^g = \{X \in \underline{g},\ g([X,\underline{p}]) = 0\} \tag{1}$$

is contained in $\underline{p}$. If $p = g|\underline{p}$, and if $\underline{p}$ is coisotropic, then $\underline{p}^g = \underline{p}(p)$. It is a subalgebra of $\underline{p}$. Let $\underline{P}(p)_0$ be the analytic subgroup of G with Lie algebra $\underline{p}(p)$. Then $P(p)_0 g$ is an open subset of $g + \underline{p}^\perp$. We say that $\underline{p}$ satisfies the Pukanszky condition if $P(p)_0 g = g + \underline{p}^\perp$.

Let Cos (g) be the set of coisotropic subalgebras which satisfy the Pukanszky condition, and which are algebraic.

(2) Example. $\underline{g} \in$ Cos (g). In particular, Cos($\underline{g}$) $\neq \emptyset$.

(3) Example. Let $\underline{p}$ be a real polarization which satisfies Pukanszky's condition. Then $\underline{p} \in$ Cos (g).

(4) Example. Let $\underline{b}$ be a polarization such that $\underline{b} + \overline{\underline{b}}$ is a subalgebra. Let $\underline{p} = (\underline{b} + \overline{\underline{b}}) \cap \underline{g}$. Then $\underline{p}(p) = \underline{b} \cap \underline{g}$ and $\underline{p}$ is coisotropic. If moreover $\underline{p} \in$ Cos (g), then $\underline{b}$ is what is called a "polarization satisfying Pukanszky's condition" in e.g. [Au-Ko].

(5) Example. If $\underline{g}$ is semisimple, coisotropic algebras are parabolic ([Di2]).

(6) Lemma. Let $\underline{p} \in$ Cos (g). Then g has a good polarization if and only if p has one.

Proof. (i) Let $\underline{b} \subset \underline{p}_{\mathbb{C}}$ be a good polarization at p. It is a good polarization at g.

(ii) Conversely, let $\underline{b} \subset \underline{g}_{\mathbb{C}}$ be a good polarization at g. We have to prove that there exists a good polarization contained in $\underline{p}$. It is obvious if dim $\underline{g}$ = 0. We prove the assertion by induction on dim $\underline{g}$. We assume dim $\underline{g}$ > 0 and the assertion proved for the algebraic algebras of strictly less dimension. Following a well known path, we consider several cases.

(a) There exists a unipotent ideal $\underline{a}$ of $\underline{g}$ contained in ker g, such that $\underline{a} \neq 0$. The result is then easily proved by reduction to $\underline{g}/\underline{a}$.

In what follows, case (a) is excluded. We let $\underline{z}$ be the unipotent part of the center of $\underline{g}$. Then dim $\underline{z} \leq 1$, and $g|\underline{z} \neq 0$ if $\underline{z} \neq \{0\}$. We let $\underline{u}$ be the unipotent radical of $\underline{g}$.

(b) $\underline{u} = \underline{z}$. Then $\underline{g}$ is a direct product $\underline{r} \times \underline{z}$, with $\underline{r}$ reductive.

Thus $\underline{g}(g)$ is a Cartan subalgebra of $\underline{g}$ and $\underline{p}$ is a parabolic subalgebra of $\underline{g}$ which contains $\underline{g}(g)$. As $\underline{p}_{\mathbb{C}}$ contains a Borel subalgebra $\underline{b}$ of $\underline{g}_{\mathbb{C}}$, the result is proved.

(c) We suppose that $\underline{g}$ contains a commutative characteristic ideal $\underline{a}$ of $\underline{g}$, such that $\underline{u} \supset \underline{a} \supset \underline{z}$, $\underline{a} \neq \underline{z}$, and $\underline{g} \neq \underline{p} + \underline{a}$. Let $a = g|\underline{a}$, $\underline{h} = \underline{g}(a)$. Then $\underline{h} \neq \underline{g}$. Let $\underline{p}' = (\underline{p} \cap \underline{h}) + \underline{a}$, $h = g|\underline{h}$. Then $\underline{p}' \in$ Cos (h). Let $\underline{b}$ be a good polarization at g, and let $\underline{b}' = (\underline{b} \cap \underline{h}_{\mathbb{C}}) + \underline{a}_{\mathbb{C}}$. It is a good polariztion at h. By induction, $\underline{p}'$ contains a good polarization $\underline{b}''$ at h. Obviously, $\underline{b}''$ is contained in $\underline{p} + \underline{a}$. Let $\underline{g}' = \underline{p} + \underline{a}$ and $g' = g|\underline{g}'$. Then $\underline{b}''$ is a good polarization at g'. As $\underline{g}' \neq \underline{g}$, the induction hypothesis, applied to $\underline{g}'$, shows that there exsists a good polarization contained in $\underline{p}$.

(d) We suppose that $\underline{u}$ is a Heisenberg algebra with center $\underline{z}$, $\underline{u} \neq \underline{z}$. Let $u = g|\underline{u}$ and $\underline{r} = \underline{g}(u)$, $r = g|\underline{r}$. Let $\underline{p}' = \underline{p} + \underline{u}$, $\underline{k} = \underline{p}' \cap \underline{r}$, $\underline{p}'' = \underline{g} + \underline{u}$. Then $\underline{k} \in$ Cos (r), $\underline{p}'' \in$ Cos (g) (cf. [Du5]). Let $\underline{b}$ be a good polarization at g. There exists a good polarization $\underline{b}'$ at g such that $\underline{b}' \cap \underline{u}_{\mathbb{C}}$ is a polarization at u and $\underline{b}' \cap \underline{r}$ a good polarization at r. By the inductive hypothesis there exists a good polarization $\underline{c}$ at r contained in $\underline{g}$. Let $\underline{b}'' = \underline{c} + (\underline{b}' \cap \underline{u}_{\mathbb{C}})$. It is a good polarization contained in $\underline{p}''_{\mathbb{C}}$. Suppose $\underline{p}' \neq \underline{g}$. We apply the inductive hypothesis to $\underline{p}'$ and find a good polarization at p.

So we suppose moreover that $\underline{g} = \underline{p} + \underline{u}$. Suppose $\underline{p} = \underline{g}$. Then there is nothing to prove. We now assume that $\underline{g} \neq \underline{p}$. Let $\underline{v} = {}^{u}\underline{p}$. Then $\underline{v} = \underline{u} \cap \underline{p}$, and $[\underline{u},\underline{v}] \subset [\underline{u},\underline{u}] \subset \underline{z} \subset \underline{p}$. Thus $\underline{v}$ is an ideal in $\underline{g}$. Let $\underline{w}$ be the centralizer of $\underline{v}$ in $\underline{u}$. Then $\underline{v} \cap \underline{w}$ is a commutative ideal. I claim that $\underline{v} \cap \underline{w} \neq \underline{z}$. Suppose that $\underline{v} \cap \underline{w} = \underline{z}$. Let $\underline{m}$ be a $\underline{p}$-invariant subspace of $\underline{w}$ such that $\underline{m} \oplus \underline{z} = \underline{w}$. Let $g' \in \underline{g}^*$ be such that $g|\underline{p} = p$, $g'|\underline{m} = 0$. By Pukanszky's hypothesis, $\underline{p} \in$ Cos (g'). As $\underline{m} \subset {}^{g'}\underline{p}$, this is a contradiction.

As $\underline{v} \cap \underline{w} \neq \underline{z}$ we can apply case (c) to $\underline{v} \cap \underline{w}$.

(e) $\underline{g} = \underline{p}$. The result is obvious.

To finish the proof of (ii), it remains to see that at least one of the cases (a), (b), (c), (d), (e) is satisfied if $\dim \underline{g} > 0$. We assume that $\dim \underline{z} \le 1$, $g|\underline{z} \ne 0$, that $\underline{u} \ne \underline{z}$, that $\underline{u}$ is not Heisenberg, and that for all commutative characteristic ideals $\underline{a}$ of $\underline{g}$ contained in $\underline{u}$, such that $\underline{a} \supset \underline{z}$, $\underline{a} \ne \underline{z}$, we have $\underline{g} = \underline{p} + \underline{a}$. We assume $\underline{g} \ne \underline{p}$. We show it is absurd.

Because $\underline{u}$ is not Heisenberg, there exists a commutative ideal $\underline{a}$ of $\underline{g}$ such that $\underline{a} \subset \underline{u}$, $\underline{a} \supset \underline{z}$, $\underline{a} \ne \underline{z}$. We choose such an ideal of minimal dimension. Consider the ideal $[\underline{u},\underline{a}]$. By minimality, $[\underline{u},\underline{a}] \subset \underline{z}$ and $\underline{p} \cap \underline{a} \subset \underline{z}$. Let $\underline{v} = {}^{u}\underline{p}$. Then $\underline{v}$ is contained in $\underline{u}$ (because $\underline{p} + \underline{u} = \underline{g}$) and any ideal of $\underline{p}$ contained in $\underline{v}$ is an ideal in $\underline{g}$. So $\underline{p}$ has no commutative ideal contained in $\underline{v}$ and strictly containing $\underline{z}$. So $\underline{v}$ is an Heisenberg algebra with center $\underline{z}$ (or $\underline{z}$ itself). Let $\underline{a}'$ be the center of $\underline{u}$. Then $\underline{a}'$ is an ideal of $\underline{g}$, commutative, and $\underline{a}' \cap \underline{v} = \underline{z}$, $\underline{a}' + \underline{v} = \underline{u}$. Replacing $\underline{a}'$ by $\underline{a}$, we assume that $[\underline{a},\underline{a}] = 0$. Let $\underline{m}$ be a subspace of $\underline{a}$ such that $[\underline{g},\underline{m}] \subset \underline{m}$, $\underline{m} \oplus \underline{z} = \underline{a}$. Let $g' \in \underline{g}^*$ be such that $g'|\underline{p} = p$, $g'|\underline{m} = 0$. By Pukanszky's condition, $\underline{p} \in \mathrm{Cos}\,(g')$. This is a contradiction because $\underline{m}$ is a non zero subspace of ${}^{g'}\underline{p}$.

Q.E.D.

Let $p \in \mathrm{Cos}\,(g)$ be an element stable under $G(g)$. Let $\overset{\circ}{\underline{\underline{P}}}$ be the irreducible subgroup of $\underline{\underline{G}}$ with Lie algebra $\underline{p}$, P_0 the analytic subgroup of G with Lie algebra $\underline{p}$, $P = G(g)P_0$, $\underline{\underline{P}}$ the algebraic closure of P/Γ in $\underline{\underline{G}}$. Because of algebraicity, $\underline{\underline{P}}(\mathbb{R})$ is closed in $\underline{\underline{G}}(\mathbb{R})$. The group P is open in the inverse image of $\underline{\underline{P}}(\mathbb{R})$ in G, and so, it is a closed subgroup of G. Note that $(P,\Gamma,\underline{\underline{P}})$ is an almost algebraic group.

Let $\underline{v} = {}^{u}\underline{p}(p)$ be the unipotent radical of $\underline{p}(p)$. The subgroup $\underline{\underline{V}}(R)$ of $\underline{\underline{G}}(R)$ which corresponds to $\underline{v}$ is simply connected. Thus the analytic subgroup V of G with Lie algebra $\underline{v}$ is closed and simply connected. Moreover, the Pukanszky condition implies

$$P(p) = G(g)V \tag{7}$$

$$V/V \cap G(g) \text{ is simply connected.} \tag{8}$$

(9) Lemma (i) The linear form g is admissible if and only if p is admissible.

(ii) Let $\tau \in X_G(g)$. There exists a unique class $\tau' \in X_P(p)$ such that

$$(\delta^g\tau)(x) = (\delta^P\tau')(x) \quad (x \in G(g)).$$

(iii) The map $\tau \to \tau'$ is a bijection of $X_G(g)$ onto $X_P(p)$ which preserves irreducibility.

(10) Remark. In the particular case of a polarization, this kind of lemma was discovered by Auslander and Kostant, and the proof is the same (except for the usual complication coming from covering groups).

The main result of this chapter is:

(11) Theorem. Suppose Γ is finite. Let $(g,\tau) \in X_G$. Let $\underline{p} \in \mathrm{Cos}\ (g)$ be $G(g)$-stable. Let $p = g|\underline{p}$, and let $\tau' \in X_P(p)$ be as in (9). Then $(p,\tau') \in X_P$ and

$$T^G_{g,\tau} = \mathrm{Ind}^G_P\ (T^P_{P,\tau'})$$

(The hypothesis "Γ finite" is certainly unnecessary. It is there because Bouaziz's result III (22) is ued in the proof - cf. the remark there.)

(12) Remark. Theorem (11) probably generalizes in some way to non almost algebraic groups. However it is certainly much more difficult. For instance, in the particular case of real polarizations satisfying

Pukanszky's condition, the result analoguous to remark 14 below has been solved for non algebraic solvable groups only recently by Fujiwara [Fu].

(13) Remark. Let $\underline{n}$ be a unipotent ideal of $\underline{g}$, G-invariant. Let us use the notations of theorem III (33). Let $\underline{p} = \underline{g}(n) + \underline{n}$. Then $\underline{p} \in$ Cos (g), $\tilde{P} = G(n)N$, $P = G(g)P_0$. It is easy to see that $\tilde{P}(p) = P(p)$, and that

$$T^{\tilde{P}}_{p,\tau'} = \mathrm{Ind}^{\tilde{P}}_{P}\,(T^{P}_{p,\tau'}) \tag{13a}$$

$$T^{P}_{p,\tau'} = T^{G_1}_{g_1,\tau_1} \otimes S_n T^N_n . \tag{13b}$$

(of course 13b is a particular case of (33), applied to $\tilde{G}$). By induction in stages:

$$T^{G}_{g,\tau} = \mathrm{Ind}^{G}_{P}\,(T^{P}_{p,\tau'}) \tag{13c}$$

This particular case of theorem 10 will be used in the proof of the general case.

(14) Remark. Let us be more explicit when $\underline{p}$ is a real polarization of $\underline{g}$ at g which is $G(g)$ invariant and $\dim \tau < \infty$. The representation $T^G_{g,\tau}$ is then equivalent to the representation obtained by left translations in the Hilbert space H obtained by completion from the space H' of C^∞ functions ϕ on G such that

$$\left.\begin{aligned} \phi(xy) &= (\rho_{\underline{p}}\tau)(y)^{-1}\phi(x) \quad (x \in G,\ y \in G(g)) \\ R(X)\phi &= (-ig(X) - \rho_{\underline{p}}(X))\phi \quad (X \in \underline{p}) \end{aligned}\right\} \tag{15}$$

$$\int_{G/P} |\phi|^2 d\mu < \infty \tag{16}$$

where μ is a positive G-invariant "measure" on G/P (cf. [Be] ch V).

In particular, the result does not depend on the choice of $\underline{p}$: this result is due to Andler ([An], theorem 1, ch II, §7).

(17) Remark. Consider the subset $G(g + \underline{p}^{\perp})$ of $\underline{g}^*$. It is easy to see that it is equal to Gg. This is the fact which corresponds to theorem 11 from the point of view of orbits in the coadjoint representation.

(18) Remark. If $\underline{g} = \underline{p}$, the theorem says:

$$T^G_{g,\tau} = \mathrm{Ind}^G_{G(g)G_0} (T^{G(g)G_0}_{g,\tau}).$$

This is true even when G is not almost algebraic, and it is easy to prove using the definition of $T_{g,\tau}$ given in III.

Proof of theorem (11). The proof is by induction on $\dim \underline{g}$ along the lines of the proof of lemma (6). We consider the same five cases (a), (b), (c), (d), (e).

(a) It is a particular case of (33).

(b) It is exactly the "Corollaire" of Bouaziz [Bou], §3.

(c) Let $\underline{a}$, $\underline{h}$, $\underline{p}'$, $\underline{g}'$ be as in the proof of lemma (6). We consider the corresponding groups $H = G(g)H_0$, $P' = G(g)P_0$, $G' = G(g)G_0$, and consider the representations T^H, $T^{P'}$, $T^{G'}$, T^P, T^G associated to these groups and to $\tau \in X_G(g)$, as in the theorem. By the induction hypothesis, $T^{G'} = \mathrm{Ind}^{G'}_P(T^P)$, $T^{G'} = \mathrm{Ind}^{G'}_{P'}(T^{P'})$. By induction in stages, it is enough to prove that $T^G = \mathrm{Ind}^G_{P'}(T^{P'})$. By the inductive hypothesis, $T^H = \mathrm{Ind}^H_{P'}(T^{P'})$. By remark (13), $T^G = \mathrm{Ind}^G_H (T^H)$. The result follows by induction in stages again.

(d) Let $\underline{p}'$, $\underline{p}''$ be as in the proof of lemma (6). We consider the corresponding groups $P' = G(g)P'_0$, $P'' = G(g)P''_0$, and the corresponding representations $T^{P'}$, $T^{P''}$. Suppose $\underline{p}' \neq \underline{g}$. By the inductive hypothesis, $T^{P'} = \mathrm{Ind}^{P'}_{P''} (T^{P''})$ and $T^{P'} = \mathrm{Ind}^{P'}_P (T^P)$. By induction in stages, it is enough to prove that $T^G = \mathrm{Ind}^G_{P''} (T^{P''})$. Let $\underline{r} = \underline{g}(u)$, $R = G(u)^{\underline{u}}$,

$\underline{k} = \underline{p}' \cap \underline{r}$, $k = g|\underline{k}$, $r = g|\underline{r}$, $K = R(r)_0K_0 \subset R$. From (g,τ) we obtain, as in III (25) and (9), elements $(r,\tau") \in X_R(r)$, $(k,\tau"') \in X_K(k)$, and, by III (33), defining representations T^R and T^K,

$$T^G = \mathrm{Ind}_{G(u)U}^G (T^R \otimes S_uT_u^U). \tag{19}$$

(The group $G(u)U$ is of finite index in G, but this does not matter.) By the inductive hypothesis

$$T^R = \mathrm{Ind}_K^R (T^K). \tag{20}$$

Consider $\underline{P}"(u)$. Then $P"(u)^{\underline{u}} = K$, and, by III (33) again (and because $P" = P"(u)U$):

$$T^{P"} = T^K \otimes S_uT_u^U. \tag{21}$$

Proposition 15 of [Du5], and (19), (20), (21), imply the desired equality $T^G = \mathrm{Ind}_{P"}^G (T^{P"})$.

We suppose now that $\underline{g} = \underline{p}'$, and $\underline{g} \neq \underline{p}$. As seen in the proof of lemma (6), we can apply case (c).

(e) $\underline{g} = \underline{p}$. See remark (18).

Q.E.D.

Let $g \in \underline{g}^*$ be an element which is admissible and has a good polarization. Let $\underline{p} \in \mathrm{Cos}\ (g)$. It is sometimes useful (as in [Au]) to consider elements $\underline{p} \in \mathrm{Cos}\ (g)$ which are not $G(g)$-invariant. By algebraicity there exists a subgroup $G(g)_1$ of $G(g)$, of finite index, which normalizes $\underline{p}$. Let τ_1 be a representation of $G(g)_1^{\underline{g}}$, such that $\tau_1(e) = -\mathrm{Id}$, $\tau_1(\exp X) = \exp ig(X)\mathrm{Id}$ for $X \in \underline{g}(g)$. (We denote the set of such classes by $X_G(g, G(g)_1)$.) We define:

$$\tau = \mathrm{Ind}_{G(g)_1^{\underline{g}}}^{G(g)^{\underline{g}}} (\tau_1) \tag{22}$$

$$P_1 = G(g)_1P_0,\ p = g|\underline{p}\ . \tag{23}$$

As in (9), we define an element $\tau_1' \in X_{P_1}(p)$. The theorem below generalizes theorem 11. As its proof is completely similar, we leave it to the reader.

(24) Theorem. Suppose Γ is finite. $T^G_{g,\tau} = \mathrm{Ind}^G_{P_1}(T^{P_1}_{p,\tau_1'})$, where τ_1, τ_1', and τ are as above.

(25) Remark. If $\underline{p}$ is a real polarization which satisfies Pukanszky's condition, this is theorem 1 of §II.7 of [An].

IV.3 Coisotropic subalgebras of unipotent type

Let $(G,\Gamma,\underline{\underline{G}})$ be an almost algebraic group, and $g \in \underline{g}^*$.

Let $\underline{p} \in \mathrm{Cos}\,(g)$. Let $\underline{v} = {}^u\underline{p}$, $v = g|\underline{v}$. We say that $\underline{p}$ is of unipotent type if

$$\underline{p} = \underline{p}(u) + \underline{v}. \tag{26}$$

There exists $\underline{p} \in \mathrm{Cos}\,(g)$ of unipotent type which are even invariant by all automorphisms of $\underline{g}$ which stabilize g. I repreat the construction of a particular one called in [Du 5], n°20, the "Canonical acceptable subalgebra". Let us denote it by $\underline{p}^c$. The construction of $\underline{p}^c$ is by induction on $\dim \underline{g}$. If $\dim \underline{g} = 0$, then $\underline{p}^c = 0$. Let $\dim \underline{g} > 0$, and assume the canonical acceptable subalgebra is defined for all algebraic subalgebras of strictly less dimension. Let $\underline{u} = {}^u\underline{g}$, $u = g|\underline{u}$, $\underline{g}_1 = \underline{g}(u)$, $g_1 = g|\underline{g}_1$. We consider two cases.

First case. $\underline{g}_1 = \underline{g}$. Then we let $\underline{g} = \underline{p}^c$.

Second case. $\underline{g}_1 \neq \underline{g}$. Let $\underline{p}_1^c$ be the canonical acceptable subalgebra at g_1. Then $\underline{p}^c = \underline{p}_1^c + \underline{u}$.

Let $\mathrm{Cos}\,u(g,G)$ be the set of elements $\underline{p} \in \mathrm{Cos}\,(g)$ of unipotent type and $G(g)$-invariant. Let $(g,\tau) \in X$, and $\underline{p} \in \mathrm{Cos}\,(u,G)$. Theorem (11) is especially interesting in this case, because the construction of

$T^P_{p,\tau'}$ is particularly simple and does not require induction.[10] Define $\underline{\underline{P}}$ as after lemma (6), and let $\underline{\underline{V}}$ be the unipotent radical of $\underline{\underline{P}}$. Let $\underline{\underline{S}}$ be a reductive factor of $\underline{\underline{G}}(g)$ and $\underline{\underline{R}}$ a reductive factor of $\underline{\underline{P}}$ which contains $\underline{\underline{S}}$, and is contained in $\underline{\underline{P}}(v)$ where $v = g|\underline{v}$. Let V be the analytic subgroup of G with Lie algebra $\underline{v}$, and $P = G(g)P_0$, $R \subset P$ as above, so that P is a semi-direct product RV, and R fixes v. We consider the group $R^{\underline{V}}$ and $r = g|\underline{r}$. It is easy to see that $R(r)$ is the inverse image of $\underline{\underline{S}}(\mathbb{R})$ in G. There is a unique element $\tau'' \in (R(r)^{\underline{V}})^{\underline{r}}$ such that:

$$(\tau\delta^g)(x) \quad = \quad ((\tau''\delta^r)\delta^V)(x) \quad (x \in G(g)). \tag{27}$$

(This is analogous to III (25).) Then $(r,\tau'') \in X_{R^{\underline{V}}}$, and the representation $T^{R^{\underline{V}}}_{r,\tau''}$ of $R^{\underline{V}}$ (which is a group with reductive Lie algebra) has been defined in III.4. The representation $T^{R^{\underline{V}}}_{r,\tau''} \otimes S_v T^V_v$ of $P = RV$ is defined in II (16). Assume Γ is finite. As a consequence of theorem (33), we obtain $T^P_{p,\tau'} = T^{R^{\underline{V}}}_{r,\tau''} \otimes S_v T^V_v$ and

$$T^G_{g,\tau} \quad = \quad \mathrm{Ind}^G_{RV}\ (T^{R^{\underline{V}}}_{r,\tau''} \otimes S_v T^V_v). \tag{28}$$

Particular cases of this formula are given in [Du5] III 20.

When $(G,\Gamma,\underline{\underline{G}})$ is an almost algebraic group, formula (28) can be used as an alternative definition of the representations $T^G_{g,\tau}$.

IV.4 More special classes of coisotropic subalgebras

Let $(G,\Gamma,\underline{\underline{G}})$ be an almost algebraic group and $(g,\tau) \in X$. If $\underline{p} \in \mathrm{Cos}\ u(g,G)$, we use the notations R, V, r, v, $\underline{r}$, $\underline{v}$ introduced before formula (28). We consider $\underline{q} = \underline{v}(v) \cap \ker g$. It is a P-invariant ideal in $\underline{p}$. Let Q be the analytic subgroup of P with Lie algebra $\underline{q}$. We consider $V_1 = V/Q$, $\underline{v}_1 = \underline{v}/\underline{q}$, $P_1 = P/Q$, $\underline{p}_1 = \underline{p}/\underline{q}$. We consider R as a subgroup of P_1. In an obvious way, we define $p_1 \in \underline{p}_1^*$, $v_1 \in \underline{v}_1^*$. Obviously, the representation $T^{R^{\underline{V}}}_{r,\tau''} \otimes S_v T^V_v$ is obtained from the representation $T^{R\underline{v}}_{r,\tau''} \otimes S_{v_1} T^{V_1}_{v_1}$ of $P_1 = RV_1$ by composing with the canonical projection $P \to P_1$. In what follows, we try to make $T^{V_1}_{v_1}$ and $T^{R\underline{v}}_{r,\tau''}$ as simple as possible.

We say that $r \in \underline{r}^*$ is standard if the following is true: we write the Cartan subalgebra $\underline{h} = \underline{r}(h)$ of $\underline{r}$ as $\underline{t} \oplus \underline{a} \oplus \underline{z}$ (cf III (4)). Let μ be the element of h^* such that

$$\mu|\underline{t} = r|\underline{t},\ \mu|\underline{a} + \underline{z} = 0. \qquad (29)$$

We ask:

$$\mu \text{ is regular in } \underline{r}^* \qquad (30)$$

(This implies that $\underline{h}$ is a fundamental subalgebra of $\underline{g}$).

(31) Lemma. There exists an element $\underline{p} \in \text{Cos } u(g,G)$ such that $\underline{v}_1$ is Heisenberg with center $\underline{v}_1(v_1)$, and r standard in $\underline{r}^*$.

Proof. We in fact construct a canonical one. Start from the canonical acceptable element $\underline{p}^c \in \text{Cos } u(g,G)$. Write (with abuse of notation) $P^c = R'V'$, $\underline{p}' = p^c = \underline{r}' + \underline{v}'$, $r' = g|\underline{r}'$, $\underline{h}' = \underline{r}'(r')$, $\underline{h}' = \underline{t}' + \underline{a}' + \underline{z}'$. Let $\gamma' \in \underline{a}'^*$ be defined by $\gamma|\underline{a}' = g|\underline{a}'$, $\gamma|\underline{t}' + \underline{z}' = 0$. Let $\underline{r}$ be the centralizer of γ in $\underline{r}'$, $r = r'|\underline{r}$. Let K be the Killing form of $[\underline{r}',\underline{r}']$, and $H_\gamma \in [\underline{r}',\underline{r}']$ the element which corresponds (by K) to γ. Let $\underline{n} \subset \underline{r}'$ be the subalgebra sum of the strictly positive eigenvalue eigensubspaces of ad H_γ. Let $\underline{p}^0 = \underline{r} + \underline{n} + \underline{v}$. It is an element of $\text{Cos } u(g,G)$ with reductive factor $\underline{r}$, unipotent radical $\underline{n} + \underline{v}$. Let $\underline{w} = \underline{n} + \underline{v}$, $w = g|\underline{n} + \underline{v}$; Penney constructed in [Pe] a canonical coisotropic subalgebra $\underline{v}''$ of $\underline{w}$ with the following properties. Let $v'' = v|\underline{v}''$, $\underline{v}''_1 = \underline{v}''/\underline{v}''(v'')$; then $\underline{v}''$, is Heisenberg. Moreover, $\underline{v}''$ is invariant by any automorphism of $\underline{v}''$ which fixes v''. We let $\underline{p}'' = \underline{r} + \underline{v}''$. It satisifies the conditions of the lemma.

Q.E.D.

Let $\underline{\underline{S}}$ be a reductive factor of the Zariski closure of $G(g)/\Gamma$ in $\underline{\underline{G}}$, and S the corresponding redutive factor of $G(g)$. Its Lie algebra $\underline{s}$ is a reductive factor of $\underline{g}(g)$. There is a unique decomposition

$$\underline{s} = \underline{s}_a \oplus \underline{s}_d$$

where $\underline{s}_a$ is the Lie algebra of the anisotropic part of $\underline{\underline{S}}$, and $\underline{s}_d$ the Lie algebra of the split part of $\underline{\underline{S}}$ (d for "deployé"). Recall that $\underline{s}$ is commutative because $\underline{g}(g)$ is solvable. I consider the following.

(32) Condition. S-centralizes $\underline{s}_d$.

For instance, (32) is always satisfied if G is connected and solvable, or connected and semisimple. Lemma (33) and Remark (35) is implicit in Lipsman [Li 3].

(33) Lemma. Suppose condition (32) is satisfied. Then there exists $\underline{p} \in \mathrm{Cos}\ u(g,G)$ such that $\underline{v}_1$ is Heisenberg with center $\underline{v}_1(v_1)$, and $\underline{s}_d$ central in $\underline{p}_1$.[(11)]

(34) Remark. Unlike in lemma (31), there is usually no canonical such object.

Proof. Let $\underline{p}', \underline{r}', \underline{v}', \underline{h}', \underline{t}', \underline{a}', \underline{z}'$ be as in the proof of lemma 31. Then $\underline{s}_d$ contains $\underline{a}'$. Let $\Theta \subset \underline{s}_d^*$ be the set of non zero roots of $\underline{s}_d$ in $\underline{p}'$, $\underline{d}$ the centralizers of $\underline{s}_d$ in $\underline{p}'$, $\underline{p}'_\theta$ the root space corresponding to a root θ. Let $F \subset \underline{s}_d$ be a subspace of codimension one such that $\Theta \cap F = \emptyset$. Let Θ^+ be the subset of Θ which are on one side of F, $\underline{e} = \sum_{\theta\in\Theta^+} \underline{p}'_\theta$. Let $\underline{f} = \underline{d} + \underline{e}$. It is easy to see that $\underline{f}$ is in $\mathrm{Cos}\ u(g,G)$. Let $\underline{j} = {}^u\underline{d}$, $j = g|\underline{j}$, $\underline{k}$ the canonical Penney coisotropic subalgebra of $\underline{j}$, $\underline{r} = \underline{d} \cap \underline{r}'$, $\underline{p} = \underline{r} + \underline{k} + \underline{e}$, $\underline{v} = \underline{k} + \underline{e}$. It is easy to see that $\underline{p}$ has the properties of lemma 33.

(35) Remark. If (32) is satisfied, there exists $\underline{p} \in \mathrm{Cos}\ (g)$, $G(g)$-

invariant, such that $P(p)$ acts compactly in $\underline{p}/\underline{p}(p)$ (by lemma (33), we can even choose $\underline{p} \in \mathrm{Cos}\, u(g,G)$).

IV.5 Harmonic induction

Let $(G,\Gamma,\underline{G})$ and (g,τ) be as in IV.4. We assume $\dim \tau < \infty$.

Let $\underline{b} \subset \underline{g}_{\mathbb{C}}$ be a polarization at g. Consider the following conditions on $\underline{b}$:

$$\underline{b} + \overline{\underline{b}} \text{ is a subalgebra of } \underline{g}_{\mathbb{C}}. \tag{36}$$

Let $\underline{e} = (\underline{b} + \overline{\underline{b}}) \cap \underline{g}$. It is an algebraic coisotropic subalgebra of $\underline{g}$.

$$\underline{e} \text{ satisfies the Pukanszky condition} \tag{37}$$

$$\underline{b} \text{ is } G(g)\text{-invariant.} \tag{38}$$

Let $\underline{d} = \underline{g} \cap \underline{b}$, D_0 the analytic subgroup of G with Lie algebra $\underline{d}$, $D = G(g)D_0$.

$$\text{Then } \mathrm{Ad}_{\underline{e}/\underline{d}}(D) \text{ is compact.} \tag{39}$$

These polarizations are called metric in Lipsman [Li 1]. From the data g,τ, and a metric $\underline{b}$, one can manufacture a complex of fiber bundles over $\underline{G}/D$, with a differential usually denoted by $\bar{\partial}$. Let L be the space of the representation τ, and consider the representation $\rho_{\underline{b}}\tau$ of $G(g)$ in L. By the usual argument, (cf [Li 1], Lemma 2.3) $\rho_{\underline{b}}\tau$ extends uniquely to a representation of D with differential $X \to ig(X) - \frac{1}{2}\operatorname{tr}\operatorname{ad} X_{\underline{g}_{\mathbb{C}}/\underline{b}}$. For $j \in \mathbb{N}$, let $L_j = \Lambda^j(\underline{e}_{\mathbb{C}}/\underline{b})^* \otimes L$, and $\mathcal{L}_j = G \times_D L_j$. Choose a D invariant hermitian metric on $\underline{e}_{\mathbb{C}}/\underline{b}$, and a "measure" μ on the space of positive Borel functions ψ on G such that

$$\psi(xy) = |\det(y_{\underline{g}/\underline{d}})|\,\psi(x) \quad (x \in G,\ y \in D). \tag{40}$$

$\mathcal{L}_j$ is an hermitian fiber bundle, and if ϕ and ϕ' are sections of $\mathcal{L}_j$, then (ϕ,ϕ') can be identified to a function ψ satisfying (40). We let $\Gamma^\infty(\mathcal{L}_j)$ be the space of C^∞ sections of $\mathcal{L}_j$. There is an order one differential operator

$$\bar{\partial}:\Gamma^{\infty}(L_j) \to \Gamma^{\infty}(L_{j+1}).$$

Let $\bar{\partial}^*:\Gamma^{\infty}(L_{j+1}) \to \Gamma^{\infty}(L_j)$ be its formal adjoint (with respect to the scalar product

$$\int_{G/D} |\phi|^2\mu, \quad \phi \in \Gamma^{\infty}(L_j),$$

ϕ with compact support. We let

$$H_2^{\prime j}(\underline{b}) = \{\phi \in \Gamma^{\infty}(L_j),\ \bar{\partial}\phi = 0,\ \bar{\partial}^*\phi = 0,\ \int_{G/D} |\phi|^2\mu < \infty\}.$$

This is a prehilbert space (and an Hilbert space if $\underline{g} = \underline{e}$). We let $H_2^j(\underline{b})$ be the completion of $H_2^{\prime j}(\underline{b})$. This is an Hilbert space, in which G acts unitarily. It can be proved that the equivalence class of the representation of G in $H_2^j(\underline{b})$ does not depend on the choice of the hermitian metric on $\underline{e}_{\mathbb{C}}/\underline{b}$ used to define ∂^*. It is proved in [He] when $\underline{e} = \underline{g}$, and it follows from this case by induction.

I introduce a special class of metric polarizations. A metric polarization is said to be of unipotent type if $\underline{e} \in \mathrm{Cos}\ u(g,G)$ and if $\underline{b}$ is good. The following lemma is a refinement of Lipsman [Li 2] th. 5.5 and solves positively the conjecture contained in remark 5.5 of [Li 2].[12]

(41) Lemma. Suppose condition (32) is satisfied. Then g has a metric polarization of unipotent type.

Proof. Let $\underline{p}$, $\underline{p}_1$, be as in lemma (33). We can suppose that $G = P_1$. Then G has the following structure. The unipotent radical $\underline{u}$ of $\underline{g}$ is an Heisenberg algebra. The center $\underline{j}$ of $\underline{u}$ is contained in the center of $\underline{g}$, and $g|\underline{j} \neq 0$ if $\underline{u} \neq \{0\}$. Let $u = g|\underline{u}$, and R be a reductive factor of G contained in $G(u)$. Then $R = G(g)R_0$, $G = RU$. Let $\underline{r}$ be the Lie algebra of R, $r = g|\underline{r}$. Then $\underline{g}(g) = \underline{r}(r) + \underline{j}$,

$\underline{r}(r)$ is a Cartan subalgebra of $\underline{r}$. Let us denote it by $\underline{h}$. It is equal to $\underline{s}$. We write $\underline{h} = \underline{t} \oplus \underline{z}$, where $\underline{z}$ is the center of $\underline{r}$, $\underline{t} = [\underline{r},\underline{r}] \cap \underline{h}$. Define $\mu \in \underline{t}^*$ as in (29). The roots Δ of $\underline{h}_{\mathbb{C}}$ in $\underline{r}_{\mathbb{C}}$ are contained in $i\underline{t}^*$. The element $i\mu$ is regular in r^* and we define $\Delta^+ = \{\alpha \in \Delta, K(\alpha, i\mu) > 0\}$. Let $\underline{n} = \sum_{\alpha \in \Delta^+} \underline{r}^{\alpha}_{\mathbb{C}}$, where $\underline{r}^{\alpha}_{\mathbb{C}}$ is the root space corresponding to α. It will be enough to prove: there exists a $G(g)$-invariant and $\underline{n}$-invariant polarization $\underline{c}$ at u in $\underline{u}_{\mathbb{C}}$, such that $\underline{c} + \overline{\underline{c}} = \underline{u}$. Indeed $\underline{b} = \underline{h}_{\mathbb{C}} \oplus \underline{n} \oplus \underline{c}$ will be a metric polarization of unipotent type.

We assume $\dim \underline{u}/\underline{z} > 0$ (otherwise it is obvious). We consider an R-invariant subspace $\underline{m} \subset \underline{u}$ such that $\underline{m} \oplus \underline{j} = \underline{u}$. Let Ψ be the set of roots of $\underline{h}_{\mathbb{C}}$ in $\underline{m}_{\mathbb{C}}$. Then Ψ is contained in $i\underline{h}^*$. If $\beta \in \Psi$, we define $K(\beta,i\mu) = K(\beta|\underline{t}^*, i\mu)$. We define

$$\Psi^+ = \{\beta \in \Psi, K(\beta, i\mu) > 0\}.$$

In an analogous way we define Ψ° and Ψ^-. We denote by $\underline{v}^+$, $\underline{m}^\circ_{\mathbb{C}}$, $\underline{v}^-$ the corresponding subspaces of $\underline{m}_{\mathbb{C}}$, and $\underline{m}^\circ = \underline{m}^\circ_{\mathbb{C}} \cap \underline{m}$. Then $\underline{m}^\circ$ is a symplectic vector space, normalized by $R(r)$. As the action of $R(r)$ in $\underline{m}^\circ$ is compact, there exists a positive totally complex lagrangian subspace $\underline{\ell}$ of $\underline{m}^\circ_{\mathbb{C}}$ which is normalized by $R(r)$. We let $\underline{c} = \underline{\ell} \oplus \underline{v}^+ \oplus \underline{j}_{\mathbb{C}}$. Obviously, it is a commutative subalgebra of $\underline{u}_{\mathbb{C}}$, normalized by $R(r)$, and $\underline{c} + \overline{\underline{c}} = \underline{u}$. Moreover $[\underline{n},\underline{c}] \subset \underline{v}^+ \subset \underline{c}$.

Q.E.D.

The interest of metric polarizations of unipotent type comes from the following result, generalizing results of Bott, Schmid [Sch], Lipsman [Li 1] and [Li 2]. Ginsburg [G] announces

(42) Theorem. Suppose Γ is finite. Let $\underline{b}$ a metric polarization of unipotent type. Let $q(\underline{b})$ be as in II (1). Then $H^j_2(\underline{b}) = 0$ if $j \neq q(\underline{b})$, and the representation of G in $H^{q(\underline{b})}_2$ is equivalent to

$T_{g,\tau}$. (For the hypothesis "Γ finite", see the comment following (11)).

Proof. By induction, and because of theorem (18) we can assume $G = E$. Let $\underline{u} = {}^u\underline{g}$, be the unipotent radical of $\underline{g}$, $u = g|\underline{u}$, $\underline{c} = \underline{b} \cap \underline{u}_{\mathbb{C}}$. We prove that $\underline{c}$ is a polarization at u. This generalizes [An], lemma 35, and I reproduce the proof. Let $\underline{b}' = \underline{b}/\underline{c}$. It is a solvable subalgebra of $\underline{g}_{\mathbb{C}}/\underline{u}_{\mathbb{C}}$, and thus còntained in a Borel subalgebra of $\underline{g}_{\mathbb{C}}/\underline{u}_{\mathbb{C}}$. This imposes a majoration of $\dim \underline{b}/\underline{c}$, and thus a minoration of $\dim \underline{c}$. This minoration gives that $\underline{c}/\underline{u}(u)_{\mathbb{C}}$ is lagrangian in $\underline{u}_{\mathbb{C}}/\underline{u}_{\mathbb{C}}(u)$, which proves the assertion. Theorem 42 follows then from Rosenberg [Ro] th. 4.8.

Q.E.D.

(43) Remark. Let $\underline{p}$ be as in remark (35). Then Dirac type equations can be used (instead of $\bar{\partial}$ in the more special case when $\underline{p} = \underline{e}$) to build representation spaces H^+ and H^- of G whose "difference" is presumably equivalent to $T_{g,\tau}$ by the Connes-Moscovici Index Theorem (cf. [Co-Mo]). However, the vanishing of one of the spaces $H^{\pm}$ would be desirable, and is not known in general, even when $\underline{g}$ is nilpotent.

IV.6 The representation $T'_{\underline{c},\tau}$

In the Plancherel formula, it will be useful to consider particular classes of (usually not irreducible) representations of G.

Let g an admissible element of $\underline{g}^*$, with a good polarization. We choose a reductive factor $\underline{s}$ of $\underline{g}(g)$ (it is commutative, because $\underline{g}(g)$ is solvable). Let $\underline{\underline{S}}$ the irreducible torus of $\underline{\underline{G}}$ with Lie algebra $\underline{s}$ and S the inverse image in $\underline{\underline{S}}(\mathbb{R})$ in G. We define:

$$G(g)_1 = SG(g)_0. \tag{44}$$

(Remark that the definition of $G(g)_1$ depends on the choice of the almost algebraic structure $(G,\Gamma,\underline{\underline{G}})$ on G.)

The group $G(g)_1$ is of finite index in $G(g)$ and does not depend on the choice of $\underline{s}$. We define $Z_G(g) = X_G(g,G(g)_1)$ (cf. the definition before formula (22)), and denote by $Z_G^{irr}(g)$ the subset of irreducible subclasses of $Z_G(g)$.

Let $\sigma \in Z_G^{irr}(g)$. Let $\tau = \mathrm{Ind}_{G(g)_1^{\underline{g}}}^{G(g)^{\underline{g}}}\sigma$. Then we can write

$$\tau = \oplus\, n_j \tau_j, \text{ with } \tau_j \in X_G^{irr}(g), \tag{45}$$

and n_j the multiplicity of τ_j in τ (it is a finite sum, and the multiplicites are finite). We define

$$T'_{g,\sigma} = T_{g,\tau}. \tag{46}$$

Then, the decomposition of $T'_{g,\sigma}$ into irreducible representations is given by:

$$T'_{g,\sigma} = \oplus\, n_j T_{g,\tau_j}, \tag{47}$$

because of theorem III (32)(i).

As $G(g)_1$ obviously centralizes $\underline{s}_d$ (cf. Condition (32)), one can mimic (41) and (42) to realize $T'_{g,\sigma}$ in spaces of harmonic forms without any restriction on g. We leave this to the reader.

V. Strongly regular forms

V.1 Conventions

In this chapter, $(G,\Gamma,\underline{G})$ is an almost algebraic group. We denote by $\underline{g}$ its Lie algebra, $\underline{u} = {}^u\underline{g}$. We fix a subalgebra $\underline{j}$ of $\underline{u}$ contained in the center of $\underline{g}$, and centralized by G. The analytic subgroup of G with Lie algebra $\underline{j}$ is denoted by J. It is closed and central in G. The group ΓJ is also closed and central in G (because its image in G/Γ is closed) and $\Gamma \cap J \subset \{1\}$.

We fix a unitary character Ξ of ΓJ with differential $i\xi \in i\underline{j}^*$. We also assume (because we can reduce to this case by dividing G by $(\ker \Xi)_0$, that ξ is injective; that is

$$\dim \underline{j} \le 1, \text{ and if } \dim \underline{j} = 1,\ \xi \neq 0. \tag{1}$$

We denote by $\hat{G}_\Xi$ the set of unitary classes $T \in \hat{G}$ whose restriction to ΓJ is a multiple of Ξ. We denote by $\underline{j}^*_\xi$ the set of $g \in \underline{g}^*$ such that $g|\underline{j} = \xi$. Let $g \in \underline{g}^*$. We denote by $X_G(g,\Xi)$ the set of $\tau \in X_G(g)$ such that $\tau(r_g(x)) = \Xi(x)\mathrm{Id}$ for $x \in \Gamma J$ (recall the definition of r_g in III.5). We say that g is Ξ-admissible if $X_G(g,\Xi)$ is not empty. This implies that g is admissible and belongs to $\underline{g}^*_\xi$.

Let $(g,\tau) \in X_G^{irr}$. It follows from III (23) that $T_{g,\tau}$ belongs to $\hat{G}_\Xi$ if and only if τ belongs to $X_G^{irr}(g,\Xi)$.

We denote by $C_c^\infty(G/\Gamma J,\Xi)$ the space of C^∞ functions ϕ on G, with compact support modulo ΓJ, such that

$$\phi(xy) = \Xi(y)^{-1}\phi(x) \quad x \in G,\ y \in \Gamma J. \tag{2}$$

We choose a Lebesgue measure dX on $\underline{g}/\underline{j}$, and denote by dx the left invariant measure on G/Γ which is tangent to dX. We define $L^2(G/\Gamma J,\Xi)$ to be the obvious Hilbert space completion of $C_c^\infty(G/\Gamma J,\Xi)$. The group G acts in $L^2(G/\Gamma J,\Xi)$ by the left regular representation.

Let $[n]$ be the Plancherel measure class of $L^2(G/\Gamma J,\Xi)$. It is supported by $\hat{G}_\Xi$. (Recall G is type one [Di1].)

If $\phi \in C_c^\infty(G/\Gamma J, \Xi)$ and π is a representation of G whose restriction to ΓJ is a multiple of Ξ, we define

$$\pi(\phi) = \int_{G/\Gamma J} \phi(x)\pi(x)dx. \tag{3}$$

Suppose that G is unimodular. Then there is a unique measure n in the class $[n]$ such that

$$\phi(1) = \int_{\hat{G}_\Xi} \operatorname{tr} \pi(\phi)dn(\pi). \tag{4}$$

(cf. [Di3]).

In this chapter, we study n when G is unimodular.

(5) <u>Remark</u>. $X_G^{irr}(g,\Xi)$ is a finite set of finite dimensional representations, because $\Gamma G(g)_0$ is of finite index in $G(g)$ by algebracity. We use on discrete groups the measure giving the mass one to each point, and still denote by Ξ the character of $(\Gamma G(g)_0)^{\underline{g}}$ such that τ restricts to $\Xi\mathrm{Id}$ on $(\Gamma G(g)_0)^{\underline{g}}$ for $\tau \in X_G^{irr}(g,\Xi)$. The Plancherel formula for $L^2(G(g)^{\underline{g}}/(\Gamma G(g)_0)^{\underline{g}},\Xi)$ reads

$$\phi(1) = (\# G(g)/\Gamma G(g)_0)^{-1} \sum_{\tau \in X_G^{irr}(g,\Xi)} \dim \tau \operatorname{tr} \tau(\phi) \tag{6}$$

for $\phi \in C_c^\infty(G^{\underline{g}}/(\Gamma G(g)_0)^{\underline{g}},\Xi)$.

(6) <u>Remark</u>. Let H be a Lie group, connected and locally isomorphic to an algebraic group. Let A be a closed subgroup of the center of H and ζ a unitary character of A. The study of $L^2(H/A,\zeta)$ can be reduced, by elementary manipulations which we describe briefly below, to the preceding situation. Obviously we may assume that H is simply connected. Let $\underline{\underline{H}}$ be a linear algebraic group defined over $\mathbb{R}$ such that H and $\underline{\underline{H}}(\mathbb{R})$ are locally isomorphic. We may assume that the unipotent radical $\underline{u}$ of $\underline{h}$ contains the center of

$\underline{g}$ (by changing $\underline{\underline{H}}$ if necessary). Let U be the analytic subgroup of H with Lie algebra $\underline{u}$. Let I be the (unique) connected closed subgroup of U containing $A \cap U$ and such that $I/A \cap U$ is compact. Let $B = AI$. It is a closed central subgroup of H. It is clear that $L^2(H/A,\zeta)$ is the direct sum of the spaces $L^2(H/B,\eta)$ with $\eta \in \hat{B}_\zeta$. Let R be a reductive factor of H (cf. IV.1). Let $\Theta = B \cap R$. It is easy to see that $\Theta \cap I = \{1\}$ and $B = \Theta I$. Fix $\eta \in \hat{B}_\zeta$. Let Q be the connected component of $\ker \eta$. Dividing by Q, and changing notations, we are reduced to the following situation: $(G,\Gamma,\underline{\underline{G}})$ is an almost algebraic group, $\underline{u} = {}^u\underline{g}$ contains the center of $\underline{g}$, G is simply connected, R is a reductive factor of G, Θ a (discrete) subgroup of R contained in the center of G. By changing $\underline{\underline{G}}$ if necessary, we may assume that Θ is contained in Γ. We fix $J \subset U$, a connected central subgroup, $\eta \in (\Theta J)^\wedge$, and we assume that $\ker \eta$ is discrete. It is then easy to see that $L^2(G/\Theta J,\eta)$ is a direct integral or sum (depending on the order of Γ/Θ) of spaces $L^2(G/\Gamma J,\Xi)$, with $\Xi \in (\Gamma J)^\wedge_\eta$.

(7) Examples. The first type of reduction (from A to B) occurs for connected but not simply connected groups with Heisenberg Lie algebra. The second type (from ΘJ to ΓJ) occurs for the simply connected covering group of $SL(2,\mathbb{R})$, or of the oscillator group.

V.2 Strongly regular forms

Recall that a linear form $g \in \underline{g}^*$ is called regular if $\dim \underline{g}(g)$ is minimum. In this case $\underline{g}(g)$ is commutative [Du - Ve]), and it has a unique reductive factor which we will denote by $\underline{s}(g)$. Then we say that g is strongly regular if g is regular and if $\dim \underline{s}(g)$ is maximum. This notion was introduced in [Du3].

We denote by $\underline{g}^*_{st}$ the subset of strongly regular forms. The set $\underline{g}^*_{st}$ is a G-invariant Zariski open set in $\underline{g}^*$. There exists a finite subset $\{\underline{s}_1, \underline{s}_2,\ldots,\underline{s}_N\}$ of $\{\underline{s}(g),\ g \in \underline{g}^*_{st}\}$ which represent the conjugacy classes. Moreover the $\underline{s}(g)_{\mathbb{C}}$ are conjugate under the complex adjoint group (see [Du3] or [Cha 1]).

Let

$$\mathcal{U} = \underline{g}^*_{st} \cap \underline{g}^*_{\xi}. \tag{8}$$

Then $\mathcal{U}$ is not empty, and the $\underline{s}_j (j = 1,\dots,N)$ are still a set of representatives of the $\underline{s}(g)$, for $g \in \mathcal{U}$, because of condition (1). The elements of $\mathcal{U}$ are the strongly regular elements of $\underline{g}^*_{\xi}$.

(9) Remark. If we abandon condition 1, $\mathcal{U}$ might be empty, and we define the strongly regular elements of $\underline{g}^*_{\xi}$ as the strongly regular elements of $\underline{g}/\ker \xi$ contained in $\underline{g}^*_{\xi}$.

We denote by $\mathcal{P}$ the set of strongly regular Ξ-admissible elements of $\underline{g}^*_{\xi}$. It is a G-invariant subset of $\mathcal{U}$(it follows from (15) and (19) below that $\mathcal{P}$ is closed in $\mathcal{U}$).

All elements $g \in \mathcal{P}$ have a good polarization ([An]; ch. II.12). We define

$$\mathcal{Y} = \{(g,\tau) \in X^{irr}(\underline{g},\Xi),\ g \in \mathcal{U}\}. \tag{10}$$

If necessary, we shall denote $\mathcal{Y}$ by $\mathcal{Y}_G(\Xi)$. It is a subset of $\mathcal{X}_G$, and the map $(g,\tau) \to T_{g,\tau}$ induces an injection,

$$\mathcal{Y}/G \to \hat{G}_{\Xi}; \tag{11}$$

and the map $(g,\tau) \to g$ a surjection

$$\mathcal{Y}/G \to \mathcal{P}/G \tag{12}$$

with finite fiber.

Let $j = 1,2,\dots,N$. We let $\mathcal{U}_j = \{g \in \mathcal{U},\ \underline{s}(g)$ is G-conjugate to $\underline{s}_j\}$. Then $\mathcal{U}_j$ is a G-invariant subset of $\underline{g}^*_{\xi}$, open for the Hausdorff topology ([Du3]). Now we set

$$\underline{h}_j = \text{centralizer of } \underline{s}_j \text{ in } \underline{g}$$

$$\begin{aligned} H_j &= \text{centralizer of } \underline{s}_j \text{ in } G \\ \underline{\underline{H}}_j &= \text{centralizer of } \underline{s}_j \text{ in } \underline{\underline{G}} \\ H'_j &= \text{normalizer of } \underline{s}_j \text{ in } G \\ W_j &= H'_j/H_j \\ \underline{g}_j &= [\underline{s}_j, \underline{g}]. \end{aligned} \tag{13}$$

Then W_j is finite, $\underline{g} = \underline{h}_j \oplus \underline{g}_j$, and we identify $\underline{h}^*_j$ and $\underline{g}^{\perp}_j$. We let

$$\begin{aligned} \underline{h}^*_{j,\xi} &= \underline{h}^*_j \cap \underline{g}^*_\xi \\ V_j &= (\underline{h}^*_j \cap U_j) \text{ (it is open and } H'_j \text{ invariant in } \underline{h}^*_{j,\xi}). \end{aligned} \tag{14}$$

Then the map $(x,g) \to xg$ from $G \times V_j$ is a submersion

$$G \times V_j \to U_j. \tag{15}$$

The following important fact supplements (15).

Two points of V_j are G conjugate if and only if they are H'_j-conjugate. (16)

We denote by Δ_j the set of non-zero roots of $\underline{s}_{j\mathbb{C}}$ in $\underline{g}_{\mathbb{C}}$, and if $\alpha \in \Delta_j$, $\underline{g}^\alpha$ the corresponding subspace . Each $\underline{g}^\alpha$ is stable under H, and $\underline{g}_{j\mathbb{C}} = \oplus\, \underline{g}^\alpha$.

Let $\underline{\underline{S}}_j$ be the algebraic connected torus of $\underline{\underline{G}}$ with Lie algebra $\underline{s}_j$. We denote by $\underline{\underline{T}}_j$ and $\underline{\underline{A}}_j$ the anisotropic and split parts, and by $\underline{t}_j$ and $\underline{a}_j$ the corresponding Lie algebras.

Fix an element $X_j \in i\underline{t}_j + \underline{a}_j$ such that $\alpha(X_j) \neq 0$ for all α. Let

$$\begin{aligned} \Delta^+_j &= \{\alpha \in \Delta_j,\ \alpha(X_j) > 0\} \\ \rho_j &= \frac{1}{2} \sum_{\alpha \in \Delta^+_j} \dim \underline{g}^\alpha\, \alpha \\ \sigma_j &= -i\rho_j|\underline{t}_j. \end{aligned} \tag{17}$$

Then $\sigma_j \in \underline{t}_j^*$.

We consider the set

$$\mathcal{Q}_j = P \cap V_j. \tag{18}$$

Then

$$\mathcal{Q}_j = \{g \in V_j,\ \sigma_j + ig|\underline{t}_j \text{ is the differential of a character of the analytic subgroup } (T_j)_0 \text{ with Lie algebra } \underline{t}_j \text{ whose restriction to } \Gamma \cap (T_j)_0 \text{ is equal to the restriction of } \Xi\}. \tag{19}$$

The injection from $\underline{h}_j^*$ into $\underline{g}^*$ induces bijective maps:

$$\begin{aligned} &V_j/H_j^! \to U_j/G \\ &\mathcal{Q}_j/H_j^! \to P_j/G. \end{aligned} \tag{20}$$

V.3 Invariant generalized functions and integral formulas

Let $j = 1,2,\ldots N$, and let the notations be as above.

We denote by $F(M)$ the set of generalized functions[13] on a manifold M. Let $\theta \in F(U_j)$ be a G-invariant element. Because of the transversality property (15), the restriction of θ to V_j is well defined (cf. e.g. the appendix to [Du-He-Ve]). We denote it by $R(\theta)$. It is clearly $H_j^!$-invariant, and the map $\theta \to R(\theta)$ is continuous for the weak topologies.

(21) Lemma. The map $\theta \to R(\theta)$ is an homeomorphism of the space of G-invariant elements of $F(U_j)$ onto the space of $H_j^!$-invariant elements of $F(V_j)$.

Proof. Let ψ be an $H_j^!$-invariant element of $F(V_j)$. Let $g_0 \in U_j$. We are looking for a G-invariant element $\theta \in F(U_j)$ such that $R(\theta) = \psi$. We construct θ in a small neighborhood of g_0 in the following way. We choose $x \in G$ such that $xg_0 \in V_j$. In a neighborhood of xg_0 in U_j there exists a unique θ^x which is locally G invariant and restricts to ψ. We define θ by $\theta(g) = \theta^x(xg)$ in a neighborhood of

g_0. To see that it is well defined, consider $y \in G$ such that $yg_0 \in V_j$. Then $y = zx$ with $z \in H_j^!$ (by (16)). Consider $\theta^x(zg)$. It restricts to $\psi(zxg)$ $(g \in V_j)$ in a neighborhood of zxg_0 (by functorial properties of the restriction map), which is equal to $\psi(xg)$ by invariance. Thus $\theta^{zx}(zg) = \theta^x(g)$ in a neighborhood of xg_0. The continuity of $\psi \to \theta$ is easily seen using the construction of θ.

Q.E.D.

To calculate $R(\theta)$, there is an integral formula similar to Weyl's and Harish Chandra's integration formulas. We need some more notations.

Recall that a measure dX on $\underline{g}/\underline{j}$ has been chosen. On $(\underline{g}/\underline{j})^*$ we use the dual measure, and on $\underline{g}_\xi^*$ (which is parallel to $(\underline{g}/\underline{j})^*$) we transport it by translation. Let dg be this measure.

We choose a measure dY on $\underline{h}_j/\underline{j}$, and a measure dZ such that $dX = dYdZ$. On $\underline{h}_{j,\xi}^*$, we use the measure $d\lambda$ dual as above to dY. Choose a basis $e_1,\ldots,e_{2d}$ of $\underline{g}_j$ such that $dZ = |e_1^*\wedge\ldots\wedge e_{2d}^*|$, where $e_1^*,\ldots,e_{2d}^*$ is the dual basis. We define an element $\omega \in S(\underline{h}_j)$ by the formula:

$$(d!)^{-1}B_\lambda^d \quad = \quad \omega(\lambda)\; e_1^* \wedge\ldots\wedge e_{2d}^* \tag{22}$$

for $\lambda \in \underline{h}^*$ where B_λ is the two form on $\underline{g}_j$ such that $B_\lambda(X,Y) = \lambda([X,Y])$ $(X,Y \in \underline{g}_j)$.

On the set of functions ϕ on G which satisfy

$$\phi(xy) \quad = \quad |\det \mathrm{Ad}\, y_{\underline{g}/\underline{h}_j}|\,\phi(x) \quad (x \in G,\ y \in H_0) \tag{23}$$

there is a well deinfed "measure" which will be denoted by dx, tangent to dZ on $\underline{g}_j \simeq \underline{g}/\underline{h}_j$.

Let $\phi \in C_c^\infty(U_j)$. We denote the value of θ on $\phi\,dg$ by
$$\int_{U_j} \theta(g)\phi(g)dg.$$

(24) Lemma. (cf. [Du4] theorem 1).

$$\int_{\mathcal{U}_j} \theta(g)\phi(g)dg = \frac{1}{c_j} \int_{G/H_j} \{|\det \mathrm{Ad}\, x|^{-1} \int_{\mathcal{U}_j} R(\theta)(\lambda)\omega(\lambda)^2 \phi(x\lambda)d\lambda\}$$

with

$$c_j = (2\pi)^{2d}\,\#\, W_j,\ 2d = \dim \underline{g}_j. \tag{25}$$

(Note that the function inside { } satisfies (23).)

Proof. Let us use on G/H_j' the "measure" tangent to dZ as above. We rewrite (24) as

$$\int_{\mathcal{U}_j} \theta(g)\phi(g)dg = (2\pi)^{-2d} \int_{G/H_j'} \{\ldots\}dx, \tag{26}$$

with $\{\ldots\}$ as above.(14)

It follows from (16) that $\mathcal{U}_j$ is homeomorphic to the open subset $G \times_{H_j'} \mathcal{U}_j$ of the affine bundle $G \times_{H_j'} \underline{h}_\xi^*$.

The two sides of (26) (considered as a function of ϕ) have the same property of tranformation under the action of G. It is then enough to prove (26) for ϕ with support in an arbitrary neighborhood of V_j in $\mathcal{U}_j$. Let $\mathcal{O}$ be an open neighborhood of 0 in $\underline{g}_j$ such that $(X,y) \to \exp Xy$ is a diffeomorphism of $\mathcal{O} \times H_j'$ on an open set in G. The map $(X,\lambda) \to \exp X\lambda$ is a diffeomorphism of $\mathcal{O} \times V_j$ on an open set $\mathcal{O}'$ of $\mathcal{U}_j$. If $\beta \in C_c(\mathcal{O}')$ we write

$$\int_{\mathcal{U}_j} \theta(g)\phi(g)dg = \int_{\mathcal{O}\times V_j} \theta(\exp X\lambda)\phi(\exp X\lambda)\psi(X,\lambda)d\lambda dX \tag{27}$$

with a Jacobian $\psi(X,\lambda)$. By invariance, the generalized function $(X,\lambda) \to \theta(\exp X\lambda)$ on $\mathcal{O} \times V_j$ is equal to $1 \otimes R(\theta)$. By Fubini's Theorem (27) is equal to

$$\int_{\mathcal{O}} \left(\int_{V_j} R(\theta)(\lambda)\phi(\exp X\lambda)\psi(X,\lambda)d\lambda \right) dX.$$

By G-invariance again, to prove (26) it is enough to prove that

$$\psi(0,\lambda) = (2\pi)^{-2d}\omega(\lambda)^2. \tag{28}$$

This computation is easy.

Q.E.D.

(29) Example. Let $\theta = 1$ in lemma (24). We get

$$\int_{U_j} \phi(g)dg = \frac{1}{c_j} \int_{G/H_j} \{|\det \mathrm{Ad}\ x|^{-1} \int_{V_j} \omega(\lambda)^2\phi(x\lambda)d\lambda\}\ dx.$$

(30) Example. Suppose G is unimodular. Let Ω be a closed G-orbit contained in U_j. Let β'_Ω be the element of $F(U_j)$ such that $\beta'_\Omega dg = \beta_\Omega$ (cf III (40)). Let $\Lambda = \Omega \cap \underline{h}_j^*$. It is a closed H'_j orbit and we define in the same manner β_Λ and β'_Λ such that $\beta'_\Lambda d\lambda = \beta_\Lambda$. We compare β'_Λ and $R(\beta'_\Omega)$:

$$\beta'_\Lambda = (2\pi)^{-d}|\omega(\lambda)|R(\beta_\Omega). \tag{31}$$

In fact (31) is equivalent to

$$\int_\Omega \phi(g)d\beta_\Omega(g) = \frac{1}{(2\pi)^d \# W_j} \int_{G/H_j} dx\{\int_\Lambda |\omega(\lambda)|\phi(x\lambda)d\beta_\Lambda(\lambda)\} \tag{32}$$

which is proved by the same kind of arguments as (24).

(33) Remark. (32) is valid also if G is non unimodular. In fact, (21) and, with a slight modification, (24) may be extended to semi-invariant generalized functions.

Consider the series of generalized functions on $\underline{h}_{j,\xi}^*$

$$P_j(\lambda) = \sum_{\substack{X\in \underline{t}_j,\\ \exp X\in\Gamma}} e^{(-i\xi+\sigma_j+i\lambda)(X)}. \tag{34}$$

(Note that $e^{\sigma_j(X)} = \pm 1$.) By Poisson's formulas, its support is the set of $g \in \underline{h}^*_j$, such that $\sigma_j + i\lambda|\underline{t}_\lambda$ is the differential of a character of $(T_j)_0$ whose restriction to $\Gamma \cap (T_j)_0$ is equal to the restriction of Ξ. It is obviously $H_j^!$-invariant. We denote by P_j the generalized function on $\mathcal{U}_j$ which is G-invariant and such that $R(p_j)$ is equal to P_j in $\mathcal{V}_j$(cf. (33)). By (18) and (19), it follows that the support of p_j is $\mathcal{P} \cap \mathcal{U}_j$.

(35) Example. If $\underline{t}_j = 0$, then $p_j = 1$.

In general, p_j seems to be the most natural generalized function on $\mathcal{U}_j$ which is G-invariant and whose support is $\mathcal{P} \cap \mathcal{U}_j$.

We let p be the generalized function on $\mathcal{U}$ whose restriction to each of the $\mathcal{U}_j$ is p_j. Note that it is G-invariant, with support in $\mathcal{P}$. Note also that it depends on G, j, Γ and Ξ, but not on the choice of measures.

Consider pdg: it is clearly a positive Radon measure on $\mathcal{U}$, with support $\mathcal{P}$. We usually consider pdg as a positive Borel measure on $\underline{g}^*_\xi$, concentrated in $\mathcal{P}$.

(36) Remark. I conjecture that pdg is tempered. But it is not even obvious that it is a Radon measure. It is known that pdg is tempered in the following cases:

$\underline{g}$ semi-simple (Harish-Chandra),

and more generally if $\underline{h}_j$ is commutative $(j = 1,\ldots,N)$ (Torasso);

$\underline{g}$ solvable (this is easy);

$\underline{g}$ complex (cf. [Du4]).

Let us recall the following result which follows from [Du-Ra], proposition 5.14 because orbits of algebraic groups are locally closed:

(37) Lemma. Let ψ be a positive Borel function on $\mathcal{U}$. The function

$g \to \int_{Gg} \psi(f)\, d\beta_{Gg}(f)$ is also a Borel function (with values in $[0,\infty]$) on $\mathcal{U}$.

Again because the orbits are locally closed, the set P/G (with the quotient Borel structure) is a countably separated Borel space.

(38) Corollary. Suppose G is unimodular. There exists a unique positive Borel measure on P/G, which we denote by m, such that

$$\int_P \psi(g)\, p(g)\, dg \;=\; \int_{P/G} \Big(\int_\Omega \psi(g)\, d\beta_\Omega(g) \Big) dm(\Omega).$$

Proof. cf. the proof of lemma 5.1.7 in [Du-Ra].

(39) Remark. When G is not unimodular, the division of $p(g)dg$ by the β_Ω can also be performed, but the result is a measure with values in a line bundle over P/G instead of an ordinary measure. (Of course this is true also of the Plancherel measure on $\hat{G}_\Xi$, and the two facts are connected--cf. [Du-Ra] for solvable exponential groups).

V.4 The Plancherel formula

In this section, we assue that G is unimodular. Notations are as in V.1, V.2 and V.3.

(40) Theorem. There exists a G-invariant[(16)] function ζ on $\mathcal{Y}$, with values in $]0,\infty[$, such that the following is true.

For all $\phi \in C_c^\infty(G/\Gamma J, \Xi)$, the function

$$\psi(g) \;=\; \sum_{\tau \in X^{irr}(g,\Xi)} \frac{\dim \tau \;\zeta(g,\tau)}{\# G(g)/G(g)_0\Gamma}\; \operatorname{tr} T_{g,\tau}(\phi) \tag{41}$$

is a G-invariant pdg -measurable function on P, and

$$\phi(1) \;=\; \int_{P/G} \psi(g)\, dm(g). \tag{42}$$

Let us mention several particular cases of theorem (40). If G is discrete, then $\zeta = 1$, and (42) is exactly (6). If G is unipotent, then $\zeta = 1$, and (42) is Kirillov's Plancherel theorem [Ki]. If G is solvable, $\zeta = 1$. This is Charbonnel's theorem [Cha 1]. If G is a complex algebraic group, $\zeta = 1$. This is Andler's theorem [An]. If G is reductive in the "Harish-Chandra class", this is Harish-Chandra. Moreover, the function ζ is explixitly computed in [Ha]. I reproduce below the formulas for ζ in this case (formulas (54)-(55)). Harish-Chandra's formula has been extended to the case of connected Lie groups by Herb and Wolf [He-Wo].

The reductive connected case is fundamental, because the general case is reduced to this one by using the Mackey and Kleppner-Lipsman method [Kl-Li] to go from G to G_0, or from G to a group of smaller dimension.

I do not give a proof of theorem (40). In the case of complex algebraic groups, it is given in [An] and it is completely similar in the general case.

In two respects, theorem (40) is not entirely satisfying. First one would like to define on $\mathcal{Y}$ a standard Borel structure with respect to which all natural maps $(g,\tau) \to g$, $(g,\tau) \to \dim \tau$, $(g,\tau) \to \zeta(g,\tau)$, $(g,\tau) \to T_{g,\tau}$, etc. are Borel. Andler defined such a structure when G is algebraic complex. It is not difficult to imagine an adaption of his definition to the present setting. But adaption of proofs is certainly painful because one has to take into account the more complicated structure of the metaplectic representation in the real case. We shall call the existence of such a structure hypothesis (H).[15]

Under hypothesis (H), theorem (40) describes the class of Plancherel measure. In fact, the unimodularity assumption is not necessary for the next corollary.

(43) <u>Corollary</u>. Assume (H) is satisfied. Then a set $E \subset \hat{G}_{\Xi}$ is of

Plancherel measure zero if and only if the set $\{g \in U$, there exists $\tau \in X^{irr}(g,\Xi)$ such that $T_{g,\tau} \in E\}$ is of measure zero with respect to pdg.

Second, one would like an explicit determination of the function ζ. The most obvious possibility (about which I am not completely enthusiastic, because it is not extremely explicit) is to reduce the computation to the semi-simple case, where the results are known. One way to do that uses a special class of coisotropic subalgebras of unipotent type, called acceptable coisotropic subalgebras, introduced in [Du5].

(44) Definition. Let $g \in \underline{g}^*$. A coisotropic subalgebra $\underline{p}$ of unipotent type at g is called acceptable if its reductive factors are of maximal dimension (when $\underline{p}$ is coisotropic of unipotent type).

(45) Example. The canonical acceptable subalgebra constructed in IV.3 is invariant under all automorphisms of G which fix g.

Let $g \in P$, and let $\underline{p}$ be a $G(g)$-invariant acceptable subalgebra at g. Let $\tau \in X_G^{irr}(g)$. We define P, R, V, $R^{\underline{V}}$, $r \in \underline{r}^*$, $\tau'' \in X^{irr}_{R^{\underline{V}}}(r)$ as in IV.3. Recall that $\underline{r}$ is reductive. Its isomorphism class is independent of $\underline{p}$ (cf. Du[5]). We consider $\zeta_{R^{\underline{V}}}(r,\tau'')$. It can be proved that it is independent of $\underline{p}$, and that:

$$\zeta_G(g,\tau) = \zeta_{R^{\underline{V}}}(r,\tau''). \tag{46}$$

V.5 Another form of the Plancherel formula

Let $g \in P$. Recall the group $G(g)_1$ and the set $Z^{irr}(g)$ defined in IV.6. Let $Z^{irr}(g,\Xi)$ be the subset of $\sigma \in Z^{irr}(g)$ such that $\sigma(r_g(x)) = \Xi(x)\mathrm{Id}$ for $x \in \Gamma J$. The group $G(g)$ acts in $Z^{irr}(g,\Xi)$. We let Z be the set of pairs (g,σ) with $g \in P$ and $\sigma \in Z^{irr}(g,\Xi)$.

Because the structure of Z is simpler than that of Y, and the structure of the representations $T_{g,\sigma}$ simpler than that of the representations $T_{g,\tau}$, the following version of the Plancherel formula is interesting.

(47) Theorem. There exists a G-invariant[17] function ζ' on $\mathcal{Z}$, with values in $]0,\infty[$, such that the following is true:

For all $\phi \in C_c^\infty(G/\Gamma J,\Xi)$, the function

$$\psi'(g) = \sum_{\sigma\in Z^{irr}(g,\Xi)/G(g)} \frac{\dim \sigma \ \zeta'(g,\sigma)}{\#G(g)/G(g)_0\Gamma} \operatorname{tr} T'_{g,\sigma}(\phi) \tag{48}$$

is a G-invariant pdg measurable function on $\mathcal{P}$, and

$$\phi(1) = \int_{\mathcal{P}/G} \psi'(g)dm(g). \tag{49}$$

If necessary, we denote ζ' by ζ'_G.

When G is algebraic and complex, then $G(g)_1 = G(g)_0$. Thus $Z^{irr}(g,\Xi)$ contains only one element σ, and its dimension is one. We write $T'_g = T'_{g,\sigma}$. Moreover $\zeta' \equiv 1$, so that formula (48) reads in this case $\psi'(g) = \frac{1}{\#G(g)/G(g)_0\Gamma} \operatorname{tr} T'_g(\phi)$. This is theorem 7' in [An].

In the general case, theorem (47) is more complicated, but still remarkably simple.

The relation between ζ and ζ' is as follows: Let $(g,\tau) \in \mathcal{Y}$, and let σ be an irreducible component of the restriciton of τ to $G(g)_1^{\mathfrak{g}}$. Then:

$$\zeta(g,\tau) = \zeta'(g,\sigma). \tag{50}$$

It is easy to see that theorem (47) is equivalent to theorem (41), plus the fact that $\zeta(g,\tau)$ depends only on the quasiequivalence class of the restriction of τ to $G(g)_1^{\mathfrak{g}}$.

Obviously, the set $\mathcal{Z}_G$ (relative to G and Ξ) and the set $\mathcal{Z}_{G_0}$ relative to G_0 and the restriction of Ξ to $\Gamma \cap G_0$ are the same. One advantage of the formulation (47) of the Plancherel theorem is the following easy lemma which reduces the computation of the Plancherel formula to connected groups (cf. the proof of proposition 18 in [An].)

(51) Lemma. The functions ζ'_G and ζ'_{G_0} are equal.

V.6 Connected reductive groups

Formulas (46), (50), (51) reduce the computation of ζ_G to the case where G is connected and $\underline{g}$ reductive.

We assume in this paragraph that G is connected and $\underline{g}$ reductive.

Let $g \in P$. Then $\underline{g}(g)$ is a Cartan subalgebra of $\underline{g}$, and $G(g)$ is equal to $G(g)_1$, and also to the centralizer of $\underline{g}(g)$ in G. We let $\underline{g}(g) = \underline{h}$, and use the notations of III.3.

A root $\alpha \in \Delta$ is called real if $\underline{g}^{\alpha} = (\underline{g}^{\alpha} \cap \underline{g})_{\mathbb{C}}$. We denote by $\Delta_{\mathbb{R}}$ the set of real roots. If $\alpha \in \Delta_{\mathbb{R}}$, we choose $X_{\alpha} \in \underline{g} \cap \underline{g}^{\alpha}$, $Y_{\alpha} \in \underline{g} \cap \underline{g}^{-\alpha}$, $H_{\alpha} \in \underline{g}$ such that $H_{\alpha} = [X_{\alpha}, Y_{\alpha}]$, $\alpha(H\ ) = 2$. We let $\gamma_{\alpha} = \exp(\pi(X_{\alpha} - Y_{\alpha}))$. It is known that γ_{α} belongs to H.

(52) Lemma. Let a be an automorphism of G, and suppose a normalizes $\underline{h}$ and $\ker \alpha \subset \underline{h}$. Then either $a(\gamma_{\alpha}) = \gamma_{\alpha}$ or $a(\gamma_{\alpha}) = \gamma_{\alpha}^{-1}$.

Proof. Let S_{α} be the analytic subgroup of G with Lie algebra $\underline{s}_{\alpha}$. It is locally isomorphic to $SL(2,\mathbb{R})$, and normalized by a. Thus we may assume that $G = S_{\alpha}$, and obviously it is enough to make the proof when G is simply connected. Then the center of G is isomorphic to $\mathbb{Z}$, and γ_{α} is a generator of it. So $a(\gamma_{\alpha}) = \gamma_{\alpha}$ or γ_{α}^{-1}.

We let $\Delta_{\mathbb{R}}^{+} = \{\alpha + \Delta_{\mathbb{R}},\ g(H_{\alpha}) > 0\}$. If $\alpha \in \Delta_{\mathbb{R}}^{+}$, we let $\varepsilon_{\alpha} = (-1)^{n_{\alpha}}$, where $2n_{\alpha} = \sum \beta(H_{\alpha})$, and the summation is over the set of roots $\beta \in \Delta$ such that $\beta|\underline{a} = c\alpha|\underline{a}$, with $c > 0$. Consider $\tau \in X^{irr}(g)$, and define the irreducible representation $\delta^{g}\tau$ of H as in III (12).

(53) Lemma. Let $\alpha \in \Delta_{\mathbb{R}}$, and let χ_{α} be an eigenvalue of $\delta^{g}\tau(\gamma_{\alpha})$. Then the eigenvalues of $\delta^{g}\tau(\gamma_{\alpha})$ belong to the set $\{\chi_{\alpha}, \bar{\chi}_{\alpha}\}$.

Proof. Let $x \in H$. By Lemma (50), $x\gamma_\alpha x^{-1} = \gamma_\alpha$ or γ_α^{-1}. This means that the cyclic group $<\gamma_\alpha>$ generated by γ_α is invariant in H. Mackey's theory shows that the eigenvalues of $\delta^g\tau(\gamma_\alpha)$ are the $\chi_\alpha(x\gamma_\alpha x^{-1})$ where $x \in H$.

Q.E.D

We see that the real part $\mathrm{Re}\,\chi_\alpha$ is well defined.

The following theorem is a way of expressing Harish Chandra's theorem [Ha] (when Γ is finite) or Herb-Wolf's theorem [He-Wo] (when Γ is arbitrary).

(59) Theorem. There exists a constant $c_H > 0$ (depending perhaps on H, but not on g such that $G(g) = H$, nor on τ) such that

$$\zeta(g,\tau) = c_H \left| \prod_{\alpha \in \Delta^+_{\mathbb{R}}} \frac{\mathrm{sh}\ g(H_\alpha)}{\mathrm{ch}\ g(H_\alpha) - \varepsilon_\alpha \ \mathrm{Re}\ \chi_\alpha} \right| .$$

The constants c_H in [Ha] and [He-Wo] are written in an explicit but somewhat complicated way. Moreover they are not written in the same way. My guess is that they are extremely simple. In fact:

(55) Theorem. Suppose Γ is finite. Then $c_H = 1$.

Sketch of proof. Look at the constants in Harish-Chandra. They involve the volume of some homogeneous compact spaces, for a specific measure. Then, compute these volumes.

(56) Remark. A particular case of (55) is given in Mneinne [Mne].

I have no doubts that (55) is valid also when Γ is not finite (for instance it's easy to verify for groups locally isomorphic to $SL(2,\mathbb{R})$, or when H is fundamental) but I did not study the constants in [He-Wo]. I do not persist more on this, or on the proof of (55), because I believe it will be possible to give a proof of the Plancherel theorem which directly gives the constants c_H in the form $c_H = 1$.

V.7 Complement

In this section, we assume that G is unimodular and that Γ is finite (to be able to use (55)).

The following proposition perhaps renders a more natural definition of $\zeta(g,\tau)$ in (41): in fact, it tends to 1 when g goes to ∞:

(57) Proposition. Let $g_0 \in P$. Then $\zeta(g,\tau)$ tends to 1 uniformly in $\tau \in X^{irr}(g)$ when g goes to ∞ in $P \cap \mathbb{R}g_0$.

This follows from (46), (55), and from the fact that the algebra $\underline{r}$ of (46) depends only on $\mathbb{R}g_0$.

Notes

(1) A given Lie group G can be endowed with several structures of an almost algebraic group.

Example 1. Let $G = \mathbb{R}$. Then for $\underline{\underline{G}}$ we can take either the additive group for which $\underline{\underline{G}}(\mathbb{R}) = \mathbb{R}$, or the multiplicative group, for which $\underline{\underline{G}}(\mathbb{R}) = \mathbb{R}^{\times}$.

Example 2. Let $G = SL(2,\mathbb{R})$. Then for $\underline{\underline{G}}$ we can take either $SL(2)$, for which $G = \underline{\underline{G}}(\mathbb{R})$, or $PSL(2)$, for which $SL(2,\mathbb{R})/\{\pm 1\}$ is of index two in $G(\mathbb{R})$.

If a Lie group G can be given a structure of an almost algebraic group, it is locally isomorphic to the group of real points of an algebraic group. The converse is not true, even if we assume G connected Consider for instance a connected non commutative nilpotent Lie group with compact center.

(2) In this paper, representation means continuous representation in a separable Hilbert space.

(3) The definition of S depends on the choice of the square root i of -1 by the formula I (3). The choice $-i$ would lead to the conjugate representation, which is not equivalent.

(4) $\underline{\ell}$ is said to be totally complex if $\underline{\ell} \cap \overline{\underline{\ell}} = \{0\}$. In this case the mapping $x \to x_{\underline{\ell}}$ is an isomorphism of $Sp(V)_{\underline{\ell}}$ onto a subgroup of real automorphisms of $\underline{\ell}$ isomorphic to $U(p,q)$, with $q = q(\underline{\ell})$, $p = \dim \underline{\ell} - q$.

(5) We do not require that $\underline{b} + \overline{\underline{b}}$ is a subalgebra.

(6) Vogan defines also non unitary pseudo-characters, but we do not have to use them.

(7) Harish-Chandra assumes that the derived group (M_o,M_o) is of finite center, but this is not a problem. Cf. e.g. [Vo2].

(8) This implies strong conditions on $\underline{m}$ and g.

(9) An algebraic algebra $\underline{h}$ such that ${}^u\underline{h} = \{0\}$ is reductive in the ordinary sense, that is it is a direct product of semi-simple and commutative algebras. The converse is not true. I hope this will not cause confusion.

(10) Since usually the construction of an element $\underline{p} \in \mathrm{Cos}(g,G)$ involves more or less the same inductive stops as those used in the construction of $T^G_{g,\tau}$ in III, this does not mean that induction is not used in this alternative construction of $T^G_{g,\tau}$.

(11) We consider $\{0\}$ as an Heisenberg Lie algebra.

(12) Lispman uses a condition which is equivalent to "S centralizes $\underline{s}$". The advantage of (32) is that it is generic for the Plancherel formula of G at least if we can apply corollary (43) of Ch. V.

(13) A generalized function is a continuous linear map on the set of C^∞-densities with compact support.

(14) One may wonder why we prefer the form (24) to (26). In fact it is easier to use in most applications because H_j is a subgroup of many interesting subgroups of G, which is less often the case for H_j'.

(15) Of course, it is hard to imagine that such an hypothesis might not be satisfied.

(16) ζ is in fact invariant by the group of automorphims of G.

(17) Note (16) is also valid for ζ'.

References

[An] M. Andler, La formule de Plancherel pour les groupes algébriques complexes unimodulaires. To appear in Acta Mathematica.

[Au-Ko] L. Auslander and B. Kostant, Polarizations and unitary representations of solvable Lie groups. Invent. Math. 14(1971), 255-354.

[Be] P. Bernat and al., Représentations des groupes de Lie résolubles Dunod, Paris 1972.

[Bou] A. Bouaziz, Sur les représentations des groupes de Lie réductifs non connexes. Preprint, Paris 1983.

[Cha 1] J.-Y. Charbonnel, La formule de Plancherel pour un groupe résoluble connexe II. Math. Annalen 250(1980), 1-34.

[Cha 2] J.-Y. Charbonnel, Sur les orbites de la représentation coadjointe. Compositio Mathematica 46(1982), 273-305.

[Che] C. Chevalley, Theory of Lie groups I. Princeton University Press.

[Co-Mo] A. Connes and H. Moscovici, The L^2-index theorem for homogeneous spaces of Lie groups. Ann. of Math. 115(1982), 291-330.

[Di 1] J. Dixmier, Sur les représentations unitaires des groupes de Lie algébriques. Ann. Inst. Fourier 7(1957), 315-328.

[Di 2] J. Dixmier, Polarisations dans les algébres de Lie. Bull. Soc. Math. France 104(1976), 145-164.

[Di 3] J. Dixmier, Les C*-algébres et leurs représentations. Gauthier-Villars, Paris 1964.

[Du 1] M. Duflo, Sur les extensions des représentations irréductibles des groupes de Lie nilpotents. Ann. Scient. Ec. Norm. Sup. 5(1972), 71-120.

[Du 2] M. Duflo, Opérateurs différentials biinvariants sur un groupe de Lie. Ann. Scient. Ec. Norm. Sup. 10(1977), 265-288.

[Du 3] M. Duflo, Construction de représentations unitaires d'un groupe de Lie. Cours d'été du C.I.M.E., Cortona 1980, published by Liguori, Napoli.

[Du 4] M. Duflo, Représentations unitaires des groupes de Lie et méthode des orbites. In G.M.E.L., Bordas 1982.

[Du 5] M. Duflo, Théorie de Mackey pour les groupes de Lie algébriques. Acta Mathematica 149(1982), 153-213.

[Du-He-Ve] M. Duflo, G. Heckman, M. Vergne, Projection d'orbites, formule de Kirillov et formule de Blattner. Preprint, Paris, 1983.

[Du-Ra] M. Duflo and M. Rais, Sur l'analyse harmonique sur les groupes de Lie résolubles. Ann. Scient. Ec. Norm. Sup 9(1976), 107-144.

[Du-Ve] M. Duflo and M. Vergne, Une propriété de la représentation coadjointe d'une algèbre de Lie. C.R. Acad. Sci. Paris 268 (1969), 583-585.

[Fu] H. Fujiwara, Polarisations réelles et représentations associées d'une groupe de Lie résoluble. Preprint, Kyushu 1983.

[Gi] V.A. Ginsburg, Fast decreasing functions and characters of real algebraic groups. Funktional'nyi Analiz i Ego Prilozheniya 16(1982), 66-69.

[Ha] Harish-Chandra, Harmonic analysis on real reductive groups III. Annals of Math. 104(1976), 117-201.

[He-Wo] R. Herb and J. Wolf, The Plancherel theorem for general semisimple groups. Preprint, Berkeley 1983.

[He] A. Hersant, Formes harmoniques et cohomologies relative des algèbres de Lie. Journal für die Reine und Angewandte Mathematik 344(1983), 71-86.

[Kha 1] M.S. Khalgui, Caractères des groupes de Lie. Journal of Functional Analysis 47(1982), 64-77.

[Kha 2] M.S. Khalgui, Extensions des représentations des groupes de Lie construites par M. Duflo. Math. Annalen 265(1983), 343-376.

[Kha 3] M.S. Khalgui, Caractères des représentations factorielles normales d'un groupe de Lie connexe. Preprint, Tunis 1983.

[Ki 1] A.A. Kirillov, Représentations unitaires des groupes de Lie nilpotents. Uspekhi Math. Nauk 17(1962), 57-110.

[Ki 2] A.A. Kirillov, Plancherel measure of nilpotent Lie groups. J. Functional Analysis and its applications 1(1967), 330-332.

[Ki 3] A.A. Kirillov, Eléments de la théorie des représentations. Ed. MIR. Moscow 1974.

[Kl-Li] A. Kleppner and R.L. Lipsman, The Plancherel formula for group extension II. Ann. Scient. Ec. Norm. Sup 6(1973), 103-132.

[Li-Ve] G. Lion and M. Vergne, The Weil representation, Maslov index and theta series, Birkhauser, Boston 1980.

[Li 1] R.L. Lipsman, Harmonic induction on Lie groups. Journal für die Reine and Angemandte Mathematik 344(1983), 120-148.

[Li 2] R.L. Lipsman, On the existence of metric polarizations. Preprint 1983.

[Li 3] R.L. Lipsman, Generic representations are induced from square integrable representations. Trans. AMS, 1984, to appear.

[Mne] R. Mneimne, Equation de la chaleur sur un espace riemannien symétrique et formule de Plancherel. A paraitre dans Bull. Sc. Math.

[Pe] R. Penney, Canonical objects in Kirillov theory on nilpotent Lie groups. Proc. Amer. Math. Sco. 66(1977), 175-178.

[Pu] L. Pukanszky, Characters of connected Lie groups. Acta Math. 133(1974), 82-137.

[Ro] J. Rosenberg, Realization of square integrable representations of unimodular Lie groups on L^2-cohomology spaces. Trans. Amer. Math. Soc. 261(1980), 1-32.

[Sch] W. Schmid, L^2-cohomology and the discrete series. Annals. of Math. 103(1976), 375-394.

[Ve] M. Vergne, Representations of Lie groups and the orbit method. Emmy Noether Collequium, Birkhauser, Boston 1983.

[Vo 1] D. Vogan, The algebraic structure of the representations of semi-simple Lie groups. Annals. of Math. 109(1979), 1-60.

[Vo 2] D. Vogan, Irreducible characters of semi-simple Lie groups, I. Duke Math. Journal 46(1979), 61-108.

Harmonic Analysis on Semisimple Symmetric Spaces
A Method of Duality

Mogens Flensted-Jensen*

Introduction

In these lectures I shall try to describe some of the basic problems, as I see them, in harmonic analysis on semisimple symmetric spaces, and to indicate how far we are in understanding these problems. Not very much is known in complete generality. However, very many special classes of such spaces have been studied in great detail. This means that there is a rich pool of knowledge to draw from, when trying to formulate and attack problems and conjectures for the general case.

In Section 1 the basic notation and structure theory of a semisimple symmetric space $X = G/H$ is treated along with some examples. In Section 2 we discuss in general terms the fundamental questions in harmonic analysis on X.

The duality principle, on which much of our analysis is based, is introduced in Section 3. The following sections contain applications of the duality principle, such as construction of discrete series and other series of representations of G related to $X = G/H$, determination of minimal K-types of such representations, multiplicity in the Plancherel formula, relationship between elementary spherical functions on real and on complex groups, along with some examples and references.

I want to thank the Department of Mathematics at University of Maryland for the invitation to give these lectures and for the great hospitality shown to me during my stay.

§ 1. Structure of Semisimple Symmetric Spaces.

A semisimple symmetric space or in short a symmetric space is a homogeneous manifold $X = G/H$, where G is a connected semisimple Lie group having a non-trivial involution τ and H is a closed subgroup of the fixed point group G^{τ} of τ containing the identity conponent, i.e.

$$G_e^{\tau} \subset H \subset G^{\tau} .$$

For simplicity of the exposition we shall for the most make the following assumptions:

(1.1) G is a real form of a simply connected complex linear group G.

(1.2) H is connected, i.e. $H = G_e^{\tau}$, and H does not contain any non-trivial connected normal subgroup of G.

(1.3) $X = G/H$ is irreducible in the following sense: Let $X^{\sim}$ be the universal covering space of X. Then $X^{\sim}$ cannot be decomposed into a product of two non-trivial symmetric spaces.

There exists a, up to H-conjugacy, unique maximal compact subgroup K of G with Cartan involution σ such that $\sigma\tau = \tau\sigma$, or equivalently $\sigma(H) = H$ or $\tau(K) = K$. We choose such a K and σ. The universal covering space $X^{\sim}$ of X can be realized as $\tilde{G}^{\sim}/H$, where $G^{\sim}$ is a covering group of G and, by abuse of notation, H in G is isomorphic to H in $G^{\sim}$. Let $K^{\sim}$ be the connected covering group of K in $\tilde{G}$.

We distinguish between the following three types of irreducible symmetric spaces $X = G/H$:

(I) X is of <u>compact - type</u>, if G is compact.

(II) X is of <u>noncompact - type</u>, if G is noncompact and H is compact.

(III) X is of <u>non-Riemannian-type</u>, if H is noncompact.

When X is not irreducible, different irreducible components may have different types.

Of course one may think of (II) as the degenerate case of (III) with $\sigma = \tau$ or $H = K$, and think of (I) as the degenerate case of (III) with $\sigma = \mathrm{id}$ or $K = G$.

Let $\mathfrak{g}_{\mathbb{C}}, \mathfrak{g}, \mathfrak{h}, \mathfrak{k}, \tau$ and σ be the corresponding notions on the Lie algebra level. Define

$$\mathfrak{p} = \{X \in \mathfrak{g} \mid \sigma(X) = -X\} \text{ and } \mathfrak{q} = \{X \in \mathfrak{g} \mid \tau(X) = -X\}.$$

Then as direct sum of vectorspaces:

$$\mathfrak{g} = \mathfrak{h} + \mathfrak{q} = \mathfrak{k} + \mathfrak{p} = \mathfrak{h} \cap \mathfrak{k} + \mathfrak{h} \cap \mathfrak{p} + \mathfrak{q} \cap \mathfrak{k} + \mathfrak{q} \cap \mathfrak{p}. \tag{1.4}$$

Since the tangent space at eH of G/H may by identified with $\mathfrak{q}$ the Killing form on $\mathfrak{q}$ defines a G-invariant pseudo-Riemannian structure on $X = G/H$. In cases (I) and (II) X is actually Riemannian.

<u>Example 1.1. A semisimple Lie group G_1 as a symmetric space</u>. Let $G = G_1 \times G_1$ and $\tau(x,y) = (y,x)$. Then H is the diagonal subgroup, i.e. $H = d(G_1) = \{(x,x) \mid x \in G_1\}$. If K_1 is a maximal compact subgroup of G_1, we take $K = K_1 \times K_1$. As a manifold $G/H = G_1 \times G_1/d(G_1)$ is isomorphic to G_1 via the mapping

$$(x,y)d(G_1) \to xy^{-1}.$$

If G_1 is compact we are in case (I). If G_1 is noncompact we are in case (III). □

Example 1.2.

(a) The unit sphere in $\mathbb{R}^{n+1}$: $S^n \simeq SO(n+1)/SO(n)$.

(b) The hyperbolic upper half plane

$$\{z\in\mathbb{C} \mid \operatorname{Im} z > 0\} \simeq SL(2,\mathbb{R})/SO(2)$$

or isomorphic to this, the unit disk

$$\{z\in\mathbb{C} \mid |z|<1\} \simeq SU(1,1)\ /S(U(1)\times U(1)). \qquad \square$$

Example 1.3. The hyperbolic spaces.

Let $n = p+q$ be nonnegative integers, $n\geq 2$. Let $\mathbb{F} = \mathbb{R}$, $\mathbb{C}$ or $\mathbb{H}$, where $\mathbb{H}$ is the quarternions.

Define

$$X'_{p,q}(\mathbb{F}) = \{x\in\mathbb{F}^{n+1} \mid |x_1|^2+\dots+|x_p|^2-|x_{p+1}|^2-\dots-|x_{n+1}|^2=-1\}_e$$

where the component containing $(0,\dots,0,1)$ is taken.

Let $X_{p,q}(\mathbb{R}) = X'_{p,q}(\mathbb{R})$, and let, for $\mathbb{F} = \mathbb{C}$ or $\mathbb{H}$, $X_{p,q}(\mathbb{F})$ be the corresponding projective space over $\mathbb{F}$. Then we have

$$X_{p,q}(\mathbb{R}) \simeq SO_e(p,q+1)/SO_e(p,q) \ , \ K = SO(p)\times SO(q+1)$$

$$X_{p,q}(\mathbb{C}) \simeq SU(p,q+1)/S(U(p,q)\times U(1)) \ , \ K = S(U(p)\times U(q+1))$$

$$X_{p,q}(\mathbb{H}) \simeq Sp(p,q+1)/Sp(p,q)\times Sp(1) \ , \ K = Sp(p)\times Sp(q+1).$$

In all three cases τ is conjugation with the diagonal matrix

$$I_{p,q+1} = \begin{pmatrix} -I_p & 0 \\ 0 & I_{q+1} \end{pmatrix}.$$

(a) For $p = 0$ we get the classical rank one Riemannian symmetric spaces of compact type.

(b) For $q = 0$ we get the classical rank one Riemannian symmetric spaces of noncompact type.

(c) For $p \neq 0$ and $q \neq 0$ we get the rest of the classical isotropic symmetric spaces of constant curvature. $\square$

We now define a <u>Cartan subspace for G/H</u> as a maximal Abelian subspace $\mathfrak{a}$ of $\mathfrak{q}$ consisting of semisimple elements in $\mathfrak{q}$. Since all Cartan subspaces have the same dimension we may define

$$\underline{\mathrm{rank}(G/H) = \dim \mathfrak{a}.}$$

Each H-conjugacy class of Cartan subspaces contains a , up to $K\cap H$-conjugacy, unique σ-invariant one. Also each σ-invariant maximal Abelian subspace of $\mathfrak{q}$ is a Cartan subspace.

In the cases (I) and (II) the notion of rank is the usual one. In Example 1.1 $\mathrm{rank}(G_1 \times G_1/d(G_1)) = \mathrm{rank}(G_1)$. In Examples 1.2 and 1.3 $\mathrm{rank}(G/H) = 1$.

<u>Example 1.4</u>. There are more rank one symmetric spaces than contained in Example 1.3. We list the remaining ones:

(a) The isotropic spaces, given by the Lie algebras:

$$(\mathfrak{g}, \mathfrak{h}) = (\mathfrak{f}_{4(-20)}, \mathfrak{o}(9)) \text{ and } (\mathfrak{f}_{4(-20)}, \mathfrak{o}(1,8)) .$$

(b) The nonisotropic spaces

$$SL(n,\mathbb{R})/S(GL^{+}(n-1,\mathbb{R}) \times GL^{+}(1,\mathbb{R})) \quad , n \geq 3$$

$$Sp(n,\mathbb{R})/Sp(n-1,\mathbb{R}) \times Sp(1,\mathbb{R}) \quad , n \geq 3$$

and

$$(\mathfrak{f}_{4(4)}, \mathfrak{o}(4,5)).$$ □

A classification of all Riemannian symmetric spaces i.e. classes (I) and (II), due to E. Cartan can be found in Helgason [20]. A classification of the symmetric spaces of nonRiemannian type is due to Berger [1]. See Wolf [41] for the isotropic spaces. See also Loos [26], and Matsuki [27] for some of the general theory of symmetric spaces.

§ 2. Harmonic Analysis on Symmetric Spaces.

With $X = G/H$ let $U(\mathfrak{g})$ be the universal enveloping algebra over $\mathbb{C}$ of $\mathfrak{g}$. We consider $U(\mathfrak{g})$ as acting on the left as differential operators on X. The centralizer $U(\mathfrak{g})^{\mathfrak{h}}$ of $\mathfrak{h}$ in $U(\mathfrak{g})$ acts (on the right) on $X = G/H$ and gives the commutative algebra $D(X) = D(G/H)$ of G-invariant differential operators on X. Let $Z(\mathfrak{g})$ be the center of $U(\mathfrak{g})$. Since $Z(\mathfrak{g}) \subset U(\mathfrak{g})^{\mathfrak{h}}$, $Z(\mathfrak{g})$ defines a subalgebra $Z(X)$ of $D(X)$. Notice that for all classical groups and for all spaces of rank one we have $Z(X) = D(X)$, and in general $D(X)$ is finitely generated over $Z(X)$, with the same quotient field, see [15].

Both G and H are unimodular so that X has an essentially unique G-invariant measure.

Let L be the (<u>quasi-</u>) <u>regular representation</u> of G on $L^2(X)$, i.e.

$$(L_g f)(x) = f(g^{-1}x) \quad , f \in L^2(X) \, , \, g \in G \, , \, x \in X \, .$$

L is clearly unitary.

Let $G^\wedge$ denote the set of equivalence classes of irreducible unitary representations of G. Then, using Harish-Chandra's result [10] that G is "Type-I", L can be decomposed over $G^\wedge$:

$$L \simeq \int_{G^\wedge} m_\pi \cdot \pi \, d\mu(\pi)$$

(2.1)

$$L^2(X) \simeq \int_{G^\wedge} V_\pi \otimes H_{m_\pi} \, d\mu(\pi)$$

where m_π is the multiplicity of π, $d\mu$ is a measure on $G^\wedge$, $(\pi, V_\pi) \in G^\wedge$ and H_{m_π} is a Hilbert space of dimension m_π. Formula (2.1) is called the Plancherel formula for $X = G/H$.

The first and basic problem in the harmonic analysis on X is to find explicitly the Plancherel formula. There are several aspects to this. First notice that it is only the measure class of $d\mu$ which is uniquely determined. So to describe the formula we need to

(i) find a parametrization (in terms of the structure of G/H) of a subset Ω of $G^\wedge$ sufficiently large to carry $d\mu$,

(ii) find m_π for $\pi\in\Omega$,

(iii) describe $d\mu$ in terms of these parameters along with the maps: $f \to f^\wedge(\pi)$ of $L^2(G/H)$ into $V_\pi \otimes H_{m_\pi}$,such that

$$f = \int_\Omega f^\wedge(\pi)\,d\mu(\pi) \quad ,$$

(iv) if possible find a canonical normalization of the maps $f \to f^\wedge(\pi)$, and thus a canonical Plancherel measure.

Example 2.1. Let G_1 be compact and simply connected. Then $G_1 \simeq G_1 \times G_1/d(G_1) = G/H$. $G_1^\wedge$ is described by the theory as dominant weights. The set Ω may be taken to be the subset $G_H^\wedge$ of $G^\wedge = G_1^\wedge \times G_1^\wedge$ consisting of representations with an $H = d(G_1)$ fixed vector, i.e.

$$\Omega = \{\delta \times \check{\delta} \mid \delta \in G_1^\wedge\} \quad .$$

Then m_π is identically one on Ω . The canonical normalization of $d\mu$ is determined by the canonical choice of normalized characters χ_δ for $\delta\in G_1^\wedge$, i.e $\chi_\delta(x) = \frac{1}{\dim(\delta)} \operatorname{Tr}(\delta(x))$. The characters are given by Weyl's characterformula and the Plancherel measure is a discrete measure given by Weyl's dimension formula. □

Example 2.2. Let G/K be a Riemannian symmetric space of non-compact type. The Plancherel formula is explicitly determined by Harish-Chandra's theory of the spherical Fourier transform on G/K. We shall indicate a few of the features of this theory. It is natural to take

$$\Omega = G_K^\wedge = \{\pi \in G^\wedge \mid \exists v_o \in V_\pi \ \forall k \in K : v_o \neq o, \pi(k)v_o = v_o\}.$$

If $\pi \in G_K^\wedge$ then $\mathbb{C}v_o$ is uniquely determined. Take $\|v_o\| = 1$ and define $\varphi(x) = \langle v_o, \pi(x)v_o \rangle$, $x \in G$. Then φ is positive definite, and φ is an elementary spherical function, that is

$$\varphi \in C^\infty(K \backslash G/K), \text{ i.e. } K\text{-biinvariant}, \ \varphi(e) = 1 \text{ and}$$

(2.2)

$$D\varphi = \chi(D)\varphi \quad \text{for all} \quad D \in D(G/K) .$$

On the other hand every positive definite elementary spherical function arises in this fashion.

The full set of elementary spherical functions is given by Harish-Chandra's integral formula in the following way:

Let $G = KAN$ be an Iwasawa decomposition of G, and correspondingly $\mathfrak{g} = \mathfrak{k} + \mathfrak{a} + \mathfrak{n}$. Define the function $x \to H(x)$ of G into $\mathfrak{a}$ by

$$x \in K\exp H(x) N.$$

Let W be the Weyl group of $(G, \mathfrak{a})$.

Theorem.2.3. (Harish-Chandra). Every elementary spherical function is of the form

$$(2.3) \qquad \varphi_\lambda(x) = \int_K e^{\langle i\lambda-\rho, H(xk)\rangle} dk = \int_K e^{\langle -i\lambda-\rho, H(x^{-1}k)\rangle} dk ,$$

and $\varphi_{\lambda_1} = \varphi_{\lambda_2}$ if and only if $\lambda_2 \in W \cdot \lambda_1$.

The uniqueness of $\mathbb{C}v_o$ for $\pi \in G_K^\wedge$ implies that m_π is one on the support of the Plancherel measure.

The canonical normalization is determined by:

$$f \to f^\sim(\lambda) = \mathrm{Tr}(\pi_\lambda(f)) = \langle \pi_\lambda(f)\, v_o^\lambda, v_o^\lambda \rangle$$

$$= \int_G f(x)\varphi_\lambda(x)\,dx, \quad f \in C_c^\infty(G/K), \ \lambda \in \mathfrak{a}_{\mathbb{R}}^* .$$

One formulation of the Plancherel formula is then:

$$(2.4) \qquad f(e) = \frac{1}{|W|} \int_{\mathfrak{a}_{\mathbb{R}}^*} f^\sim(\lambda) \ |c(\lambda)|^{-2} d\lambda \ , f \in C_c^\infty(G/K) \ ,$$

where $c(\lambda)$ is the so-called c-function related to the asymptotic behaviour of φ_λ and explicitly known. See Helgason [21] for more details. □

For the non-compact group $G_1 \simeq G_1 \times G_1/d(G_1)$ of Example 1.1 the Plancherel formula is also known explicitly by the work of Harish-Chandra [13].

For the hyperboloids $X_{p,q}(\mathbb{F})$ of Example 1.3 Faraut [5] proved a Plancherel formula. See also Strichartz [39] and the references in these two papers for special cases.

In all the above examples the multiplicity m_π is one on the support of $d\mu$, except for the cases $X_{1,q}(\mathbb{R})$, $q>0$, where the multiplicity is two on the continuous spectrum and one on the discrete spectrum. In general there may be higher multiplicities, as we shall discuss later.

In the above examples, in particular in Harish-Chandra's proof of the Plancherel formula for G_1, it is an important step to classify the discrete series, which we define in the context of G/H in the following way.

An irreducible unitary representation π of G is said to belong to the <u>discrete series for G/H</u> if π is equivalent to an irreducible subrepresentation of the regular representation of G in $L^2(G/H)$. This means that precisely the discrete series occur discretely in the Plancherel formula.

<u>Example 2.4</u>.(a). The discrete series for $G/H = G_1 \times G_1/d(G_1)$ is given by the set $\{\pi_1 \otimes \check{\pi}_1 \mid \pi_1$ is a discrete series representation for $G_1\}$.

(b). In the noncompact Riemannian case G/K , there is no discrete series for G/K , cf. (2.4).

(c). For the hyperbolic spaces $X_{p,q}(\mathbb{F})$ there is a non-empty discrete series whenever $q>o$. □

For the non-Riemannian hyperboloids, (i.e. $p \neq 0$ and $q \neq 0$), there does not seem to be any good candidate for a canonical normalization of the Plancherel measure. This would be necessary if one wanted to generalize to G/H the notion of a <u>formal degree</u> of a discrete representation.

For $\pi \in G^\wedge$ to play a role in the Plancherel measure it is necessary that π has a non-trivial <u>H-spherical distribution vector</u>, by which we mean the following.

Let V_∞ be the C^∞-vectors for π , with the standard topology. Let $V_{-\infty}$ be the space of distribution vectors, i.e. the dual space to V_∞ . Define the set of H-spherical distribution vectors by

$$V_{-\infty}^H = \{v \in V_{-\infty} \mid \pi_{-\infty}(h)v = v, \text{ for all } h \in H\} .$$

The set $G_H^\wedge$, for which $V_{-\infty}^H \neq \{0\}$, and the analogue for some non-unitary representations is our main object of study. The examples we have mentioned suggest two methods in an attempt to parametrize (parts of) $G_H^\wedge$:

1. Method: Spherical distributions.

Let $\pi \in G_H^\wedge$, $0 \neq v^* \in V_{-\infty}^H$. Then

$$T_{v^*}: \ f \to \langle v^*, \pi_{-\infty}(f) v^* \rangle , \qquad f \in C_c^\infty(G)$$

is a well-defined distribution on G biinvariant under H, or equivalently an H-invariant element in $\mathcal{D}'(G/H)$.
Since π is irreducible T_{v^*} is an eigendistribution for $Z(\mathfrak{g})$, and of positive type.

We shall call $T \in \mathcal{D}'(G/H)$ an H-spherical distribution if

(i) T is H-invariant

(ii) T is an eigendistribution for each $D \in D(G/H)$.

Notice that we actually put a slightly stronger condition on T than we had on T_{v^*} , since $Z(G/H)$ may be smaller than $D(G/H)$. (However it may be true that $v^* \in V_{-\infty}^H$ always can be chosen such that T_{v^*} is a spherical distribution.)

Example 2.5. (a). In the group case, i.e. $G/H = G_1 \times G_1/d(G_1)$, we have $G_H^\wedge = \{ \pi_1 \otimes \check{\pi}_1 \mid \pi_1 \in G_1^\wedge \}$. For each $\pi = \pi_1 \otimes \check{\pi}_1$ we have that $\dim(V_{-\infty}^H) = 1$. The choice of $v^* \in V_{-\infty}^H$ is normalized by

$$\langle T_{v^*}, f \rangle = \mathrm{Tr}(\pi_1(f))$$

for $f \in C_c^\infty(G/H) \simeq C_c^\infty(G_1)$. This means that the H-spherical distribution related to $\pi_1 \otimes \check{\pi}_1$ is just the distribution character of π_1 . These play a fundamental role in the work of Harish-Chandra.

(b). The noncompact Riemannian case G/K . In this case every K-spherical distribution T is up to a constant an elementary spherical function, i.e.

$$\langle T, f \rangle = \int_G f(x) \varphi_\lambda(x)\, dx , \quad f \in C_c^\infty(G) ,$$

for some $\lambda \in \mathfrak{a}_{\mathbb{C}}^*$.

(c). <u>The hyperbolic spaces</u> $X_{p,q}(\mathbb{F})$, $q \neq o$. The H-spherical distributions form the main object of study in Faraut [5], see also Kosters [24], Kengmana [23] and references there. □

Harish-Chandra showed, cf. [13], that any invariant eigendistribution of $Z(\mathfrak{g}_1)$ on G_1 is a locally integrable function. This is no longer the case for the hyperbolic spaces. One may think that this will make the general study of H-spherical distributions more difficult.

<u>2. Method</u>: <u>Minimal K-types</u>.

Let again $\pi \in \hat{G}_H$ and let V_K be the space of K-finite vectors in V , the Hilbert space of π. Denote by $C_K^\infty(G/H)$ the space of K-finite C^∞ functions on G/H. For any nonzero $v^* \in V_{-\infty}^H$ we may realize the Harish-Chandra module (or $(\mathfrak{g},K)$-module) V_K in $C_K^\infty(G/H)$ by

$$V_K \ni v \quad \to \quad \varphi_v(x) = \langle v, \pi_{-\infty}(x) v^* \rangle \ , \ x \in G.$$

(φ_v is an eigendistribution for an elliptic operator on G/H . Thus φ_v is C^∞).

Let $V_K = \oplus_{\delta \in K^\wedge} V_\delta$ with $V_\delta = m_\delta \cdot E_\delta$ be the decomposition of V_K over $K^\wedge$. We say that δ is a K-type of V if $m_\delta \neq 0$.
Let now $K^\wedge$ be parametrized by the dominant weights, $K^\wedge = \{\delta_\mu \mid \mu \in \Lambda\}$, and let $2\rho_c$ be the sum of the corresponding positive roots. Vogan showed in [40] that if $\mu_0 \in \Lambda$ is minimal in V_K in the sense that

$$\| \mu_0 + 2\rho_c \| = \min\{ \| \mu + 2\rho_c \| \mid m_\mu \neq 0\} ,$$

then $\delta^\sim = \delta_{\mu_0}$ occurs with multiplicity one, i.e. $m_{\mu_0} = 1$.
This means that $U(\mathfrak{g})^{\mathfrak{k}}$ acts on $\delta^\sim$ as a scalar.
Summing up: For any $v^\sim \in V_{\delta^\sim}$ and any $v^* \in V_{-\infty}^H$ the function

$\tilde{\varphi}(x) = \langle \tilde{v}, \pi_{-\infty}(x) v^* \rangle$ belonging to $C_K^\infty(G/H)$ is of K-type $\tilde{\delta}$ and an eigenfunction of $Z(G/H)$. Furthermore the $U(\mathfrak{g}) \times K$-module generated by $\tilde{\varphi}$ in $C_K^\infty(G/H)$ contains $\tilde{\delta}$ with multiplicity one. This implies in particular that $\tilde{\varphi}$ is an eigenfunction for each u in $U(\mathfrak{g})^{\mathfrak{h}}$.

Example 2.6. (a). Let G_1 be as before, and $\pi_1 \in \hat{G}_1$. The above realization (2.5) of the Harish-Chandra module of $\pi_1 \otimes \check{\pi}_1$ in $C^\infty_{K_1 \times K_1}(G_1 \times G_1/d(G_1))$ is the same as the realization of $\pi_1 \otimes \check{\pi}_1$ on the K_1-finite matrix coefficients of π_1. Let δ_1 be a minimal K_1-type in π_1. Then $\tilde{\delta} = \delta_1 \otimes \check{\delta}_1$ is a minimal $K_1 \times K_1$-type for $\pi_1 \otimes \check{\pi}_1$. Notice in particular that $\tilde{\delta} \in \hat{K}_{H \cap K}$, i.e. $\tilde{\delta}$ has a non-trivial fixed vector for $d(K_1)$, the diagonal in $K_1 \times K_1$.

(b). Let G/K be a Riemannian symmetric space and let $\pi \in \hat{G}_K$. The minimal K-type of π is the trivial representation δ_o of K.

This example suggests the following

Conjecture 2.7. Let $\pi \in \hat{G}_H$. Then there exists a minimal K-type of π which belongs to $\hat{K}_{K \cap H}$.

Definition 2.8. A function $\varphi \in C_K^\infty(G/H)$ is called a minimal spherical function if

(i) $\varphi(e) = 1$, $\varphi(kx) = \varphi(x)$ for all $k \in K \cap H, x \in G$, and φ is of some irreducible K-type δ, $\delta \in \hat{K}_{K \cap H}$.

(ii) φ is a joint eigenfunction for $D(G/H)$, (right action).

(iii) φ is a joint eigenfunction for $U(\mathfrak{g})^{\mathfrak{h}}$ (left action).

(iv) The $U(\mathfrak{g}) \times K$-module V_K^φ generated by φ in $C_K^\infty(G/H)$ contains δ as one of its minimal K-types.

Notice again that the discussion before the definition suggests a slightly weaker definition, with $Z(G/H)$ instead of $D(G/H)$ and without the requirement that $\delta \in \hat{K}_{K \cap H}$. However we feel that the definition is conceptually the right one, and furthermore as mentioned before $Z(G/H) = D(G/H)$ for all classical groups, and in the group case, i.e. when $G/H \simeq G_1$, as discussed above, every representation $\pi \in G_1^{\wedge}$ (or even every irreducible admissible representation) has for every minimal K-type δ a unique minimal spherical function associated to it.

We are now ready to state the problem which shall be our main concern in the rest of these lectures:

Problem: Find all minimal spherical functions for G/H. In particular sort out those which are in $L^2(G/H)$.

We would like to give the answer in a form similar to Harish-Chandra's solution of the problem for the Riemannian symmetric spaces of noncompact type., cf. Harish-Chandra's integral formula (2.3). To motivate our approach, let us look at the compact spaces.

Example 2.7. The symmetric spaces of compact type.

Let G^*/K be such a space and $\mathfrak{g}^* = \mathfrak{k} + i\mathfrak{p}$. Then $\mathfrak{g} = \mathfrak{k} + \mathfrak{p}$ is another real form of $\mathfrak{g}_{\mathbb{C}}$. Let G be the analytic subgroup of $G_{\mathbb{C}}$ with Lie algebra $\mathfrak{g}$. Then G/K is a Riemannian symmetric space of noncompact type. This is the classical duality between G/K and G^*/K used already by Cartan in his classification.

Following essentially Helgason [19] and [16] we know that any non-zero eigenfunction of $D(G^*/K)$ on G^*/K generates an irreducible (finite-dimensional) subrepresentation V_μ of $L^2(G^*/K)$, with some dominant weight μ. V_μ contains a non-zero K-fixed vector, which normalized is nothing else than an elementary spherical function φ^*. Since any finite-dimensional representation of G^* has a holomorphic extension to $G_{\mathbb{C}}$, φ^* has an analytic continuation $\varphi^*_{\mathbb{C}}$ to all of $G_{\mathbb{C}}$. Furthermore φ^* is uniquely determined by the restriction φ of $\varphi^*_{\mathbb{C}}$ to G. Now it is obvious that φ is an elementary spherical function on G/K, which is moreover G-finite.

Theorem 2.8. (Helgason) The set of elementary spherical functions on G^*/K is via analytic continuation the same as the set of G-finite elementary spherical functions on G/K. These are given by Harish-Chandra's formula

$$(2.6) \qquad \varphi_\lambda(x) = \int_K e^{\langle -i\lambda-\rho, H(x^{-1}k)\rangle} dk = \int_K e^{\langle i\lambda-\rho, H(xk)\rangle} dk \ , x \in G \ ,$$

where $\lambda \in \mathfrak{a}^*_{\mathbb{C}}$ and $\mu = i\lambda - \rho$ satisfies

$$\frac{\langle \mu, \alpha\rangle}{\langle \alpha, \alpha\rangle} \in \mathbb{Z}^+ \text{ for all } \alpha \in \Delta^+ .$$

One should notice that if $\operatorname{rank}(G) \neq \operatorname{rank}(G/K)$ then there is a slight abuse of notation in that "μ" in Theorem 2.8 is a linear form on $\mathfrak{a}_{\mathbb{C}}$ where as the dominant weight "μ" of the corresponding irreducible representation is a linear form on a full Cartan subalgebra containing $\mathfrak{a}_{\mathbb{C}}$. The dominant weight is obtained from μ of Theorem 2.8 by extending it to be zero on the orthogonal complement of $\mathfrak{a}_{\mathbb{C}}$. It now follows immediately that Theorem 2.8 precisely gives all the minimal spherical functions for G^*/K. One just has to notice that in this case G^* is in itself compact. Thus K in the Definition 2.8 is equal to G^*. □

The aim of the next section is to define a notion of duality for K-finite functions on G/H generalizing and combining the ideas involved in (2.3) and (2.6).

§ 3. The Duality principle.

Let again G/H be a general semisimple symmetric space. With the notation from (1.4) we define inside $\mathfrak{g}_{\mathbb{C}}$ the real form $\mathfrak{g}^o$ by

$$\mathfrak{g}^o = \mathfrak{h}\cap\mathfrak{k}+i(\mathfrak{h}\cap\mathfrak{p})+i(\mathfrak{q}\cap\mathfrak{k})+\mathfrak{q}\cap\mathfrak{p}, \tag{3.1}$$

where $i = \sqrt{-1}$.

Also define $\mathfrak{h}^o$, $\mathfrak{k}^o$ and $\mathfrak{g}_o$ by

$$\mathfrak{g}_o = \mathfrak{g}^o\cap\mathfrak{g} = \mathfrak{h}\cap\mathfrak{k}+\mathfrak{q}\cap\mathfrak{p}$$

$$\mathfrak{h}^o = \mathfrak{h}_{\mathbb{C}}\cap\mathfrak{g}^o = \mathfrak{h}\cap\mathfrak{k}+i(\mathfrak{h}\cap\mathfrak{p})$$

$$\mathfrak{k}^o = \mathfrak{k}_{\mathbb{C}}\cap\mathfrak{g}^o = \mathfrak{h}\cap\mathfrak{k}+i(\mathfrak{q}\cap\mathfrak{k}) .$$

Let G^o, G_o , H^o and K^o be the corresponding analytic subgroups of $G_{\mathbb{C}}$.

Notice that $G_o \subset G\cap G^o$, that $K\cap H = K^o\cap H^o$ is connected and that $(H/H\cap K\ ,\ H^o/H\cap K)$ and $(K/K\cap H\ ,\ K^o/K\cap H)$ are dual pairs in the classical sense cf. Example 2.6 (disregarding the fact that K and H may be only reductive and not semisimple). The involutions τ and σ extend holomorphically to $G_{\mathbb{C}}$ and thus in particular to G^o . This means that (G,H,τ,K,σ) and $(G^o,K^o,\sigma,H^o,\tau)$ satisfy the same general conditions, the first related to the symmetric space G/H , the second related to the symmetric space G^o/K^o . It is this pairing we call the duality. We prefer to think of it in the following way: <u>The dual space for the symmetric space $X = G/H$ of non-Riemannian type is the Riemannian symmetric space $X^o = G^o/H^o$ with the associated (noncompact) symmetric subgroup K^o .</u>

<u>Example 3.1</u>.(a).Let G_1 be a noncompact real form of a simply connected complex semisimple Lie group $G_{1\mathbb{C}}$, and let U_1 be a suitably chosen compact real form. Then $K_1 = G_1\cap U_1$ is maximal compact in G_1 . We identify as usual G_1 with the symmetric space

$G/H = G_1 \times G_1/d(G_1)$ and choose $K = K_1 \times K_1$. The dual $(G^o/H^o, K^o)$ of G/H is in this case isomorphic to $(G_{1\mathbb{C}}/U_1, K_{1\mathbb{C}})$.

(b). In the degenerate cases (II), i.e. $H = K$, the dual of G/K is again $(G/K,K)$.

(c) Let G^*/H be a symmetric space of compact type, cf. Example 2.6, i.e. the degenerate case (I) with $K = G^*$. Then the dual is $(G/H, G)$, where G^*/H is just the classical dual of G/H.

(d). The duals of the hyperbolic spaces $X_{p,q}(\mathbb{F})$ are the following, cf. Example 1.3:

$$X_{p,q}(\mathbb{R})^o = (X_{p+q,o}(\mathbb{R}),\ SO(p)\times SO_e(q,1)),$$

$$X_{p,q}(\mathbb{C})^o = (X_{p+q,o}(\mathbb{C}),\ S(U(p)\times U(q,1))),$$

$$X_{p,q}(\mathbb{H})^o = (X_{p+q,o}(\mathbb{H}),\ Sp(p)\times Sp(q,1)).$$

(e). Let $X = G/H = SL(n,\mathbb{R})/SO_e(p,q)$ with $p+q = n$. Then the dual is $(SL(n,\mathbb{R})/SO(n),\ SO_e(p,q))$. □

Since, cf. Mostow [28] and Loos [26], G is diffeomorphic to $K \times \mathfrak{q}\cap\mathfrak{p} \times \mathfrak{h}\cap\mathfrak{p}$ via the mapping

$$(k,X,Y) \to k\exp X \exp Y,$$

one may consider G/H as the principal fiberbundle $(K/K\cap H, \mathfrak{q}\cap\mathfrak{p})$, where $K\cap H$ acts on $\mathfrak{q}\cap\mathfrak{p}$ by the adjoint action or in other words

$$G/H \simeq K\times_{K\cap H} G_o/K\cap H, \tag{3.2}$$

where $(k, yK\cap H) \sim (k_1, y_1K\cap H)$ iff there exists a $k_2 \in K\cap H$ such that $K_1 = kk_2$ and $y_1 \in k_2^{-1} yK\cap H$. From this it follows easily that

$$G = KG_oH \quad \text{and} \quad G = K\bar{B}^+H, \tag{3.3}$$

where $\mathfrak{b}$ is maximal Abelian in $\mathfrak{q}\cap\mathfrak{p}$, $\bar{\mathfrak{b}}^+$ is a closed positive Weyl chamber for $(\mathfrak{b}, \mathfrak{g}_o)$ and $\bar{B}^+ = \exp(\bar{\mathfrak{b}}^+)$. For the dual $(G^o/H^o, K^o)$ (3.3) takes the form

(3.4) $G^o = K^o G_o H^o$ and $G^o = K^o \overline{B}^+ H^o$.

Now generalizing the arguments using analytic continuation in Example 2.6 we get, cf. [8].

<u>Theorem 3.2</u>. <u>The duality principle</u>

There is a unique isomorphism $f \to f^o$ of $C^\infty_{K^\sim}(G^\sim/H)$ onto $C^\infty_{K^o}(G^o/H^o)$ considered as $U(\mathfrak{g})\times U(\mathfrak{g})^{\mathfrak{h}}$ -[(left, right) - action] - modules such that

$$f(x) = f^o(x) \text{ , for all } f\in C^\infty_{K^\sim}(G^\sim/H) \text{ , } x\in G_o \text{ .}$$

Using the duality principle we may reformulate our problem of finding the minimal spherical functions for G/H in the following way.

<u>Find all solutions</u> $\varphi^o \in C^\infty_{K^o}(G^o/H^o)$ to

(i) $\varphi^o(e) = 1$, $\varphi^o(kx) = \varphi^o(x)$ for all $k\in K\cap H$, $x\in G$ and φ^o is of some irreducible finite-dimensional K^o-type δ .

(ii) φ^o is a joint eigenfunction for $D(G^o/H^o)$,(right action).

(iii) φ^o is a joint eigenfunction for $U(\mathfrak{g})^{\mathfrak{k}}$, (left action).

(iv) The $U(\mathfrak{g})\times K^o$-module generated by φ^o in $C^\infty_{K^o}(G^o/H^o)$ contains δ as a minimal K^o-type.

In particular we are interested in finding all solution φ^o to (3.5) such that the dual function φ is contained in $L^2(G/H)$. From the formula for the Haar measures related to (3.3) and (3.4), cf [8, § 2] , this may be detected by studying the asymptotic behaviour of

$$\varphi^o \quad \text{on} \quad \overline{B^+}.$$

In the next section we shall discuss how to attack problem (3.5) using the Poisson transform on the Riemannian symmetric space G^o/H^o . We close this section by giving a very simple application of the duality principle.

Example 3.3. Elementary Spherical functions.

(This example is a very short account of some ideas and results from [7]). In this example we assume for a certain technical reason that G is a classical group, (cf. condition (5.10) in [7]).

Theorem 3.4. Let $1 \leq p \leq \infty$. The duality $f \to f^{o}$ defines a norm-isomorphism of $L^p(K;G/H)$, (the K-invariant L^p-functions on G/H), onto $L^p(K^o \backslash G^o; H^o)$, (the H^o-invariant L^p-functions on $K^o \backslash G^o$). This follows from the duality principle and inspection of the decomposition of Haar measures related to (3.3) and (3.4).

Corollary 3.5. (G a classical group.) The restriction of the Plancherel measure for $L^2(G/H)$ to $\hat{G}_K \cap \hat{G}_H$, (i.e. the K-spherical representations of G contained in $\hat{G}_H$) is the same as the restriction of the Plancherel measure for $L^2(K^o \backslash G^o)$ to $\hat{G}_{H^o} \cap \hat{G}_{K^o}$, (i.e the H^o-spherical representations of G^o contained in $\hat{G}_{K^o}$).

Notice that this gives a pairing of certain positive-definite elementary spherical functions on G w.r.t. K to certain positive-definite elementary spherical functions on G^o, w.r.t. H^o.

Conjecture. The duality defines a bijection between $\hat{G}_K \cap \hat{G}_H$ and $\hat{G^o}_{K^o} \cap \hat{G^o}_{H^o}$.

If this conjecture holds, then in particular the problem of classifying the positive-definite elementary spherical functions on G_1/K_1 would be the same as that of classifying the positive-definite elementary spherical functions on $G_{1\mathbb{C}}/U_1$, for which a $K_{1\mathbb{C}}$-spherical distribution exists. For the case of a split group G_1, f.ex. $G_1 = SL(n,\mathbb{R})$,this conjecture must in some way be related to the so-called base-change, relating representations of G_1 to representations of $G_{1\mathbb{C}}$, see Clozel [2].

Let now G_1 be non-compact and as usual $G_1 \simeq G/H = G_1 \times G_1/d(G_1)$ with $K = K_1 \times K_1$. Let φ_λ be an elementary spherical function on G_1, w.r.t. K_1 . Then $\varphi_\lambda^\sim$ defined by

$$\varphi_\lambda^\sim(x,y) = \varphi_\lambda(xy^{-1})$$

is contained in $C^\infty(K;G/H)$. Thus the dual function $\varphi_\lambda^o = (\varphi_\lambda^\sim)^o$ belongs to $C^\infty(K_{1\mathbb{C}} \backslash G_{1\mathbb{C}};U_1)$cf. Example 3.1(a), and Theorem 3.2. Now let $\mathfrak{g}_1 = \mathfrak{k}_1 + \mathfrak{a}_1 + \mathfrak{n}_1$ and extend $\mathfrak{a}_1$ to a Cartan subalgebra $\mathfrak{a}_{\mathbb{C}}$ of $\mathfrak{g}_{1\mathbb{C}}$. Let $\mathfrak{a} = \mathfrak{a}_{\mathbb{C}} \cap i\mathfrak{u}_1$. Choose compatible orderings of the roots, and let ρ_m be half the sum of the positive roots vanishing on $\mathfrak{a}_1$.

<u>Theorem.3.6</u> The function Φ_Λ defined by

$$(3.6) \qquad \Phi_\Lambda(x) = \int_{U_1} \varphi_\lambda^o(ux)\,du \ , \ x \in G_{1\mathbb{C}}$$

defines an elementary spherical function on $G_{1\mathbb{C}}$ w.r.t. U_1 with Harish-Chandra parameter $\Lambda = 2(\lambda - i\rho_m)$.

<u>Theorem 3.7</u> Let $\lambda \in \mathfrak{a}_{1\mathbb{C}}^*$. Then φ_λ is bounded if and only if Φ_Λ is bounded, where $\Lambda = 2(\lambda - i\rho_m)$.

If the above conjecture is true, then that means in this case that "bounded" may be exchanged with "positive-definite" in Theorem 3.7.

It is also possible to relate the spherical Fourier transform on G_1/K_1 to that on $G_{1\mathbb{C}}/U_1$. The spherical Fourier transform is much simpler on $G_{1\mathbb{C}}/U_1$, since the elementary spherical functions are given essentially by Weyl's character formula and the Plancherel measure by Weyl's dimension formula.

When G_1 is a split group it is possible to reduce the proof of the Plancherel formula, the inversion formula and the Paley-Wiener theorem to that of $G_{1\mathbb{C}}/U_1$. For example the Plancherel measure comes out to be given by:

$$(3.7) \qquad d\mu(\lambda) = \text{const.}\cdot(\pi(2\lambda)\int_{K_{1\mathbb{C}}} \Phi_{2\lambda}(y)\,dy)\,d\lambda \quad \text{over} \quad \mathfrak{a}^*_{\mathbb{R}}\,,$$

where $\pi(\Lambda)d\Lambda$ is the Plancherel measure for $G_{1\mathbb{C}}/U_1$.
In particular since we know that $d\mu(\lambda) = |c(\lambda)|^{-2}d\lambda$, the formula (3.7) is an expression for the "real" c-function in terms of the "complex"-spherical Fourier transform. □

Construction of H-finite eigenfunctions of $D(G/K)$
by means of the Poisson transform.

§ 4. The Poisson transform and Helgason's conjecture.

In this section we shall exchange the roles of (G,H,τ,K,σ) and $(G^o,K^o,\sigma,H^o,\tau)$. We do this because we want to study the problem (3.5) by means of known techniques for the Riemannian symmetric space G^o/H^o. So for the notation to be more familiar we name the space G/K.

So to recapitulate our problem in this notation: G is a non-compact real form of a complex simply connected semisimple Lie group; K is a maximal compact subgroup of G; τ is an involution of G such that $\tau(K) = K$; H is the connected component of the identity of the fixed points for τ. We look for functions $\psi \in C^\infty(G/K)$ satisfying

(i) $\psi(e) = 1$, $\psi(kx) = \psi(x)$ for all $k \in K \cap H, x \in G$ and ψ is of some irreducible finite-dimensional H-type δ.

(4.1) (ii)-(iii) ψ is an eigenfunction of $U(\mathfrak{g})^{\mathfrak{h}}$ by the left action and of $D(G/K)$ (by the right action).

(iv) The $U(\mathfrak{g}) \times H$-module in $C^\infty_H(G/K)$ generated by ψ contains δ as a minimal H-type.

We are particularly interested in finding solutions for which the dual function ψ^o belongs to $L^2(G^o/K^o)$.

Example 4.1. Let $G = G_{1\mathbb{C}}, K = U_1$ and $H = K_{1\mathbb{C}}$ in our usual notation for G_1, K_1, U_1. Then $G^o/K^o = G_1 \times G_1/d(G_1) \simeq G_1$ and $H^o = K_1 \times K_1$. □

We shall briefly recall a few properties of the Poisson transform for G/K, see Helgason [16] or [21].

Let $G = KAN$ be an Iwasawa decomposition of G, and let the notation be as in Example 2.2. Let M be the centralizer of A in K. Then $P = MAN$ is a minimal parabolic subgroup. The coset space

$K/M \simeq G/P$ is called the boundary of G/K. Let for $\lambda \in \mathfrak{a}^*_{\mathbb{C}}$ $\chi_\lambda : D(G/K) \to \mathbb{C}$ give the eigenvalues for the elementary spherical function $\varphi_\lambda(x) = \int_{K/M} e^{\langle -i\lambda-\rho, H(x^{-1}k)\rangle} dk$.

It is known, cf. Helgason [16, page 94], that $x \to e^{\langle -i\lambda-\rho, H(x^{-1}k)\rangle}$ is itself an eigenfunction with eigenvalue $\chi_\lambda(D)$ for each $D \in D(G/K)$ and $k \in K$. Therefore given any function, distribution or hyperfunction (i.e. analytic functional) T on K/M it makes sense to form

$$(4.2) \qquad \psi(x) = \int_{K/M} e^{\langle -i\lambda-\rho, H(x^{-1}k)\rangle} T(k)\, dk ,$$

and ψ will be an eigenfunction of $D(G/K)$, with eigenvalues χ_λ. We shall call $\psi(\cdot)$ the <u>Poisson transform</u> $P_\lambda(T,\cdot)$ of T. This may also be thought of more in the spirit of the spherical principal series for G in the following way.

Let for $\lambda \in \mathfrak{a}^*_{\mathbb{C}}$

$$\mathcal{B}(G/P, L_\lambda) = \left\{ f \in \mathcal{B}(G) \;\middle|\; \begin{array}{l} f(gman) = e^{\langle i\lambda-\rho, H(a)\rangle} f(g) \\ \forall g \in G,\ \forall man \in P \end{array} \right\}$$

where $\mathcal{B}$ stands for hyperfunctions, and we use notation as if f was an ordinary function. Notice that $f \in \mathcal{B}(G/P, L_\lambda)$ is determined by its restriction to K/M. Actually in this way $\mathcal{B}(G/P, L_\lambda)$ is isomorphic to $\mathcal{B}(K/M)$. Since K/M is a compact analytic manifold, the hyperfunctions on K/M are the same as the analytic functionals on K/M.

Let $\mathcal{D}'_\lambda = \mathcal{D}'(G/P, L_\lambda)$ be the subspace of $\mathcal{B}(G/P, L_\lambda)$ consisting of distributions. Notice that the standard model for the spherical principal series is obtained by considering the Hilbert subspace $\mathcal{H}_\lambda$ of $\mathcal{D}'_\lambda$ consisting of L^2-functions on K/M.

All these spaces $\mathcal{B}(G/P,L_\lambda)$, $\mathcal{D}'_\lambda$ and $\mathcal{H}_\lambda$ are invariant under the action of G , and thus define representations of G . They all contain a K-invariant element, namely the constant function f_o on K/M . Now in the spirit of (2.3) and (2.6) we may obtain representations of G in $C^\infty(G/K)$ by the intertwining operator

$$f \to P_\lambda(f,\cdot) \text{ , where}$$

$$P_\lambda(f,x) = \langle f,\pi(x)f_o\rangle = \int_K f(xk)dk \text{ ,}$$

which is also equal to, cf. (2.3),

$$P_\lambda(f,x) = \int_K e^{\langle -i\lambda-\rho,H(x^{-1}k)\rangle} f(k)dk \text{ ,}$$

only using the restriction of f to K/M .

From this last formula and our previous discussion we have that $P_\lambda(f,\cdot)$ belongs to the space

$$\mathcal{E}_\lambda(G/K) = \{f\in C^\infty(G/K) \mid Df = \chi_\lambda(D)f \text{ , } D\in D(G/K)\}.$$

<u>Example 4.2</u> . Notice that Harish-Chandra's integral formula for the elementary spherical function φ_λ is just $\varphi_\lambda(x) = P_\lambda(f_o,x)$. □

The following theorem was proved by Helgason in special cases, cf.[18], and is known under the name "<u>Helgason's Conjecture</u>":

Let $\mathfrak{a}^*_+ = \{\lambda\in \mathfrak{a}^*_{\mathbb{C}} \mid \mathrm{Re}\langle i\lambda,\alpha\rangle \geq 0, \forall\alpha\in\Delta^+\}$. Note that $W\cdot \mathfrak{a}^*_+ = \mathfrak{a}^*_{\mathbb{C}}$.

<u>Theorem 4.3</u>. (Kashiwara, Kowata, Minemura, Okamoto, Oshima, Tanaka).

Let $\lambda\in \mathfrak{a}^*_+$ (or just $\frac{2\langle\lambda,\alpha\rangle}{\langle\alpha,\alpha\rangle}\notin\{-1,-2,\cdots\}$ for every $\alpha\in\Delta^+$). Then P_λ is a G-equivariant isomorphism of $\mathcal{B}(G/P,L_\lambda)$ onto $\mathcal{E}_\lambda(G/K)$.

In Oshima, Sekiguchi [33] it is shown that a certain natural growth condition on functions on G/K characterize the image of $\mathcal{D}'_\lambda$. The inverse mapping is called the normalized boundary value map:

$$\beta^\sim_\lambda = \mathcal{E}_\lambda(G/K) \to \mathcal{B}(G/P, L_\lambda) .$$

(The boundary value map β_λ is defined for almost all λ, and if $c(\lambda)$ is Harish-Chandra's c-function, then

$$\beta_\lambda \circ P_\lambda(f) = c(\lambda)f \quad \text{and} \quad \beta^\sim_\lambda = c(\lambda)^{-1}\beta_\lambda ,$$

whenever well-defined.)

Let now φ belong to $\mathcal{E}_\lambda(G/K) \cap C^\infty_H(G/H)$ and assume that λ is such, that $\beta^\sim_\lambda$ is well defined. Then we define the <u>λ-support of φ on $G/P = K/M$</u> to be:

$$\operatorname{supp}_\lambda(\varphi) = \operatorname{supp}(\beta^\sim_\lambda(\varphi)) .$$

Notice that $\operatorname{supp}_\lambda(\varphi)$ is $U(\mathfrak{g})$-stable and also H-stable. This means that <u>$\operatorname{supp}_\lambda(\varphi)$ is a union of H-orbits in G/P</u>.

<u>Remark</u>. I think it can be proved using [33], that

$$\beta^\sim_\lambda(\mathcal{E}_\lambda(G/K) \cap C^\infty_H(G/K)) \subset \mathcal{D}'_\lambda . \qquad \square$$

Matsuki [27] has, generalizing the Bruhat decomposition for the case $G_1 \times G_1/d(G_1)$ and previous works of Aomoto and Wolf, classified all orbits of H in G/P. In particular he shows that there is a finite number of orbits and he describes the open and the closed orbits. See also Rossmann [34].

Let now $\mathcal{O} \subset G/P$ be an H-orbit. Let $\mathcal{D}'_{\lambda,H}(\mathcal{O})$ denote the H-finite distributions in $\mathcal{D}'_\lambda$ with support contained in $\overline{\mathcal{O}}$, the closure of $\mathcal{O}$. Clearly $\mathcal{D}'_{\lambda,H}(\mathcal{O})$ is an $U(\mathfrak{g}) \times H$-module. For the harmonic analysis of G^o/K^o, we feel that, recalling the duality, the study of these modules is very important.

<u>Example 4.4</u>. $G/K = SL(2,C)/SU(2)$, $H = SO(2,\mathbb{C})$. Notice that $G^o/K^o = SL(2,\mathbb{R})\times SL(2,\mathbb{R})/d(SL(2,\mathbb{R})) \simeq SL(2,\mathbb{R})$.

G/P is the 2-sphere S^2 and there are 3 H-orbits: $\mathcal{O}_1$ the "north pole", $\mathcal{O}_2$, the "south pole", and $\mathcal{O}_3$, the rest. A simple calculation shows that if $\delta_i, i = 1,2$ is the Dirac measure at $\mathcal{O}_i$, then $\overset{o}{\psi}_{\bar{\lambda},i}$, the dual functions to $\psi_{\lambda,i} = P_\lambda(\delta_i)$, for $i = 1,2$ and $i\lambda>0$ provide the matrix coefficients of the minimal $SO(2)$-types of all the (relative) discrete series representations for the universal covering group $SL(2,\mathbb{R})^\sim$ of $SL(2,\mathbb{R})$. In a similar fashion $\mathcal{O}_3$ is related to the minimal $SO(2)$-types of the principal series for $SL(2,\mathbb{R})^\sim$. □

In the next sections we are going to construct solutions to our problem (4.1) by means of suitably chosen distributions in $\mathcal{D}'_{\lambda,H}(\mathcal{O})$. First, in Section 5, for the closed orbits we show that G^o/K^o has discrete series if it has a compact Cartan subspace, i.e. if rank (G^o/K^o) = rank $(H^o/H\cap K)$. Then, in Section 6, we give a general construction of solutions to (4.1) (i)-(iii) involving a large class (but not all) of the orbits. We do not prove property (4.1)(iv) for these functions, but the formula for the H-type is very similar to Knapp's formula for the minimal K-types [25]. In Section 7 we shall, using the open orbits, indicate why there may be multiplicity in the Plancherel formula.

§ 5. Existence of discrete series - The closed orbits.

For Example 4.1 Harish-Chandra showed [12] that G_1 has discrete series if and only if $\operatorname{rank}(G_1) = \operatorname{rank}(K_1)$. Translated to our notation for the symmetric space $G^o/K^o \simeq G_1$ this means

$$(5.1) \qquad \operatorname{rank}(G^o/K^o) = \operatorname{rank}(H^o/K\cap H) .$$

In this section we assume this condition satisfied. Therefore it is possible to choose a Cartan subspace $\mathfrak{a}$ for G/K such that $\mathfrak{a}\subset\mathfrak{h}\cap\mathfrak{p}$. We could say that $i\mathfrak{a}$ is a compact Cartan subspace for G^o/K^o. Fix a positive system Δ^+_c, for the (restricted) root system $\Delta_c = \Delta(\mathfrak{a},\mathfrak{h})$, (the "compact" roots). Let $\Delta = \Delta(\mathfrak{a},\mathfrak{g})$. Matsuki [27] shows in this case that the closed H-orbits in G/P are parametrized by the possible choices of Δ^+, which are compatible with Δ^+_c, i.e. such that $\Delta^+_c\subset\Delta^+$.

Fix such an orbit $\mathcal{O}$, i.e. orderings $\Delta^+_c\subset\Delta^+$. Let $G = KAN$ and $H = (K\cap H)A(N\cap H)$ be the corresponding Iwasawa decompositions of G and H, and $P = MAN$ and $P_c = P\cap H = M_cAN_c$. Then $\mathcal{O}$ may be described by $\mathcal{O} = HP/P\subset G/P$ and clearly $\mathcal{O}$ is isomorphic to H/P_c and again to $K\cap H/M_c$.

We look at the Poisson transform of the $K\cap H$-invariant measure on $\mathcal{O}$:

$$(5.2) \qquad \psi_\lambda(x) = \psi_\lambda^{\mathcal{O}}(x) = \int_{K\cap H} e^{\langle -i\lambda-\rho,H(x^{-1}k)\rangle}dk , \quad x\in G , \ \lambda\in\mathfrak{a}^*_{\mathbb{C}} .$$

Theorem 5.1. Let $\lambda\in\mathfrak{a}^*_{\mathbb{C}}$ and let ψ_λ be defined by (5.2). Then

(i) ψ_λ is an eigenfunction of $D(G/K)$, with eigenvalues χ_λ.

(ii) For every $u\in U(\mathfrak{g})^{\mathfrak{h}}$ there exists $u_o\in U(\mathfrak{a})$ such that $u - u_o\in U(\mathfrak{g})(\mathfrak{m}+\mathfrak{n})_{\mathbb{C}}$ and $u\psi_\lambda = u_o(-i\lambda-\rho)\psi_\lambda$.

(iii) If $y\in H$, $x\in G$ and $\mu_\lambda = i\lambda+\rho-2\rho_c$, then

$$\psi_\lambda(y^{-1}x) = \int_{K\cap H} e^{<\mu_\lambda,\, H(yk)>} e^{<-i\lambda-\rho, H(x^{-1}k)>} dk .$$

In particular ψ_λ is H-finite of irreducible H-type $\check{\mu}_\lambda$ if

(5.3) $\frac{<\mu_\lambda,\alpha>}{<\alpha,\alpha>} \in \mathbb{Z}^+$, for all $\lambda \in \Delta_c^+$.

The proof of this theorem is rather easy, see [8].

Theorem 5.2. Assume λ satisfies (5.3) and that

(5.4) $<i\lambda,\alpha> > 0$ for all $\alpha \in \Delta^+$.

Then ψ_λ^o belongs to $L^2(G^o/K^o)$, and thus generates a discrete series representation for G^o/K^o .

Proof: In [8] it was proved by a rather easy estimate of $e^{<-i\lambda-\rho,H(x^{-1}k)>}$ that ψ_λ^o is bounded provided $<i\lambda+\rho,\alpha> \geq 0$ for all $\alpha \in \Delta^+$ and that it belongs to $L^2(G^o/K^o)$ provided $<i\lambda,\alpha> > C$, for all $\alpha \in \Delta^+$, for a certain constant $C \geq 0$. Oshima proved using the methods that went into the proof of Helgason's Conjecture, that actually $C = 0$. For this see the announcement [32] and the notes by Schlichtkrull [37]. □

Remarks (a). For $G_1 = G_1 \times G_1/d(G_1)$ this gives the full discrete series for G_1 , as was already established in [8].

(b). For the hyperbolic spaces $X_{p,q}(\mathbb{F})$, when $p \neq 0$ and q is somewhat bigger than p , a finite number of discrete series representations are missing. We shall return briefly to this in Section 8.

(c). Oshima and Matsuki have announced that they can prove that the rank condition (5.1) is also necessary for the existence of a discrete series for G^o/K^o , see [32].

(d). Schlichtkrull proves in [35] that all the representations constructed in Theorem 5.2 have $\check{\mu}_\lambda$ as the unique minimal H^o-type.

Also he finds the Langlands parameters for most of these representations. An important feature used in his proof is the explicit knowledge of the action of $U(\mathfrak{g})^{\mathfrak{h}}$ on the H^o-type $\check{\mu}_\lambda$.

(e). Schlichtkrull generalizes in [36] Theorems 5.1 and 5.2 to deal with $L^2(G^o/K^o;\pi)$, where π is a finite dimensional unitary representation of K^o. For example if

$$G^o/K^o = SO_e(p+m,q)/SO(m)\times SO_e(p,q) \quad , \; H^o = SO(p+m)\times SO(q),$$

then any $\pi \in SO(m)^{\wedge}\otimes 1$ can be used. It is shown by examples that Theorem 5.2 and its generalization to $L^2(G^o/K^o;\pi)$ exhibit <u>new unitary representations</u>. □

§ 6. The general orbits.

Any orbit of H in G/P is determined by a τ-invariant Cartan subspace $\mathfrak{a}$ for G/K and an ordering of the roots $\Delta = \Delta(\mathfrak{a},\mathfrak{g})$. So let $\mathfrak{a} = \mathfrak{t}+\mathfrak{b} \in \mathfrak{h}\cap\mathfrak{p}+\mathfrak{q}\cap\mathfrak{p}$ be any fixed τ-invariant Cartan subspace. We are able to construct H-finite elements in $\mathcal{D}'_\lambda$ for suitable λ's whenever the ordering is such that

$$(6.1) \qquad \forall \alpha\in\Delta,\ \alpha_{|\mathfrak{t}} > 0 \ \Rightarrow\ \alpha\in\Delta^+ ,$$

i.e. a root is positive whenever its restriction to $\mathfrak{t}$ is positive.

Let in the rest of this section $\mathcal{O}$ be a fixed orbit determined such that (6.1) holds. We extend $\mathfrak{t}$ to a "fundamental" Cartan subspace in the following way. First extend $\mathfrak{t}$ to a maximal Abelian subspace $\mathfrak{t}^\sim = \mathfrak{t} + \mathfrak{t}'$ in $\mathfrak{h}\cap\mathfrak{p}$ and then to a Cartan subspace $\mathfrak{a}^\sim = \mathfrak{t}^\sim + \mathfrak{b}'$, where $\mathfrak{b}'\subset\mathfrak{b}$.

Let $\mathfrak{m}_\mathfrak{b}$ and $\mathfrak{l}$ be defined by the centralizers of $\mathfrak{b}$ and $\mathfrak{t}$

$$\mathfrak{g}^\mathfrak{b} = \mathfrak{m}_\mathfrak{b} + \mathfrak{b} \text{ and } \mathfrak{g}^\mathfrak{t} = \mathfrak{l} + \mathfrak{t} ,$$

where the sum is direct and $\mathfrak{m}_\mathfrak{b}$ and $\mathfrak{l}$ are σ and τ invariant. Let $M_\mathfrak{b}$ and L be the analytic subgroups of G correspondingly. We may think of

$$(M_\mathfrak{b}\cap H\ ,\ M_\mathfrak{b}/M_\mathfrak{b}\cap K) \quad\text{and}\quad (L\cap H\ ,\ L/L\cap K) \quad\text{as}$$

the dual spaces for certain "symmetric subspaces"

$$M_\mathfrak{b}^o/M_\mathfrak{b}^o\cap K^o \quad\text{and}\quad L^o/L^o\cap K^o \quad\text{of}\quad G^o/K^o ,$$

where the first has $i\mathfrak{t}$ as a compact Cartan subspace and the second has $\mathfrak{b}$ as a purely non-compact Cartan subspace.

Example 6.1. Let G^o/K^o be a group G_1, i.e. $G^o/K^o = G_1\times G_1/d(G_1)$. Let $\mathfrak{a}_1 = \mathfrak{t}_1+\mathfrak{b}_1$ be a Cartan subalgebra of $\mathfrak{g}_1$. Then

$M^o_{\mathfrak{b}}/M^o_{\mathfrak{b}} \cap K^o$ is the connected component of the identity of M_1, where M_1B_1 is the reductive part of a cuspidal parabolic subgroup of G_1, such that $\mathfrak{t}_1$ is a compact Cartan subalgebra for M_1.

Similarly $L^o/L^o \cap K^o$ is a reductive split subgroup of G_1 with the noncompact Cartan subalgebra $\mathfrak{b}_1$. □

The root systems for $\mathfrak{m}_{\mathfrak{b}}$ and $\mathfrak{l}$ may be identified with subsets of Δ:

$$\Delta_{\mathfrak{m}_{\mathfrak{b}}} = \Delta(\mathfrak{t}, \mathfrak{m}_{\mathfrak{b}}) = \{\alpha \in \Delta \mid \alpha|_{\mathfrak{b}} = 0\} ,$$

$$\Delta_{\mathfrak{l}} = \Delta(\mathfrak{b}, \mathfrak{l}) = \{\alpha \in \Delta \mid \alpha|_{\mathfrak{t}} = 0\} .$$

We also define

$$\Delta_c = \Delta(\mathfrak{t}^{\sim}, \mathfrak{h}) , \quad \Delta_{c,\mathfrak{l}} = \{\alpha \in \Delta_c \mid \alpha|_{\mathfrak{t}} = 0\}$$

$$\text{and } \Delta^{\sim} = \Delta(\tilde{\mathfrak{a}}, \mathfrak{g}) .$$

Recall that we have chosen Δ^+. This gives natural choices of $\Delta^+_{\mathfrak{m}_{\mathfrak{b}}}$ and $\Delta^+_{\mathfrak{l}}$. Choose $(\Delta^{\sim})^+$, Δ^+_c and $\Delta^+_{c,\mathfrak{l}}$ compatible with each other and such that

(6.2) $\quad \forall \alpha \in \Delta^{\sim} : \alpha|_{\mathfrak{t}} > 0 \Rightarrow \alpha \in (\Delta^{\sim})^+ .$

Let $\rho, \rho^{\sim}, \rho_c$ etc. be the corresponding linear forms, i.e. half the sum of the positive (restricted) roots counted with multiplicity. (In [9] there is an extra condition (9) on the choice of Δ^+_c, but that condition is always satisfied).

Let $G = KAN$ be the Iwasawa decomposition corresponding to Δ^+, and $P = MAN$ the related minimal parabolic subgroup. The orbit $\mathcal{O}$ we are dealing with is

$$\mathcal{O} = HP/P \subset G/P .$$

$\mathcal{O}$ is isomorphic to $H/H\cap P$. We now try to define the Poisson transform of a naturally chosen H-quasiinvariant measure on $\mathcal{O}$. This leads to

$$(6.3) \qquad \psi_\lambda(x) = \psi_\lambda^{\mathcal{O}}(x) = \int_{H\cap L}\int_{K\cap H} e^{\langle -i\lambda-\rho, H(x^{-1}ky)\rangle} dk\, dy, \; x\in G ,$$

for any $\lambda\in\mathfrak{a}_{\mathbb{C}}^*$, for which there is absolute, uniform convergence on compact subsets of G.

For $\lambda\in\mathfrak{a}_{\mathbb{C}}^*$ we write $i\lambda = (i\lambda_o, i\nu)$, where $\lambda_o\in\mathfrak{t}_{\mathbb{C}}^*$ and $\nu\in\mathfrak{b}_{\mathbb{C}}^*$.

We think of $i\lambda_o$ as our parameter for a discrete series representation of $M_{\mathfrak{b}}^o/M_{\mathfrak{b}}^o\cap K^o$, cf. Section 5. If we do that, then the minimal $M_{\mathfrak{b}}^o\cap H^o$-type should be:

$$\mu_{\lambda_o} = i\lambda_o + (\rho-2\rho_c)|_{\mathfrak{t}} .$$

(We use here that by our choices of orderings $\rho-2\rho_c$ restricted to $\mathfrak{t}$ is the same as the $\rho-2\rho_c$ for $M_{\mathfrak{b}}$).

Let now $\mu_\lambda\in(\tilde{\mathfrak{t}}_{\mathbb{C}})^*$ be defined, using the canonical embedding of $\mathfrak{t}^*$ into $(\mathfrak{t}^\sim)^*$, by

$$6.4) \qquad \mu_\lambda = \mu_{\lambda_o} - \kappa = (i\lambda+\rho-2\rho_c)|_{\mathfrak{t}} - E(2\rho_c) + 2\rho_{c,\ell'}$$

where E denotes the projection from $(\mathfrak{t}^\sim)^*$ to $(\mathfrak{t}')^*$, i.e. restriction from $\mathfrak{t}^\sim$ to $\mathfrak{t}'$.

We shall also write ψ_{ν,μ_λ} instead of ψ_λ .

<u>Theorem 6.2</u>. Let $\lambda\in\mathfrak{a}_{\mathbb{C}}^*$. Then

(i) there exists $C>o$ such that $\psi_\lambda = \psi_{\nu,\mu_\lambda}$ is well-defined whenever

$$(6.5) \qquad \mathrm{Re}\langle i\nu,\alpha\rangle > C \qquad \text{for all } \alpha\in\Delta_\ell^+ .$$

Now we assume that λ is chosen such that ψ_λ is well-defined.

(ii) ψ_λ is an eigenfunction of $D(G/K)$ with eigenvalues χ_λ.

(iii) ψ_λ is an eigenfunction of $U(\mathfrak{g})^{\mathfrak{h}}$ with the eigenvalues determined in the following way (where $\mathfrak{n}$ is the Lie algebra of N and $\mathfrak{m}$ is the centralizer of $\mathfrak{a}$ in $\mathfrak{k}$). For every $u\in U(\mathfrak{g})^{\mathfrak{h}}$ there exists $u_o\in U(\mathfrak{a})$ such that

$$u-u_o\in(\mathfrak{l}\cap\mathfrak{h})_{\mathbb{C}}U(\mathfrak{g})+U(\mathfrak{g})(\mathfrak{m}+\mathfrak{n})_{\mathbb{C}} .$$

Then $u\psi_\lambda = u_o(-i\lambda-\rho)\psi_\lambda$.

(iv) ψ_λ is H-finite of irreducible type $\check{\mu}_\lambda$ if

$$\frac{\langle\mu_\lambda,\alpha\rangle}{\langle\alpha,\alpha\rangle}\in\mathbb{Z}^+ \quad \text{for all} \quad \alpha\in\Delta_c^+ .$$

This result is announced in [9]. The full proof is still to appear. Actually one can do a little better. As indicated in [9] one can similarly to the notion of "fine-K_1-type" in the example of a split group G_1, define a "nice $L^o\cap H^o$-type for $L^o/L^o\cap K^o$" .If $\mu_L\in(\mathfrak{t}')^*$ is such, then (6.3) may be generalized to

$$(6.6) \qquad \psi_\lambda^{\mu_L}(x) = \int_{H\cap L} e^{\langle\mu_L,H^\sim(y)\rangle}\int_{K\cap H} e^{\langle -i\lambda-\rho,H(x^{-1}ky)\rangle}dkdy,\ x\in G ,$$

where $H^\sim(\cdot)$ is defined as $H(\cdot)$, but related to the Iwasawa decomposition corresponding to $(\Delta^\sim)^+$.

Let now

$$(6.7) \qquad \mu = \mu_\lambda+\mu_L = \mu_{\lambda_o}-E(2\rho_c^\sim)+2\rho_{c,\mathfrak{l}}+\mu_L .$$

Whenever well-defined and whenever μ satisfies

$$\frac{\langle\mu,\alpha\rangle}{\langle\alpha,\alpha\rangle}\in\mathbb{Z}^+ \quad \text{for all} \quad \alpha\in\Delta_c^+ ,$$

then $\psi_\lambda^{\mu_L}$ is an eigenfunction of $D(G/K)$ of H-type $\check{\mu}$. Furthermore $\psi_\lambda^{\mu_L}$ is a finite linear combination of eigenfunctions of $U(\mathfrak{g})^{\mathfrak{h}}$.

Remark. The last expression for μ in (6.7) is written in a form such that the similarity with Knapp's formula for the minimal-K-type, cf.[25], and Vogan's results, [40], should be apparent. □

Remark. Notice that $\mathcal{O}$ is a closed orbit whenever, with $\mathfrak{a} = \mathfrak{t} + \mathfrak{b}$, $\mathfrak{t}$ is maximal Abelian in $\mathfrak{h} \cap \mathfrak{p}$, and (6.1) holds. In this case condition (6.5) is automatically satisfied for all λ since $\Delta_{\mathfrak{l}}$ is empty. □

§ 7. The open orbits. - Examples of multiplicity.

Matsuki [27] describes the open orbits of H in G/P in the following way. Let $\mathfrak{a} = \mathfrak{t} + \mathfrak{b}$ be a τ-invariant Cartan subspace, and let Δ^+ be a choice of positive roots. Then the corresponding orbit is open if and only if

(7.1)
(i) $\mathfrak{b}$ is maximal Abelian in $\mathfrak{q} \cap \mathfrak{p}$, and

(ii) $\forall \alpha \in \Delta \quad \alpha_{|\mathfrak{b}} > 0 \Rightarrow \alpha \in \Delta^+$.

If G^o/H^o is irreducible (i.e. satisfies (1.3)), then conditions (7.1) and (6.1) can only be satisfied at the same time if $\mathfrak{t} = \{0\}$, or in other words if

(7.2) $\quad \operatorname{rank}(G/K) = \operatorname{rank}(G_o/K \cap H)$.

Assuming condition (7.2) the open orbits are parametrized in the following way. Fix a choice of positive roots Δ_o^+ for $\Delta_o = \Delta(\mathfrak{a}, \mathfrak{g}_o)$. Then different open orbits correspond to different possible choices of Δ^+ compatible with Δ_o^+ , i.e. such that $\Delta_o^+ \subset \Delta^+$. The number of such orbits is then equal to the number r of cosets in $W(\mathfrak{a}, \mathfrak{g}_o) \backslash W(\mathfrak{a}, \mathfrak{g})$.

Example 7.1. Let $G = SL(n, \mathbb{R})$, $H = SO_e(p,q)$ and $K = SO(n)$, where $p+q = n$. In this case $r = \frac{n!}{p!q!}$. By the way this is an example of self duality in the sense that $G^o \simeq G$, $H^o \simeq K$ and $K^o \simeq H$. Notice furthermore in relation to the next theorem that $\operatorname{rank}(G) = \operatorname{rank}(G/H) = \operatorname{rank}(G/K) = n-1$,and that $\operatorname{rank}(H) = [\frac{n}{2}]$, (i.e. the biggest integer less than or equal to $\frac{n}{2}$) , and that $\operatorname{rank}(H/H \cap K) = \min\{p,q\}$. Thus in particular $\operatorname{rank}(H) = \operatorname{rank}(H/H \cap K)$ if $p = [\frac{n}{2}]$. $\square$

Let now, again under the assumption (7.2), $\mathcal{O}$ be an open orbit. In the notation of § 6 we have $\ell = \mathfrak{a}$ and $L = G$. Therefore the formula (6.3) for $\lambda \in \mathfrak{a}_{\mathbb{C}}^*$ reduces to the form

$$(7.3) \qquad \psi_\lambda(x) = \psi_\lambda^{\mathcal{O}}(x) = \int_H e^{\langle -i\lambda-\rho, H(x^{-1}y)\rangle} dy \ , x \in G \ ,$$

and $\mu_\lambda = 0$, cf. (6.4). This means according to Theorem 6.2 that ψ_λ is H-finite of the trivial H-type, whenever well-defined.

Let now $\mathcal{E}_\lambda(H;G/K)$ denote the subspace of H-invariant functions in $\mathcal{E}_\lambda(G/K)$, cf. § 4. From the above discussion and Theorem 6.2 we can get

<u>Proposition 7.2.</u> Assume that (7.2) is satisfied. There is a constant $C>0$, such that

$$\dim \mathcal{E}_\lambda(H;\ G/K) = \dim \mathcal{E}_\lambda(H^o;\ G^o/K^o) \geq r$$

for every $\lambda \in \mathfrak{a}_{\mathbb{C}}^*$ satisfying

$$\mathrm{Re} \langle i\lambda, \alpha\rangle > C \quad \text{for all} \quad \alpha \in \Delta^+ \ .$$

<u>Proof</u>: Fix λ satisfying the condition, where C is the biggest constant in Theorem 6.2 (i) coming from any of the open orbits. For each of the open orbits formula (7.3) then defines an element in $\mathcal{E}_\lambda(H;\ G/K)$. It follows from Theorem 4.3 that these elements are linearly independent. Since r is the number of orbits the proposition follows using the duality. □

<u>Example 7.3</u>. Let $H = K_\varepsilon$ in the notation of Oshima-Sechiguchi [33]. Then condition (7.2) is satisfied. From Proposition 4.2 in [33] it follows that $\dim \mathcal{E}_\lambda(H;\ G/K) = r$ for every $\lambda \in \mathfrak{a}_{\mathbb{C}}^*$. Notice that there are many cases where condition (7.2) is satisfied but where H is not of the type K_ε . For example all the non-Riemannian hyperbolic spaces. □

Under the extra condition that

(7.4) $$\begin{aligned} &\operatorname{rank}(G/H) = \operatorname{rank}(G)\ , \text{ and} \\ &\operatorname{rank}(H/H\cap K) = \operatorname{rank}(H)\ , \end{aligned}$$

we can improve the proposition.

Theorem 7.4. Assume that (7.2) and (7.4) hold .

(i) Let $\mathcal{O}$ be any open H-orbit, and let Δ^+ be the corresponding system of positive roots. If $\lambda \in \mathfrak{a}^*_{\mathbb{C}}$ is such that

$$\operatorname{Re}\langle i\lambda,\alpha\rangle > 0 \quad \text{for all} \quad \alpha\in\Delta^+ ,$$

then $\psi_\lambda = \psi_\lambda^{\mathcal{O}}$ in (7.3) is well-defined. Furthermore ψ_λ is H-finite of the trivial H-type and ψ_λ is a joint eigenfunction for $D(G/K)$ and for $U(\mathfrak{g})^{\mathfrak{h}}$.

(ii) For every $\lambda \in \mathfrak{a}^*_{\mathbb{C}}$ we have

$$\dim \mathcal{E}_\lambda(H;\ G/K) = \dim \mathcal{E}_\lambda(H^o;\ G^o/K^o) \geq r\ .$$

Proof: (i) follows rather easily from Theorem 6.2 using Theorem 8.1 of [7]. (ii) follows from Proposition 7.2. using that $W\cdot\mathfrak{a}^*_+ = \mathfrak{a}^*_{\mathbb{C}}$. □

Remarks (a). From Example 7.1 it is seen that (7.2) and (7.4) may hold simultaneously with r as big as one should like.

(b). It is natural to expect that for $\lambda \in \mathfrak{a}^*_{\mathbb{R}}$ the functions $(\psi_\lambda^{\mathcal{O}})^o$ should be related to the Plancherel formula for $L^2(G^o/K^o)$. It is therefore to be expected that the Plancherel formula should have multiplicity greater than or equal to r , at least over $(G^o)^\wedge_{H^o} \cap (G^o)^\wedge_{K^o}$, cf. Example 3.3. See in this connection also Oshima [31]. □

Example 7.5. For the hyperbolic spaces $G^o/K^o = X_{1,q}(\mathbb{R})$,$q>1$, we have condition (7.2) satisfied with rank equal one. In this case $r = 2$. There is multiplicity two in the Plancherel formula for the continuous spectrum as it already follows from Strichartz [39]; (the spaces of even and odd functions on $X_{1,q}(\mathbb{R})$ are both preserved under the group action). On the discrete spectrum there is by the way multiplicity one. Notice that condition (7.4) is only satisfied for $q = 1$. □

§ 8. Exceptional discrete series for G/H - A few remarks.

Already in [8, Section 8] it was clear that the construction in § 5 did not exhaust the discrete series for G^o/K^o in general. This followed since it is easy to see that for $X_{p,q}(\mathbb{F})$, with $p \neq o$ and q somewhat bigger than p, there is a finite set of representations missing.

We shall call such representations the exceptional discrete series for G/H. To see how they occur one should, given the rank condition (5.1) and a closed orbit, look at the two conditions (5.3) and (5.4) again:

(8.1) $\quad \langle i\lambda,\alpha\rangle > 0 \quad$ for all $\alpha \in \Delta^+$,

(8.2) $\quad \dfrac{\langle \mu_\lambda,\alpha\rangle}{\langle \alpha,\alpha\rangle} \in \mathbb{Z}^+ \quad$ for all $\alpha \in \Delta_c^+$,

where $\mu_\lambda = i\lambda+\rho-2\rho_c$.

Whereas (8.1) ensures that the asymptotics of ψ_λ are good enough for ψ_λ^o to be in $L^2(G^o/K^o)$, (8.2) implies that ψ_λ is H-finite, such that ψ_λ^o is well-defined.

Now for the hyperbolic spaces, if one relates the parameter λ to the eigenvalue of the Casimir operator, one can see that the full discrete series occur if we allow λ to take values such that (8.1) is satisfied and such that

(8.3) $\quad \dfrac{\langle \mu_\lambda,\alpha\rangle}{\langle \alpha,\alpha\rangle} \in \mathbb{Z} \quad$ for all $\alpha \in \Delta_c^+$.

(For $X_{p,q}(\mathbb{F})$ Δ^+ consists only of one element!)

We see that in these cases the exceptional discrete series correspond to values of λ for which $i\lambda$ is dominant (i.e. (8.1) holds) and for which μ_λ is integral but not dominant, (i.e. (8.3) holds, but not (8.2)).

Oshima and Matsuki have announced [32] that the full discrete series for the general case is determined by the rank condition (5.1), which is necessary and sufficient for the existence, and three conditions on λ : (8.1), (8.3) and the following

(8.4) $\quad \langle i\lambda-\rho,\alpha\rangle \geq 0$ for every simple root

$\alpha\in\Delta^+$ for which $\mathfrak{g}_\alpha \subset \mathfrak{h}$.

That something like (8.4) is necessary can easily be seen by looking at the compact case, cf. Theorem 2.8.

For a λ corresponding to an exceptional discrete series representation the function ψ_λ defined in § 5 is not H-finite. Recall that ψ_λ is the Poisson transform of the normalized $K\cap H$-invariant measure on a closed orbit $\mathcal{O}$. The remaining problem, which does not seem to be solved in [32], is to construct the $K\cap H$-invariant distributions T supported on $\mathcal{O}$ for which the Poisson transform $\psi = P_\lambda(T,\cdot)$ via the duality defines a minimal spherical function for a discrete series representation for G^o/K^o.

In a paper to appear the author has in collaboration with Okamoto solved this problem for the hyperbolic spaces.

REFERENCES

[1] Berger, M., Les espaces symétriques non compacts. Ann. Sci. École Norm. Sup., 74 (1957), 85-177.

[2] Clozel, L., Changement de base pour les représentations tempérées des groupes réductive réels. Ann. Sc. E.N.S.(4), 15 (1982) , 45-115.

[3] Dixmier, J., Algebras Enveloppantes, Gauthier-Villars, Paris, 1974.

[4] Enright,T.J., Varadarajan,V.S., On an infinitesimal characterization of the discrete series. Ann. of Math., 102 (1975), 1-15.

[5] Faraut,J., Distributions sphériques sur les espaces hyperbolic. J.Math. pures et appl. 58 (1979), 369-444.

[6] Flensted-Jensen, M., Spherical functions on rank one symmetric spaces and generalizations. Proceedings of symposia in pure mathematics, vol XXVI, (Amer. Math. Soc. 1973), 339-342.

[7] Flensted-Jensen, M., Spherical functions on a real semisimple Lie group. A method of reduction to the complex case. J. Funct. Anal. 30 (1978), 106-146.

[8] Flensted-Jensen, M., Discrete series for semisimple symmetric spaces. Ann. of Math. 111 (1980), 253-311.

[9] Flensted-Jensen, M., K-finite joint eigenfunctions of $U(g)^K$ on a non-Riemannian semisimple symmetric space G/H . In: Noncommutative harmonic analysis and Lie groups, Proceedings, Marseille-Luminy 1980, ed. Carmona, J. and Vergne, M..Lecture Notes in Math. 880 (1981) 91-101.

[10] Harish-Chandra, Representations of a semisimple Lie group on a Banach space,I. Trans. Amer. Math. Soc. 75 (1953), 185-243.

[11] Harish-Chandra, Spherical functions on a semisimple Lie group I and II. Amer. J. Math. 80 (1958), 241-310 and 553-613.

[12] Harish-Chandra, Discrete series for semisimple Lie groups, I, II. Acta Math. 113 (1965), 241-318, 116 (1966), 1-111.

[13] Harish-Chandra, Harmonic analysis on semisimple Lie groups. Bull. Amer. Math. Soc. 78 (1970), 529-551.

[14] Helgason, S. Differential geometry and symmetric spaces. Academic Press, New York 1962.

[15] Helgason, S., Fundamental solutions of invariant differential operators on symmetric spaces. Amer. J. Math. 86 (1964), 565-601.

[16] Helgason, S., A duality for symmetric spaces with applications to group representations. Adv. Math. 5 (1970),1-154.

[17] Helgason, S., Eigenspaces of the Laplacian; integral representations and irreduciblity. J. Functional Analysis 17 (1974), 328-353.

[18] Helgason.S., A duality for symmetric spaces with applications to group representations II. Differential equations and eigenspace representations. Adv. Math. 22 (1976), 187-219.

[19] Helgason, S., Some results on eigenfunctions on symmetric spaces and eigenspace representations. Math. Scand. 41 (1977), 79-89.

[20] Helgason, S., Differential geometry, Lie groups and symmetric spaces. Academic Press, New York-San Francisco-London 1978.

[21] Helgason, S., Groups and geometric analysis I. To appear Academic Press.

[22] Kashiwara, M., Kowata, A., Minemura, K., Okamoto, K., Oshima, T. and Tanaka, M., Eigenfunctions of invariant differential operators on a symmetric space. Ann. of Math., 107 (1978), 1-39.

[23] Kengmana, T.,Characters of the discrete series for pseudo-Riemannian symmetric spaces. Preprint, Harvard University, 1983.

[24] Kosters, M.T., Spherical distributions on rank one symmetric spaces. Thesis, University of Leiden, 1983.

[25] Knapp, T., Minimal K-type formula. Proceedings of the 1982-Marseille-Luminy Conference.

[26] Loos, O., Symmetric spaces, I: General theory. New York-Amsterdam, W.A. Benjamin, Inc., 1969.

[27] Matsuki, T., The orbits of affine symmetric spaces under the action of minimal parabolic subgroups, J. Math. Soc. Japan 31 (1979), 331-357.

[28] Mostow, G.D., On covariant fiberings of Klein spaces. Amer. J. Math. 77 (1955), 247-278.

[29] Olafsson, G., Die Langlands-parameter für die Flensted-Jensensche Fundamentale Reihe. To appear.

[30] Oshima, T., Poisson transformation of affine symmetric spaces. Proc. Jap. Acad. Ser. A 55 (1979), 323-327.

[31] Oshima, T., Fourier analysis on semisimple symmetric spaces. In: Non commutative harmonic analysis and Lie groups. Proceedings, Marseille-Luminy 1980, ed. Carmona,J. and Vergne, M. Lecture Notes in Math. 880, (1981), 357-369.

[32] Oshima, T. and Matsuki, T., A complete description of discrete series for semisimple symmetric spaces. Preprint.

[33] Oshima, T. and Sekiguchi, J.: Eigenspaces of invariant differential operators on an affine symmetric space, Inventiones Math. 57 (1980), 1-81.

[34] Rossmann, W., The structure of semisimple symmetric spaces. Can. J. Math. 31 (1979), 157-180.

[35] Schlichtkrull, H. , The Langlands parameters of Flensted-Jensen' s discrete series for semisimple symmetric spaces, J. Func. Anal. 50 (1983), 133-150.

[36] Schlichtkrull,H., A series of unitary irreducible representations induced from a symmetric subgroup of a semisimple Lie group, Invent. Math. 68 (1982), 497-516.

[37] Schlichtkrull, H., Applications of hyperfunction Theory to representations of semisimple Lie groups. Rapport 2 a-b, Dept. of Math., University of Copenhagen, April 1983.

[38] Schlichtkrull, H., One dimensional K-types in finite dimensional representations of semisimple Lie groups. to appear Math. Scand.

[39] Strichartz, R.S., Harmonic analysis on hyperboloids. J. Funct. Anal. 12 (1973), 341-383.

[40] Vogan, D., Algebraic structure of irreducible representations of semisimple Lie groups. Ann. of Math. 109 (1979), 1-60.

[41] Wolf, J.A., Spaces of constant curvature. McGraw-Hill, New York, 1967.

[42] Wolf, J.A., Fineteness of orbit structure for real flag manifolds. Geom. Dedicata 3 (1974), 377-384.

[43] Warner, G., Harmonic analysis on semi-simple Lie groups I and II. New York-Heidelberg-Berlin, Springer-Verlag, 1972.

<u>Remark</u> 1°. After these notes were written I have received the following preprint, which contains proofs of results from the announcement [32].

Oshima, T. and Matsuki, T., A description of discrete series for semisimple symmetric spaces.

2°. I also discovered an example in which Conjecture 2.7 does not hold.

*Dept. of Mathematics and Statistics
Royal Veterinary- and Agricultural University
Thorvaldsensvej 40
DK-1871- Copenhagen, Denmark

PARTIAL DIFFERENTIAL EQUATIONS ON NILPOTENT GROUPS

B. Helffer

Université de Nantes

§0. Introduction

In these talks, I want to present a survey on recent results on hypoellipticity, analyticity and local solvability for left invariant operators on graded nilpotent groups. Though this theory was in great part constructed (at least for hypoellipticity) as a first step for the study of general partial differential equations (see for example the basic work of Rothschild-Stein [R.S]), we shall restrict ourselves in these lectures to the context of nilpotent groups and refer to a forthcoming book written in collaboration with J. Nourrigat ($[HE-NO]_7$) for generalizations or details (see also $[HE-NO]_8$, [NO]). However we have tried to give details for the example of $\mathbb{N}_4$ (Lie algebra of rank of nilpotency 3 and of dimension 4). The paper is organized as follows:

I want to thank the organizers of this Lie Groups Year at the University of Maryland for their kind invitation and June Slack for her excellent typing.

§1. Operators on nilpotent groups and representation theory

Let $\mathfrak{G}$ be a nilpotent graded Lie algebra of rank r: we have:

(1.1) $$\mathfrak{G} = \bigoplus_{i=1}^{r} \mathfrak{G}_i$$

$$\text{with} \quad [\mathfrak{G}_i, \mathfrak{G}_j] \subset \mathfrak{G}_{i+j} \quad \text{if} \quad i + j \leq r$$

$$[\mathfrak{G}_i, \mathfrak{G}_j] = 0 \quad \text{if} \quad i + j > r.$$

We define $\mathfrak{G}^i$ as:

(1.2) $$\mathfrak{G}^i = \sum_{j=1}^{i} \mathfrak{G}_j .$$

$\mathfrak{G}$ is provided with a natural family of dilations $\delta_t (t \in \mathbb{R}^+)$:

(1.3) $$\delta_t(a) = t^i a \quad \text{if} \quad a \in \mathfrak{G}_i .$$

Via the exponential map, $\mathfrak{G}$ is identified with the associated simply-connected, connected group G.

We denote by $\mathfrak{U}(\mathfrak{G})$ the complexified enveloping algebra which is identified with the space of the left invariant operators on G. We have a natural extension of δ_t to $\mathfrak{U}(\mathfrak{G})$ and for $m \in \mathbb{N}$, $\mathfrak{U}_m(\mathfrak{G})$ would be:

(1.4) $$\mathfrak{U}_m(\mathfrak{G}) = \{A \in \mathfrak{U}(\mathfrak{G}),\ \delta_t(A) = t^m \cdot A\}$$

i.e. the set of homogeneous operators of order m in $\mathfrak{U}(\mathfrak{G})$.

Let $\mathfrak{G}^*$ be the dual space of $\mathfrak{G}$. δ_t^* is just the transpose of δ_t acting on $\mathfrak{G}^*$.

We know that G acts on $\mathfrak{G}^*$ via the coadjoint action. We denote this action like this:

(1.5) $$G \times \mathfrak{G}^* \ni (g, \ell) \to g \cdot \ell \in \mathfrak{G}^* .$$

The Kirillov theory says that there exists a bijection $\hat{\beta}$ between

$G\backslash\mathfrak{G}^*$ (the space of the orbits for the coadjoint action in $\mathfrak{G}^*$) and $\hat{G}$ (the set of equivalence classes of irreducible unitary representations). If π_0 is the projection of $\mathfrak{G}^*$ on $G\backslash\mathfrak{G}^*$, we define β with the following diagram:

$$(1.6) \qquad \begin{array}{ccc} \mathfrak{G}^* & \xrightarrow{\beta} & \hat{G} \\ \downarrow{\scriptstyle \pi_0} & \nearrow{\scriptstyle \hat{\beta}} & \\ G\backslash\mathfrak{G}^* & & \end{array}$$

Moreover, Kirillov's theory gives us a constructive method to realize β by the introduction of induced representations.

Let us say a few words on induced representations because it would be useful later.

If $\ell \in \mathfrak{G}^*$, let B_ℓ be the bilinear form defined by:

$$(1.7) \qquad B_\ell([a,b]) = \ell([a,b]), \qquad \forall a \in \mathfrak{G} \quad \forall b \in \mathfrak{G}$$

and let $\mathfrak{H}$ be a subalgebra of $\mathfrak{G}$ which is isotropic for B_ℓ. Let H be the associated subgroup of G via the exponential map:

$$(1.8) \qquad H = \exp \mathfrak{H}.$$

Then the unitary representation of H defined by:

$$\exp h \to e^{i<\ell,h>} \qquad \forall h \in \mathfrak{H}$$

induces a representation on $L^2(H\backslash G)$ (this is the right regular representation restricted to measurable functions on G satisfying: $f(e^h \cdot g) = e^{i<\ell,h>} f(g)$ and s.t $|f(g)|$ (which is defined on $H\backslash G$) is $L^2(H\backslash G)$ integrable).

We denote this representation by: $\pi_{\ell,\mathfrak{H}}$.

Recall an explicit realization of $\pi_{\ell,\mathfrak{H}}$ in $L^2(\mathbb{R}^k)$, where k is the codimension of $\mathfrak{H}$ in $\mathfrak{G}$.

It is classical (see [PU]) that we can construct a sequence of subalgebras of $\mathfrak{G}$ s.t.:

(1.9) $$\mathfrak{H} = \mathfrak{A}_0 \subset \mathfrak{A}_1 \subset \ldots \subset \mathfrak{A}_{k-1} \subset \mathfrak{A}_k = \mathfrak{G}$$

with $\mathfrak{A}_i$ an ideal of codimension 1 in $\mathfrak{A}_{i+1}$.

Choosing for each $i = 1,\ldots,k$, $e_i \neq 0$ in $\mathfrak{A}_i$ s.t.

(1.10) $$\mathfrak{A}_{i-1} \oplus \mathbb{R} e_i = \mathfrak{A}_i$$

we can write: $\forall x = (x_1,\ldots,x_k) \in \mathbb{R}^k$, $\forall a \in \mathfrak{G}$

(1.11) $$\exp(x_1e_1)\cdot\exp(x_2e_2)\cdot\ldots\cdot\exp(x_ke_k)\cdot\exp a =$$
$$= \exp(h(x,a))\cdot\exp(\sigma_1(x,a)e_1)\cdot\ldots\cdot\exp(\sigma_k(x,a)\cdot e_k)$$

with $h(x,a) \in \mathfrak{H}$, $\sigma_i(x,a) \in \mathbb{R}$.

We then realize $\pi_{(\ell,\mathfrak{H})}$ in $L^2(\mathbb{R}^k)$ by:

(1.12) $\forall a \in \mathfrak{G}$, $\forall f \in L^2(\mathbb{R}^k)$

$$(\pi_{(\ell,\mathfrak{H})}(\exp a)f)(x) = e^{i<\ell,h(x,a)>} f(\sigma(x,a))$$

where $\sigma(x,a) = (\sigma_1(x,a),\sigma_2(x,a),\ldots,\sigma_k(x,a))$.

If $\mathfrak{H}$ is of maximal dimension $(= \frac{\dim \ker B_\ell}{2} + \frac{\dim \mathfrak{G}}{2})$, $\pi_{\ell,\mathfrak{H}}$ is irreducible and depends only, up to unitary equivalence, on the orbit of ℓ. In this case, we denote simply $\pi_{\ell,\mathfrak{H}}$ by π_ℓ (or with the notation (1.6): $\beta(\ell)$).

§2. Hypoellipticity

The starting points of the study are the works of L. Hörmander [HO] on $\sum X_i^2$, of L. P. Rothschild-E. M. Stein [R.S] and of C. Rockland [ROC], which gives a theoretical group interpretation of old results of Grušin [GRU] (and other specialists in P.D.E.) on hypoellipticity.

Recall first that a differential operator P on an open set of $\mathbb{R}^n$ is called <u>hypoelliptic</u> if we have:

$$\left.\begin{array}{l} \forall\ u \in \mathcal{D}'(\Omega) \\ \forall\ \omega \text{ open set in } \Omega \\ \quad Pu \in C^\infty(\omega) \end{array}\right\} \Rightarrow u \in C^\infty(\omega)$$

The following conjecture of C. Rockland [ROC] was proved by Helffer-Nourrigat ($[\text{HE-NO}]_{1,2}$):

<u>Theorem 2.1.</u> Let $P \in \mathfrak{U}_m(\mathfrak{G})$; then the following properties are equivalent:

i) P is hypoelliptic.

ii) For each nontrivial irreducible representation of G ($\Leftrightarrow \forall \pi \in \hat{G}\setminus\{0\}$), Rockland's condition is satisfied:

$$(\text{RO})\begin{cases} \pi(P) \text{ is injective in } \mathcal{S}_\pi, \text{ where } \mathcal{S}_\pi \text{ is the space of } C^\infty \\ \text{vectors of } \pi. \end{cases}$$

The implication i) $\Rightarrow$ ii) was proved before in Beals [BE] and Rothschild-Stein [R.S.].

When $\mathfrak{G}$ is stratified (i.e. generated by $\mathfrak{G}_1$), we have also equivalence with:

$$(2.1)\quad \text{iii)} \begin{cases} P \text{ is maximally hypoelliptic, i.e.:} \\ \text{For each } A \in \mathfrak{U}_m(\mathfrak{G}),\ \exists C_A > 0 \text{ such that} \\ \forall u \in C_0^\infty(G) \quad \text{we have:} \\ \|Au\|^2_{L^2(G)} \leq C_A \|Pu\|^2_{L^2(G)}. \end{cases}$$

The fact that the inequality (2.1) implies hypoellipticity was relatively standard in the P.D.E. literature (see [FED], $[TR]_1$, [U]), if you remark that if we denote by:

(2.2) $$(Y_1, \ldots, Y_{p_1}) \quad \text{a basis of } \mathfrak{G}_1,$$

these vector fields satisfy the so-called Hörmander's condition at each point of G, which implies (see [R.S])

(2.3) $$\begin{gathered} \forall K \subset G,\ \exists C_K \text{ such that } \forall u \in C_0^\infty(K) \\ \|u\|^2_{1/r} \le C_K \left(\sum_{i=1}^{p_1} \|Y_i u\|_0^2 + \|u\|_0^2\right) \end{gathered}$$

where $\| \ \|_s$ is the classical norm of Sobolev spaces H^s.

Let us mention some steps in the proof of the theorem which shall be useful in the following sections.

As in many theorems in the Kirillov theory, the proof is by induction on the rank of nilpotency.

Admitting the theorem for nilpotent groups of rank $<r$, we prove in a first step that if:

(2.4) (RO-DE) $$\begin{cases} \text{For each } \pi \in \hat{G}\backslash\{0\} \text{ which is trivial on} \\ G_r \equiv \exp \mathfrak{G}_r,\ \pi(P) \text{ is injective in } \mathcal{S}_\pi \end{cases}$$

is satisfied ((RO-DE) for RO-degenerate), then we have the following inequalities with remainder terms:

(2.5) $$\begin{cases} \text{For each } A \in \mathfrak{U}_m(\mathfrak{G}),\ \exists C_A \text{ such that} \\ \forall u \in C_0^\infty(G_r\backslash G),\ \forall \ell \in \mathfrak{G}^* \text{ such that } |\ell_r| = |\ell_{/\mathfrak{G}_r}| = 1 \\ \|\pi_{\ell,\mathfrak{G}_r}(A)u\|^2_{L^2(G_r\backslash G)} \le C_A(\|\pi_{\ell,\mathfrak{G}_r}(P)u\|^2_{L^2(G_r\backslash G)} + \|u\|^2_{L^2(G_r)}). \end{cases}$$

To simplify the notations, let us introduce a fixed base B_j^m of $\mathfrak{U}_m(\mathfrak{G})$ for each m, and define for each couple $(\ell,\mathfrak{H})$ defined in §1, the semi-norms:

(2.6) $$\|u\|^2_{m,\pi_{(\ell,\mathfrak{H})}} = \sum_j \|\pi_{\ell,\mathfrak{H}}(B^m_j)u\|^2_{\mathfrak{H}_{\pi(\ell,\mathfrak{H})}},$$

which are defined at least for $u \in \mathcal{S}_{\pi(\ell,\mathfrak{H})}$. Here,

(2.7) $\mathfrak{H}_{\pi_{\ell,\mathfrak{H}}}$ is the space of the representation $\pi_{(\ell,\mathfrak{H})}$.

So we can rewrite (2.5) like this (we use some density argument to replace $C_0^\infty(G_r\backslash G)$ by $\mathcal{S}_{\pi(\ell_r,\mathfrak{G}_r)}$):

(2.8) $$\|u\|^2_{m,\pi_{(\ell_r,\mathfrak{G}_r)}} \leq C[\|\pi_{(\ell_r,\mathfrak{G}_r)}(P)u\|^2_0 + \|u\|^2_0],\ \forall u \in \mathcal{S}_{\pi(\ell_r,\mathfrak{G}_r)}.$$

This inequality implies (see §4, Prop. 4.6).

(2.9) $$\begin{cases} \exists C > 0, \text{ such that } \forall \ell \in \mathfrak{G}^* \text{ such that } |\ell_r| = 1, \\ \forall u \in \mathcal{S}_{\pi_\ell}: \\ \|u\|^2_{m,\pi_\ell} \leq C(\|\pi_\ell(P)u\|^2_0 + \|u\|^2_0). \end{cases}$$

($\|\ \|_0$ is used here for $\|\ \|_{0,\pi_\ell}$).

Inequality (2.9) is basic to relate the injectivity of $\pi_\ell(P)$ in $\mathcal{S}_{\pi_\ell}$ to the existence of a left inverse for $\pi_\ell(P)$ which goes from $\mathfrak{H}_\pi(\equiv \mathfrak{H}^0_\pi)$ to $\mathfrak{H}^m_\pi$, where $\mathfrak{H}^m_\pi$ is by definition:

(2.10) $$\mathfrak{H}^m_\pi = \{u \in \mathfrak{H}_\pi,\ \pi(A)u \in \mathfrak{H}_\pi,\ \forall A \in \mathfrak{U}_j(\mathfrak{G}),\ j \leq m\}.$$

We take on $\mathfrak{H}^m_\pi$ the following norm:

(2.11) $$\|u\|^2_{\mathfrak{H}^m_\pi} = \sum_{j=0}^m \|u\|^2_{j,\pi}.$$

Then the (RO)-condition gives the inequality:

$$(2.12)\quad \begin{cases} \forall \ell \in \mathfrak{G}^*, \text{ such that } |\ell_r| = 1, \ \exists C_\ell \text{ such that} \\ \forall u \in \mathcal{S}_{\pi_\ell} \\ \|u\|^2_{\mathcal{H}^m_{\pi_\ell}} \leq C_\ell (\|\pi_\ell(P)u\|_0^2). \end{cases}$$

The last step (and probably the most difficult) is to prove that the constant C_ℓ can be chosen independently of ℓ (see §4).

Remark 2.2. In $[MEL]_{1,2}$, A. Melin has given a proof which is a partial alternative to our proof of Theorem 2.1 using parametrix constructions.

Remark 2.3. Let us mention that all the restrictions (which have been made during our original proof in $[HE\text{-}NO]_{1,2}$) on the order of P can be removed in the case of stratified groups using the interpolation theory for the $\mathcal{H}^m_\pi$ spaces (see $[FO]_2$, [MO]).

§3. <u>An example of application: study of</u> $\sum_j Y_j^2 + \sum_{j,k} C_{jk}[Y_j,Y_k]$

In this section, we want to explain, by studying an example, why Theorem 2.1 has not only a theoretical interest. The result presented here is a complement to Rothschild-Stein results in [R.S] and is due to B. Helffer, G. Métivier and J. Nourrigat (cf. $[HE]_5$).

Let $\mathfrak{G}$ be stratified, $(Y_1,\ldots,Y_{p_1})$ be a basis of $\mathfrak{G}_1$; we want to consider:

(3.1) $$L = \sum_{j=1}^{p_1} Y_j^2 + \frac{i}{2} \sum_{j,k} b_{jk}[Y_j,Y_k],$$

where (b_{jk}) is a real antisymmetric matrix.

Let us introduce some definitions. If ρ is a real antisymmetric matrix, we can define its trace-norm by:

(3.2) $$\|\rho\|_1 = \sum_j |\rho_j|,$$

where the ρ_j are the eigenvalues of ρ.

We denote by S the subspace of the real antisymmetric matrices such that

(3.3) $$\sum_{j,k} s_{jk}[Y_j,Y_k] = 0.$$

The theorem is the following:

<u>Theorem 3.1.</u> (Rothschild-Stein [R.S] and Helffer-Métivier-Nourrigat $[HE]_5$). Suppose that $\mathfrak{G}$ is not an Heisenberg algebra. Then L is hypoelliptic if and only if:

(3.4) $$\sup_{\substack{\|\rho\|_1 \le 1 \\ \rho \in S^\perp}} |\mathrm{tr}(b\cdot\rho)| < 1.$$

(Here $S^\perp$ is the set of ρ such that $\mathrm{tr}(s\cdot\rho) = 0$, $\forall s \in S$.)

The sufficiency part is proved in [R.S]. The necessity part is also proved in [R.S.] under the additional hypothesis that:

(3.5) $\mathfrak{G}/\mathfrak{G}^3$ is not an Heisenberg algebra.

Let us remark that in the case where $\mathfrak{G}$ is an Heisenberg algebra the characterization of the hypoellipticity is well known, but of different nature (see $[HE]_5$ for references).

According to these results, the last case to consider is the case where $\mathfrak{G}$ is effectively of rank >2 and with the property that: $\mathfrak{G}/\mathfrak{G}^3$ is an Heisenberg algebra. Then we have to prove that if:

(3.6)
$$\sup_{\substack{\|\rho\|_1 \leq 1 \\ \rho \in S^\perp}} |\mathrm{tr}(b\cdot\rho)| \geq 1.$$

L is not hypoelliptic.

We need first the following easy lemma:

<u>Lemma 3.2</u>. Let $\mathfrak{G}$ be a stratified algebra such that $\mathfrak{G}/\mathfrak{G}^3$ is a Heisenberg algebra. If $\mathfrak{G}$ is effectively of rank >2, then $\dim \mathfrak{G}_1 = p_1 = 2$.

Applying Lemma 3.2 and taking eventually a quotient of $\mathfrak{G}$ by an ideal of $\mathfrak{G}$, we can reduce the proof to the case where $\mathfrak{G}$ is the algebra $\mathfrak{N}_4$ of dimension 4 which admits a basis (Y_1, Y_2, Z, W) satisfying the following relations:

(3.7)
$$\left\{\begin{array}{l} [Y_1, Y_2] = Z, \\ [Y_1,[Y_1,Y_2]] = W, \\ [Y_2,[Y_1,Y_2]] = 0, \end{array}\right.$$
$$\mathfrak{N}_4 = \mathfrak{G}_1 \oplus \mathfrak{G}_2 \oplus \mathfrak{G}_3$$
$$\text{with} \left\{\begin{array}{l} \mathfrak{G}_1 = \mathbb{R}\, Y_1 + \mathbb{R}\, Y_2, \\ \mathfrak{G}_2 = \mathbb{R}\, Z, \\ \mathfrak{G}_3 = \mathbb{R}\, W. \end{array}\right.$$

L is now the operator:

(3.8) $$L_\alpha = Y_1^2 + Y_2^2 + i\alpha Z, \qquad \alpha \in \mathbb{R}.$$

We now have to prove that if $|\alpha| \geq 1$, L_α is not hypoelliptic. According to Theorem 2.1 (necessary part), we have to find π in $\hat{G}\backslash 0$ such that $\pi(L_\alpha)$ is not injective.

Let us describe briefly the nature of $\hat{G}$ (for $G = \exp \mathfrak{N}_4$). We prefer to write the expressions of the associated representations of $\mathfrak{G}$.

1. Representations degenerate on $\exp(\mathfrak{G}_2 \oplus \mathfrak{G}_3)$ parametrized by $\mathbb{R}^2$.

(3.9) $$\begin{aligned} \pi_{\eta_1,\eta_2}(Y_1) &= i\eta_1 & \qquad \pi_{\eta_1,\eta_2}(Z) &= 0 \\ \pi_{\eta_1,\eta_2}(Y_2) &= i\eta_2 & \qquad \pi_{\eta_1,\eta_2}(W) &= 0. \end{aligned}$$

2. Representations degenerate on $\exp(\mathfrak{G}_3)$ parametrized by $\mathbb{R}\backslash 0$.

(3.10) $$\begin{aligned} \pi_\lambda(Y_1) &= \partial_t & \qquad \pi_\lambda(Z) &= i\lambda \\ \pi_\lambda(Y_2) &= i\lambda t & \qquad \pi_\lambda(W) &= 0. \end{aligned}$$

3. Nondegenerate representations.

$$\begin{aligned} \pi_{\beta,\gamma}(Y_1) &= \partial_t & \qquad \pi_{\beta,\gamma}(Z) &= it\gamma \\ \pi_{\beta,\gamma}(Y_2) &= i\left(\frac{t^2}{2}\gamma - \beta\right) & \qquad \pi_{\beta,\gamma}(W) &= i\gamma . \end{aligned}$$

In what follows, we are only interested in $\pi_{\pm 1}$ which we denote simply $\pi_\pm$ and by $\pi_{\beta,2}$ which we denote simply π_β. We shall now study:

(3.12) $$\begin{cases} \pi_\pm(-L_\alpha) = -\partial_t^2 + t^2 \pm \alpha \\ \pi_\beta(-L_\alpha) = -\partial_t^2 + (t^2-\beta)^2 + 2\alpha t, \qquad \beta \in \mathbb{R}. \end{cases}$$

The study of the injectivity of $\pi_{\pm}(L_\alpha)$ gives the answer if:

$$\alpha = 2p+1 \quad (p \in \mathbb{Z}).$$

Looking at the symmetry of $\pi_\beta(-L_\alpha)$ $(t \to -t)$, we see that it is sufficient to study the case where $\alpha < -1$. Then the following lemma gives the answer in the remaining cases.

Lemma 3.3. For each $\alpha < -1$, there exists $\beta \in \mathbb{R}$ such that

$$P_{\alpha,\beta} = -\partial_t^2 + (t^2-\beta)^2 + 2\alpha t$$

is not injective in $\mathcal{S}(\mathbb{R})$.

Proof. The operator $P_{0,\beta}$ is clearly an essentially self-adjoint operator on $\mathcal{S}(\mathbb{R})$ and its self-adjoint extension A is positive, injective and with compact inverse.

For β fixed, we consider the spectral problem in α for the operator $A + \alpha B$ where B is the operator $2t$ which is well defined on a domain of definition of A and $A^{1/2}$. It is easier (and equivalent) to study:

$$I + \alpha\, A^{-1/2}\, BA^{-1/2}.$$

$A^{-1/2}\, BA^{-1/2}$ is a compact operator, self-adjoint, whose spectrum is contained in $[-1,+1]$ (because $P_{\alpha\beta}$ is invertible for $|\alpha|<1$, symmetric with respect to 0; and there is a bijection between the values of α for which P is noninjective and the nonzero eigenvalues λ of $A^{-1/2}\, BA^{-1/2}$: $\alpha \to \lambda = -1/\alpha$. The minimax principle applied to the operator $(A^{-1/2}\, BA^{-1/2})$ gives that, if $\alpha(\beta)$ is the greatest strictly negative eigenvalue of $(A+\alpha B)$, we have:

$$(3.14)\quad \frac{-1}{\alpha(\beta)} = \sup_{\substack{u\in\mathcal{S}\\ u\neq 0}} \left(\frac{((A^{-1/2}BA^{-1/2})u,u)}{\|u\|^2}\right) = \sup_{\substack{u\in\mathcal{S}\\ u\neq 0}} \frac{(tu,u)}{(-\partial_t^2+(t^2-\beta)^2u,u)}.$$

Looking at (3.14), it is easy to see that $\alpha(\beta)$ depends continuously on $\beta \in]-\infty, \infty[$.

Lemma 3.3 is then a consequence of the following lemma.

(3.15) $$\lim_{\beta\to-\infty} \alpha(\beta) = -\infty,$$

(3.16) $$\lim_{\beta\to+\infty} \alpha(\beta) = -1.$$

Proof of 3.15. We start from the following maximal inequality for A (which is just inequality (2.12) for $Y_1^2 + Y_2^2 = P$ (see §4) for the representation π_ℓ introduced in (3.11)):

(3.17) $$\begin{cases} \|\partial_t^2 u\|^2 + \|(t^2-\beta)^2 u\|^2 \le C(\|(-\partial_t^2 + (t^2-\beta)^2)u\|^2 \\ \qquad \forall u \in \mathcal{S}(\mathbb{R}). \end{cases}$$

Then an easy computation shows the existence of $\varepsilon_0 > 0$ such that if $|\alpha| \le \varepsilon_0|\beta|$, $\beta < 0$, we have:

(3.18) $$\|\partial_t^2 u\|^2 + \|t^4 u\|^2 + \beta^4\|u\|^2 \le 4C\|P_{\alpha\beta}u\|^2.$$

This proves the injectivity of $P_{\alpha\beta}$ if $|\alpha| \le \varepsilon_0|\beta|(\beta<0)$, and then (3.15)

Proof of 3.16. We transform the problem with the help of unitary transformations which preserve the eigenvalues (we follow here the ideas developed in the proof of Theorem 2.1). We suppose here that β is >0.

By the unitary transformation $\mathcal{T}$ associated to the change of variables:

(3.19) $$s = \sqrt{2}\,\beta^{1/4}\,(t-\sqrt{\beta})$$

we find that:

(3.20) $$\mathcal{F}^{-1} P_{\alpha\beta}\mathcal{F} = 2\sqrt{\beta}\,[Q_\varepsilon^\alpha]$$

where

(3.21) $$\varepsilon = (\sqrt{2})^{-3}\,\beta^{-3/4},$$

(3.22) $$Q_\varepsilon^\alpha(s,\partial_s) = -\partial_s^2 + (\varepsilon s^2+s)^2 + \alpha(2\varepsilon s+1).$$

We recognize here a singular perturbation of the harmonic oscillator. The determination of $\alpha(\beta)$ when β tends to ∞ can be reduced to the determination of the greatest strictly negative eigenvalue of the spectral problem in α associated to the family Q_ε^α when ε tends to zero. We have:

(3.23) $$-\frac{1}{\tilde{\alpha}(\varepsilon)} = \sup_{\substack{u\neq 0\\ u\in\mathcal{S}}} \frac{((2\varepsilon s+1)u,u)}{((-\partial_s^2+(\varepsilon s^2+s)^2)u,u)}.$$

We know also that $\frac{1}{\tilde{\alpha}(\varepsilon)} \in [-1,+1]$. So we have:

(3.24) $$1 \geq -\frac{1}{\tilde{\alpha}(\varepsilon)} \geq \frac{((2\varepsilon s+1)u,u)}{(-\partial_s^2+(\varepsilon s^2+s)^2u,u)} \quad \forall u \in \mathcal{S},\quad u \neq 0.$$

Taking $u = e^{-s^2/2}$ (which is the eigenvector of the harmonic oscillator associated to the eigenvalue 1), we see that:

$$\lim_{\varepsilon\to 0} -\frac{1}{\tilde{\alpha}(\varepsilon)} = -1 \Leftrightarrow \lim_{\beta\to\infty} \alpha(\beta) = -1.$$

The lemma is proved.

§4. Operators in homogeneous spaces: the analytic approach (after Helffer-Nourrigat)

We shall first recall the notions related to the spectrum (or support) of a representation. Let us consider a couple $(\ell,\mathfrak{H})$ where $\ell \in \mathfrak{G}^*$, and $\mathfrak{H}$ is isotropic for B_ℓ:

$$(4.1) \qquad \ell([\mathfrak{H},\mathfrak{H}]) = 0.$$

We introduce the set:

$$(4.2) \qquad \Omega_{(\ell,\mathfrak{H})} = G\cdot[\mathfrak{H}^{\perp}+\ell].$$

This is the set of the points $\mathcal{I} \in \mathfrak{G}^*$ whose orbit meets a point ℓ' whose restriction to $\mathfrak{H}$ is equal to that of ℓ.

Definition 4.1. We shall call the spectrum of $\pi_{(\ell,\mathfrak{H})}$, denoted by $Sp(\pi_{\ell,\mathfrak{H}})$, the set:

$$(4.3) \qquad Sp(\pi_{\ell,\mathfrak{H}}) = \beta(\overline{\Omega}_{\ell,\mathfrak{H}}).$$

Remark 4.2. For the natural topology on $\hat{G}$ (see [BR], [CON]), $\hat{\beta}$ is a homeomorphism from $G\backslash\mathfrak{G}^*$ onto $\hat{G}$. Then the spectrum is a closed set in $\hat{G}$ for this topology. This notion coincides with the notion of support of $\pi_{\ell,\mathfrak{H}}$. (See Dixmier [DI], Guivarch [GU].)

If $\mathfrak{H}$ is a graded subalgebra of $\mathfrak{G}$ and $P \in \mathfrak{U}_m(\mathfrak{G})$ we want to study the maximal hypoellipticity of $\pi_{(0,\mathfrak{H})}(P)$ in the sense of (2.1). Then we have the following theorem:

Theorem 4.3. (Helffer-Nourrigat [HE-NO]). Let $\mathfrak{G}$ be a stratified algebra, $\mathfrak{H}$ a graded subalgebra of $\mathfrak{G}$ and suppose that:

$$(4.4) \qquad \overline{\Omega}_{(0,\mathfrak{H})} \subset [\mathfrak{G}^2,\mathfrak{G}^2]^{\perp}.$$

Let $P \in \mathfrak{U}_m(\mathfrak{G})$. Then the following properties are equivalent:

i) $\forall \pi \in Sp\pi_{(0,\mathfrak{H})} \setminus \{0\}$,

$\pi(P)$ is injective in $\mathcal{S}_\pi$,

ii) $\exists C > 0$ such that $\forall u \in \mathcal{S}_{\pi_{(0,\mathfrak{H})}}$

$$\|u\|^2_{m,\pi_{(0,\mathfrak{H})}} \leq C\|\pi_{(0,\mathfrak{H})}(P)u\|^2_0 .$$

Remark 4.4. The condition (4.4) is always satisfied when $r = 2$, $r = 3$. This condition is probably only of a technical nature. As we shall see later, the part ii) $\Rightarrow$ i) is proved without condition (4.4).

One step in the proof is the following proposition:

Proposition 4.5. (Helffer-Nourrigat [HE-NO]). Let (A_i) $(i = 1,\dots,k)$, (B_j) $(j = 1,\dots,\ell)$ two families in $\mathfrak{U}(\mathfrak{G})$. Let E be the set of the unitary representations π of G such that for each u in $\mathcal{S}_\pi$ we have:

$$(4.5) \qquad \sum_{i=1}^{k} \|\pi(A_i)u\|^2_{\mathfrak{H}_\pi} \leq \sum_{j=1}^{\ell} \|\pi(B_j)u\|^2_{\mathfrak{H}_\pi} .$$

Let $\mathfrak{H}$ be a subalgebra of $\mathfrak{G}$ and let $\ell \in \mathfrak{G}^*$ such that: $\ell([\mathfrak{H},\mathfrak{H}]) = 0$. Then the following properties are equivalent:

i) $\pi_{(\ell,\mathfrak{H})} \in E$,

ii) $Sp(\pi_{\ell,\mathfrak{H}}) \subset E$.

This proposition is essentially an iteration of Proposition 2.1 in $[HE\text{-}NO]_2$ and of the following lemma:

Lemma 4.6. Let ℓ_n be a sequence in $\mathfrak{G}^*$, and $\ell \in \mathfrak{G}^*$ such that $\pi_0(\ell_n) \to \pi_0(\ell)$ (see 1.6). Then, if $\beta(\ell_n) \in E$ $\forall n \geq 0$, $\beta(\ell) \in E$.

In other words, $(E \cap \hat{G})$ is closed in $\hat{G}$. The proof of this lemma follows the proof of Theorem 8.2 in [KI].

Proof of ii) $\Rightarrow$ i). We deduce from ii) and Proposition 4.5 that for every $\pi \in Sp\,\pi_{(0,\mathfrak{H})}$ we have: $\forall u \in \mathcal{S}_\pi$

$$(4.6) \qquad \|u\|^2_{m,\pi} \leq C\|\pi(P)u\|^2_0.$$

In particular, there exists $\tilde{C}$ such that

$$(4.7) \qquad \sum_{i=1}^{p_1} \|\pi(Y_i^m)u\|_0^2 \leq \tilde{C}\|\pi(P)u\|_0^2, \quad \forall u \in \mathcal{S}_\pi.$$

If u satisfies: $\pi(P)u = 0$, then we get immediately:

$$(4.8) \qquad \pi(Y_i^m)u = 0, \quad i = 1,\dots,p_1,$$

which implies

$$(4.9) \qquad \pi(Y_i)u = 0, \quad i = 1,\dots,p_1.$$

Now $\mathfrak{G}$ is stratified, so (4.9) implies:

$$(4.10) \qquad \pi(a)u = 0, \quad \forall a \in \mathfrak{G}.$$

If π is not trivial, this implies: $u = 0$. □

As explained in §2, the proof of the converse is more difficult. To formalize the problem, we first observe that, if $\mathfrak{H}$ is a graded subalgebra, the spectrum of $\pi_{(0,\mathfrak{H})}$ is a closed cone in $\hat{G}$, or equivalently, $\overline{\Omega}_{(0,\mathfrak{H})}$ is a closed conic G-stable subset of $\mathfrak{G}^*$ (by conic, we mean stable by $\delta_t^*(t \in \mathbb{R}^+)$). Then the proof of i) $\Rightarrow$ ii) can be deduced, keeping in mind Proposition 4.5, from the following theorem:

Theorem 4.7 (Helffer-Nourrigat). Let Γ be a closed G-stable cone of $\mathfrak{G}^*$ contained in $[\mathfrak{G}^2,\mathfrak{G}^2]^\perp$. Let us assume that $\mathfrak{G}$ is stratified. Then the two following conditions are equivalent for $P \in \mathfrak{U}_m(\mathfrak{G})$:

(4.11) i) $\forall \pi \in \beta(\Gamma\backslash 0)$, $\pi(P)$ is injective in $\mathcal{S}_\pi$

(4.12) ii) $\exists C, \forall \pi \in \beta(\Gamma), \quad \forall u \in \mathcal{S}_\pi$:

$$\|u\|^2_{m,\pi} \leq C\|\pi(P)u\|^2_0.$$

As in §2 the proof of (4.11) ⇒ (4.12) is articulated in three steps:

Step 1. Proof of (4.12) with remainder term:

$$\|u\|^2_{m,\pi_\ell} \leq C[\|\pi_\ell(P)u\|^2_0 + |\ell_r|^{2m}\|u\|^2_0]. \tag{4.13}$$

For this step, (4.11) with Γ replaced by $\Gamma \cap \mathfrak{G}_r^\perp$ is sufficient.

Step 2. Proof of (4.12) with a constant C_π possibly depending on π.

Step 3. Proof that C_π can be chosen independently of π.

The most technical part is Step 1. We refer to a future paper for the details ([HE-NO]). We want to present here some aspects of the proof which can be useful in other contexts.

Definition 4.8. Let ℓ^n be a sequence in $\mathfrak{G}^*$; we define $\mathfrak{L}(\{\ell^n\})$ as the set of the ℓ in $\mathfrak{G}^*$ such that there exists a subsequence n_i and a sequence $g_i \in G$ such that $g_i \cdot \ell^{n_i} \to \ell$ in $\mathfrak{G}^*$.

In other words, $\pi_0(\mathfrak{L}(\{\ell^n_m\}))$ is the set of the adherent points of $\pi_0(\ell^n_m)$ in $G\backslash\mathfrak{G}^*$.

With this new terminology, Lemma 4.6 says:

$$\begin{cases} \text{If } \ell^n \text{ is a sequence in } \mathfrak{G}^* \text{ such that } \beta(\ell^n) \in E, \text{ then} \\ \beta(\mathfrak{L}(\{\ell^n\}) \subset E. \end{cases} \tag{4.14}$$

The following theorem can be seen as a partial converse:

Theorem 4.9 (Helffer-Nourrigat). Let P in $\mathfrak{U}_m(\mathfrak{G})$ ($\mathfrak{G}$ stratified) and let $(\ell^n)_{n\in\mathbb{N}}$ be a sequence in $[\mathfrak{G}^2,\mathfrak{G}^2]^\perp$. We suppose that $\exists C_0$ such that:

(4.15) $\forall \ell \in \mathcal{L}(\{\ell^n\})$, $\forall u \in \mathcal{S}_{\pi_\ell}$ we have:

$$\|u\|^2_{m,\pi_\ell} \leq C_0 \|\pi_\ell(P)u\|^2_0.$$

(4.16) $\forall \varepsilon > 0$, $\exists C(\varepsilon)$ such that $\forall n \in \mathbb{N}$, $\forall u \in \mathcal{S}_{\pi_{\ell^n}}$

$$\|u\|^2_{m,\pi_{\ell^n}} \leq (C_0+\varepsilon/2)\|\pi_{\ell^n}(P)u\|^2_0 + C(\varepsilon)\|u\|^2_0.$$

(4.17) $$\inf_{g \in G}\|g\ell^n\| \geq 1/2.$$

Then for each $\varepsilon > 0$, $\exists N(\varepsilon)$ such that $\forall n \geq N(\varepsilon)$, $\forall u \in \mathcal{S}_{\pi_{\ell^n}}$ we have:

(4.18) $$\|u\|^2_{m,\pi_{\ell^n}} \leq (C_0+\varepsilon)\|\pi_{\ell^n}(P)u\|^2_0.$$

For $\ell \in \mathfrak{G}^*$, $\|\ell\|$ is the homogeneous quasinorm on $\mathfrak{G}^*$ defined by:

(4.19) $$\|\ell\| = \sum_{j=1}^{r} |\ell_j|^{1/j}_{\mathfrak{G}^*_j}$$

where $|\ |$ is a fixed Euclidean norm on $\mathfrak{G}^*_j$.

<u>Sketch of the proof of (4.7).</u> We admit Theorem 4.9 and the proof of Step 1.

We prove Theorem 4.7 by induction. Suppose that it is proved for stratified algebras of rank $<r$ and let us prove it for rank r. Let $\Gamma_0 = \Gamma \cap \{\ell_r = 0\}$. From Theorem 4.7 (rank r-1), we deduce:

(4.20) $$\begin{cases} \exists C_0,\ \forall \ell \in \Gamma_0,\ \forall u \in \mathcal{S}_{\pi_\ell} \\ \|u\|^2_{m,\pi_\ell} \leq C_0\|\pi_\ell(P)u\|^2. \end{cases}$$

Perhaps after changing C_0, Step 1 gives:

$$(4.21)\quad \begin{cases} \forall\varepsilon > 0, \exists C(\varepsilon) \text{ such that } \forall \ell \in \Gamma \text{ such that } |\ell_r| = 1, \forall u \in \mathcal{S}_{\pi_\ell} \\ \|u\|^2_{m,\pi_\ell} \le (C_0+\varepsilon)\|\pi_\ell(P)u\|^2_0 + C(\varepsilon)\|u\|^2_0. \end{cases}$$

Then as in (2.12), (4.11) and (4.21) give (this is Step 2):

$$(4.22)\quad \begin{cases} \forall \ell \in \Gamma \text{ such that } |\ell_r| = 1, \exists C(\ell) > 0 \text{ such that } \forall u \in \mathcal{S}_{\pi_\ell}: \\ \|u\|^2_{m,\pi_\ell} \le C(\ell)(\|\pi_\ell(P)u\|^2_0). \end{cases}$$

It remains to prove:

(4.23) $C(\ell)$ can be chosen independently of ℓ.

For $j = 1,\dots,r$ and for $(\ell_j,\dots,\ell_r) \in (\mathfrak{G}^j)^*$ such that $|\ell_r| = 1$, define:

$$(4.24)\quad \Gamma^j(\ell_j,\dots,\ell_r) = \{\tilde{\ell} \in \Gamma, \exists g \in G \text{ such that } g\cdot\tilde{\ell}_{\restriction\mathfrak{G}_j} = (\ell_j,\dots,\ell_r)\}.$$

For $j = 1$, $\Gamma(\ell_1,\dots,\ell_n)$ is just the orbit of ℓ if $\ell \in \Gamma$ and $\emptyset$ if $\ell \notin \Gamma$.

For $j = r + 1$, we take the convention that:

$$(4.25)\quad \Gamma^{(r+1)} = \Gamma \cap \{|\ell_r| = 1\}.$$

It is not difficult to see that all these sets are closed in $\mathfrak{G}^*$.

Then we get (4.23) from (4.22) and from the following lemma:

<u>Lemma 4.10 (Helffer-Nourrigat</u>. Let $1 \le j \le r$ and suppose that (4.21) is satisfied.

Suppose we have proved the property:

(P_j) For all $(\ell_j,\dots,\ell_r)$ such that $|\ell_r| = 1$, $\exists C(\ell_j,\dots,\ell_r)$ such that $\forall\tilde{\ell} \in \Gamma^j(\ell_j,\dots,\ell_r)$, $\forall u \in \mathcal{S}_{\pi_{\tilde{\ell}}}$

$$(4.26)_j \qquad \|u\|^2_{m,\pi_{\tilde{\ell}}} \leq C(\ell_j,\ldots,\ell_r)\,\|\pi_{\tilde{\ell}}(P)u\|^2_0 .$$

Then we have the property (P_{j+1}).

(Remark that (P_1) is (4.22) and that (P_{r+1}) is (4.23).)

Proof of Lemma 4.10. Let $(\ell_{j+1},\ldots,\ell_r)$ be fixed with $|\ell_r| = 1$ (if $j = r$, we have nothing to choose). We prove the lemma by contradiction.

Suppose that there exists a sequence ℓ^n in $\Gamma^{j+1}(\ell_{j+1},\ldots,\ell_r)$ and $u_n \in \mathcal{S}_{\pi_{\ell^n}}$ such that:

$$(4.27) \qquad 1 = \|u_n\|^2_{m,\pi_{\ell^n}} \geq n\|\pi_{\ell^n}(P)u_n\|^2 .$$

We apply Theorem 4.9. We have two cases:

(a) $\mathcal{L}(\{\ell^n\})$ is empty.

In this case, all the hypotheses of Theorem 4.9 are satisfied: (4.15) is empty; (4.21) $\Rightarrow$ (4.16); $|\ell^n_r| = 1 \Rightarrow$ (4.17). We deduce from (4.18) that for n big enough:

$$(4.28) \qquad \|u_n\|^2_{m,\pi_{\ell^n}} \leq 2C_0\|\pi_{\ell^n}(P)u_n\|^2_0$$

and we get the contradiction with (4.27).

(b) $\mathcal{L}(\{\ell^n\}) \neq \emptyset$.

Let $\tilde{\ell} \in \mathcal{L}$. Then there exists a sequence of integers $n_i \geq i$ and for each $i \in \mathbb{N}$, $\tilde{\ell}^i \in G\cdot\ell^{n_i}$ such that:

$$\tilde{\ell}^i \underset{i\to\infty}{\longrightarrow} \tilde{\ell}.$$

It is possible to prove (see [PU]) that, in this case, there exists a sequence $\tilde{g}_i \in G$ converging to g in G such that:

$$\tag{4.29} \begin{aligned} \tilde{\ell}^i \restriction \mathfrak{g}^{j+1} &= \tilde{g}_i \cdot (\ell_{j+1}, \ldots, \ell_r) \\ \tilde{\ell} \restriction \mathfrak{g}^{j+1} &= g \cdot (\ell_{j+1}, \ldots, \ell_r). \end{aligned}$$

Let us now define ℓ_j as:

$$\tag{4.30} \ell_j = g^{-1} \cdot \tilde{\ell} \restriction \mathfrak{g}_j .$$

Then we can prove that the set $\mathfrak{L}(\{\tilde{\ell}^i\})$ (which contains of course $\tilde{\ell}$) is contained in $\Gamma^j(\ell_j, \ell_{j+1}, \ldots, \ell_r)$. Then by property P_{j+1}, we have for each $\mu \in \mathfrak{L}(\{\tilde{\ell}^i\})$

$$\tag{4.31} \|u\|^2_{m,\pi_\mu} \leq C\|\pi_\mu(P)u\|^2_0 .$$

All the hypotheses of Theorem 4.9 are satisfied for the sequence $\tilde{\ell}^i$, so we get for $i \geq i_0$ and for each u in $\mathfrak{S}_{\pi_{\tilde{\ell}^i}}$:

$$\tag{4.32} \|u\|^2_{m,\pi_{\tilde{\ell}^i}} \leq 2C\|\pi_{\tilde{\ell}^i}(P)u\|^2_0 .$$

But if you have (4.32) for $\tilde{\ell}^i$, you have the same estimate for $\ell \in G \cdot \tilde{\ell}^i$. In particular, we get:

$$\tag{4.33} \|u\|^2_{m,\pi_{\ell^{n_i}}} \leq 2C\|\pi_{\ell^{n_i}}(P)u\|^2$$

$$\forall u \in \mathfrak{S}_{\pi_{\ell^{n_i}}} , \quad \forall i \geq i_0 .$$

We then get easily the contradiction between (4.33) and (4.27). □

To finish this section, let us explain some variants of Theorem 4.7 which can perhaps be useful in other contexts.

Let Δ be a G-stable closed set in $\mathfrak{g}^*$. For each $\ell \in \Delta$ we define $V_\Delta(\ell)$ as:

(4.34) $$V_{\Delta}(\ell) = \{\tilde{\ell} \in \Delta, \exists \ell^n \in \Delta (n \in \mathbb{N}), g_n \in G (n \in \mathbb{N}) \text{ such that } \ell^n \to \ell, \; g_n \ell^n \to \tilde{\ell}\}.$$

If $\Delta = \mathfrak{G}^*$, we write simply $V(\ell)$. In a dense subset of $\mathfrak{G}^*$, $V(\ell)$ is just the orbit $G \cdot \ell$, but the notion is important at the other points (for example at points where the rank of B_ℓ is not constant in a neighborhood).

Suppose now that: $\Delta \subset [\mathfrak{G}^2, \mathfrak{G}^2]^\perp \cap \{\ell_r = 1\}$ and that:

(4.35) $\exists C_0, C_1$ such that $\forall \ell \in \Delta$, $\forall u \in \mathcal{S}_{\pi_\ell}$:
$$\|u\|^2_{m,\pi_\ell} \le C_0(\|\pi_\ell(P)u\|_0^2) + C_1\|u\|_0^2.$$

Then we have the following proposition:

<u>Proposition 4.11</u> (localized version of Theorem 4.7). Under hypothesis (4.35), the following properties are equivalent for $\ell \in \Delta$.

(i) $\pi_{\tilde{\ell}}(P)$ is injective, $\forall \tilde{\ell} \in V_\Delta(\ell)$,

(ii) $\forall \tilde{\ell} \in V_\Delta(\ell)$, $\exists C(\tilde{\ell})$ such that $\forall u \in \mathcal{S}_{\pi_{\tilde{\ell}}}$,

(4.36) $$\|u\|^2_{m,\pi_{\tilde{\ell}}} \le C(\tilde{\ell})\|\pi_{\tilde{\ell}}(P)u\|_0^2,$$

(iii) $\exists C$, $\forall \tilde{\ell} \in V_\Delta(\ell)$, $\forall u \in \mathcal{S}_{\pi_{\tilde{\ell}}}$,

(4.37) $$\|u\|^2_{m,\pi_{\tilde{\ell}}} \le C\|\pi_{\tilde{\ell}}(P)u\|^2,$$

(iv) $\exists$ a neighborhood $W(\ell)$ of ℓ in Δ, $\exists \tilde{C}$ such that $\forall \tilde{\ell} \in W(\ell)$, $\forall u \in \mathcal{S}_{\pi_{\tilde{\ell}}}$:

(4.38) $$\|u\|^2_{m,\pi_{\tilde{\ell}}} \le \tilde{C}\|\pi_{\tilde{\ell}}(P)u\|^2.$$

<u>Proof</u>. (i) $\Rightarrow$ (ii) is just proved like Theorem 4.7, Step 2.

(ii) $\Rightarrow$ (iii) is just proved like Theorem 4.7, Step 3.

(iii) $\Rightarrow$ (iv).

Suppose (iv) is not true. Then there exists a sequence in ℓ^n in Δ such that $\ell^n \Rightarrow \ell$ and $u_n \in \mathcal{S}_{\pi_{\ell^n}}$ such that (4.27) is satisfied. Then the proof is like the proof of Lemma 4.10 if we remark that $\Sigma(\{\ell^n\}) \subset V_\Delta(\ell)$.

(iv) $\Rightarrow$ (iii). It is clear that (4.38) is also true for $G \cdot W(\ell)$. Applying Lemma 4.6, we get (iii).

(iii) $\Rightarrow$ (i). (See the proof of ii) $\Rightarrow$ i) in Proposition 4.5.) $\square$

§5. Operators on homogenous spaces: the algebraic approach

In this section, we want to present an alternative approach to the proof of Theorem 4.3 in the case where $\overline{\Omega}_{(0,\mathfrak{H})}$ is an algebraic subset of $\mathfrak{G}^*$. Details for this chapter can be found in $[HE]_2$ or in $[HE\text{-}NO]_9$. Keeping the notations of §4, we introduce:

(5.1) $\quad \widetilde{\Omega}_{(0,\mathfrak{H})}$ is the Zariski-closure of $G \cdot \mathfrak{H}^{\perp}$ in $\mathfrak{G}^*$.

We now call the <u>algebraic spectrum</u> of $\pi_{(0,\mathfrak{H})}$ the set:

$$\widetilde{Sp}\pi_{(0,\mathfrak{H})} = \beta(\widetilde{\Omega}_{(0,\mathfrak{H})}). \tag{5.2}$$

Of course, we have the inclusions:

$$\overline{\Omega}_{(0,\mathfrak{H})} \subset \widetilde{\Omega}_{(0,\mathfrak{H})}; \quad Sp(\pi_{(0,\mathfrak{H})}) \subset \widetilde{Sp}\pi_{(0,\mathfrak{H})} \tag{5.3}$$

but these inclusions can be strict as in:

<u>Example 5.1</u>. (See 3.7). $\mathfrak{G} = \mathfrak{N}_4$, $\mathfrak{H} = \mathbb{R}Z$. Then we have:

$$\overline{\Omega}_{(0,\mathfrak{H})} = \{\ell = (\eta_1,\eta_2,\zeta,w^*),\ 2w^* \cdot \eta_2 - \zeta^2 \leq 0\}$$

and

$$\widetilde{\Omega}_{(0,\mathfrak{H})} = \mathfrak{G}^*.$$

We have the following theorem:

<u>Theorem 5.2</u>. Let P be in $\mathfrak{U}_m(\mathfrak{G})$. Suppose that:

(5.4) $\pi(P)$ is injective in $\mathcal{S}_\pi$ for all π in $\widetilde{Sp}\pi_{(0,\mathfrak{H})}$ (π nontrivial).

Then $\pi_{(0,\mathfrak{H})}(P)$ is maximally hypoelliptic.

<u>Remark 5.3</u>. When $\widetilde{Sp}\pi_{(0,\mathfrak{H})} = Sp\pi_{(0,\mathfrak{H})}$, Theorem 5.2 gives the proof of Theorem 4.3 under other hypotheses.

In the case of Example 5.1, it does not give anything new. In fact, the theorem says only that if P is hypoelliptic, $\pi_{(0,\mathfrak{A})}(P)$ is hypoelliptic.

<u>Sketch of the proof of Theorem 5.2.</u> Let I be the ideal (two-sided) defined by:

(5.5) $$I = \{A \in \mathfrak{U}(\mathfrak{G});\ \pi_{(0,\mathfrak{A})}(A) = 0\}.$$

Then, using classical results (see Dixmier [DI]$_3$), we can construct a system of generators for I: A_i^j $(i = 1,\dots,k, j = 1,\dots,n_i)$ with the following properties:

(5.6) $$A_i^j \text{ is homogeneous for } \delta_t;$$

if

(5.7) $$I_\ell \text{ is the ideal generated by the } A_i^j \text{ for } i \leq \ell$$

then

(5.8) $$[\mathfrak{U}(\mathfrak{G}), A_\ell^j] \subset I_{\ell-1} \text{ for } \ell = 1,\dots,k$$

(with the convention that $I_0 = 0$).

The crucial point is to have another characterization of (5.2). Using results of Duflo [DU] and R. Rentschler [RE]$_{1,2}$, we can prove:

<u>Proposition 5.4.</u>

(5.9) $$\widetilde{\mathrm{Sp}}\ \pi_{(0,\mathfrak{A})} = \{\pi \in \hat{G},\ \pi(I) = 0\}.$$

Then Theorem 5.2 is easy to prove. Consider the system:

$$\mathfrak{P} = (P, A_i^j).$$

According to a natural extension of Theorem 2.1 to systems, we have just to prove that the system $\mathfrak{P}$ verifies (RO) which means here:

$\forall \pi \in \hat{G}\setminus\{0\}$

$$(5.10) \qquad \left.\begin{array}{l} u \in \mathcal{S}_\pi \\ \pi(P)u = 0 \\ \pi(A_i^j)u = 0 \end{array}\right\} \Rightarrow u = 0.$$

Suppose we have proved (5.10) for π verifying:

$$\pi(I_{\ell-1}) = 0.$$

Then, according to (5.8), we see that $\pi(A_\ell^j)$ is a scalar.

If $\pi(A_\ell^j) \neq 0$ for some j, we get immediately $u = 0$; in the other case we have $\pi(I_\ell) = 0$.

By induction, we reduce the problem to the case where $\pi(I) = 0$.

According to Proposition 5.4, (5.4) gives (5.10) in this case. □

§6. Local solvability: results of P. Levy-Bruhl

The aim of this section is not to present a complete survey on local solvability. For a general survey, you can see the survey of L. Corwin in this conference. We just want to show how results of local solvability are connected with the preceding ones.

Recall first that a differential operator on Ω is called locally solvable if it has the following property:

$$(\mathrm{L.S.})\begin{cases}\forall x_0 \in \Omega,\ \exists V_{x_0} \text{ (open neighborhood of } x_0) \text{ such that} \\ \text{for each } f \in C_0^\infty(V_{x_0}),\ \exists T \in \mathcal{D}'(V_{x_0}) \text{ such that} \\ PT = f \text{ in } V_{x_0}.\end{cases}$$

A classical result of functional analysis says that if the formal adjoint P^* is hypoelliptic then P is locally solvable. So we get immediately from Theorem 5.2:

$$(6.1)\begin{cases}\text{Let } P \text{ be in } \mathfrak{U}_m(\mathfrak{G}), \text{ and suppose that for each } \pi \text{ in } \hat{G}\setminus\{0\}, \\ \pi(P^*) \text{ is injective in } \mathcal{S}_\pi; \text{ then } P \text{ is locally solvable.}\end{cases}$$

This result is in fact too weak. On one side, it is well known that all non-zero bi-invariant operators are locally solvable. On the other side, well known examples (on the Heisenberg group) show that local solvability is not true for all left invariant operators.

In the last few years, some effort has been made to understand better the problem of local solvability for left invariant operators ($[\mathrm{CO}]_{1,2}$, [C.R], $[\mathrm{LE}]_{1,2,3}$, [LI], $[\mathrm{ROT}]_2$, [R.T]).

The first result I want to present is a result of Corwin-Rothschild [C.R], recently improved by P. Levy-Bruhl $[\mathrm{LE}]_3$.

Theorem 6.1. Let P in $\mathfrak{U}_m(\mathfrak{G})$ be such that

i) P^* satisfies [RO-DE] (see §2).

ii) $\operatorname{Ker} \pi(P^*) \neq \{0\}$ for all π in an open set $(\neq \emptyset)$ in $\hat{G}$.

Then P is not locally solvable.

In the case where RO-degenerate is not satisfied, the situation is not clear, as we can see from the two following examples:

Example 6.2. ([C.R]). Let us consider $\mathfrak{G} = \mathfrak{N}_4$ (see §3) and the operator: $L = Y_i + iY_2$. (Y_i+iY_2) is not invertible but $\pi_{\beta,\gamma}(L)$ is injective for all $(\beta,\gamma) \in \mathbb{R} \times \mathbb{R}\setminus\{0\}$ in $\mathcal{S}(\mathbb{R})$. It is easy to see that (RO-DE) is not satisfied.

Example 6.3. ([C.R]). Let us consider the three dimensional Heisenberg group.

$$H_3 = \mathbb{R}X \oplus \mathbb{R}Y \oplus \mathbb{R}Z$$

$$\text{with } [X,Y] = Z.$$

$P = X[Y^2-iZ]$ is locally solvable, but we can find an open subset in $\hat{G}$ where $\pi(P^*)$ is not injective in $\mathcal{H}_\pi$ (but injective in $\mathcal{S}_\pi$).

In example (6.3), the problem is that when (RO-DE) is not satisfied, we can lose the property:

$$\operatorname{Ker}\pi(P^*) \cap \mathcal{S}_\pi = \operatorname{Ker} \pi(P^*) \cap \mathcal{S}'_\pi$$

(which is proved in $[\text{HE-NO}]_{1,2}$)

and the notion of injectivity in $\mathcal{S}_\pi$ does not coincide with a property of left invertibility.

Then the problem of finding sufficient conditions for local solvability can be posed like this:

"Does there exist a conic closed set $\hat{F}$ in $\hat{G}$ (as small as possible!) such that the injectivity of $\pi(P^*)$ for π in $\hat{F}\setminus\{0\}$ implies local solvability for P?"

As mentioned at the beginning of the section, we know that we can take $\hat{F} = \hat{G}$ but the result is not very interesting. The examples given before suggest that it is better to assume that $\hat{F}$ contains $\beta(\mathfrak{G}_r^{\perp})$. Let us present a result of this type obtained by P. Levy-Bruhl $[LE]_2$.

Theorem 6.4. (P. Levy-Bruhl). Let $\mathfrak{G} = \mathfrak{G}_1 \oplus \mathfrak{G}_2$. Let Ω be the open set in $\mathfrak{G}_2^*$ where B_{ℓ_2} restricted to $\mathfrak{G}_1 \times \mathfrak{G}_1$ is of maximal rank and let d be the dimension of the kernel of B_{ℓ_2} in $\mathfrak{G}_1$.

Assume that

$$d \geq 1. \tag{6.2}$$

Let P in $\mathfrak{U}_m(\mathfrak{G})$ be such that

$$\pi_\ell(P^*) \text{ is injective for each } \ell \in \mathfrak{G}_1^* \times {}^c\Omega\backslash\{0\}. \tag{6.3}$$

Then P is locally solvable.

Remark 6.5. If $[\mathfrak{G}_1,\mathfrak{G}_1] = 0$, no condition appears (we are in the constant coefficient case).

Remark 6.6. If $[\mathfrak{G}_1,\mathfrak{G}_1] \neq \{0\}$, then $0 \notin \Omega$ and the condition (6.3) contains (RO-DE).

Remark 6.7. If B_ℓ is of constant rank for $\mathfrak{G}_2 \neq 0$, then $\Omega = \mathfrak{G}_2^*\backslash\{0\}$ and the condition (6.3) is reduced to (RO-DE).

When $d = 0$, the situation is different. Let us introduce a definition:

Definition 6.8 (See $[MET]_1$). Let $\mathfrak{G} = \mathfrak{G}_1 \oplus \mathfrak{G}_2$. We shall say that G (or $\mathfrak{G}$) is of type (H) if $\Omega = \mathfrak{G}_2^*\backslash\{0\}$ and $d = 0$.

In the case where $\mathfrak{G}$ is of type (H), we have a quite satisfactory result:

Theorem 6.9. (Rothschild-Tartakoff [R.T]). Let P be in $\mathfrak{U}_m(\mathfrak{G})$, $\mathfrak{G}$ of type (H), and assume that RO-DE is satisfied. Then the following conditions are equivalent:

i) P is locally solvable,

ii) There is no open set $U \subset \mathfrak{G}^*$ such that $\pi_\lambda(P^*)$ has a non-trivial kernel for all $\lambda \in U$.

To finish this presentation, we just give an example corresponding to the case: $r = 3$. (See P. Levy-Bruhl [Ll]$_2$ for more general results)

<u>Theorem 6.10</u>. (P. Levy-Bruhl). Let $\mathfrak{G} = \mathfrak{N}_4$ (see §3) and $P \in \mathfrak{U}_m(\mathfrak{G})$; then if P^* satisfies [RO-DE], P is locally solvable. (Compare this with Example 6.2.)

The proof of Theorems 6.4 and 6.10 is based on the Plancherel Formula. Let us explain the technique in the case of Theorem 6.10.

If $\varphi \in L^1(G)$ and $\pi \in \hat{G}$, we can define:

$$\pi(\varphi) = \int \varphi(g)\pi(g^{-1})dg \tag{6.4}$$

where dg is a Haar measure.

With the notations of §3, (3.11) we have:

$$(2\pi)^4 \int_{\mathbb{R}^2} |\gamma| \operatorname{tr} \pi_{\beta,\gamma}(\varphi)d\beta d\gamma = \varphi(e) \tag{6.5}$$

where e is the unit in G.

Formally, if we define $E(\varphi)$ by:

$$E(\varphi) = (2\pi)^4 \int \operatorname{tr}(\pi^{-1}_{\beta,\gamma}(P) \cdot \pi_{\beta,\gamma}(\varphi))|\gamma|d\beta d\gamma, \tag{6.6}$$

then we have:

$$\langle PE,\varphi\rangle = \varphi(e) \quad \text{for each } \varphi \in \mathcal{D}(G). \tag{6.7}$$

In fact, there are many difficulties in giving a sense to (6.6), in particular $\pi^{-1}_{\beta,\gamma}(P)$ is not always invertible. We suppose to simplify things that P is self adjoint.

We modify Formula (6.6) like this: for each $\varphi \in \mathcal{D}(G)$ we define a distribution F_φ such that:

(6.8) $$\overline{P}\cdot F_\varphi = L\cdot\varphi$$

where $L = W^n$, n great enough

by

(6.9)

$$<F_\varphi,\psi>_{\mathcal{D}',\mathcal{D}} = \int_{|\beta|\geq K|\gamma|^{1/3}} \mathrm{Tr}[\pi_{\beta,\gamma}(\psi)(\pi_{\beta,\gamma}(P))^{-1}\pi_{\beta,\gamma}(L\varphi)]|\gamma|d\beta d\gamma$$

$$+\sum_j\int_{|\beta|\leq K|\gamma|^{1/3}}\psi_j(\beta,\gamma)\cdot\mathrm{tr}[(\pi_{\beta,\gamma}(\psi)(\pi_{\beta,\gamma}(P)\,{}^{\dagger}W_j(\beta,\gamma))^{-1}(I-\mathbb{P}_j(\beta,\gamma))\cdot$$

$$\pi_{\beta,\gamma}(L\varphi)]|\gamma|d\beta d\gamma$$

$$+\sum_j\int_{|\beta|\leq K|\gamma|^{1/3}}\psi_j(\beta,\gamma)\frac{1}{2i\pi}\int_{\Gamma_j(\beta,\gamma)}\mathrm{Trace}[\pi_{z,\gamma}(\psi)(\pi_{z,\gamma}(P)\,{}^{\dagger}V_j(z,\gamma))^{-1}$$

$$(\mathbb{P}_{j(z,\gamma)})\pi_{z,\gamma}(L\varphi)]\frac{dz}{z-\beta}\cdot|\gamma|d\beta d\gamma$$

$$= (I) + (II) + (III).$$

Let us explain the different terms which appear in (6.9).

<u>Study of (I)</u>. (I) has a good sense because $(\pi_{(\beta,\gamma)}(P))^{-1}$ is invertible for $|\beta| \geq K|\gamma|^{1/3}$ with K big enough. This is an immediate consequence of (RO-DE) which gives:

(6.10) $$\|D_t^m u\|^2 + \|(\pm\frac{t^2}{2}-\beta)^m u\|^2$$
$$+ |\beta|^{\frac{2m}{3}}\|u\|^2 \leq C[\|\pi_{\beta,\pm 1}(P)u\|^2 + \|u\|^2]$$

with C independent of β.

For $|\beta| \geq K$, this gives the invertibility of $\pi_{\beta,\pm 1}(P)$. We then conclude by a homogeneity argument.

<u>Study of II</u>. $\mathbb{P}_j(\beta,\gamma)$ is a projector associated to a convenient contour $\tilde{\Gamma}_j(\beta,\gamma)$ in $\mathbb{C}$.

(6.11) $$\mathfrak{P}_j(\beta,\gamma) = \frac{1}{2i\pi}\int_{\tilde{\Gamma}_j(\beta,\gamma)} (\pi_{\beta,\gamma}(P)-s)^{-1}\, ds.$$

The contour is chosen such that, if $W_j(\beta,\gamma) = \mathrm{Im}(I-\mathfrak{P}_j(\beta,\gamma))$, the spectrum of $\pi_{\beta,\gamma}(P) \restriction W_j(\beta,\gamma)$ satisfies

(6.12) $$\text{spectrum } \pi_{\beta,\gamma}(P)\restriction_{W_j(\beta,\gamma)} \subset \{z \in \mathbb{C} \mid |z| \geq C|\gamma|^{m/3}\}.$$

$\psi_j(\beta,\gamma)$ is a partition of unity introduced for technical reasons.

<u>Study of III</u>. Let $V_j(\beta,\gamma) = \mathrm{Im}(\mathfrak{P}_j(\beta,\gamma))$; then $\pi_{\beta,\gamma}(P)\restriction_{V_j(\beta,\gamma)}$ is not necessarily invertible.

Here the idea is to go into the complex plane (complexification of β) and to find a contour $\Gamma_j(\beta,\gamma)$ in $\mathbb{C}$ such that $\pi_{z,\gamma}(P)\restriction_{V_j(z,\gamma)}$ is invertible for $z \in \Gamma_j(\beta,\gamma)$.

Note that according to Formulas (3.11), there is a natural way to define $\pi_{z,\gamma}$ for $z \in \mathbb{C}$.

In fact you need a more precise information on the distance between the spectrum of $\pi_{z,\gamma}(P)\restriction_{V_j(z,\gamma)}$ and 0. The "good" properties of the family $\pi_{z,\gamma}(P)$ are derived also from RO-degenerate.

<u>End of the proof</u>. We remark that L is locally solvable (it is an operator with constant coefficients); then (6.8) gives the local solvability for $\overline{P}$ and then for P.

<u>Philosophy of the proof</u>. The choice of the set $\hat{F}$ (in Theorem 6.10 it is just $\beta(\mathfrak{G}_3^{\perp})$) permits us to reduce the problem of inverting $\pi(P)$ $(\pi \in \hat{G})$ to a problem of invertibility for a family of operators depending smoothly on a parameter lying in a compact set K of $G\backslash\mathfrak{G}^* \cap |\ell_r| = 1$ where the orbits are regular.

In a particular case, this is what appears in our study of (I). The choice of the set $\hat{F}$ permits us also to obtain good properties for the family $\pi_\lambda(P)$ for $\lambda \in \bigcup_{t>0} (\delta_t K)$.

Another point is to have something to complexify. This is the sense of hypothesis (6.2) in Theorem 6.4. Theorem 6.9 shows us that this condition is not only technical.

§7. Hypoanalyticity. In this section, we want to make a survey of the known results on hypoanalyticity for elements in $\mathfrak{U}_m(\mathfrak{G})$ and discuss some open problems. (See the survey of L. Rothschild in this conference for other results.) Recall first that a differential operator P on $\Omega \subset \mathbb{R}^n$ is called hypoanalytic if we have:

$$\text{(H.A.)} \qquad \left.\begin{array}{l} \forall\, u \in \mathcal{D}'(\Omega) \\ \forall\, \omega \ \text{ open set in } \ \Omega \\ \quad Pu \in \mathcal{A}(\omega) \end{array}\right\} \Rightarrow u \in \mathcal{A}(\omega).$$

The first general result in this direction is the following theorem of G. Métivier $[\mathrm{MET}]_1$:

Theorem 7.1 (G. Métivier). Let G be a nilpotent group of type (H) (see definition 6.8) and let P be in $\mathfrak{U}_m(\mathfrak{G})$. Then the following conditions are equivalent:

(i) P is hypoanalytic.

(ii) P is hypoelliptic,

(iii) For all π in $\hat{G}\backslash 0$, $\pi(P)$ is injective in $\mathcal{S}_\pi$,

(iv) For all π in $\hat{G}\backslash 0$, $\pi(P)$ is injective in E_π,

where E_π is the set of entire vectors of π.

E_π is the set in $\mathcal{H}_\pi$ of the v such that

$$g \to \pi(g)v$$

extends to an entire function with value in $\mathcal{H}_\pi$ on $G_{\mathbb{C}}$ (the complexified group of $G = \exp \mathfrak{G}$).

This result generalizes preliminary results on the Heisenberg group and results of F. Trèves $[\mathrm{TR}]_2$ and D. S. Tartakoff [TA], and is the starting point of more general results in P.D.E. of G. Métivier ($[\mathrm{MET}]_3$), J. Sjöstrand (to appear), A. Grigis - J. Sjöstrand (in preparation). We want also to mention recent results of A. Grigis-

L. P. Rothschild ([G.S] and this conference). When condition (H) is not satisfied, it has been proved by G. Métivier ($[\mathrm{MET}]_2$) that non-hypoanalytic, hypoelliptic operators appear in $\mathfrak{U}_2(\mathfrak{G})$ (with $\mathfrak{G}$ of rank 2). Just interpreting Métivier's Proof, we give in $[\mathrm{HE}]_3$ the following theorem:

Theorem 7.2. Let $\mathfrak{G} = \mathfrak{G}_1 \oplus \mathfrak{G}_2$ and P in $\mathfrak{U}_m(\mathfrak{G})$ a hypoelliptic operator. Suppose that:

(7.1) $\exists \ell_2 \in \mathfrak{G}_2^*\backslash 0$ such that B_{ℓ_2} (restricted to $\mathfrak{G}_1 \times \mathfrak{G}_1$) is degenerate.

Let R_{ℓ_2} be the kernel of B_{ℓ_2} (in $\mathfrak{G}_1$). Then if there exists $\ell_1 \in (\mathbb{R}_{\ell_2}^*)^{\mathbb{C}}$ such that

(7.2) $\pi_{\ell_2,\ell_1}(P)$ is not injective in $E_{\pi(\ell_2,\ell_1)}$,

P is not hypoanalytic.

Here $\pi_{(\ell_2,\ell_1)}$ is naturally defined (as a nonunitary representation) by complexifying ℓ_1 (see the analogy with techniques which appear in the proof of Theorems 6.4 and 6.10).

As an immediate corollary (by looking at properties of harmonic oscillators), we find Métivier's Theorem $[\mathrm{MET}]_2$ in a particular case:

Corollary 7.3 (Métivier). Let $\mathfrak{G} = \mathfrak{G}_1 \oplus \mathfrak{G}_2$ and $P = \sum_{i=1}^{p_1} Y_i^2$ (where $(Y_1,\ldots,Y_{p_i})$ is a basis of $\mathfrak{G}_1$). If $\mathfrak{G}$ does not satisfy condition (H), then P is not hypoanalytic.

Theorem 7.2 suggests that the hypoanalyticity of P is related to injectivity of $\pi(P)$ for a family of not necessarily unitary representations associated to $G_{\mathbb{C}}$ orbits in $(\mathfrak{G}_{\mathbb{C}})^*$. A step in this direction has been made in $[\mathrm{HE}]_4$:

Theorem 7.4. (B. Helffer). If $\mathfrak{G}$ is not stratified, there do not exist hypoanalytic operators in $\mathfrak{U}_m(\mathfrak{G})$ (for $m > 0$).

Theorem 7.4 is a natural extension of the fact that the heat equation is not hypoanalytic.

Theorem 7.5 (B. Helffer). Let $\mathfrak{G}$ a stratified algebra of rank $r \geq 3$. Then if $\dim \mathfrak{G}_2 = 1$, the operator: $\sum_{i=1}^{p_1} Y_i^2$ is not hypoanalytic.

Let us give the proof of Theorem 7.5 (which has many analogies to the proof of Theorem 3.1 and Theorem 6.10).

By Theorem 7.2 and the condition: $\dim \mathfrak{G}_2 = 1$, we can reduce the problem to the case where $\mathfrak{G}/\mathfrak{G}^3$ is the Heisenberg algebra. Then by Lemma 3.2, we are reduced to the case where:

$$P = Y_1^2 + Y_2^2$$

on the algebra N_4 defined in (3.7).

Let us consider (cf 3.11 , 3.12) the representation π_β, and consider

$$\pi_\beta(Y_1^2+Y_2^2) = \partial_t^2 - (t^2-\beta)^2, \quad \beta \in \mathbb{C}.$$

As in the proof of Theorem 6.10, we have complexified β. Suppose we have proved the following lemma:

Lemma 7.6. (Pham the Lai-D. Robert). $\exists \beta \in \mathbb{C}$ and $u \in \mathcal{S}(\mathbb{R})$, $u \neq 0$ such that

$$(7.3) \qquad (\partial_t^2 - (t^2-\beta)^2)u = 0.$$

(The proof of this lemma is a corollary of results of Pham the Lai-D. Robert (Sur un problème aux valeurs propres non linéaire, Israel Journal of Mathematics, Vol. 36, n°2 (1980)).) Then we shall contradict the hypoanalycity of $(Y_1^2+Y_2^2)$ by constructing locally a non-analytic solution in the kernel of $(Y_1^2+Y_2^2)$.

In adapted exponential coordinates identifying G and $\mathbb{R}^4$, $(Y_1^2+Y_2^2)$ can be written as the following operator:

$$Y_1^2 + Y_2^2 = \partial_{y_1}^2 + (\partial_{y_2} + y_1\,\partial_z + \frac{y_1^2}{2}\cdot\partial_w)^2.$$

Then we introduce the following C^∞ function in the neighborhood of 0 in $\mathbb{R}^4$:

$$v(y_1,y_2,z,w) = \int_0^{+\infty} e^{-\rho}\, e^{i[2\rho^3 w - \rho\beta y_2]}\, u(\rho y_1)\,d\rho.$$

then, for $|y_2|$ small enough, v is a C^∞ function satisfying:

$$(Y_1^2 + Y_2^2)v = 0.$$

Then, because it is not difficult to prove that $u(0)$ or $u'(0)$ is different from 0, we see easily that $v(0,0,0,w)$ or $(\partial_{y_1} v)(0,0,0,w)$ is not analytic in the neighborhood of 0.

<u>Remark 7.7.</u> The same proof gives the nonhypoanalyticity of the following operators which can be considered as operators on homogeneous spaces of $\mathbb{N}_4$:

$$\partial_{y_1}^2 + (\frac{y_1^2}{2}\,\partial_w + \partial_{y_2})^2 \quad \text{in } \mathbb{R}^3$$

$$\partial_{y_1}^2 + (\frac{y_1^2}{2}\,\partial_w + y_1\partial_z)^2 \quad \text{in } \mathbb{R}^3.$$

But we know also that the following operators are hypoanalytic:

(7.4) $\partial_{y_1}^2 + (y_1\partial_z)^2$ in $\mathbb{R}^2$ (Grušin [GRU]$_1$)

$\partial_{y_1}^2 + \left(\frac{y_1^2}{2}\,\partial_w\right)^2$ in $\mathbb{R}^2$ (Grušin [GRU]$_1$)

$\partial_{y_1}^2 + (y_1\partial_z + \partial_{y_2})^2$ in $\mathbb{R}^3$ (Theorem 7.1)

$\partial_{y_1}^2 + \partial_{y_2}^2$ in $\mathbb{R}^2$ (Elliptic case).

It would be interesting to have a group theoretical criterion for the hypoanalyticity of homogeneous operators on homogeneous spaces as in Theorem 4.3.

Let us just make a remark in this direction for example 7.4. Let $\mathfrak{H}$ be the subalgebra in $\mathfrak{N}_4$ generated by $\mathbb{R}Y_2 \oplus \mathbb{R}Z$. Then $\overline{G\cdot\mathfrak{H}^{\perp}}$ is given in $\mathbb{R}^4 \approx \mathfrak{G}^*$ by:

$$(7.5) \quad \overline{G\cdot\mathfrak{H}^{\perp}} = \{\ell = (\eta_1, \eta_2, \zeta, w^*), \quad 2w^* \cdot \eta_2 = \zeta^2\}.$$

In Section 5, we have seen that we can deduce the hypoellipticity of $\pi_{(0,\mathfrak{H})}(Y_1^2 + Y_2^2) = \partial_{y_1}^2 + \frac{y_1^4}{4}\partial_w^2$ from the hypoellipticity of the system:

$$(7.6) \quad \begin{cases} Y_1^2 + Y_2^2 \\ 2WY_2 - Z^2 \end{cases}$$

(of course, the C^∞ result is well known (see [HO] or [GRU]$_2$)). Then we can hope to deduce the hypoanalyticity of $\pi_{(0,\mathfrak{H})}(Y_1^2 + Y_2^2)$ from the hypoanalyticity of the system (7.6).

Let us just remark here, that the system:

$$(7.7) \quad \begin{cases} \pi_\beta(Y_1^2 + Y_2^2) \\ \pi_\beta[2WY_2 - Z^2] \end{cases}$$

is injective for all $\beta \in \mathbb{C}$. Indeed, this is simply the system:

$$(7.8) \quad \begin{cases} \partial_t^2 - (t^2 - \beta)^2 \\ 2\beta \end{cases}$$

Equivalently, we have the property that:

$$(7.9) \quad \pi_\ell(Y_1^2 + Y_2^2) \text{ is injective for } \ell \in \widehat{G_{\mathbb{C}}\cdot\mathfrak{H}_{\mathbb{C}}^{\perp}} \cap \{(\mathfrak{G}_1^*)^{\mathbb{C}} \times (\mathfrak{G}^2)^*\}.$$

This example suggests a new way for studying hypoanalyticity for homogeneous operators on homogeneous spaces in the case where $\overline{G\cdot H^{\perp}}$ is algebraic. But actually, we do not know any results on hypoanalyticity for systems when $\mathfrak{G}$ is <u>not</u> of type (H).

§8. References

[BE] R. Beals: Opérateurs invariants hypoelliptiques sur un groupe de Lie nilpotent. Séminaire Goulaouic-Schwartz 76-77, exposé n°19.

[BR] I. D. Brown: Dual topology of a nilpotent Lie group. Ann. Scient. Ec. Normale Sup. 4ème série, t.6 (1973), p. 407-441.

[CON] N. Conze - Berline: Espace des idéaux primitifs de l'algèbre enveloppante d'une algèbre nilpotente. J. of Algebra 34, 1975, p. 444-450.

[COR] L. Corwin: [1] A representation theoretic criterion for local solvability of left invariant differential operators in nilpotent Lie groups. Trans. of the A.M.S. 246 (1981) p. 113-120.
[2] Criteria for solvability of left invariant operators on Nilpotent Lie groups. Trans. of the A.M.S. (to appear).

[C.R.] L. Corwin - L. P. Rothschild: Necessary conditions for local solvability of homogeneous left invariant operators on Nilpotent Lie groups. Acta Mathematica 147(1981), #3-4, p.265-288.

[DI] J. Dixmier: [1] Représentations irréductibles des algèbres de Lie nilpotentes. Anais. Acad. Brasil 35, 1963, p. 491-519.
[2] Les C*-algèbres et leurs représentations, Paris (Gauthiers-Villars).
[3] Algèbres enveloppantes, Paris (Gauthiers-Villars).

[DU] M. Duflo: Sur les extensions des représentations irréductibles des groupes de Lie nilpotents. Ann. Scient. Ec. Norm. Sup. 4ème série, t.5, 1972, p. 71-120.

[FED] Fedii: On a criterion of hypoellipticity, Math - Sbornik, vol. 14, (1971), n°1., 15-46.

[FEL] J. M. G. Fell: The dual spaces of C*-algebras. Trans. of the A.M.S., Mars 1960, Vol. 94 n°3, p. 365-403.

[FO] G. B. Folland: [1] A fundamental solution for a subelliptic operator Bull. Amer. Math. Soc. 79 (1973) p. 373-376.
[2] Subelliptic estimates and function spaces on nilpotent Lie groups. Ark. F. Mat. 13 (1975), 161-207.
[3] On the Rothschild-Stein lifting theorem. Comm. in P.D.E. 2(2) p. 165-191 (1977),

[GO] R. Goodman: Nilpotent Lie groups. Lecture Notes in Math. n° 562.

[G.S.] A. Grigis - L. P. Rothschild: A criterion for analytic hypoellipticity of a class of differential operators with polynomial coefficients. Ann. Math. (to appear).

[GRU] V. V. Grusin: [1] On a class of elliptic p.d.o. degenerate on a submanifold. Math. Sbornik, Tom. 84,(126) (1971) n°2, p.163-195.
[2] On a class of hypoelliptic operators. Math. Sbornik, Tom. 83 (1970) p. 456-473.

[GU] Y. Guivarc'h: Croissance polynomiale et périodes des fonctions harmoniques. Bull. Soc. Math. France 101 (1973) p. 333-379.

[HE] B. Helffer: [1] Hypoellipticité pour des opérateurs différentiels sur des groupes nilpotent. (Cours du C.I.M.E. 1973).
[2] Hypoellipticité pour des opérateurs quasihomogènes à coefficients polynomiaux II. Preprint de l'école polytechnique, Mai 1979.
[3] Remarques sur des résultats de G. Métivier sur la non-hypoanalyticité Séminaire de l'Université de Nantes - Exposé n° 9, 1978-79.
[4] Conditions nécessaires d'hypoanalyticité pour des opérateurs invariants à gauche homogènes sur un groupe nilpotent gradué. Journal of differential equations, Vol.44 n°3, Juin 1982, p.460-481.
[5] On the hypoellipticity of operators of the type $\sum_{j=1}^{n} Y_j^2 + \frac{1}{2}\sum_{j,k} C_{jk}[Y_j,Y_k]$. Séminaire de l'Université de Nantes - Exposé n° 1. 1981-82.

[HE-NO] B. Helffer - J. Nourrigat: [1] Hypoellipticité pour des groupes de rang 3. Comm. in P.D.E. 3(8) (1978) p. 643-743.
[2] Caractérisation des opérateurs hypoelliptiques homogènes invariants à gauche sur un groupe nilpotent gradué. Comm. in P.D.E. 4(9) (1979) p. 899-958.
[3] Hypoellipticité pour des opérateurs quasi-homogènes à coefficients polynomiaux. Preprint de l'Ecole Polytechnique Mars 1979.
[4] Approximation d'un système de champs de vecteurs et applications à l'hypoellipticité. Arkiv för Matematik (1979) n° 2, p. 237-254.
[5] Hypoellipticité maximale pour des opérateurs polynômes de champs de vecteurs (une nouvelle démonstration d'un théorème de L. P. Rothschild). Preprint 1980.

[6] Hypoellipticité pour des opérateurs quasi-homogènes à coefficients polynomiaux. (Actes du Colloques de Saint Cast, Juin 1979).

[7] Hypoellipticité maximale pour des opérateurs polynômes de champs de vecteurs. Note aux C. R. Acad. Sc. (Octobre 1979), Manuscrit et actes du colloque de Saint-Jean-de-Monts, Juin 1980.

[8] Hypoellipticité maximale pour des opérateurs polynômes de champs de vecteurs. Rendiconti del Sem. Math. Univ. Polit. Torino (à paraître).

[9] Book in preparation.

[HO] L. Hörmander: Hypoelliptic second order differential equations. Acta Math. 119 (1967) p. 147-171.

[KI] Kirillov: Unitary representations of nilpotent groups. Russian Math. Survery 17 (1962) p. 53-104.

[LE] P. Levy-Bruhl: [1] Résolubilite locale d'opérateurs invariants a gauche sur des groupes nilpotents d'ordre 2. 104 (1980) p. 369-391.

[2] Application de la formule de Plancherel a la résolubilité d'opérateurs invariants à gauche sur des groupes nilpotents.

. CRAS 19-1-81.

. Journées E.D.P. St Jean de Monts. Juin 1982, Exposé n°18.

. To appear.

[3] Conditions nécessaires de résolubilité locale d'opérateurs invariants à gauche sur des groupes nilpotents. To appear in Comm. in P.D.E.

[LI] G. Lion: Hypoellipticité et résolubilité d'opérateurs differentiels sur les groupes de rang 2. Note aux CRAS 1980.

[MEL] A. Melin: [1] Parametrix constructions for some classes of right invariant differential operators on the Heisenberg group.

[2] Parametrix constructions for some classes of right invariant operators on nilpotent groups. Annals of global analysis and geometry (1982).

[MET] G. Métivier: [1] Hypoanalyticity on nilpotent groups of rank 2. Duke Math. J. 47 (1980), p.195-223.

[2] Une classe d'opérateurs nonhypoelliptiques analytiques. Indiana J. Math. 29 (1980) p. 823-860.

[3] Analytic hypoellipticity for operators with multiple characteristics. Comm. in P.D.E. (1981),6(#1), p. 1-90.

[4] Equations aux dérivées partielles sur les groupes de Lie nilpotents. Exposé aux séminaire Bourbaki 1981-82.

MO] N. Moukadem: Interpolation pour des espaces de Sobolev avec poids associés à des représentations de groupe de Lie. Note aux C.R.A.S. t. 294, n°17, 10 Mai 1982.

NO] J. Nóurrigat: Cours à Récife (Brésil) Août 1982.

PU] L. Pukanszky: Leçon sur les représentations des groupes. Dunod (1967).

RE] R. Rentschler: [1] L'injectivité de l'application de Dixmier pour les algèbres résolubles Inventiones Math. 23, p. 49-71 (1974).

[2] Propriétés fonctorielles de l'application de Dixmier pour des algèbres de Lie résolubles (preprint).

ROC] C. Rockland: Hypoellipticity on the Heisenberg group. Representation theoretic criteria. Trans. of the A.M.S., vol. 240, n°517, p. 1-52 (1978).

ROT] L. P. Rothschild: [1] Non-existence of optimal estimates for the boundary Laplacian operator on certain weakly pseudo-convex domains. Comm. in P.D.E. 5 (1980) p. 897-912.

[2] Local solvability of second order differential operators on nilpotent Lie groups. Arkiv. F. Mat. vol. 19 (1981), n°2, p. 145-175.

[R.S.] L. P. Rothschild - E. M. Stein: Hypoelliptic differential operators and nilpotent groups. Acta Mathematica 137 p. 248-315.

[R.T.] L. P. Rothschild - D. S. Tartakoff: Inversion of analytic matrices and local solvability of some invariant differential operators on nilpotent Lie groups. Comm. in P.D.E., 6 (1981) p. 625-650.

[TA] D. S. Tartakoff: The analytic hypoellipticity of $\Box_b$ and related operators on C.R. Manifolds, Acta. Math.145(1980),p.177-204.

[TRE] F. Trèves: [1] An invariant criterion of hypoellipticity. Amer. J. Math., vol. 83, 1961, p. 645-668.

[2] Analytic hypoellipticity of a class of p.d.o. Comm. in P.D.E. 3 (1978), p. 475-642.

[U] A. Unterberger: Commuting with mollifiers without terms. Ann. Inst. Fourier (Grenoble) 26 (1976) n°2, p. 35-54.

WAVE EQUATIONS ON HOMOGENEOUS SPACES

By Sigurdur Helgason

§1. Introduction. It is a familiar fact from daily life that propagation of waves is very different in 2 and in 3 dimensions. When a pebble falls in water at a certain point P the initial ripple on a circle around P will be followed by subsequent ripples. Thus a given point Q will be hit by residual waves.

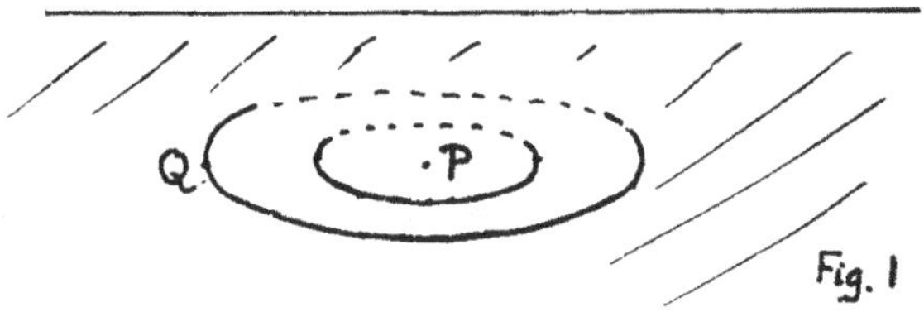

Fig. 1

In three dimensions the situation is quite different. A flash of light at a point has an effect on the surface of a sphere around the point after a certain time interval but then no more. There are no residual waves as those present on the water surface. The same is the case with sound waves, one has a pure propagation without residual waves; thus music can exist in $\mathbb{R}^3$.

The propagation of waves in $\mathbb{R}^n$ is governed by the wave equation

$$(1)\qquad \frac{\partial^2 u}{\partial x_1^2} + \cdots + \frac{\partial^2 u}{\partial x_n^2} = \frac{\partial^2 u}{\partial t^2}$$

with the special initial data

$$(2)\qquad u(x,0) = 0, \qquad \frac{\partial u}{\partial t}(x,0) = f(x).$$

The major qualitative difference between the cases $n = 2$, $n = 3$ can be deduced from the explicit solution formula

$$(3)\qquad u(x,t) = \frac{1}{(n-2)!} \frac{\partial^{n-2}}{\partial t^{n-2}} \int_0^t (M^r f)(x)\, r\, (t^2-r^2)^{\frac{1}{2}(n-3)}\, dr,$$

where $(M^r f)(x)$ is the mean value of f on the sphere $S_r(x)$ of radius r and center x (Tedone [27], Àsgeirsson [1], John [19]; the case $n = 3$ was already done by Poisson). We can write

$$(M^r f)(x) = \frac{1}{\Omega_n r^{n-1}} \int_{S_r(x)} f(y)\,d\omega(y),$$

where Ω_n is the area of the unit sphere.

<u>The case</u> $n = 3$. In this case we have

$$(4) \qquad u(x,t) = t(M^t f)(x)$$

so $u(x,t)$ is determined by the values of f on $S_t(x)$.

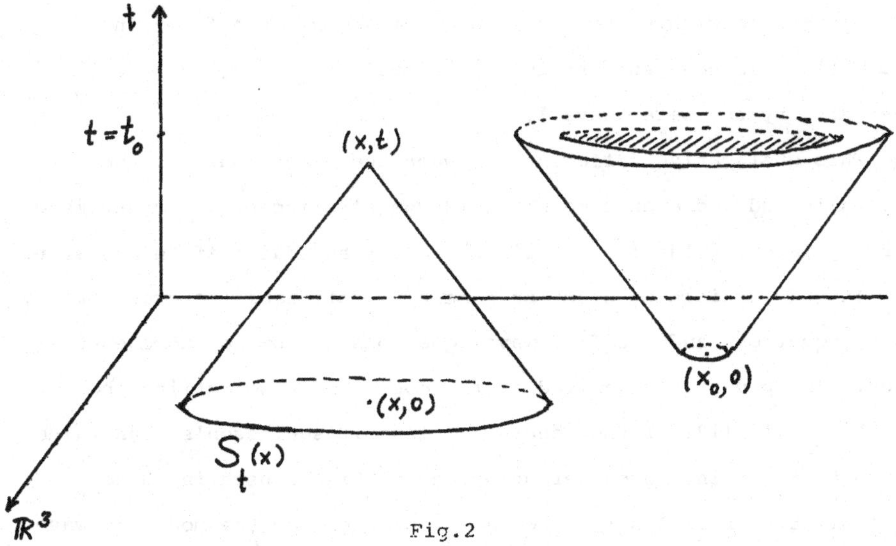

Fig.2

This is indicated on the left part of Fig.2. We can also say that if $f \in C^\infty(\mathbb{R}^3)$ has support in a small neighborhood N of $x_0 \in \mathbb{R}^3$, then the solution $u(x,t)$ has support inside the conical shell S indicated on the right in Fig.2. The shell is given by $S = \cup_{x \in \bar{N}} C_x$ where $\bar{N}$

is the closure of N and C_x denotes the support of the forward light cone with vertex x. This means that for each $t = t_0$, the function $x \to u(x,t_0)$ is 0 outside a certain shell in $\mathbb{R}^3$ around x_0. Also, for a given x, u(x,t) vanishes except when t is in a certain short time interval. This means pure propagation without residual effects.

The case n = 2. Here we have

$$u(x,t) = \int_0^t (M^r f)(x) r (t^2 - r^2)^{-\frac{1}{2}} dr$$

so, in contrast to the case n = 3, u(x,t) depends on the values of f in the solid disk $|y-x| \leq t$; also if f has support in a small neighborhood N of x_0 then we can only say that u(x,t) has support in the conical solid $B = \cup_{x \in \overline{N}} D_x$ when D_x is the forward solid light cone with vertex x. For a given x, the function $t \to u(x,t)$ will have support in an interval $t \geq |x-x_0|-\varepsilon$ (as illustrated by the water waves).

This qualitative difference between the dimensions 3 and 2 persists in odd and even dimensions: Huygens' principle, defined more generally later, holds for (1) if n is odd and fails if n is even.

In this article we describe some applicationsof Lie group theory to wave equations and modified wave equations on special homogeneous spaces. The paper is organized as follows. In §2 we outline the proof from [11],[12] of the generalization of Àsgeirsson's mean value theorem to two-point homogenerous spaces. This is used in §5 to obtain solution formulas for the Cauchy problem for the modified wave equation on a hyperbolic space. In §3 we describe some basic results of Hadamard for the general Cauchy problem for second order hyperbolic equationsand define Huygens'principle (Hadamard's "minor premise"). In §4 we consider the possibility of Huygens' principle for modified wave equations on symmetric spaces; deriving explicit solution formulas we give detailed proofs of some results from [14] and [16], where the proofs were only sketched. §6 is an exposition of some known results about the conformal invariance of Huygens' principle.

Since this article constitutes notes from lectures within the framework of analysis on Lie groups the exposition is purposefully rather elementary so far as differential equations and tools from Riemannian geometry is concerned. I am indebted to Bent Ørsted for helpful discussions about the material in §6, and to R. Schimming for sending me a copy of his informative review article [24].

§2. Ásgeirsson's Mean Value Theorem Generalized.

Let X be a two-point homogeneous space, G the identity component of the group of all isometries of X taken with the compact-open topology. We fix a point $o \in X$ and let $K = \{g \in G: g\cdot o = o\}$ so $X = G/K$. Let dk be a normalized Haar measure on the compact group K and let $L = L_X$ denote the Laplace-Beltrami operator on X. If f is a function on X let $(M^r f)(x)$ denote the mean value

(1) $$(M^r f)(x) = \frac{1}{A(r)}\int_{S_r(x)} f(y)\,d\omega(y)$$

of f over the sphere $S_r(x)$ of radius r with center x, $A(r)$ denoting the area of $S_r(x)$. We can then write

(2) $$(M^r f)(g\cdot o) = \int_K f(gkh\cdot o)\,dk$$

if $g,h \in G$ and r is the distance $d(o,h\cdot o)$ between o and $h\cdot o$ in X.

Theorem 2.1. *Let $u \in C^2(X \times X)$ satisfy the differential equation*

(2) $$L_1 u = L_2 u.$$

Then for each $r \geq 0$,

(3) $$M_1^r u = M_2^r u.$$

Here the subscripts 1 *and* 2 *indicate the action on the first and second variable, respectively.*

We outline the proof from [11], [12] of this result; the proof is inspired by one given by Àsgeirsson [1] for the case $X = \mathbb{R}^n$.

Let f be a radial function on X, i.e., $f(x) = F(d(o,x))$. Then

(4) $$(Lf)(x) = \frac{d^2F}{dr^2} + \frac{A'(r)}{A(r)}\frac{dF}{dr}, \qquad r = d(o,x),$$

as one sees easily by expressing L in geodesic polar coordinates, [12], Ch. IV. Next we note the commutativity

(5) $$LM^r = M^r L$$

and that if $F(x,y) = (M^{d(o,y)}f)(x)$

then

(6) $$L_1 F = L_2 F.$$

To see this define $\tilde{f}$ on G by $\tilde{f}(g) = f(g\cdot o)$, $(g \in G)$. Since G is either semisimple or the semidirect product of K with a Euclidean group there exists on G a second order differential operator Ω which is bi-invariant, that is invariant under all left and right translations on G. The operator Ω_0 on $C^\infty(X)$ defined by $(\Omega_0 f)^\sim = \Omega\tilde{f}$ is then a G-invariant differential operator on X and must be proportional to L; we may choose Ω such that $\Omega_0 = L$.

Hence by (2) and the bi-invariance of Ω,

$$(L_1F)(g\cdot o) = (LM^r f)(g\cdot o) = \Omega_g\left(\int_K \tilde{f}(gkh)\,dk\right) =$$

$$\int_K (\Omega\tilde{f})(gkh)\,dk = \Omega_h\left(\int_K \tilde{f}(gkh)\,dk\right) = (L_2F)(g\cdot o).$$

This proves both (5) and (6).

Now fix $(x_0,y_0) \in X \times X$ and put for $r,s \geq 0$,

$$U(r,s) = (M_1^r M_2^s u)(x_0,y_0) = (M_2^s M_1^r u)(x_0,y_0).$$

Assuming u satisfies (2) we obtain using (4), (5), (6), and the commutativity of two operators acting on different arguments,

$$\frac{\partial^2 U}{\partial r^2} + \frac{A'(r)}{A(r)}\frac{\partial U}{\partial r} = (L_1 M_1^r M_2^s u)(x_0,y_0) =$$

$$(M_1^r L_1 M_2^s u)(x_0,y_0) = (M_1^r M_2^s L_1 u)(x_0,y_0) =$$

$$(M_1^r M_2^s L_2 u)(x_0,y_0) = (M_1^r L_2 M_2^s u)(x_0,y_0) =$$

$$(L_2 M_2^s M_1^r u)(x_0,y_0) = \frac{\partial^2 U}{\partial s^2} + \frac{A'(s)}{A(s)}\frac{\partial U}{\partial s}.$$

Putting $F(r,s) = U(r,s) - U(s,r)$ we have

$$(7) \qquad \frac{\partial^2 F}{\partial r^2} + \frac{A'(r)}{A(r)}\frac{\partial F}{\partial r} - \frac{\partial^2 F}{\partial s^2} - \frac{A'(s)}{A(s)}\frac{\partial F}{\partial s} = 0,$$

$$(8) \qquad F(r,s) = -F(s,r).$$

Multiplying the first equation by $2A(r)\frac{\partial F}{\partial s}$ we obtain after some manipulation

$$-A(r)\frac{\partial}{\partial s}\left[(\frac{\partial F}{\partial r})^2 + (\frac{\partial F}{\partial s})^2\right] + 2\frac{\partial}{\partial r}(A(r)\frac{\partial F}{\partial r}\ \frac{\partial F}{\partial s}) = 2\frac{A(r)}{A(s)}\ A'(s)(\frac{\partial F}{\partial s})^2.$$

Now we integrate this over the region OMN; for the left hand side we use the divergence theorem

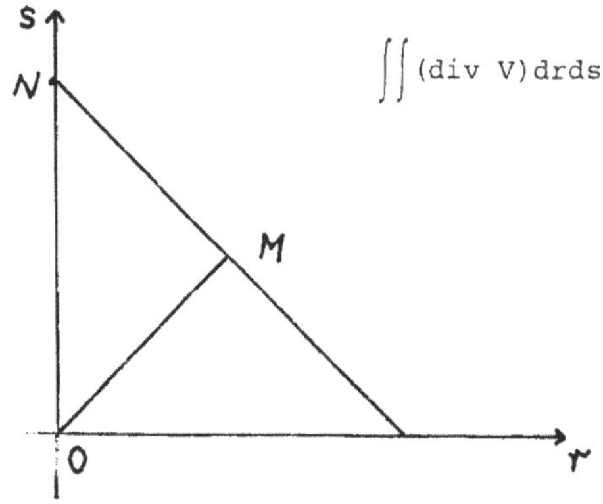

$$\iint(\operatorname{div} V)drds = \int(V\cdot\bar{n})d\ell,$$

$d\ell$ being the element of arc length, $\bar{n}$ the outgoing unit normal. After some computation this gives the formula

$$(9)\quad \frac{1}{\sqrt{2}}\int_{MN} A(r)\left(\frac{\partial F}{\partial r} - \frac{\partial F}{\partial s}\right)^2 d\ell + \iint_{OMN} 2\frac{A(r)}{A(s)}\ A'(s)\left(\frac{\partial F}{\partial s}\right)^2 drds = 0.$$

If X is noncompact one can prove $A'(s) > 0$ for all $s > 0$ so (9) implies $F \equiv 0$, so $U(r,0) = U(0,r)$ which is the conclusion of the theorem. If X is compact we have $A'(s) > 0$ on a certain interval $0 < s < L$ so as before $M_1^r u = M_2^r u$ for $0 \leq r \leq L$. If u is analytic it follows that (3) holds for all r. Let u be an arbitrary C^2 solution to (2). Let ϕ,ψ be two analytic functions on the compact Lie group G and consider the convolution

$$u_{\phi,\psi}(x,y) = \int_G\int_G u(g_1^{-1}\cdot x, g_2^{-1}\cdot y)\phi(g_1)\psi(g_2)dg_1dg_2.$$

Then $u_{\phi,\psi}$ satisfies (2) and since it is analytic it satisfies (3) for all r. By approximation it follows that u satisfies (3) for all r.

Consider now the Cauchy problem of solving the wave equation

(10) $$\frac{\partial^2 u}{\partial x_1^2} + \cdots + \frac{\partial^2 u}{\partial x_n^2} = \frac{\partial^2 u}{\partial t^2}$$

with initial data

(11) $$u(x,0) = u_0(x) \quad , \quad \frac{\partial u}{\partial t}(x,0) = u_1(x).$$

As shown by Àsgeirsson [1] a solution can be written down by using Theorem 2.1 for $X = \mathbb{R}^n$ on the functions

$$V(x_1,\ldots,x_n,y_1,\ldots,y_n) = u(x_1,\ldots,x_n,t), \qquad (y_1 = t)$$

$$W(x_1,\ldots,x_n,y_1,\ldots,y_n) = \frac{\partial u}{\partial t}(x_1,\ldots,x_n,t), \qquad (y_1 = t).$$

The identity (3) then gives the following explicit solution to (10) and (11). For n odd we have for $r > 0$

(12) $$u(x,r) = \frac{\Omega_n}{2(\frac{n-3}{2})!\,\Omega_{n-1}} \left[\frac{\partial}{\partial r}\left(\frac{\partial}{\partial (r^2)}\right)^{\frac{1}{2}(n-3)} \{r^{n-2}(M^r u_0)(x)\} + \left(\frac{\partial}{\partial (r^2)}\right)^{\frac{1}{2}(n-3)} \{r^{n-2}(M^r u_1)(x)\} \right].$$

For n even and $r > 0$ we have a similar formula (but now the formula involves integration),

(13) $$u(x,r) = \frac{1}{(\frac{n-2}{2})!} \left[\frac{\partial}{\partial r}\int_0^r t(r^2-t^2)^{-\frac{1}{2}} \left(\frac{\partial}{\partial (t^2)}\right)^{\frac{1}{2}(n-2)} \{t^{n-2}(M^t u_0)(x)\}dt + \int_0^r t(r^2-t^2)^{-\frac{1}{2}} \left(\frac{\partial}{\partial (t^2)}\right)^{\frac{1}{2}(n-2)} \{t^{n-2}(M^t u_1)(x)\}dt \right].$$

For the case $u_0 = 0$ these formulas can be shown to be equivalent to (3) §1.

§3. Results of Hadamard. Huygens' Principle

Let X' be a manifold with a pseudo-Riemannian structure g of signature $(1,n)$ (one + sign, n - signs). Such a manifold is called a Lorentzian manifold. A tangent vector Y to X' at a point $x \in X$ is called timelike, isotropic, or spacelike if $g_x(Y,Y)$ is positive, 0, or negative, respectively. Let $X \subset X'$ be an open subset with the property that any two points $x,y \in X$ can be joined by a "unique" geodesic in X. The isotropic geodesics through a point $x \in X$ constitute the light cone $C_x \subset X$ with vertex x. A submanifold $S \subset X$ is called spacelike if each of its tangent vectors is spacelike. Let $L = L_X$ be the Laplace-Beltrami operator and L' a differential operator on X such that $L' - L$ has order ≤ 1. Suppose now a Cauchy problem is posed for the equation $L'u = 0$ with initial data given on a spacelike surface $S \subset X$. From Hadamard's theory [8] it is known that the value $u(x)$ of the solution at $x \in X$ only depends on the initial data on the piece $S^* \subset S$ which lies inside the light cone C_x. (Formulas (12), (13) in §2 give an example of this.)

Huygens' principle is said to hold for $L'u = 0$ if for each x and S the value $u(x)$ only depends on the initial data in an arbitrarily small neighborhood of the edge s of S^*, $s = C_x \cap S$. Hadamard showed that Huygens' principle can not hold if $\dim X$ is odd. On the other hand formula (12) in §2 shows that if n is odd Huygens' principle does hold for the wave equation (10) which is the equation $L = 0$ for the manifold $\mathbb{R} \times \mathbb{R}^n$ with the flat Lorentzian structure $dt^2 - dx_1^2 - \cdots - dx_n^2$. Hadamard proved a necessary and sufficient criterion for Huygens' principle, the absence of a logarithmic term in his "elementary solution"; in particular it is immaterial which spacelike surface is considered. Hadamard also raised the question whether any Huygensian differential equation $L'u = 0$ is equivalent to the flat wave equation by a change of variables combined with a replacement of

L' by an operator $\phi L' \circ \psi$, ϕ and ψ being nonvanishing functions. Examples to the contrary were given by Stellmacher [26] and Günther [7].

The simplest examples to test for Huygens' principle are of course manifolds of constant curvature. If X is a Riemannian manifold of constant curvature $\neq 0$ the wave equation

$$(1) \qquad \left(\frac{\partial^2}{\partial t^2} - L_X\right)u = 0$$

does not satisfy Huygens' principle (Hölder [18]); neither does the equation

$$(2) \qquad L_X u = 0$$

on a Lorentzian manifold X of constant curvature (Helgason [12]). We shall, however, see in §5 that positive results hold for certain natural modifications of equations (1) and (2).

§4. Wave Equations on Symmetric Spaces.

Let G be a connected semisimple Lie group with finite center, K a maximal compact subgroup and X the symmetric space G/K with the Riemannian structure induced by the Killing form B of the Lie algebra $\mathfrak{g}$ of G. Let $\mathfrak{g} = \mathfrak{k} + \mathfrak{p}$ be the corresponding Cartan decomposition and $\mathrm{Exp}: \mathfrak{p} \longrightarrow X$ the diffeomorphism given by sending rays through the origin 0 in $\mathfrak{p}$ onto geodesics through the origin $o = \{K\}$ in X. Let $\mathrm{Log}: X \longrightarrow \mathfrak{p}$ denote the inverse map. Let dx denote the volume element on X and dY the (Euclidean) volume element on $\mathfrak{p}$. Let J denote the Jacobian given by

$$(1) \qquad \int_{G/K} f(x)\,dx = \int_{\mathfrak{p}} f(\mathrm{Exp}\, Y) J(Y)\,dY, \qquad f \in C_c^\infty(X).$$

Then, vertical bar denoting restriction, it is well known that

$$(2) \qquad J(Y) = \det\left(\frac{\sinh \operatorname{ad}Y}{\operatorname{ad}Y}\bigg|_{\mathfrak{p}}\right), \qquad Y \in \mathfrak{p},$$

where $\operatorname{ad}Y(Z) = [Y,Z]$. Let L_X and $L_{\mathfrak{p}}$ denote the Laplace-Beltrami operators on X and $\mathfrak{p}$, respectively. If ϕ is a diffeomorphism of a manifold M onto a manifold N and if $f,g \in C_c^\infty(N)$, T a distribution on M, D a differential operator on M we put $g^{\phi^{-1}} = g\circ\phi$, $T^\phi(f) = T(f^{\phi^{-1}})$, $D^\phi(g) = (D(g^{\phi^{-1}}))^\phi$. Then $g^{\phi^{-1}} \in C_c^\infty(M)$, T^ϕ is a distribution on N and D^ϕ is a differential operator on N, the image of D under ϕ. With this notation we have the following result ([15], Ch. II, §1).

Theorem 4.1. *Let F be a K-invariant function on $\mathfrak{p}$. Then*

$$(3) \qquad L_X^{\operatorname{Log}} F = (L_{\mathfrak{p}} + \operatorname{grad}(\log J))\, F.$$

Since the proof in the above reference is rather brief we give some additional details here. Let V be a Riemannian manifold, H a closed unimodular subgroup of the group $I(V)$ of isometries of V. Assume a submanifold $W \subset V$ is an orthogonal transversal submanifold in the sense that for each $w \in W$,

$$W \cap (H\cdot w) = \{w\}, \quad V_w = (H\cdot w)_w \oplus W_w,$$

(orthogonal direct sum) the subscripts denoting tangent space. Then the radial part $\Delta(L_V)$ of the Laplace-Beltrami operator L_V is given by

$$(4) \qquad \Delta(L_V) = D^{-\frac{1}{2}} L_W \circ D^{\frac{1}{2}} - D^{-\frac{1}{2}} L_W(D^{\frac{1}{2}}),$$

(loc. cit. Ch.II, §2). Here D is the <u>density function</u>

$$(5) \qquad D(w) = \frac{d\sigma(H\cdot w)}{d\dot{h}}, \qquad w \in W,$$

where $d\sigma(H\cdot w)$ is the Riemannian volume element on the orbit $H\cdot w$ and $d\dot{h}$ is an invariant measure on the homogeneous space H/H_w where $H_w \subset H$ is the isotropy subgroup at w. This measure is determined by a Haar measure dh on H (fixed once for all) and the normalized Haar measure on the compact group H_w.

Let $\mathfrak{a} \subset \mathfrak{p}$ be a maximal abelian subspace and $A \subset G$ the corresponding analytic subgroup. Let $\mathfrak{a}'$ and A' be the corresponding sets of regular elements, $\mathfrak{a}^+ \subset \mathfrak{a}'$ a fixed Weyl chamber and $A^+ = \exp \mathfrak{a}^+$. For the action of K on X, $A^+\cdot o$ serves as an orthogonal transversal submanifold. With δ the corresponding density function we have by (4)

$$(6) \qquad \Delta(L_X) = \delta^{-\frac{1}{2}} L_A \circ \delta^{\frac{1}{2}} - \delta^{-\frac{1}{2}} L_A(\delta^{\frac{1}{2}}).$$

For K acting on $\mathfrak{p}$, $\mathfrak{a}^+$ serves as an orthogonal transversal submanifold and

$$(7) \qquad \Delta(L_{\mathfrak{p}}) = \delta_0^{-\frac{1}{2}} L_{\mathfrak{a}} \circ \delta_0^{\frac{1}{2}} - \delta_0^{-\frac{1}{2}} L_{\mathfrak{a}}(\delta_0^{\frac{1}{2}}),$$

δ_0 being the new density function. Also $J = (\delta\circ\mathrm{Exp})/\delta_0$. Now if F is a K-invariant function on $\mathfrak{p}$ and $f = F\circ\mathrm{Log}$ we have for $H \in \mathfrak{a}^+$ using (6) and (7),

$$(L_X^{\mathrm{Log}} F)(H) = (L_X f)(\mathrm{Exp} H) = (\delta^{-\frac{1}{2}} L_A(\delta^{\frac{1}{2}} f) - \delta^{-\frac{1}{2}} L_A(\delta^{\frac{1}{2}}) f)(\mathrm{Exp}\, H);$$

$$(J^{-\frac{1}{2}} L_{\mathfrak{p}}(J^{\frac{1}{2}} F))(H) - (J^{-\frac{1}{2}} L_{\mathfrak{p}}(J^{\frac{1}{2}}) F)(H) =$$

$$J^{-\frac{1}{2}}(H) \left[\delta_0^{-\frac{1}{2}} L_{\mathfrak{a}}(\delta_0^{\frac{1}{2}} J^{\frac{1}{2}} F) - \delta_0^{-\frac{1}{2}} L_{\mathfrak{a}}(\delta_0^{\frac{1}{2}}) J^{\frac{1}{2}} F\right](H)$$

$$-J^{-\frac{1}{2}}(H) \left[\delta_0^{-\frac{1}{2}} L_{\mathfrak{a}}(\delta_0^{\frac{1}{2}} J^{\frac{1}{2}}) - (\delta_0^{-\frac{1}{2}} L_{\mathfrak{a}}(\delta_0^{\frac{1}{2}})) J^{\frac{1}{2}}\right](H) F(H).$$

Here the second and fourth term cancel and $J\delta_0 = \delta \circ \mathrm{Exp}$. Also Exp is an isometry of $\mathfrak{a}$ onto $A\cdot o$ so $(L_{\mathfrak{a}}F)(H) = (L_A f)(\mathrm{Exp}H)$. Our expression therefore reduces to

$$\left[\delta^{-\frac{1}{2}}L_A \circ \delta^{\frac{1}{2}} - \delta^{-\frac{1}{2}}L_A(\delta^{\frac{1}{2}})\right](f)(\mathrm{Exp}H).$$

This proves by continuity that

$$L_X^{\mathrm{Log}}F = \left[J^{-\frac{1}{2}}L_{\mathfrak{p}} \circ J^{\frac{1}{2}} - J^{-\frac{1}{2}}L_{\mathfrak{p}}(J^{\frac{1}{2}})\right](F) \tag{8}$$

and now the theorem follows since

$$L_{\mathfrak{p}}(uv) = uL_{\mathfrak{p}}(v) + 2\,\mathrm{grad}\,u(v) + vL_{\mathfrak{p}}(u),$$

and since $2J^{-\frac{1}{2}}\mathrm{grad}(J^{\frac{1}{2}}) = \mathrm{grad}(\log J)$.

Consider now the Cauchy problem for the wave equation on $\mathfrak{p}$,

$$\left(\frac{\partial^2}{\partial t^2} - L_{\mathfrak{p}}\right)u = 0, \quad u(Y,0) = 0, \qquad u_t(Y,0) = f(Y). \tag{9}$$

According to (12) and (13) in §2 the solution is given by

$$u(Y,t) = f * E_t(Y) = \int_{\mathfrak{p}} f(Y-Z)\,dE_t(Z),$$

where E_t is a distribution on $\mathfrak{p}$ given as follows:

For $n = \dim X$ odd,

$$E_t(\phi) = c_1\left(\frac{\partial}{\partial(t^2)}\right)^{\frac{n-3}{2}} (t^{n-2}(M^t\phi)(0)) \tag{10}$$

and for $n = \dim X$ even

$$E_t(\phi) = c_2\int_0^t r(t^2-r^2)^{-\frac{1}{2}}\left(\frac{\partial}{\partial(r^2)}\right)^{\frac{n-2}{2}} \{r^{n-2}(M^r\phi)(0)\}dr. \tag{11}$$

The constants c_1 and c_2 are given by (12) and (13) in §2. Then (9) can be stated

$$(12) \qquad \left(\frac{\partial^2}{\partial t^2} - L_{\mathfrak{p}}\right)E_t = 0, \ \lim_{t\to 0} E_t = 0, \ \lim_{t\to 0} \frac{\partial}{\partial t}E_t = \delta.$$

The family E_t is called the <u>fundamental solution for the Cauchy problem</u> (9).

Let now $\Omega \in C^\infty(G/K)$ be defined by

$$\Omega(\operatorname{Exp} Y) = J^{-\frac{1}{2}}(Y)L_{\mathfrak{p}}(J^{\frac{1}{2}})(Y).$$

Then (8) can be written

$$(13) \qquad (L_X + \Omega)(F^{\operatorname{Exp}}) = (J^{-\frac{1}{2}}L_{\mathfrak{p}}(J^{\frac{1}{2}}F))^{\operatorname{Exp}}$$

for F K-invariant.

<u>Lemma 4.2. Let</u> E <u>be a</u> K-<u>invariant distribution on</u> $\mathfrak{p}$ <u>and define the distribution</u> $\mathcal{E}_E$ <u>on</u> X <u>by</u>

$$(14) \qquad \mathcal{E}_E(f) = E(J^{\frac{1}{2}}(f\circ\operatorname{Exp})) \qquad f \in C_c^\infty(X).$$

<u>Then</u>

$$(L_X + \Omega)\mathcal{E}_E = \mathcal{E}_{L_{\mathfrak{p}}E}.$$

<u>Proof</u>. With F K-invariant we have (since L_X is symmetric)

$$((L_X + \Omega)\mathcal{E}_E)((J^{-\frac{1}{2}}F)^{Exp}) = \mathcal{E}_E[(L_X + \Omega)((J^{-\frac{1}{2}}F)^{Exp})] = \mathcal{E}_E[(J^{-\frac{1}{2}}L_{\mathfrak{p}}(F))^{Exp}]$$

$$= E[J^{\frac{1}{2}}(J^{-\frac{1}{2}}L_{\mathfrak{p}}(F))^{Exp}\circ Exp] = E(L_{\mathfrak{p}}(F)) = (L_{\mathfrak{p}}E)(F) = (J^{\frac{1}{2}}(L_{\mathfrak{p}}E))(J^{-\frac{1}{2}}F)$$

$$= (J^{\frac{1}{2}}L_{\mathfrak{p}}E)^{Exp}((J^{-\frac{1}{2}}F)^{Exp}).$$

Since both $(L_X + \Omega)\mathcal{E}_E$ and $(J^{\frac{1}{2}}L_{\mathfrak{p}}E)^{Exp}$ are K-invariant the derivation above shows that they are equal. But if $f \in C_c^\infty(X)$ then

$$(J^{\frac{1}{2}}L_{\mathfrak{p}}E)^{Exp}(f) = (L_{\mathfrak{p}}E)(J^{\frac{1}{2}}(f\circ Exp)) = \mathcal{E}_{L_{\mathfrak{p}}(E)}(f)$$

so the lemma is proved.

Theorem 4.3. *Let $X = G/K$ be a symmetric space of the noncompact type. With E_t as in (10) and (11) the distribution $\mathcal{E}_t = \mathcal{E}_{E_t}$ on X satisfies*

$$\text{(15)} \qquad \frac{\partial^2 \mathcal{E}_t}{\partial t^2} = (L_X + \Omega)\mathcal{E}_t, \quad \lim_{t\to 0} \mathcal{E}_t = 0, \quad \lim_{t\to 0} \frac{\partial \mathcal{E}_t}{\partial t} = \delta_o.$$

In fact,

$$\frac{\partial^2}{\partial t^2}(\mathcal{E}_t(f)) = \frac{\partial^2}{\partial t^2}(E_t(J^{\frac{1}{2}}(f\circ Exp))) = (L_{\mathfrak{p}}E_t)(J^{\frac{1}{2}}(f\circ Exp)) = \mathcal{E}_{L_{\mathfrak{p}}E_t}(f)$$

$$= ((L_X + \Omega)\mathcal{E}_t)(f)$$

and the limit relations follow from (12).

Remarks. In analogy with (12) one can view the family $\mathcal{E}_t$ satisfying (15) as a fundamental solution of the Cauchy problem for the operator

$$\text{(16)} \qquad \frac{\partial^2}{\partial t^2} - L_X - \Omega.$$

In the case when $n = \dim X$ is odd we see from (10) and (14) that $\&_t$ has support on the sphere $S_t(o)$ in X (with radius t, center o). In this sense the operator (16) satisfies Huygens' principle when n is odd, (cf. [16]). However, the operator (16) is not invariant under G, except when Ω is a constant, so only then does this (by convolution) imply Huygens' principle for the wave equation

$$\frac{\partial^2 u}{\partial t^2} = (L_X + \Omega)u$$

where u is a function on $\mathbb{R} \times X$ (see Theorem 4.4 below).

If we diagonalize the operators $(\mathrm{ad}H)^2 (H \in \mathfrak{a})$ on $\mathfrak{p}$ we see that

$$J(H) = \prod_{\alpha \in \Sigma^+} \left(\frac{\sinh \alpha(H)}{\alpha(H)}\right)^{m_\alpha}, \qquad H \in \mathfrak{a}, \tag{17}$$

where Σ^+ is the set of the positive restricted roots and m_α is the multiplicity of $\alpha \in \Sigma^+$. Let $\tilde{\&}_t$ be the lift of the distribution $\&_t$ to G, i.e.,

$$\int_G F(g)\,d\tilde{\&}_t(g) = \int_{G/K} \left(\int_K F(gk)\,dk\right) d\&_t(gK) \tag{18}$$

for $F \in C_c^\infty(G)$.

THEOREM 4.4. Let $X = G/K$ be a symmetric space of the noncompact type and assume G complex. Let $f \in C_c^\infty(X)$. Then the Cauchy problem for the modified wave equation on X

$$\frac{\partial^2 u}{\partial t^2} - \left(L_X + \frac{\dim X}{48}\right)u = 0, \quad u(x,0) = 0\,, \quad u_t(x,0) = f(x) \tag{19}$$

has the solution

$$u(gK,t) = \int_G f(gh^{-1}\cdot o)\,d\tilde{\varepsilon}_t(h).$$

If dim X *is odd, Huygens' principle holds for* (19).

Proof. For the case when G is complex each multiplicity m_α is 2 and

$$\prod_{\alpha\in\Sigma^+}(e^{\alpha}-e^{-\alpha}) = \sum_{s\in W}\det(s)e^{s\rho}$$

where W is the Weyl group for X and 2ρ is the sum of the roots in Σ^+ with multiplicity; here each $m_\alpha = 2$ so $\rho = \sum_{\alpha\in\Sigma^+}\alpha$ (cf. [10]). Since $\delta_0(H)$ in (7) now equals $\prod_{\alpha\in\Sigma^+}\alpha(H)^2$ $(H \in \mathfrak{a})$ we have $L_{\mathfrak{a}}\delta_0 = 0$ and by (7) Ω is the constant $B(\rho,\rho)$, where B denotes the Killing form. To evaluate this constant extend $\mathfrak{a}$ to a Cartan subalgebra $\mathfrak{h}^c$ of the complexification $\mathfrak{g}^c$, and let $\Delta(\mathfrak{g}^c,\mathfrak{h}^c)$ be the corresponding set of nonzero roots. We can order the roots in such a way that if β is a positive root its restriction $\bar{\beta} = \beta|\mathfrak{a}$ is positive and in fact, $\bar{\beta} \in \Sigma^+$, cf. [17] p.274. Let P_+ denote the set of positive elements in $\Delta(\mathfrak{g}^c,\mathfrak{h}^c)$ and let $\tilde{\rho} = \frac{1}{2}\sum_{\beta\in P_+}\beta$. If θ is the Cartan involution of G with respect to K, the map $\beta \longrightarrow -\theta\beta$ is a permutation of P_+. Hence $\tilde{\rho} = -\theta\tilde{\rho}$ so $\tilde{\rho}$ vanishes on $\mathfrak{h}^c \cap \mathfrak{k}$. Since each $\alpha \in \Sigma^+$ has multiplicity two it therefore follows that

$$\tilde{\rho} = 2\rho.$$

Let B^c denote the Killing form of the complex Lie algebra $\mathfrak{g}$. Then by [17] p.180, $B = 2\,\mathrm{Re}\,B^c$. (Re = real part.) On the other hand, if G is simple,

$$B^c(\tilde{\rho},\tilde{\rho}) = \frac{1}{24}\dim_{\mathbb{C}} G = \frac{1}{24}\dim X$$

(Freudenthal-de Vries [4], §47). Consequently,

$$(20)\qquad B(\rho,\rho) = 2B^{C}(\tfrac{1}{2}\tilde{\rho},\tfrac{1}{2}\tilde{\rho}) = \frac{1}{48}\dim X,$$

if G is simple. However, in the general semisimple case, X is a product and both sides of (20) are additive; so (20) holds in general.

Let $\tilde{u}(g,t) = u(g\cdot o,t), \tilde{f}(g) = f(g\cdot o)$; then $\tilde{u}(\cdot,t) = \tilde{f} * \tilde{\mathcal{E}}_t$ so if C is the Casimir operator on G,

$$C_g(\tilde{u}(g,t)) = \tilde{f}*(C\tilde{\mathcal{E}}_t)(g).$$

Since C induces the Laplacian on X this equals

$$\tilde{f}*(L_X\mathcal{E}_t)^{\sim} = \tilde{f}*\left(\frac{\partial^2\mathcal{E}_t}{\partial t^2} - \frac{\dim X}{48}\tilde{\mathcal{E}}_t\right) = \left(\frac{\partial^2}{\partial t^2} - \frac{\dim X}{48}\right)\tilde{u}(g,t)$$

so the differential equation (19) holds. The initial conditions $u(x,0) = 0$, $u_t(x,0) = f(x)$ also follow from (15).

For the last assertion of the theorem we note that

$$u(o,t) = \int_G f(h^{-1}\cdot o)\,d\tilde{\mathcal{E}}_t(h) = \int_G f(h\cdot o)\,d\tilde{\mathcal{E}}_t(h)$$

$$= \mathcal{E}_t(f) = \int_{\mathfrak{p}} f(\operatorname{Exp} X)\,J^{\frac{1}{2}}(X)\,dE_t(X)$$

$$= c\left(\frac{\partial}{\partial(t^2)}\right)^{\frac{n-3}{2}}\left(t^{n-2}\{M^t((f\circ\operatorname{Exp})J^{\frac{1}{2}})\}(0)\right),$$

where $c = c_1$ for n odd and $c = c_2$ for n even. Thus $u(o,t)$ depends only on the values of f in an arbitrary neighborhood of $S_t(o)$. By G-invariance, $u(x,t)$ only depends on the values of f in an arbitrary neighborhood of $S_t(x)$. Thus Huygens' principle holds.

Next we consider the dual compact symmetric space U/K; here U

is a compact simply connected semisimple Lie group with an involutive automorphism σ with fixed group K. Let $\mathfrak{u} = \mathfrak{k} + \mathfrak{p}_*$ be the corresponding eigenspace decomposition of the Lie algebra $\mathfrak{u}$ of U and let $\mathfrak{a}_* \subset \mathfrak{p}_*$ be a maximal abelian subspace. Let $B_R(0) \subset \mathfrak{p}_*$ and $B_R(o) \subset X_* = U/K$ be balls around the origins (measured by the negative of the Killing form B_* of $\mathfrak{u}$) such that the mapping $\mathrm{Exp}: \mathfrak{p}_* \longrightarrow U/K$ is a diffeomorphism of $B_R(0)$ onto $B_R(o)$. In analogy with (1) let J_* be its Jacobian and then we have in analogy with (8)

$$(21) \qquad L_{X_*}{}^{\mathrm{Log}}F = \left[J_*^{-\frac{1}{2}} L_{\mathfrak{p}_*} \circ J_*^{\frac{1}{2}} - J_*^{-\frac{1}{2}} L_{\mathfrak{p}_*}(J_*^{\frac{1}{2}})\right](F)$$

if $F \in C_c^\infty(\mathfrak{p}_*)$ is K-invariant with support in $B_R(0)$. We define Ω_* by

$$(22) \qquad \Omega_*(\mathrm{Exp}\, Y) = J_*^{-\frac{1}{2}}(Y) L_{\mathfrak{p}_*}(J^{\frac{1}{2}})(Y) \qquad Y \in B_R(0)$$

and for a K-invariant distribution E on $\mathfrak{p}_*$ we put

$$(23) \qquad \mathcal{E}_E(f) = E(J_*^{\frac{1}{2}}(f \circ \mathrm{Exp})), \qquad f \in C_c^\infty(B_R(o))$$

Then we have the analogue of Lemma 4.2,

$$(24) \qquad (L_{X_*} + \Omega_*)\mathcal{E}_E = \mathcal{E}_{L_{\mathfrak{p}_*}E} \qquad \text{on } B_R(o),$$

and the following analogue of Theorem 4.3.

<u>THEOREM 4.5.</u> <u>With</u> E_t <u>as in</u> (10) <u>and</u> (11), <u>the distribution</u> $\mathcal{E}_t = \mathcal{E}_{E_t}$ <u>defined by</u> (23), <u>we have on</u> $B_R(o)$, $(0 < t < R)$,

$$\frac{\partial^2 \mathcal{E}_t}{\partial t^2} = (L_{X_*} + \Omega_*)\mathcal{E}_t\,, \quad \lim_{t\to 0} \mathcal{E}_t = 0, \quad \lim_{t\to 0} \frac{\partial \mathcal{E}_t}{\partial t} = \delta.$$

We consider now the case when U/K is a symmetric space $(K \times K)/\Delta K$ (ΔK = diagonal in $K \times K$) which under the map $(k_1,k_2)\Delta K \to k_1k_2^{-1}$ is diffeomorphic to K. Taking K with the bi-invariant metric induced by its Killing form the metrics on $(K \times K)/\Delta K$ and on K differ by a factor of 2. Because of this it is better just to work with K directly. Let $\mathfrak{t} \subset \mathfrak{k}$ be a maximal abelian subalgebra $\mathfrak{t}^c \subset \mathfrak{k}^c$ their complexifications, $\Delta(\mathfrak{k}^c, \mathfrak{t}^c)$ the corresponding set of roots and ρ the sum of the corresponding positive roots. Let $R > 0$ be such that $\exp\colon B_R(0) \longrightarrow B_R(e)$ is a diffeomorphism. Under the action of K on itself by conjugacy with $T = \exp \mathfrak{t}$ as transversal submanifold we have

$$\Delta(L_K) = \delta^{-\frac{1}{2}} L_T \circ \delta^{\frac{1}{2}} + |\rho|^2,$$

where

$$\delta^{\frac{1}{2}}(\exp H) = \sum (\det s) e^{(s\rho)(H)}, \qquad (H \in \mathfrak{t})$$

and $|\rho|^2$ is the (positive) norm of ρ with respect to the Killing form B of K ([15], Ch.I §2). Under the adjoint action of K with $\mathfrak{t}$ as transversal submanifold we have (Harish-Chandra [9])

$$\Delta(L_{\mathfrak{k}}) = \pi^{-1} L_{\mathfrak{t}} \circ \pi,$$

where π is the product of the positive roots. Combining these relations we have

$$L_K{}^{\mathrm{Log}}F = \left(J^{-\frac{1}{2}} L_{\mathfrak{k}} \circ J^{\frac{1}{2}}\right) F + |\rho|^2 F$$

if $F \in C_c^\infty(B_R(0))$ is invariant under conjugation. Here J is the volume element ratio given by

$$\int_K f(k)\,dk = \int_{B_R(0)} f(\exp T)J(T)\,dT$$

if $f \in C_c^\infty(B_R(e))$. Again if E is a distribution on $\mathfrak{k}$ invariant under $\mathrm{Ad}(K)$ and if

$$\mathcal{E}_E(f) = E(J^{\frac{1}{2}}(f\circ\exp)) \qquad f \in C_c^\infty(K)$$

then

$$(L_K - |\rho|^2)\mathcal{E}_E = \mathcal{E}_{L_{\mathfrak{k}}E} \qquad \text{on} \quad B_R(e).$$

Since by the quoted result, $|\rho|^2 = \frac{1}{24}\dim K$ we can deduce the following analog of Theorem 4.4 which shows that Huygens' principle holds for the modified wave equation on any odd-dimensional compact semisimple Lie group.

THEOREM 4.6. *Let* K *be a simply connected compact semisimple Lie group with the bi-invariant metric given by the negative of the Killing form. Let* $f \in C_c^\infty(B_R(e))$. *Then the Cauchy problem for the modified wave equation on* K,

$$\frac{\partial^2 u}{\partial t^2} - \left(L_K - \frac{\dim K}{24}\right)u = 0,\ u(k,0) = 0, \qquad u_t(k,0) = f(k), \tag{25}$$

has the solution

$$u(k,t) = \int_K f(kh^{-1})\,d\mathcal{E}_t(h).$$

If $\dim K$ *is odd Huygens' principle holds for* (25).

Here $\mathcal{E}_t = \mathcal{E}_{E_t}$ where E_t is given by (10) and (11).

5. The Case of a Hyperbolic Space.

For the three-dimensional hyperbolic space the group of isometries is locally isomorphic to $SL(2,\mathbb{C})$ and therefore has a complex structure. Thus the Cauchy problem (19) §4 satisfies Huygens' principle in this case. We shall now generalize this to all odd dimensions by obtaining an explicit solution by means of Theorem 2.1. The procedure is analogous to Àsgeirsson's method of proving (12) §1.

Let $\mathbb{H}^n$ denote the n-dimensional hyperbolic space with the Riemannian structure

$$(1) \qquad ds^2 = y_n^{-2}(dy_1^2+\cdots+dy_n^2), \qquad y_n > 0.$$

We shall solve the modified wave equation on $\mathbb{H}^n \times \mathbb{R}$,

$$(2) \qquad (L+(\tfrac{n-1}{2})^2)u = \frac{\partial^2 u}{\partial s^2}, \quad u(x,0) = u_0(x), \qquad u_s(x,0) = u_1(x).$$

For (1) the Laplace-Beltrami operator is given by

$$(3) \qquad L = y_n^2\left(\frac{\partial^2}{\partial y_1^2}+\cdots+\frac{\partial^2}{\partial y_{n-1}^2}\right) + y_n^2\frac{\partial^2}{\partial y_n^2} - (n-2)y_n\frac{\partial}{\partial y_n},$$

which by the substitution $y_n = e^s$ can be written

$$L = e^{2s}\left(\frac{\partial^2}{\partial y_1^2}+\cdots+\frac{\partial^2}{\partial y_{n-1}^2}\right) + e^{\frac{1}{2}(n-1)s}\left(\frac{\partial^2}{\partial s^2} - (\tfrac{n-1}{2})^2\right)\circ e^{-\frac{1}{2}(n-1)s}.$$

With $y_n = e^s$ put

$$(4) \qquad v(x,y) = v(x_1,\ldots,x_n,y_1,\ldots,y_n) = e^{\frac{1}{2}(n-1)s}u(x_1,\ldots,x_n,s).$$

Then, using (2)

$$(5) \qquad L_x(v(x,y)) = e^{\frac{1}{2}(n-1)s}\left(\frac{\partial^2 u}{\partial s^2} - \left(\frac{n-1}{2}\right)^2 u\right) = L_y(v(x,y)).$$

We now use Theorem 2.1 for the point $(x_0,y_0) = (x,o)$, where o is the origin $(0,\ldots,0,1)$ in $\mathbb{H}^n$. Since $v(x,o) = u(x,0)$ we obtain

$$(6) \qquad \int_{S_r(x)} u(z,0)\,d\omega(z) = \int_{S_r(o)} v(x,y)\,d\omega(y).$$

The non-Euclidean sphere $S_r(o)$ is a Euclidean sphere with center $(0,\ldots,0,\operatorname{ch} r)$ and radius $\operatorname{sh} r$. On the plane $y_n = e^s$ the metric is $e^{-2s}(dy_1^2+\cdots+dy_{n-1}^2)$ so the intersection of this plane with the sphere $S_r(o)$ is a non-Euclidean sphere with Euclidean radius $(2e^s(\operatorname{ch} r - \operatorname{ch} s))^{\frac{1}{2}}$; its non-Euclidean $(n-2)$-dimensional area is

$$\Omega_{n-1}\{2e^{-s}(\operatorname{ch} r - \operatorname{ch} s)\}^{\frac{1}{2}(n-2)}$$

and the function $y \longrightarrow v(x,y)$ is constant on it. The arc-element on the circle

$$y_1 = (2e^s(\operatorname{ch} r - \operatorname{ch} s))^{\frac{1}{2}},\ y_2 = \cdots = y_{n-1} = 0,\ y_n = e^s,$$

is

$$y_n^{-1}(dy_1^2 + dy_n^2)^{\frac{1}{2}} = \operatorname{sh} r\,(2e^s(\operatorname{ch} r - \operatorname{ch} s))^{\frac{1}{2}} ds.$$

Thus (6) takes the form

$$(7) \qquad \Omega_n \operatorname{sh}^{n-1} r\,(M^r u_0)(x)$$

$$= \Omega_{n-1} \operatorname{sh} r \int_{-r}^{r} u(x,s)(2\operatorname{ch} r - 2\operatorname{ch} s)^{\frac{1}{2}(n-3)} ds.$$

The function $w(x,s) = u_s(x,s)$ satisfies (2) with $u_0(x)$ replaced by $u_1(x)$. Thus by (7)

$$(8) \qquad \Omega_n \operatorname{sh}^{n-1} r\,(M^r u_1)(x)$$

$$= \Omega_{n-1} \operatorname{sh} r \int_{-r}^{r} u_s(x,s)\,(2\operatorname{ch} r - 2\operatorname{ch} s)^{\frac{1}{2}(n-3)}\,ds.$$

<u>Assume now</u> n <u>odd</u>. Then apply $\left[\partial/\partial(2 \operatorname{ch} r)\right]^{\frac{1}{2}(n-3)}$ to (7) and (8). This gives

$$(9) \qquad \int_{-r}^{r} u(x,s)\,ds = \frac{\Omega_n}{(\frac{n-3}{2})!\,\Omega_{n-1}} \left(\frac{\partial}{\partial(2 \operatorname{ch} r)}\right)^{\frac{1}{2}(n-3)} (\operatorname{sh}^{n-2} r\,(M^r u_0)(x)),$$

$$(10) \qquad \int_{-r}^{r} u_s(x,s)\,ds = \frac{\Omega_n}{(\frac{n-3}{2})!\,\Omega_{n-1}} \left(\frac{\partial}{\partial(2 \operatorname{ch} r)}\right)^{\frac{1}{2}(n-3)} (\operatorname{sh}^{n-2} r\,(M^r u_1)(x)).$$

Applying $\partial/\partial r$ to (9) and integrating (10) we obtain by addition the desired formula

$$(11) \qquad u(x,s) = \frac{1}{2(\frac{n-3}{2})!}\,\frac{\Omega_n}{\Omega_{n-1}} \left[\frac{\partial}{\partial s}\left(\frac{\partial}{\partial(2\operatorname{ch} s)}\right)^{\frac{1}{2}(n-3)} \{\operatorname{sh}^{n-2} s\,(M^s u_0)(x)\} + \left(\frac{\partial}{\partial(2\operatorname{ch} s)}\right)^{\frac{1}{2}(n-3)} \{\operatorname{sh}^{n-2} s\,(M^s u_1)(x)\}\right]$$

from which the validity of Huygens' principle is obvious.

This formula is derived for $n = 3$ in Günther [6], p.23, and for general n by Kiprijanov and Ivanov [20], using different methods.

§6. <u>Conformal Invariance</u>.

Let M be a manifold with pseudo-Riemannian structure g and curvature tensor R. Let $p \in M$ and X,Y two vector fields on M. Let M_p denote the tangent space to M at p and X_p the value of

X at p. The trace of the endomorphism

$$L \longrightarrow R_p(Y_p, L)(X_p) \qquad L \in M_p$$

is denoted $r_p(X_p, Y_p)$ and the tensor field r on M given by

$$r(X,Y)(p) = r_p(X_p, Y_p)$$

is called the <u>Ricci curvature</u> of the M. (Sometimes $-r$ is called the Ricci curvature.) In local coordinates $\{x_1, \ldots, x_n\}$ with $\partial_i = \partial/\partial x_i$ we write as usual

$$g(\partial_i, \partial_j) = g_{ij}, \quad r(\partial_i, \partial_j) = r_{ij},$$

$$R(\partial_i, \partial_j)(\partial_\ell) = \sum_k R^k_{\ell ij} \partial_k,$$

so

(1) $$r_{ij} = \sum_k R^k_{ijk}.$$

The <u>scalar curvature</u> is then defined by

$$K = \sum_{i,j} g^{ij} r_{ij}$$

if (g^{ij}) is the inverse of the matrix (g_{ij}).

In the case when g is Riemannian and $e_1 \in M_p$ a unit vector then (1) shows that $-r_p(e_1, e_1)$ is the sum of the sectional curvatures for two-planes $\mathbb{R}e_1 + \mathbb{R}e_i$, $(e_1, \ldots, e_n)$ being an orthonormal basis of M_p.

Now let two manifolds M_1 and M_2, respectively, have pseudo-Riemannian structures g_1 and g_2, scalar curvatures K_1 and K_2

and Laplace-Beltrami operators L_1 and L_2. Let

$$T: M_1 \longrightarrow M_2$$

be a <u>conformal diffeomorphism</u>, i.e.,

$$T^*g_2 = \tau^2 g_1, \quad \text{where} \quad \tau \in C^\infty(M_1).$$

The following quasi-invariance of the Laplace-Beltrami operator is known (cf. Friedlander [5], Ørsted [30] and Kosman [21]; the case n=4 is well known in general relativity).

<u>THEOREM 6.1</u>. <u>With the notation above suppose</u> $T: M_1 \longrightarrow M_2$ <u>is a conformal diffeomorphism</u>. <u>Let</u>

$$c_n = \frac{n-2}{4(n-1)} \quad \underline{\text{where}} \quad n = \dim M_1 = \dim M_2 .$$

<u>Then</u>

$$(L_1 + c_n K_1)\left(\tau^{\frac{n}{2}-1}(f\circ T)\right) = \tau^{\frac{n}{2}+1}\left((L_2 + c_n K_2)f\right)\circ T \quad \underline{\text{for}} \quad f \in C^\infty(M_2).$$

For a sketch of the proof let a manifold M have pseudo-Riemannian structure g and put $h = \tau^2 g$. Then

$$h_{ij} = \tau^2 g_{ij} \qquad h^{ij} = \tau^{-2} g^{ij}$$

and the new Christoffel symbols are

$$(2) \quad \Gamma^{*k}_{ij} = \Gamma^k_{ij} + \delta^k_i \partial_j(\log\tau) + \delta^k_j \partial_i(\log\tau) - g_{ij} \sum_\ell g^{\ell k} \partial_\ell(\log\tau).$$

Since L can be written

$$L = \sum_{i,j} g^{ij}(\partial_i\partial_j - \sum_k \Gamma^k_{ij}\,\partial_k),$$

one can use (2) to prove

$$(3)\qquad L^*f = \tau^{-\frac{n}{2}-1}\left[L\left(\tau^{\frac{n}{2}-1}f\right) - L\left(\tau^{\frac{n}{2}-1}\right)f\right].$$

On the other hand the new scalar curvature K^* satisfies

$$(4)\qquad L(\tau^{\frac{n}{2}-1}) = c_n(K^*\tau^{\frac{n}{2}+1} - K^{\frac{n}{2}-1})$$

(Yamabe [29]). The theorem now follows from (3) and (4) since both L and K are invariant under isometries.

Let g be a pseudo-Riemannian structure on a manifold M. A curve $t \to \gamma(t)$ on M is said to be <u>isotropic</u> if each of its tangent vectors is isotropic.

<u>Proposition 6.2. Let g and h be two conformally equivalent pseudo-Riemannian structures on M, say $h = \tau^2 g$. Let $t \to \gamma(t)$ be an isotropic geodesic on (M,g). If the parameter change $t = f(s)$ satisfies</u>

$$(5)\qquad \tau^2(\gamma(f(s))f'(s) = 1$$

<u>then the curve $s \longrightarrow \gamma(f(s))$ is an isotropic geodesic on the pseudo-Riemannian manifold (M,h).</u>

While this well known result from relativity (e.g. Sachs-Wu [23] p.132, Beem-Ehrlich [2] p.232) is most naturally proved in the framework of symplectic geometry (see Fefferman [3]) the following version kindly communicated by B. Ørsted fits better into the present context.

Writing $\dot{x}_i = dx_i/dt$, $\ddot{x}_i = d^2x_i/dt^2$ the differential equation for geodesics is

(6) $$\ddot{x}_k + \sum_{i,j} \Gamma^k_{ij}\, \dot{x}_i \dot{x}_j = 0, \qquad \text{(each } k\text{)},$$

where the Christoffel symbols satisfy

(7) $$2\sum_{\ell} g_{i\ell}\Gamma^{\ell}_{jk} = \partial_j\, g_{ik} + \partial_k g_{ij} - \partial_i\, g_{jk}.$$

Thus (6) is equivalent to

(8) $$\tfrac{1}{2} \sum_{i,j} (\partial_k g_{ij})\dot{x}_i \dot{x}_j = \sum_j g_{kj}\ddot{x}_j + \sum_{j,\ell} (\partial_\ell g_{kj})\dot{x}_j \dot{x}_\ell.$$

The isotropy condition amounts to

(9) $$\sum_{i,j} g_{ij}\dot{x}_i \dot{x}_j = 0$$

along the curve. Now put

$$y_i(s) = x_i(f(s)) \quad , \quad y'_i = dy_i/ds \qquad \text{etc.}$$

By (8) and (9)

(10) $$\sum_j g_{kj}\ddot{x}_j + \sum_{j,\ell} (\partial_\ell g_{kj})\dot{x}_k \dot{x}_\ell = 0$$

and we shall prove the same relation for y_i with respect to h. Since

$$\sum_{i,j} h_{ij} y'_i y'_j = \tau^2 f'(s)^2 \sum_{i,j} g_{ij}\dot{x}_i \dot{x}_j = 0$$

this will prove the proposition. Using

$$y'_i = f' x_i, \qquad y''_i = (f')^2\, \ddot{x}_i + f'' \dot{x}_i$$

we have for the new right hand side in (10),

$$\sum_j \tau^2 g_{kj} y_j'' + \sum_{j,\ell} \frac{\partial}{\partial y_\ell} (\tau^2 g_{kj}) y_j' y_\ell'$$

$$= \tau^2 \sum_j g_{kj}((f')^2 \ddot{x}_j + f'' \dot{x}_j) + (f')^2 \sum_{j,\ell} (\tau^2 \frac{\partial g_{kj}}{\partial x_\ell} + \frac{\partial \tau^2}{\partial x_\ell} g_{kj}) \dot{x}_j \dot{x}_\ell .$$

Using (8) and (9) this becomes

$$\tau^2 f'' \sum_j g_{kj} \dot{x}_j + (f')^2 \sum_{j,\ell} \frac{\partial \tau^2}{\partial x_\ell} g_{kj} \dot{x}_j \dot{x}_\ell$$

$$= (\tau^2 f'' + (f')^2 \frac{d(\tau^2)}{dt}) \sum_j g_{kj} \dot{x}_j = 0$$

by (5).

<u>Proposition 6.3.</u> <u>Let</u> (M_1, g_1) <u>and</u> (M_2, g_2) <u>be two Lorentzian manifolds and</u> $T: M_1 \longrightarrow M_2$ <u>a conformal diffeomorphism. Then Huygens' principle holds for the operator</u> $L_1 + c_n K_1$ <u>if and only if it holds for</u> $L_2 + c_n K_2$.

This result is in fact an immediate consequence of Theorem 6.1 and Prop. 6.2. The result implies that if a Lorentzian manifold (M,g) is conformally equivalent to $\mathbb{R}^n$ (n even) with the standard Lorentzian structure then the differential equation $(L + c_n K)u = 0$ satisfies Huygens' principle. The vanishing of the so-called conformal curvature tensor is a necessary (Weyl [28]) and a sufficient (Schouten [25]) condition for the conformal equivalence of (M,g) to $\mathbb{R}^n$ $(n > 3)$, locally.

Günther's examples mentioned in §3 showed that Huygens' principle can hold for the equation $Lu = 0$ even if the space is not conformally flat. Consider in fact the Lorentzian structure

$$(11) \qquad 2dx_1 dx_2 + \sum_{i,j=3}^{n} g_{ij}(x_1) dx_i dx_j ,$$

where $n \geq 4$ is even and (g_{ij}) positive definite depending only on x_1. Using Hadamard's logarithm criterion, Günther showed that the equation $Lu = 0$ always satisfies Huygens' principle. Yet as a differential equation in the g_{ij}, the condition $C = 0$ is not in general satisfied, so the Lorentzian structure (11) is not in general conformally flat.

Consider now the quadric

$$Q_+(X) \equiv x_0^2 + x_1^2 - x_2^2 - \ldots - x_n^2 = 1 \qquad X = (x_0,\ldots,x_n)$$

in $\mathbb{R}^{n+1}$ with the pseudo-Riemannian structure induced by Q_+. The quadric can be identified with the coset space $\mathbb{O}(2,n-1)/\mathbb{O}(1,n-1)$. It has the Lorentzian signature $(1,n-1)$, and sectional curvature $+1$ (cf. [13]). The scalar curvature is $-n(n-1)$.

On the other hand the quadric

$$Q_-(X) \equiv x_0^2 - x_1^2 - \ldots - x_n^2 = -1, \qquad X = (x_0,\ldots,x_n)$$

with the pseudo-Riemannian structure induced by Q_- has Lorentzian signature $(1,n-1)$, sectional curvature -1 and scalar curvature $n(n-1)$. It can be identified with $\mathbb{O}(1,n)/\mathbb{O}(1,n-1)$.

It was mentioned in §3 that the equation $Lu = 0$ on these spaces does not satisfy Huygens' principle. The following positive result was however observed by Ørsted [30].

Proposition 6.4. Let n be even. Then

(i) The operator $L - \frac{1}{4}n(n-2)$ on $\mathbb{O}(2,n-1)/\mathbb{O}(1,n-1)$ satisfies Huygens' principle.

(ii) The operator $L + \frac{1}{4}n(n-2)$ on $\mathbb{O}(1,n)/\mathbb{O}(1,n-1)$ satisfies Huygens' principle.

It can be readily verified that in these two cases the conformal curvature tensor C vanishes so the result follows from Schouten's theorem mentioned together with Prop. 6.3. In Ørsted [31] a conformal equivalence is explicitly constructed. It generalizes the stereographic projection $\mathbb{R}^n \longrightarrow S^n$; for the space (i) it is given by

$$Ty = \frac{(1-\frac{1}{4}(y,y),y)}{1+\frac{1}{4}(y,y)},$$

where $y = (y_1,\ldots,y_n)$, $(y,y) = y_1^2 - y_2^2 - \ldots - y_n^2$. Then T maps the space $\mathbb{R}^n$ with the Lorentzian structure $dy_1^2 - dy_2^2 - \ldots - dy_n{}^2$ conformally into the quadric

$$x_0^2 + x_1^2 - x_2^2 - \ldots - x_n^2 = 1.$$

For a final example we take a second look at the hyperbolic space $\mathbb{H}^n$ from §5. We consider the product $\mathbb{R} \times \mathbb{H}^n$ with the Lorentzian structure

$$dt^2 - ds^2 = dt^2 - y_n^{-2}(dy_1^2 + \ldots + dy_n^2).$$

Since scalar curvature is additive under the taking of products of manifolds, $\mathbb{H}^n \times \mathbb{R}$ has scalar curvature $-n(n-1)$. Writing

$$dt^2 - ds^2 = y_n^{-2}[d(ty_n)^2 - dy_1^2 - \ldots - dy_n^2]$$

we see that $\mathbb{R} \times \mathbb{H}^n$ is conformally equivalent to a flat space $\mathbb{R}^{n+1}$. Thus we deduce from Prop. 6.3 that the operator

$$\frac{\partial^2}{\partial t^2} - L_{\mathbb{H}^n} - \left(\frac{n-1}{2}\right)^2$$

satisfies Huygens' principle for n odd. This we had also observed on the basis of (11) §5. Similarly, for the sphere S^n (n odd) the operator

$$\frac{\partial^2}{\partial t^2} - L_{S^n} + \left(\frac{n-1}{2}\right)^2$$

satisfies Huygens' principle (cf. Günther [6] for $n = 3$, Lax-Phillips [22] and Ørsted [30] in general).

Bibliography

1. L. Àsgeirsson, Über eine Mitterwertseigenschaft von Lösungen homogener linearer partieller Differentialgleichung 2. Ordnung mit konstanten Koefficienten. Math. Ann. 113(1936), 321-346.

2. J. Beem and P.Ehrlich, Global Lorentzian Geometry, Marcel Dekker, New York 1981.

3. C.L. Fefferman, Monge-Ampère equations, the Bergman kernel, and geometry of pseudoconvex domains. Ann. of Math. 103(1976), 395-416.

4. H. Freudenthal and de Vries, H. Linear Lie Groups. Academic Press, 1969.

5. F.C. Friedlander, The Wave Equation in Curved Space-Time. Cambridge Univ. Press, 1975,

6. P. Günther, Über einige spezielle Probleme aus der Theorie der linearen partiellen Differentialgleichungen zweiter Ordnung. Ber. Verh. Sächs. Akad. Wiss. Leipzig 102(1957), 1-50.

7. ________, Ein Beispiel einer nichttrivialen Huygensschen Differentialgleichung mit vier unabhängigen Variablen. Arch. Rat. Mech. Anal. 18(1965), 103-106.

8. J. Hadamard, Lectures on Cauchy's Problem in Linear Partial Differential Equations, 2nd Ed. Dover 1952.

9. Harish-Chandra, Differential operators on a semisimple Lie algebra. Amer. J. Math. 79(1957), 87-120.

10. ____________, Spherical Functions on a semisimple Lie group. Amer. J. Math. 80(1958), 241-310.

11. S. Helgason, Partial differential equations on Lie groups. Scand. Math. Congress XIII, Helsinki 1957, 110-115.

12. __________, Differential operators on homogeneous spaces. Acta Math. 102(1959), 239-299.

13. __________, Some remarks on the exponential mapping for an affine connection. Math. Scand. 9(1961), 129-146.

14. __________, Fundamental solutions of invariant differential operators on symmetric spaces. Amer. J. Math. 86(1964), 565-601.

15. __________, Analysis on Lie Groups and Homogeneous Spaces. Conf. Board Math. Sci. Series, No. 14, Amer. Math. Soc. Providence, 1972.

16. __________, Solvability questions for invariant differential operators, pp.517-527, in Proc. 5th Int. Colloq. on "Group Theoretical Methods in Physics" Montreal 1976, Academic Press 1977.

17. __________, Differential Geometry, Lie Groups and Symmetric Spaces. Academic Press, 1978.

18. E. Hölder, Poissonsche Wellenformel in nichteuclidischen Räumen. Ber. Verh. Sächs. Akad. Wiss. Leipzig 90(1938), 55-66.

19. F. John, Plane Waves and Spherical Means. Interscience, New York, 1955.

20. I.A. Kiprijanov, and L.A. Ivanov, The Euler-Poisson-Darboux equation in a Riemannian space. Soviet Math. Dokl. 24 (1981), 331-335.

21. Y. Kosman, Sur les degrées conformes des opérateurs differentiels, C.R. Acad. Sci. Paris. 280(1975), A 229-232.

22. P.D. Lax and R.S. Phillips, An example of Huygens' principle. Comm. Pure Appl. Math. 31(1978), 415-423.

23. R.K. Sachs and H. Wu, General Relativity for Mathematicians. Springer Verlag, 1977.

24. R. Schimming, A review of Huygens' principle for linear hyperbolic differential equations. Proc. IMU Symposium, "Group Theoretical Methods in Mechanics" Novosibirsk, 1978.

25. J.A. Schouten, Über die konforme Abbildung n-dimensionaler Mannigfaltigkeiten mit quadratischer Massbestimmung auf eine Mannigfaltigkeit mit Euklidischer Massbestimmung. Math. Z. 11(1921), 58-88.

26. K. Stellmacher, Ein Beispiel einer Huygensschen Differentialgleichung. Gött. Nachr. 1953, 133-138.

27. O. Tedone, Sull'integrazionne dell'equazione $\partial^2\Phi^2/\partial t^2-\Sigma_i\ \partial^2\Phi/\partial x_i^2 = 0$. Ann. di Mat. 3(1898), 1-24.

28. H. Weyl, Reine Infinitesimalgeometrie.Math. Z. 2(1918), 384-411.

29. H.Yamabe, On a deformation of Riemannian structures on compact manifolds, Osaka Math. J. 12(1960), 21-37.

30. B. Ørsted, The conformal invariance of Huygens' principle. J. Differential Geometry 16(1981), 1-9.

31. _________, Conformally invariant differential equations and projective geometry. J. Funct. Analysis. 44(1981), 1-23.

SYMBOL MAPPINGS FOR CERTAIN NILPOTENT GROUPS

by Roger Howe, Gail Ratcliff, and Norman Wildberger
Yale University

§1. Introduction: One of the most effective techniques in the theory of partial differential equations is that of the symbolic calculus. For constant coefficient differential operators on $\mathbb{R}^n$, this symbolic calculus is easy to define and its usefulness is easy to understand. We will recall the basic facts concerning symbols of constant coefficient operators.

We need some standard notation. Let $\alpha = (\alpha_1, \alpha_2, \ldots, \alpha_n)$ be a multi index-an n-tuple of non-negative integers. The typical point in $\mathbb{R}^n$ is $x = (x_1, x_2, \ldots, x_n)$. Let

$$\partial^\alpha = \left(\frac{\partial}{\partial x_1}\right)^{\alpha_1} \cdots \left(\frac{\partial}{\partial x_n}\right)^{\alpha_n} \tag{1.1}$$

be a standard monomial partial differential operator. A general constant-coefficient differential operator D is then a sum

$$D = \sum a_\alpha \partial^\alpha \tag{1.2}$$

where the a_α are complex numbers. (Only finitely many a_α are non-zero.)

The typical point in $(\mathbb{R}^n)^*$, the dual of $\mathbb{R}^n$, which of course is isomorphic to $\mathbb{R}^n$, is denoted $\zeta = (\zeta_1, \ldots, \zeta_n)$. For D as in equation (1.2) we define $\sigma(D)$, the <u>symbol</u> of D to be the polynomial on $(\mathbb{R}^n)^*$ given by the formula

$$\sigma(D) = \sum a_\alpha (-2\pi i \zeta)^\alpha \tag{1.3}$$

where

$$\zeta^\alpha = \zeta_1^{\alpha_1} \cdots \zeta_n^{\alpha_n} \tag{1.4}$$

in parallel with (1.1).

The symbol $\sigma(D)$ is useful because of the existence of the Fourier transform. Let $S(\mathbb{R}^n)$ be the usual Schwartz space of smooth, rapidly decreasing functions on $\mathbb{R}^n$. The Fourier transform is the map

$$\hat{}: S(\mathbb{R}^n) \to S((\mathbb{R}^n)^*)$$

given by the familiar formula

$$\hat{f}(\zeta) = \int_{\mathbb{R}^n} f(x)\chi(\zeta\cdot x) \qquad f \in S(\mathbb{R}^n) \tag{1.5}$$

where

$$\chi(t) = e^{2\pi i t} \tag{1.6}$$

and $\zeta\cdot x = \sum \zeta_i x_i$. In this context, we have for D as in (1.2) the well-known formula

$$(Df)\hat{} = \sigma(D)\hat{f} \qquad f \in S(\mathbb{R}^n); \tag{1.7}$$

in other words, Fourier transform converts D into a multiplication operator, specifically, to multiplication by $\sigma(D)$.

We should also recall some group-theoretic aspects of the symbol mapping. Of course $\mathbb{R}^n$ is a Lie group, and the constant coefficient differential operators are just the differential operators which commute with the action of $\mathbb{R}^n$ on itself by right translations - the right - invariant differential operators for short. As such they can be represented as convolution operators:

$$D(f) = \delta_D * f \qquad f \in S(\mathbb{R}^n) \tag{1.8}$$

where δ_D is an appropriate distribution on $\mathbb{R}^n$ and $*$ indicates convolution of distributions on $\mathbb{R}^n$. (In fact, δ_D is supported at the orgin in $\mathbb{R}^n$.) We may extend Fourier transform to a map

$$\hat{}: S^*(\mathbb{R}^n) \to S^*((\mathbb{R}^n)^*)$$

where $S^*(\mathbb{R}^n)$ is the space of tempered distributions on $\mathbb{R}^n$. Then we have the relation

$$\sigma(D) = (\delta_D)\hat{} \tag{1.9}$$

so that formula (1.7) is a special case of the basic relation

$$(\Delta * f)\hat{} = \hat{\Delta}\hat{f} \qquad f \in S(\mathbb{R}^n), \Delta \in S^*(\mathbb{R}^n). \tag{1.10}$$

In particular if we set

$$\sigma(\Delta) = \hat{\Delta} \qquad \Delta \in S^*(\mathbb{R}^n) \tag{1.11}$$

then we see that the symbolic calculus for constant coefficient differential operators on $\mathbb{R}^n$ is part of a larger symbolic calculus for convolution operators on $\mathbb{R}^n$.

It would be desirable to have a symbolic calculus, analogous to that just described for $\mathbb{R}^n$, for non-abelian groups. This paper discusses some aspects of the problem of constructing such a symbolic calculus for nilpotent Lie groups.

A notion of symbol for nilpotent Lie groups is already implicit in the fundamental paper of Kirillov [K]. Let N be a connected, simply connected nilpotent Lie group with Lie algebra $\mathcal{N}$. Let $\mathcal{N}^*$ be the vector space dual to $\mathcal{N}$. The exponential map

$$\exp:\mathcal{N} \to N$$

is well-known [M] to be a global diffeomorphism. Define $S(N)$, the Schwartz space of N, by transport of structure from $\mathcal{N}$:

$$S(N) = (\exp^{-1})^*(S(\mathcal{N})) = \{f:f\circ\exp \in S(\mathcal{N})\}.$$

Then for $f \in S(N)$, we define $\kappa(f)$, the Kirillov symbol of f, by the recipe

$$\kappa(f) = (f\circ\exp)^{\wedge} \tag{1.12}$$

where now $\hat{}$ is Fourier transform from $S(\mathcal{N})$ to $S(\mathcal{N}^*)$.

The symbol $\kappa(f)$ has a number of desirable properties. First we should note that if N is abelian, then $\exp:\mathcal{N} \to N$ is an isomorphism of groups and can be used to identify $\mathcal{N}$ and N; then the symbol κ of formula (1.12) becomes the standard symbol σ defined in formula (1.3). Second, one has the elegant Kirillov character formula. Let Ad denote the action of N on $\mathcal{N}$ by conjugation and let Ad^* denote the contragredient action of N on $\mathcal{N}^*$ (the co-adjoint action). One knows after [K] that there is a bijection between (equivalence classes of) irreducible unitary representations of N and Ad^*N orbits in $\mathcal{N}^*$. If ρ is an irreducible unitary representation of N, the corresponding orbit will be denoted by $\mathcal{O}_\rho$, and similarly the representation corresponding to an orbit $\mathcal{O}$ will be denoted $\rho_{\mathcal{O}}$. If ρ is an irreducible unitary representation of N, then $\rho(f)$ is of trace class for all $f \in S(N)$, and the trace of $\rho(f)$ is given by [K],[Pk2]

$$\operatorname{tr} \rho(f) = \int_{\mathcal{O}_\rho} \kappa(f)(\lambda)\,d\lambda \tag{1.13}$$

where $d\lambda$ is an Ad^*N-invariant measure on $\mathcal{O}_\rho$. (There is only one such measure up to multiples, and the appropriate normalization of $d\lambda$ has been specified by Pukanszky [Pk2].)

Third, the symbol mapping κ behaves quite well on the convolution algebra of central (Ad-invariant) distributions. For example, κ defines an isomorphism between the center of the von Neumann algebra generated by the left regular representation of N on $L^2(N)$ and the algebra of bounded Ad^*N-invariant Borel functions (modulo almost everywhere equality) under pointwise multiplication [Hw1].

Fourth, when N is the Heisenberg group, then κ essentially coincides with the symbol mapping of the Weyl symbolic calculus, an equally viable variant of the widely used Kohn-Nirenberg symbolic calculus for pseudo-differential operators. See [Hw2] for an account of these connections, and also [Bl],[Br],[GLS] for related discussion.

Given these facts, it seem reasonable to attempt to develop a symbolic calculus based on the symbol mapping κ. Such an attempt was begun in [Hw3] (see also [Mi] and [S]) where it was shown one could obtain L^2 estimates on certain classes of convolution operators in terms of the Kirillov symbol. The arguments of [Hw2] are based directly on group- and representation-theoretic considerations and avoid using results from the general theory of pseudo-differential operators.

However, further investigation reveals some drawbacks of the map κ as a symbol map for dealing with non-central distributions. A major shortcoming, which is somewhat ironic, is that it fails to be compatible with the Kirillov correspondence between representations and orbits. The incompatibility is of the following nature: for $f \in \mathcal{S}(N)$ and an irreducible unitary representation ρ of N, it may happen that $\kappa(f)$ vanishes identically on $\mathcal{O}_\rho$, but still $\rho(f) \neq 0$, (or vice versa). This was observed already by Ludwig [L]. Thus although for suitable f one may estimate the operator norm $\sup_\rho(\|\rho(f)\|)$ in terms of $\kappa(f)$, there is no way to localize this result to obtain an estimate on $\rho(f)$ based solely on the behavior of $\kappa(f)$ around $\mathcal{O}_\rho$ for a given irreducible ρ. Another aspect of this failure is that while the relation

$$\kappa(\Delta_1 * \Delta_2) = \kappa(\Delta_1)\kappa(\Delta_2)$$

holds for central distributions Δ_i with smooth symbols $\kappa(\Delta_i)$, it does not necessarily hold if only one of the Δ_i is central.

Thus one is led to seek a notion of symbol that preserves the good features of κ described above and that is also compatible with the Kirillov correspondence in the senses just described. In particular, for $f \in S(N)$, we would like the symbol of f to be in $S(N^*)$ and to have the property that if the symbol of f vanishes on an Ad^*N orbit $O \subseteq N^*$, then $\rho_O(f) = 0$.

In [Wi] an approach to constructing such a symbol map is described. So far, however, there are a number of unresolved issues that require further study before the symbol mappings of [Wi] can become a general and effective tool for handling convolution operators. Among these are questions of uniqueness, of smoothness and of computability. Nevertheles the examples of [Wi] suggest this symbol mapping will be nicely behaved in many interesting cases, and the results of [R] show in the case of 3-step nilpotent groups that symbols of the type of [Wi] can be used effect ively to obtain estimates on $\rho(f)$ for individual irreducible representa tions ρ.

The purpose of this paper is to set the stage for the extension of the results of [R] by showing that an easily computable symbol of the ty of [Wi] exists for a class of groups that includes many interesting examp Our approach to this symbol mapping is through the well-established standard symbolic (Weyl) calculus for the Heisenberg group. This approach was suggested by [R] and relies on an elegant formula for the operators of the oscillator representation of the symplectic group.

Here is how the paper is organized. In §2 we give a detailed acco of the 4-dimensional, 3-step nilpotent group, the simplest group that illustrates the phenomena we are discussing. We show explicitly how the symbol κ of formula (1.12) fails to behave well with respect to coadjoint orbits. Out of this study there appears a simple modification of κ which does respect the orbits. In §3 we show this modification is part of a systematic phenomenon involving the oscillator representation of the symplectic group, or more properly the semidirect product of the Heisenberg group and the symplectic group. For this representation, we obtain an explicit and simple notion of symbol. This symbol involves the classical Cayley transform. In §4 we show how to use the computation of §3 to obtain a simple symbolic calculus having, at least for generic representations, the properties we have stipulated above; we can constr this symbol for a fairly rich class of nilpotent groups, including the ni radicals of most parabolic subgroups of semisimple groups. Finally in § we indicate some applications of our results and discuss possible furthe developments.

§2. In this section we discuss the Kirillov symbol κ of formula (1.12) on the simplest 3-step nilpotent group, and show how it can be modified to satisfy the criteria discussed in §1.

First we need to look at the symbol κ on the 3-dimensional Heisenberg group, which we denote here by H. We denote by h the Lie algebra of H. We will use coordinates (x,y,z) on h in which the commutator bracket is given by

$$[(x,y,z),(x',y',z')] = (0,0,xy'-x'y) \tag{2.1}$$

Let us write

$$h = \mathcal{W} \oplus \mathcal{Z} \tag{2.2}$$

where

$$\mathcal{W} = \{(x,y,0)\} \qquad \mathcal{Z} = \{(0,0,z)\} \qquad x,y,z \in \mathbb{R}.$$

From the commutation relations (2.1) we see that $\mathcal{Z}$ is the center and commutator ideal of h.

Let (μ,ν,λ) be the coordinates on h^* dual to (x,y,z) on h. It is easy to compute that the Ad^*H orbits in h^* are of 2 types:

i) Points $(\mu,\nu,0)$ (2.3)

ii) Planes $\mathcal{O}_\lambda = \{(\mu,\nu,\lambda): \mu,\nu \in \mathbb{R}\}$ for each $\lambda \neq 0$.

Let ρ_λ be the representation of H corresponding to $\mathcal{O}_\lambda$. We may realize ρ_λ on $L^2(\mathbb{R})$ by the formula

$$\rho_\lambda(\exp(x,y,z))(\phi)(t) = \chi(\lambda(yt+(\tfrac{1}{2})xy+z))\phi(x+t) \qquad \phi \in L^2(\mathbb{R}). \tag{2.4}$$

Here χ is as in formula (1.6).

We can extend ρ_λ to a representation of the convolution algebra $S(H)$ by the usual formula:

$$\rho_\lambda(f) = \int_H f(h)\rho_\lambda(h)\,dh$$

where dh is Haar measure on H. In fact we can do something considerably more refined in this case.

Define the operator L_h of left translation by $h \in H$ in the usual way:

$$L_h f(h') = f(h^{-1}h') \qquad f \in S(H). \tag{2.5}$$

Define L_h on $S^*(H)$, the tempered distributions on H by duality. Denote by $S^*(H,\lambda)$ the subspace of distributions Δ on H such that

$$L_{\exp(0,0,z)}(\Delta) = \chi(\lambda z)(\Delta). \tag{2.6}$$

Denote by $S(H,\lambda)$ the space of smooth functions f satisfying (2.6) and such that $(f \circ \exp)|W$ is in $S(W)$. (Here W is as in (2.2).) The space $S(H,\lambda)$ is naturally the conjugate dual space of $S^*(H,\lambda)$. Since the map $f \to (f \circ \exp)|W$ clearly defines an isomorphism from $S(H,\lambda)$ to $S(W)$, the space $S^*(H,\lambda)$ is naturally identified to $S^*(W)$.

There is a projection

$$p_\lambda: S(H) \to S(H,\lambda)$$

given by

$$p_\lambda(f)(h) = \int_{\mathbb{R}} f(\exp(0,0,z)h)\chi(\lambda z)dz. \tag{2.7}$$

It is easy to check that if $p_\lambda(f) = 0$, then $\rho_\lambda(f) = 0$, so that ρ_λ factors through p_λ to define a map

$$\rho_\lambda: S(H,\lambda) \to L(L^2(\mathbb{R}))$$

where $L(L^2(\mathbb{R}))$ is the space of bounded operators on $L^2(\mathbb{R})$. In fact ρ_λ defines an isomorphism from $S(H,\lambda)$ to the integral operators with kernels in $S(\mathbb{R}^2)$. Using duality and the Schwartz Kernel Theorem, we can extend ρ_λ to an isomorphism

$$\rho_\lambda: S^*(H,\lambda) \simeq \mathrm{Hom}(S(\mathbb{R}), S^*(\mathbb{R})). \tag{2.8}$$

In particular, every operator in $L(L^2(\mathbb{R}))$ is $\rho_\lambda(D)$ for a suitable distribution $D \in S^*(H,\lambda)$. This is explained in greater detail in [Hw2].

Now consider the 4-dimensional, 3-step nilpotent group which we will denote by N. Its Lie algebra will be denoted N. We coordinatize N with 4-tuples (s,x,y,z) in terms of which the commutator bracket looks like this:

$$[(s,x,y,z),(s',x',y',z')] = (0,0,sx'-s'x,xy'-x'y) \tag{2.9}$$

From formula (2.9) we see that the points $(0,x,y,z)$ comprise an algebra isomorphic to h. We identify it with h. The center of h is also the center of N.

Let the coordinates on N^* dual to (s,x,y,z) be (τ,μ,ν,λ). The inclusion $h \subseteq N$ induces a projection

$$p: N^* \to h \tag{2.10}$$

$$p(\tau,\mu,\nu,\lambda) = (\mu,\nu,\lambda).$$

The inverse image $p^{-1}(\mathcal{O}_\lambda)$ of a coadjoint orbit $\mathcal{O}_\lambda \subseteq h$ is of course 3-dimensional; it fibers into a one-parameter family of Ad^*N orbits

$$\mathcal{O}_{a,\lambda} = \{(a-\nu^2/2\lambda,\mu,\nu,\lambda): \mu,\nu \in \mathbb{R}\} \tag{2.11}$$

each of which is a parabolic cylinder lying over the plane $\mathcal{O}_\lambda$. For details behind this and other facts stated below, see [R].

Let $\rho_{a,\lambda}$ be the representation of N corresponding to the coadjoint orbit $\mathcal{O}_{a,\lambda}$. Then we may also realize $\rho_{a,\lambda}$ on $L^2(\mathbb{R})$, by the formula

$$\rho_{a,\lambda}(\exp(s,x,y,z))(\phi)(t) = \chi(r)\phi(t+x) \qquad \phi \in L^2(\mathbb{R}) \tag{2.12}$$

where r is given by

$$r = as + \lambda(-(\tfrac{1}{2})'st^2+(y-(\tfrac{1}{2})sx)t - (\tfrac{1}{6})sx^2 + (\tfrac{1}{2})xy + z). \tag{2.13}$$

Just as we have $h \subseteq N$, we have $H \subseteq N$. From formula (2.12) we can read off that $\rho_{a,\lambda}|H = \rho_\lambda$, with ρ_λ as in formula (2.4). According to the discussion surrounding the isomorphism (2.8), for each $n \in N$, there is a distribution $g_n^{a,\lambda} \in S^*(H,\lambda)$ such that

$$\rho_{a,\lambda}(n) = \rho_\lambda(g_n^{a,\lambda}). \tag{2.14}$$

The distribution $g_n^{a,\lambda}$ can be computed explicitly. For $n = \exp(s,x,y,z)$ it is

$$g_n^{a,\lambda}(\exp(x',y',z')) = \lambda\chi(-\lambda(\tfrac{1}{2}xy'+z'))(\chi(r))^{\wedge-1}(\lambda y')\delta_x(x'). \tag{2.15}$$

Here $(\chi(r))^{\wedge-1}$ denotes the inverse Fourier transform with respect to t of the function $\chi(r)$, and $\delta_x(x')$ is the Dirac delta at $x \in \mathbb{R}$. Evidently formula (2.14) may be extended to $S(N)$ by linearity:

$$\rho_{a,\lambda}(f) = \int_N f(n)g_n^{a,\lambda}dn = \rho_\lambda(g_f^{a,\lambda}) \qquad f \in S(N) \tag{2.16}$$

where dn is Haar measure on N. The second equality defines $g_f^{a,\lambda}$.

We wish to compare the Kirillov symbols of the functions f and $g_f^{a,\lambda}$ involved in formula (2.16). Actually it is simplest to do this for $n \in N$ and $g_n^{a,\lambda}$. The projection (2.10) defines a bijection

$$\rho_{a,\lambda}:\mathcal{O}_{a,\lambda} \to \mathcal{O}_\lambda \tag{2.17}$$

$$\rho_{a,\lambda}(a-\nu^2/2\lambda,\mu,\nu,\lambda) = (\mu,\nu,\lambda).$$

We will use formula (2.15) to compare $\kappa(\delta_n)|\mathcal{O}_{a,\lambda}$, where δ_n is the point mass to $n = \exp(s,x,y,z)$ in N, with $\kappa(g_n^{a,\lambda})|\mathcal{O}_\lambda$. By the definition of κ we see that

$$\kappa(\delta_n)(\tau,\mu,\nu,\lambda) = \chi((\tau s+\mu x+\nu y+\lambda z)). \tag{2.18}$$

On the other hand a computation shows that

$$\kappa(g_n^{a,\lambda})(\mu,\nu,\lambda) = \chi((a-\nu^2/2\lambda)s+\mu x+\nu y+\lambda(z-(\tfrac{1}{24})sx^2)). \tag{2.19}$$

Comparing formulas (2.18) and (2.19) in the light of (2.17) reveals the relation

$$p_{a,\lambda}^*(\kappa(g_n^{a,\lambda})) = \chi(-(\tfrac{1}{24})\lambda sx^2))\kappa(\delta_n). \tag{2.20}$$

Here $p_{a,\lambda}^*$ is the pull-back map on functions associated to the map $p_{a,\lambda}$.

Thus $\kappa(\delta_n)$ and $p_{a,\lambda}^*(\kappa(g_n^{a,\lambda}))$ are almost the same-they differ only by the scalar factor. However, if we integrate over N, following formula (2.16), then the relation between $\kappa(f)|\mathcal{O}_{a,\lambda}$ and $p_{a,\lambda}^*(\kappa(g_f^{a,\lambda}))$ becomes completely disguised. Let us denote

$$\varepsilon_\lambda(n) = \varepsilon_\lambda(s,x,y,z) = \chi(-(\tfrac{1}{24})\lambda sx^2) \tag{2.21}$$

Then combining formulas (2.16) and (2.20) gives us

$$p_{a,\lambda}^*(\kappa(g_f^{a,\lambda})) = \int_N f(n)\varepsilon_\lambda(n)\kappa(\delta_n)dn = \kappa(f\varepsilon_\lambda) = \kappa(f)*(\varepsilon_\lambda)\hat{} \tag{2.22}$$

where ^ here denotes Fourier transform from $S^*(N)$ to $S^*(N^*)$ and $*$ denotes the usual abelian convolution on N^*. We note that ε_λ is independent of z, and of y also, so that $(\varepsilon_\lambda)\hat{}$ will be supported on the plane $\lambda = \nu = 0$. In particular, the convolution indicated in formula (2.22) effectively takes place within the hyperplanes $p^{-1}(\mathcal{O}_\lambda)$. However, within that 3-dimensional space, the convolution will mix up the orbits $\mathcal{O}_{a,\lambda}$. Indeed we can explicitly compute

$$(\varepsilon_\lambda)\hat{}(\tau,\mu,\nu,\lambda) = \delta_o(\lambda)\delta_o(\nu)(-24/\lambda\tau)^{\frac{1}{2}}\cos(2\pi\mu(-(24\tau)/\lambda)^{\frac{1}{2}}) \tag{2.23}$$

if $\lambda\tau < 0$, and it is zero when $\lambda\tau > 0$. Thus $(\varepsilon_\lambda)\hat{}$ has support in the full half plane $\{(\tau,\mu,0,0): \lambda\tau < 0\}$. This means that $\kappa(g_f^{a,\lambda})$, hence $\rho(f)$, depends on how $\kappa(f)$ behaves not only on $\mathcal{O}_{a,\lambda}$, but on how $\kappa(f)$ behaves on $\mathcal{O}_{b,\lambda}$ for all b such that $\lambda(b-a) \geq 0$. In particular $\kappa(f)$ could vanish in a neighborhood of $\mathcal{O}_{a,\lambda}$ and we might still have $\rho(f) \neq$

However, the formula (2.18) also suggests a way to modify κ in order to eliminate this undesirable feature. If we write out the integral in formula (2.22) in terms of the coordinates (s,x,y,z), we see

$$p^*_{a,\lambda}(\kappa(g_f^{a,\lambda}))(\tau,\mu,\nu,\lambda)$$

$$= \int_{\mathbb{R}^4} (f\circ\exp)(s,x,y,z)\chi(-(\tfrac{1}{24})\lambda s x^2)\chi(\ (\tau s+\mu x+\nu y+\lambda z)\, ds dx dy dz$$

$$= \int_{\mathbb{R}^4} (f\circ\exp)(s,x,y,z)\chi(\ (\tau s+\mu x+\nu y+\lambda(z-(\tfrac{1}{24})sx^2))\, ds dx dy dz$$

$$= \int_{\mathbb{R}^4} (f\circ\exp)(s,x,y,z-(\tfrac{1}{24})sx^2)\chi(-(\tau s+\mu x+\nu y+\lambda z))\, ds dx dy dz$$

$$= (f\circ\exp\circ d)^\wedge \tag{2.24}$$

where $d: \mathcal{N} \to \mathcal{N}$ is given by

$$d(s,x,y,z) = (s,x,y,z+(\tfrac{1}{24})sx^2). \tag{2.25}$$

Thus if we define a "deformed symbol" κ_d by

$$\kappa_d(f) = (f\circ\exp\circ d)^\wedge \tag{2.26}$$

the formula (2.24) reads

$$p^*_{a,\lambda}(\kappa(g_f^{a,\lambda})) = \kappa_d(f)\,|\,\mathcal{O}_{a,\lambda} \tag{2.27}$$

In otherwords, if we precede the exponential map from $\mathcal{N}$ to N by the deformation (2.25) of $\mathcal{N}$ (which just amounts to sliding each element of $\mathcal{N}$ parallel to the center of $\mathcal{N}$), then procede as before we get a symbol κ_d such that for all $f \in \mathcal{S}(N)$, the behavior of $\rho_{a,\lambda}(f)$ is completely determined by the behavior of $\kappa_d(f)$ on the orbit $\mathcal{O}_{a,\lambda}$. In the succeeding sections we will see that this pleasant discovery is not special to this group, but is part of a phenomenon that holds for a large class of nilpotent groups.

§3. In this section we will compute the Weyl symbols of the operators of the oscillator representation of the symplectic groups. This will provide a framework for understanding the example in §2 and a basis for constructing symbols in more general situations.

Let H_m be the real Heisenberg group of dimension $2m+1$, and let $\mathfrak{h}_m$ be its Lie algebra. Let X_i, Y_i, and Z be a standard basis for

h_m, with associated coordinates x_i, y_i and z. This basis satisfies the Canonical Commutation Relations:

$$[X_i,X_j] = [Y_i,Y_j] = [X_i,Z] = [Y_i,Z] = 0$$

$$[X_i,Y_j] = \delta_{ij}Z \tag{3.1}$$

where δ_{ij} is Kronecker's delta. Let W be the span of the X_i and Y_i. Then the commutation relations (3.1) define a symplectic form, to be denoted $\langle\ ,\ \rangle$, on W. Let $Sp(W) = Sp_{2m}(\mathbb{R}) = Sp_{2m}$ be the isometry group of this form. Then $Sp(W)$ acts on H_m by automorphisms.

Let X_j be the span of X_i for $i \leq j$. Let U_m be the subgroup of Sp_{2m} which preserves the X_j for all $j \leq m$, and which acts trivially on the quotients X_j/X_{j-1}. Then U_m is a maximal unipotent subgroup of Sp_{2m}. We may form the semi-direct product $U_m \ltimes H_m$. We will see explicitly somewhat later that there is an isomorphism

$$U_m \ltimes H_m \simeq U_{m+1}. \tag{3.2}$$

We note that the group N of §2 is (isomorphic to) U_2.

Let $\mathfrak{sp}_{2m}$ be the Lie algebra of Sp_{2m}, and let $\mathfrak{n}_m$ be the Lie algebra of U_m. We may think of $\mathfrak{sp}_{2m}$ as the "infinitesimal isometries" of the form $\langle\ ,\ \rangle$ on W. This means that $T \in \mathrm{End}\,W$ is in $\mathfrak{sp}_{2m}$ if and only if

$$\langle Tw,w'\rangle + \langle w,Tw'\rangle = 0 \tag{3.3}$$

which is to say, the bilinear form

$$B_T(w,w') = \langle Tw,w'\rangle \tag{3.4}$$

is symmetric in w and w'.

We have

$$\mathfrak{n}_{m+1} = \mathfrak{n}_m \oplus h_m = \mathfrak{n}_m \oplus W \oplus \mathbb{R}Z,$$

Define $Z^* \in \mathfrak{n}^*_{m+1}$ by the recipe

$$Z^*(\mathfrak{n}_m) = 0 = Z^*(W) \qquad Z^*(Z) = 1. \tag{3.5}$$

Let $\mathcal{O}$ denote the Ad^*U_{m+1} orbit in $\mathfrak{n}^*_{m+1}$ through Z^*. It is clear from the definition (3.5) that Ad^*U_m stabilizes Z^*. Thus

$$\mathcal{O} = Ad^*(H_m)(Z^*) = Ad^*(\exp W)(Z^*). \tag{3.6}$$

We can compute explicitly, for $u \in \mathfrak{n}_m$, and $w, w' \in W$, that

$$Ad^*(\exp w')(Z^*)(u + w + zZ) = Z^*(\exp(-ad\ w')(u + w + zZ)$$

$$= Z\ (u + w + zZ + u(w') + \langle w,w'\rangle Z + \tfrac{1}{2}B_u(w',w')Z)$$

$$= z + \langle w,w'\rangle + \tfrac{1}{2}B_u(w'w'). \tag{3.7}$$

Let

$$p: \mathfrak{n}^*_{m+1} \to h^*_m \tag{3.8}$$

be the projection dual to the inclusion of h_m in $\mathfrak{n}_{m+1}$ given by equation (3.2). Since H_m is normal in U_{m+1}, the map p is Ad^*U_{m+1}-equivariant. Let $0_o \subseteq h^*_m$ be the Ad^*H_m orbit of $p(Z^*)$. As is well-known, 0_o is just the affine space

$$0_o = W^{\perp} \oplus p(Z^*). \tag{3.9}$$

The formula (3.7) makes it clear that the map

$$p: 0 \to 0_o \tag{3.10}$$

is an (Ad^*U_{m+1}-equivalent) diffeomorphism.

Let ω be the representation of U_{m+1} corresponding to the orbit 0. Then the restriction

$$\rho = \omega|H_m \tag{3.11}$$

is irreducible and of course corresponds to the orbit $0_o \subseteq h^*_m$. It is the unique irreducible representation of H_m with central character χ, in the sense that

$$\rho(\exp zZ) = \chi(z)I \tag{3.12}$$

where I is the identity operator on the space where ρ is realized.

Let $S^*(H,\chi)$ be the space of tempered distribution Δ on H such that

$$L_{\exp\, zZ}(\Delta) = \chi(z)\Delta. \tag{3.13}$$

The analogy with equation (2.6) is obvious. Just as in §2, we know that if L is any operator on the Hilbert space where ρ is realized, there is a unique distribution $\Delta_L \in S^*(H,\chi)$ such that

$$L = \rho(\Delta_L). \tag{3.14}$$

The Kirillov symbol $\kappa(\Delta_L)$ will then be a distribution on h^*, supporte on $\mathcal{O}_o$. As mentioned in the introduction, $\kappa(\Delta_L)$ is essentially the same as the Weyl symbol of L. See [Hw2], §2.4 for a precise discussion.

We wish to compute the symbols $\kappa(\Delta_{\omega(u)})$ for $u \in U_{m+1}$. More precisely, we shall see that these symbols are smooth functions times the canonical invariant measure on $\mathcal{O}_o$. We will give a formula for the pull-backs $p^*(\kappa(\Delta_{\omega(u)}))$ of these symbols to the orbit $\mathcal{O} \subseteq \mathfrak{u}^*_{m+1}$.

The answer is in terms of the classical Cayley transform. For $T \in \mathrm{End}\,\mathcal{W}$ recall that $c(T)$, the <u>Cayley transform</u> of T is defined by

$$c(T) = \frac{1-T}{1+T} \tag{3.15}$$

provided that $(1+T)$ is invertible. It is easy to compute that

$$c(c(T)) = T. \tag{3.16}$$

It is well-known [We] that $c(T)$ defines a bijection between Zariski-open subsets of Sp_{2m} and $\mathfrak{sp}_{2m}$. It is clear from formula (3.15) that if $V \subseteq \mathcal{W}$ is a subspace invariant under T, then it is also invariant under $c(T)$. Therefore we see from the definition of U_m (preceding formula 3.2) that the map c will define a bijection between $\mathfrak{u}_m$ and U_m.

<u>Theorem 3.1:</u> With notations as above

$$\kappa(\Delta_{\omega(u)})(p(\lambda)) = \chi(-2\lambda(c(u))) \quad u \in U_{m+1},\ \lambda \in \mathcal{O}. \tag{3.17}$$

Remarks: a) This formula is in fact a direct generalization of formula (2.20). For if we identify the group N of §2 with U_2, and realize U_2 as matrices with respect to the basis of $\mathcal{W}$ of formula (3.1) then

$$U_2 = \left\{ \begin{bmatrix} 0 & x & 2z & y \\ 0 & 0 & y & s \\ 0 & 0 & 0 & 0 \\ 0 & 0 & -x & 0 \end{bmatrix} \right\} \tag{3.18}$$

where (s,x,y,z) are the coordinates on N used in §2. If we denote the matrix in (3.18) by T, then we see easily that

$$T^2 = \begin{bmatrix} 0 & 0 & 2xy & xs \\ 0 & 0 & -xs & 0 \\ 0 & 0 & 0 & 0 \\ 0 & 0 & 0 & 0 \end{bmatrix} \qquad T^3 = \begin{bmatrix} 0 & 0 & -x^2s & 0 \\ 0 & 0 & 0 & 0 \\ 0 & 0 & 0 & 0 \\ 0 & 0 & 0 & 0 \end{bmatrix} \tag{3.19}$$

and $T^\ell = 0$ for $\ell \geq 4$. Hence, comparing the formulas

$$\exp T = 1 + T + T^2/2 + T^3/6$$

and

$$c(-(T/2)) = 1 + T + T^2/2 + T^3/4,$$

we see that

$$\exp T = c(-((T - (T^3/12))/2)) \tag{3.20}$$

Comparing formulas (3.20), (3.19) and (3.18), we see that the deformation d of formula (2.25) precisely compensates for the discrepancy between $\exp T$ and $c(-(T/2))$.

b) In fact formula (3.17), suitably interpreted, remains true for all $g \in Sp_{2n}$(or more properly, in $\widetilde{Sp}_{2n}$, the 2-fold cover of Sp_{2m}). The full formula fits naturally into the general theory of the oscillator representation.

Proof of Theorem 3.1: We will establish formula (3.17) by a direct computation using the Schrodinger model [Ca], [Ge] for ρ. We will first compute the Kohn-Nirenberg symbol of $\omega(u)$, and then compute the Weyl symbol using the formula of [Hw2], §2.3 relating them. (In [Hw2], the isotropic symbol plays the role of the Weyl symbol, while the polarized symbol is the the Kohn-Nirenberg symbol. Again we note that the connection between the Kirillov symbol for H and the Weyl symbol is described precisely [Hw2], §2.4.) Our calculations will be formal, but there is nothing very delicate involved in making them rigorous.

With respect to the basis $\{X_i, Y_i\}$ of formula (3.1) for W, the Lie algebra $\mathfrak{n}_m$ consists of matrices of the form

$$\begin{bmatrix} A & B \\ 0 & -A^t \end{bmatrix} \tag{3.21}$$

where A, B are $m \times m$ matrices with A upper triangular, and B symmetric. When $\mathfrak{n}_{m+1}$ is written like this, the decomposition (3.2) looks like

$$\begin{bmatrix} 0 & x & 2z & y \\ 0 & A & y^t & B \\ \bigcirc & & 0 & 0 \\ & & -x^t & -A^t \end{bmatrix} \tag{3.22}$$

where $x = (x_1,\ldots,x_m)$ and $y = (y_1,\ldots,y_m)$ are m-vectors, and $z \in \mathbb{R}$, and form the coordinates of $\mathfrak{h}_m$ with respect to the basis of (3.1), while A and B are as in (3.21).

As is well-known, we can realize the representation ρ on $L^2(\mathbb{R}^n)$ according to the following formulas. In these formulas we regard the general element of $\mathfrak{n}_{m+1}$ as being a sum of components as indicated in the matrix (3.22), and we will indicate an element of $\mathfrak{n}_{m+1}$ in which only one component is non-zero simply by the name of that component. Thus A will denote the matrix

$$\begin{bmatrix} 0 & 0 & \bigcirc & \\ 0 & A & & \\ \bigcirc & & 0 & 0 \\ & & 0 & -A^t \end{bmatrix}$$

and so forth. With this convention we have

$$\begin{aligned}
&a) \quad \omega(\exp A)f(x') = f(x' \exp A) \qquad f \in L^2(\mathbb{R}^n)\\
&b) \quad \omega(\exp x)f(x') = f(x' + x)\\
&c) \quad \omega(\exp y)f(x') = \chi(x'y^t)f(x')\\
&d) \quad \omega(\exp 2z)f(x') = \chi(z)f(x')\\
&e) \quad \omega(\exp B)f(x') = \chi(\tfrac{1}{2}x'Bx'^t)f(x').
\end{aligned} \tag{3.23}$$

In these formulas, x,y and x' are regarded as row vectors. In (3.23)c), the vector y^t is the transpose of y, so is a column vector, and $x'y^t$ is the usual matrix product, which in this case is just a number. The other matrix products indicated are similar.

From the formulas (3.23) we can compute the Kohn-Nirenberg symbols of these five types of operators. The Kohn-Nirenberg symbol σ_L of an operator L on $L^2(\mathbb{R}^n)$ is defined by the formula (see [KN], [Hw2])

$$L(f)(x') = \int_{\mathbb{R}^n} \hat{f}(y)\sigma_L(x',y')\chi(xy'^t)dy' \tag{3.24}$$

We consider the operators of (3.23) in turn. If $L = \omega(\exp A)$, then we want the equation

$$f(x' \exp A) = \int_{\mathbb{R}^n} \hat{f}(y')\sigma_L(x'y')\chi(x'y'^t)dy'.$$

The Fourier Inversion formula tells us that

$$\sigma_L(x',y) = \chi((x' \exp A - x')y'^t)$$

will work. This is a special case of a well-known formula for the symbol of a diffeomorphism. A similar formula holds for the translation operators of (3.23)b). The last three operators of (3.23) are multiplication operators, and multiplication operators are their own symbols. Thus the Kohn-Nirenberg symbols of the operators (3.23), in order, are

$$\begin{aligned}
&a)\quad \chi((x' \exp A - x')y'^t)\\
&b)\quad \chi(xy'^t)\\
&c)\quad \chi(x'y^t)\\
&d)\quad \chi(z)\\
&e)\quad \chi(\tfrac{1}{2}x'Bx'^t)\,.
\end{aligned} \tag{3.25}$$

It is well-known (cf. [Hw2], §2.3) that if one follows any operator with a multiplication operator or precedes it with a convolution operator, then the symbol of the product is the product of the symbols. Thus using formulas (3.25) we can see that the Kohn-Nirenberg symbol for $\omega(\exp B \exp 2z \exp y \exp A \exp x)$ is

$$\chi((x' \exp A - x')y'^t + xy'^t + x'y^t + z + \tfrac{1}{2}x'Bx'^t) \tag{3.26}$$

We can also compute that the matrix of this 5-fold product looks like

$$\begin{bmatrix} 1 & x & 2z-y(xD^{-1})^t & y(D^{-1})^t \\ 0 & D & y^t-B(xD^{-1})^t & B(D^{-1})^t \\ & & 1 & 0 \\ \multicolumn{2}{c}{\bigcirc} & -(xD^{-1})^t & (D^{-1})^t \end{bmatrix} \tag{3.27}$$

where $D = \exp A$.

Next we convert from the Kohn-Nirenberg symbol of formula (3.26) to the Weyl symbol. Applying Fourier Inversion to [Hw2], formula (2.38) and dilating by 2 as indicated by [Hw2]§2.3 we see that the Weyl symbol of an operator with Kohn-Nirenberg symbol σ_L is given by the convolution

$$\int \sigma_L(a,b)\chi(2(x'-a)(y'-b)^t)\,da\,db. \tag{3.28}$$

If for σ_L we use expression (3.26) and compute formally we obtain

$$\int \chi(a(D-I)b^t + xb^t + ay^t + z + (\tfrac{1}{2})aBa^t)\chi(2(x'-a)(y'-b)^t)\,da\,db$$

$$= \int(\int \chi((a(D-I) + x - 2(x'-a))b^t db)\chi(ay^t + z + (\tfrac{1}{2})aBa^t + 2(x'-a)y'^t)\,da$$

$$= \int \delta(a(D+I) - (2x'-x))\chi(ay^t + z + (\tfrac{1}{2})aBa^t + 2(x'-a)y'^t)\,da$$

$$= (z - x(D+I)^{-1}y^t + \tfrac{1}{2}x(D+I)^{-1}B(D^t+I)^{-1}x^t + 2x(D+I)^{-1}y'^t + 2x'(D+I)^{-1}y^t$$

$$- 2x(D+I)^{-1}x'^t + 2x'(D+I)^{-1}B(D^t+I)^{-1}x't - 2x'c(D)y'^t) \tag{3.29}$$

as the Weyl or Kirillov symbol of $\omega(u)$ with u as in (3.27). The δ occurring in this calculation is Dirac's delta.

On the other hand, we may use standard matrix algebra to compute the Cayley transform of the matrix (3.27). The result is

$$\begin{bmatrix} 0 & -x(I+D)^{-1} & -(z-x)(I+D)^{-1}y^t + (\tfrac{1}{2})xEx^t) & xE - y(I+D^t)^{-1} \\ 0 & c(D) & Ex^t - (I+D)^{-1}y^t & -2E \\ 0 & 0 & 0 & 0 \\ 0 & 0 & (I+D^t)^{-1}x^t & -c(D)^t \end{bmatrix} \tag{3.30}$$

where $E = (I+D)^{-1}B(I+D^t)^{-1}$.

The formula (3.29) gives the left hand side of formula (3.17) when u is given by (3.27). If we combine (3.30) with formula (3.7) to compute the right hand side of (3.17), we see that, indeed, formula (3.17) is valid.

§4: The goal of this section is to take formula (3.17) and parlay it into a symbol mapping for representations of other groups than U_m. To do this, consider a nilpotent group N and an irreducible representation σ of N. Suppose there is a homomorphism

$$\phi : N \to U_{m+1} \tag{4.1}$$

such that

$$\sigma = \omega \circ \phi \tag{4.2}$$

Let $\mathcal{N}$ be the Lie algebra of N, and let

$$d\phi : \mathcal{N} \to \mathfrak{n}_{m+1} \tag{4.3}$$

be the differential of ϕ. Let

$$d\phi^* : \mathfrak{n}_{m+1}^* \to \mathcal{N}^* \tag{4.4}$$

be the map contragredient to $d\phi$. We have the diagram

$$\mathcal{O}_\sigma \xleftarrow{\;d\phi^*\;} \mathcal{O} \xrightarrow{\;p\;} \mathcal{O}_0 \tag{4.5}$$

where p is as in formula (3.10) and $\mathcal{O}_\sigma \subsetneq \mathcal{N}^*$ is the orbit corresponding to σ. The maps $d\phi^*$ and p are of course Ad^*N-equivariant, and our assumption (4.2) implies that Ad^*N acts transitively on $\mathcal{O}$, so that $d\phi^*$ and p are diffeomorphisms from $\mathcal{O}$ to $\mathcal{O}_\sigma$ and $\mathcal{O}_0$ respectively.

By assumption (4.2) we see that

$$\sigma(n) = \omega(\phi(n)) = \rho(\Delta_{\phi(n)}) \qquad n \in N \tag{4.6}$$

where $\Delta_{\phi(n)}$ is defined in formula (3.14). Let us define a symbol s_ϕ on $\mathcal{O}_\sigma$ by the formula

$$s_\phi(n)(\nu) = \chi(-2(d\phi^*)^{-1}(\nu)(c(\phi(n)))) \qquad n \in N, \nu \in \mathcal{O}_\sigma. \tag{4.7}$$

Here $(d\phi^*)^{-1}(\nu)$ is taken in $\mathcal{O}$, according to diagram (4.5), and c is the Cayley transform of formula (3.15). Note that $c(\phi(\exp x))$ will depend polynomially on $x \in \mathcal{N}$, and $(d\phi^*)^{-1}(\nu)$ will depend polynomially on $\nu \in \mathcal{O}_\sigma$, so that the right hand side of formula (4.7) is of the general form $\chi(P(x,\nu))$ where P is a polynomial on $\mathcal{N} \times \mathcal{N}^*$.

We can integrate formula (4.7) to obtain a symbol mapping on $S(N)$ or suitable spaces of distributions. For $f \in S(N)$ we would define

$$s_\phi(f) = \int_N f(n) s_\phi(n)\, dn. \tag{4.8}$$

From formula (3.17) and the well known properties of κ for H_m(cf.

[Hw2]) we see that the symbol map s_ϕ has the following properties. In these formulas f, f_1, f_2 are in $S(N)$.

a) $\sigma(f) = 0$ if and only if $s_\phi(f) = 0$ on $\mathcal{O}_\sigma$.

b) trace $\sigma(f) = \int_{\mathcal{O}_\sigma} s_\phi(f)(\nu)d\nu$ where $d\nu$ is the canonical invariant measure on $\mathcal{O}_\sigma$.

c) trace $(\sigma(f_1)\sigma(f_2)^*) = \int_{\mathcal{O}_\sigma} s_\phi(f_1)(\nu)\overline{s_\phi(f_2)(\nu)}d\nu$ where T^* is the operator adjoint of T, and $\bar{\ }$ indicates complex conjugate. Thus s_ϕ allows one to capture the Hilbert-Schmidt structure of the operators $\sigma(f)$. Note that c) implies a) and b).

d) If z is in the center of the universal enveloping algebra of N, then
$s_\phi(z{*}f) = \sigma(z)s_\phi(f) = s_\phi(z)s_\phi(f)$
where $s_\phi(z)$ is of course a constant on $\mathcal{O}_\sigma$.

e) In general $s_\phi(f_1{*}f_2)$ may be computed in terms of $s_\phi(f_1)$. and $s_\phi(f_2)$ by transfering the symbolic calculus of the Weyl symbol to $\mathcal{O}_\sigma$.

f) The symbol s_ϕ is compatible with conjugation
$s_\phi(\mathrm{Ad}n_1(n_2))(\nu) = s_\phi(n_2)(\mathrm{Ad}^*n_1^{-1}(\nu))$.

g) Somewhat more generally, if $\phi':N \to U_{m+1}$ is another homomorphism, related to ϕ by
$\phi' = \phi\circ\mathrm{Ad}g$
where $g \in Sp_{2(m+1)}$ is such that $\mathrm{Ad}g(\phi(N)) \subseteq U_{m+1}$, then
$$s_{\phi'}(n)(\nu) = s_\phi(n)(\nu) \qquad n \in N, \nu \in \mathcal{O}_\sigma. \tag{4.9}$$

Thus the symbol s_ϕ has some very nice properties. On the negative side is the highly extrinsic nature of s_ϕ, including the possibility that ϕ may not exist, or that different ϕ may exist and give different symbols. Also this s_ϕ is defined only for the representation σ and we would like a symbol map to be defined for families of representations, if possible for all representations.

We will describe a class of groups for which these undesirable features are minimal. First let us observe that for suitable ϕ, the formula (4.9) becomes substantially simpler.

<u>Definition 4.1</u>: A homomorphism

$$\phi:N \to U_{m+1}$$

is called Cayley-stable if $c(\phi(N)) = d\phi(\mathcal{N})$, where $c: U_{m+1} \to \mathfrak{u}_{m+1}$ is the Cayley transform (cf. formula (3.15)).

If ϕ is Cayley stable, then we may define

$$c_\phi : N/J \to \mathcal{N}/\mathcal{J}, \tag{4.10}$$

where $J = \ker \phi$ and $\mathcal{J}$ is its Lie algebra, by the obvious formula

$$c_\phi(n) = d\phi^{-1}(c(\phi(n))).$$

If ϕ of formula (4.2) is Cayley-stable, then $c(\phi(n)) \in \mathcal{N}$ for $n \in \mathcal{N}$, and so $(d\phi^*)^{-1}(\nu)(c(\phi)(n)) = \nu(d\phi^{-1}(c(\phi(n))))$, since $d\phi^*(\lambda)$ is just the restriction of $\lambda \in U_{m+1}$ to $d\phi(\mathcal{N})$. Thus we see

Proposition 4.2: If ϕ of formula (4. 2) is Cayley-stable then

$$s_\phi(n) = \chi(-2\nu(c_\phi(n))). \tag{4.11}$$

We will now look at a class of groups for which ϕ's as in formula (4.2) may be readily found. There has appeared in the literature [Me] a certain kind of two-step group known as an (H)-group or groups of type (H). We use a slightly less restricted class of groups which we accordingly call (wH)-groups.

Definition 4.3: A (wH)-group is a two-step nilpotent group N whose Lie algebra $\mathcal{N}$ splits into a direct sum

$$\mathcal{N} = \mathcal{N}_1 \oplus \mathcal{N}_2 \tag{4.12}$$

such that $\mathcal{N}_2$ is the center and $[\mathcal{N}_1, \mathcal{N}_1] = \mathcal{N}_2$, and such that, for a Zariski-dense set of hyperplanes $\mathfrak{a} \subseteq \mathcal{N}_2$, the quotient algebra $\mathcal{N}/\mathfrak{a}$ is a Heisenberg Lie algebra (whose center is then just $\mathcal{N}_2/\mathfrak{a}$).

In an (H)-group the final condition must hold for all hyperplanes $\mathfrak{a} \subseteq \mathcal{N}$. We note that a direct sum of (wH)-groups is again a (wH)-group. In particular a direct sum of Heisenberg groups is a (wH)-group.

We are interested in towers of successive extensions of (wH)-groups. Consider an extension

$$1 \to N \to E \to Q \to 1. \tag{4.13}$$

Definition 4.4: The group E in the sequence (4.13) is an (OKP)-group if and only if

i) E is a semidirect product: $E \simeq Q \ltimes N$

ii) N is a (wH)-group

iii) If N is decomposed as in (4.12) then the action of $\mathrm{Ad}Q$ on N by conjugation preserves N_1 and N_2 and leaves N_2 pointwise fixed.

iv) Q is an (OKP)-group

Remarks: a) Evidently (OKP)-groups will be a fairly special class of groups. On the other hand, there are a lot of them. Suppose Q is an (OKP)-group and let $\tau : Q \to U_m$ be a unipotent symplectic representation of Q. Such τ can always be found, faithful and of arbitrarily high dimension. Then form the semidirect product $Q \ltimes H_m \subseteq U_{m+1}$. Evidently this is again an (OKP)-group, and one can continue.

b) In particular, remark a) and formula (3.2) show that U_m is an (OKP)-group. In fact all unipotent radicals of parabolic subgroups of $Sp_{2m}(\mathbb{R})$ are (OKP)-groups, as are all unipotent radicals for $U_{p,q}$, $Sp_{p,q}$, O^*_{2n} and $Sp_{2m}(\mathbb{C})$. Many, but not all, unipotent radicals of parabolics in $GL_n(F)$, for $F = \mathbb{R}, \mathbb{C}$, or $\mathbb{H}$ (the quaternions), and in $O_{p,q}$ and $O_n(\mathbb{C})$, are also (OKP)-groups.

Let E be an (OKP)-group. We want to get a picture of the generic representations of E, i.e., of a family of representations whose associated orbits in E^* form a Zariski-open set. Express E as semidirect product $Q \ltimes N$ as in definition 4.4. Write $N = N_1 \oplus N_2$ as in definition 4.3. Fix a linear functional γ on N_2 such that the quotient algebra $N/\ker\gamma$ is Heisenberg. We can choose an isomorphism

$$d\phi' : /\ker\gamma \simeq h_\ell \tag{4.14}$$

for $2\ell = \dim N_1$. Clearly we can normalize $d\phi'$ so that γ defines the standard coordinate on the center of h_ℓ, equivalently, so that if $\phi : N \to H_\ell$ is the exponentiation of ϕ', and if ρ is the standard representation of H_ℓ (cf. formula (3.11)), then

$$\rho_\gamma = \rho \circ \phi' \tag{4.15}$$

is the representation of N with central character $\chi \circ \gamma \circ \exp^{-1}$.

We can and do assume that $d\phi'(N_1) = W$, the standard complement to the center in H_ℓ. Via $d\phi'$ we can transfer the action of $\mathrm{Ad}Q$ on N_1 to an action on W. From condition (4.13)iii) we see this action is symplectic. Since E is nilpotent, the transfered AdQ action is also unipotent. Hence the homomorphism $\phi':N \to H_\ell$ extends uniquely to a homomorphism

$$\phi: E \to U_{\ell+1} \tag{4.16}$$

such that $\phi(N) = H_{\ell+1}$ and $\phi(Q) \subsetneq U_\ell$. Then if we set

$$\omega_\gamma = \omega \circ \phi \tag{4.17}$$

we know that ω_γ is an extension to E of the representation of ρ_γ of N.

If we apply Mackey's theory of representations of group extensions [Ma] to this situation, we may conclude that if σ is any irreducible representation of E whose restriction to N_2 is a multiple of the character $\chi \circ \gamma \circ \exp^{-1}$, then there is an irreducible representation σ_1 of Q such that σ decomposes as a tensor product

$$\sigma \simeq \sigma_1 \otimes \omega_\gamma. \tag{4.18}$$

Since N is a (wH)-group, the generic irreducible representation of N is of the form ρ_γ for generic $\gamma \in N_2^*$. Thus the generic representation of E will have the form (4.18) where σ_1 is a generic representation of Q. Since Q is again an (OKP)-group, the decomposition (4.18) constitutes the inductive step in a precise overall picture of the generic representations of E.

To make this precise, suppose E is a tower of k (wH)-groups N^1, $N^2,\dots,N^k$, where $N^k = N$ is the bottommost one which appears explicitly in definition 4.4. Denote the successive quotients of E by Q^i, with the indexing so that

$$Q^{i+1} \simeq Q^i \ltimes N^{i+1}. \tag{4.19}$$

Then $Q^k = E$, and Q^{k-1} is what we have been calling Q. Also $Q^0 = \{1\}$ and $Q^1 = N^1$. Set

$$M = \prod_{i=1}^{k} N^i \qquad P = \prod_{i=1}^{k-1} Q^i, \tag{4.20}$$

Then M is again a (wH)-group. Indeed if $\mathcal{M}$ is the Lie algebra of M, we may write

$$M = M_1 \oplus M_2 \tag{4.21}$$

where

$$M_j = \sum_{i=1}^{k} N_j^i \qquad j = 1,2. \tag{4.22}$$

The semidirect products (4.19) may be combined to give us a direct product semidirect product

$$P \ltimes M \simeq \prod_{i=1}^{k} Q^i. \tag{4.23}$$

For each i, we have the quotient map

$$q_i : E \to Q^i. \tag{4.24}$$

Taking the direct product of these maps gives us an injection

$$\alpha : E \to P \ltimes M. \tag{4.25}$$

Consider a linear functional $\lambda \in M_2^*$. Write

$$\lambda = \sum_{i=1}^{k} \gamma^i \qquad \gamma^i \in (N_2^i)^*. \tag{4.26}$$

The Lie algebra $M/\ker \lambda$ is a quotient of the algebras $N^i/\ker \gamma^i$; clearly it will be Heisenberg if and only if each $N^i/\ker \gamma^i$ is Heisenberg. This will hold for generic λ. Assuming it, there will be a unique representation ρ_λ of M whose restriciton to M_2 is a multiple of the character $\chi \circ \gamma \circ \exp^{-1}$. Evidently ρ_λ will be an outer tensor product

$$\rho_\lambda \simeq \bigotimes_{i=1}^{k} \rho_{\gamma i} \tag{4.27}$$

where $\rho_{\gamma i}$ is the representation of N^i whose restriction to N_2^i is a multiple of $\chi \circ \gamma^i \circ \exp^{-1}$.

Each $\rho_{\gamma i}$ may be extended in the standard way to the representation $\omega_{\gamma i}$ of Q^i. The outer tensor product

$$\omega_\lambda \simeq \bigotimes_{i=1}^{k} \omega_{\gamma i} \tag{4.28}$$

then defines the standard extension of ρ_λ from M to $P \ltimes M$. If we isolate the last factor of this tensor product, writing

$$\omega_\lambda \simeq \left(\bigotimes_{i=1}^{k-1} \omega_{\gamma i}\right) \otimes \omega_{\gamma k}$$

then we can compare (4.28) with (4.18). Doing so, we see that if we assume i) that $\bigotimes_{i=1}^{k-1} \omega_{\gamma i}$ is irreducible when restricted to $\alpha(E)$, and ii) that as the γ^i vary we obtain the generic representations of $Q^{k-1}=Q$, then the discussion surrounding equation (4.18) that the representations $\omega_\lambda \circ \alpha$, for generic $\lambda \in M_2^*$, are all irreducible on $\alpha(E)$, and form the generic representations of E. In summary, we have shown

<u>Proposition 4.5</u>: If E is an (OKP)-group constructed from a tower of (wH)-groups N^i with centers N_2^i, then a generic family of representations of E is parametrized by the generic characters of $\prod_{i=1}^{k} N_2^i$. Further, all the representations are of the form $\omega_\lambda \circ \alpha$, where $\lambda \in \sum_{i=1}^{k} (N_2^i)^*$ is generic, with ω_λ as in (4.28) and α as in (4.25).

Remark: Let $\mathcal{O}_{\gamma i} \subseteq \mathfrak{Q}^{i*}$ be the $\mathrm{Ad}^* Q^i$ orbit corresponding to the representation $\omega_{\gamma i}$. The quotient mapping $q_i: E \to Q^i$ of (4.24) dualizes to give an injection of $\mathfrak{Q}^{i*}$ into $\mathfrak{E}^*$, the dual of $\mathfrak{E}$, the Lie algebra of E. Thus $\mathcal{O}_{\gamma i}$ may be regarded as an $\mathrm{Ad}^* E$ orbit in $\mathfrak{E}^*$. Its corresponding representation will be $\omega_{\gamma i} \circ q_i$. If $\mathcal{O}_\lambda$ is the orbit in $\mathfrak{E}^*$ corresponding to the representation $\omega_\lambda \circ \alpha$ of E, then the tensor product decomposition (4.28) implies that $\mathcal{O}_\lambda$ is a vector sum

$$\mathcal{O}_\lambda = \mathcal{O}_{\gamma 1} + \mathcal{O}_{\gamma 2} + \dots + \mathcal{O}_{\gamma k} \tag{4.29}$$

Moreover the decomposition (4.29) defines a diffeomorphism between the product $\prod_i \mathcal{O}_{\gamma i}$ and $\mathcal{O}_\lambda$.

We note that since for each generic $\lambda \in M_2^*$, the quotient $M/\ker \lambda$ is Heisenberg, each representation $\omega_\lambda \circ \alpha$ of E has the form (4.2), and therefore we can construct a symbolic calculus for these representations as described in formula (4.7) and the surrounding discussion. Thus for generic representations of (OKP)-groups, one has a reasonable symbol mapping. But we would like to be able to extend symbols from functions on the open set of (points in) generic orbits in $\mathfrak{E}^*$ to all of $\mathfrak{g}^*$. To do this we must study the behavior of the symbols around singular representations; an explicit formula for the symbol that could be expressed without reference to the mapping ϕ that occurs in formula (4.7) would be useful in this connection. For investigating this matter, the variant form (4.11) of the symbol formula is very helpful. For it to apply, however, we need to specialize our class of groups further. Recall the Cayley transform c from formula (3.15).

Definition 4.6: Let $G \subseteq GL(V)$ be a Lie group of linear transformations of the vector space V. Let $\mathfrak{g}$ be the Lie algebra of G. We say G is Cayley-stable if the map $x \to c(\exp x)$ maps a neighborhood of 0 in $\mathfrak{g}$ into $\mathfrak{g}$; equivalently if c maps a neighborhood of 0 in $\mathfrak{g}$ to a neighborhood of 1 in G and vice-versa.

This definition does not immediately apply to (OKP)-groups since they are not given as linear groups. To arrive at an appropriate definition of Cayley-stability for (OKP)-groups, we study some properties of the Cayley transform.

Let W be a symplectic vector space. Consider a decomposition

$$W = W_1 \oplus X \oplus Y \tag{4.30}$$

where X and Y are isotropic subspaces of W, and $X \oplus Y$ is orthogonal to W_1 with respect to the symplectic form on W. Let N_X be the nilradical of the parabolic subgroup of Sp(W) which stabilizes X. It is well-known that N_X is two-step nilpotent, in fact a (wH)-group. Explicitly we have

$$(N_X)_1 \simeq \mathrm{Hom}(Y,W_1) \qquad (N_X)_2 \simeq S^{2*}(Y) \tag{4.31}$$

where $S^{2*}(Y)$ is the space of symmetric bilinear forms on Y. We may also characterize N_X as the subgroup of Sp(W) whose elements stabilize X and $W_1 \oplus X$ and act trivially on X, on $(W_1 \oplus X)/X$ and on $W/(W_1 \oplus X)$. More precisely, we have

$$(N_X)_1(Y) \subseteq W_1 \quad (N_X)_1(W_1) \subseteq X \quad (N_X)_1(X) = \{0\} \quad (N_X)_2(Y) \subseteq X$$

$$(N_X)_2(W_1 \oplus X) = \{0\}. \tag{4.32}$$

If we consider $Sp(W_1)$ as the subgroup of Sp(W) whose elements leave $X \oplus Y$ pointwise fixed, then it is clear that $Sp(W_1)$ normalizes

N_X and that conjugation by $Sp(W_1)$ acts trivially on $(N_X)_2$.

Lemma 4.7: a) N_X is Cayley stable. In fact, the map $n \to c(\exp n)$ is the identity from N_X to itself so that all subgroups of N_X are Cayley-stable.

b) Let N be any subgroup of N_X, and let Q be a Cayley-stable subgroup of $Sp(W_1)$. Suppose Q normalizes N. Then the semidirect product $Q \ltimes N$ is Cayley stable. In fact we have the formula

$$c(c(q)c(n)) = \text{Ad} \exp n(q) + n \quad q \in \ , n \in N. \tag{4.33}$$

Proof: It is clear from relations (4.32) that $n^3 = 0$ for any $n \in N_X$. Hence part a) follows from formula (3.20). Since the elements $c(q)c(n)$ will fill out a neighborhood of 1 in $Q \ltimes N$, the formula (4.33) will imply the rest of part b). Without making any special assumptions on q and n, one can compute that

$$c(c(q)c(n)) = (1+n)(1+qn)^{-1}(q+n)(1+n)^{-1} \tag{4.34}$$

From relations (4.32) and the fact that q preserves W_1 and annihilates $X \oplus Y$, one sees that $qn^2 = n^2q = qnq = 0$. Therefore $(qn)^2 = 0$, whence $(1+qn)^{-1} = (1-qn)$, and $(1+qn)^{-1}(q+n) = (1-qn)(q+n) = q+n$. Also $(\exp n)q = (1+n)q$ and $q \exp(-n) = q(1-n) = q(1+n)^{-1}$. Using these special facts about n and q, one sees that the formula (4.34) reduces to (4.33).

Consider now a (wH)-group N. Let $A(N) = A$ be the group of automorphisms a of N such that

$$\begin{array}{ll} \text{i)} & a \text{ preserves the decomposition } N = N_1 \oplus N_2, \text{ and} \\ \text{ii)} & a \text{ acts as the identity of } N_2. \end{array} \tag{4.35}$$

Lemma 4.8: $A(N)$ is a Cayley-stable subgroup of $GL(N_2)$ (or of $GL(N)$).

Proof: Indeed, an element a of $GL(N_2)$ will belong to A if and only if a is an isometry of each antisymmetric form $\lambda \circ [\ ,\ ]$ where $[\ ,\ ]$ is the Lie bracket map from $N_1 \times N_1$ to N_2, and $\lambda \in N_2^*$. But the group of isometries of any bilinear form is Cayley-stable [We], and the intersection of Cayley stable groups is Cayley-stable. Hence A is Cayley-stable.

On the basis of this lemma and formula (4.33), we define a Cayley transform on $A \times N$ as follows

Definition 4.9: Let N be a (wH)-group and $A(N) = A$ the group defined by conditions (4.35). Define

$$c_N : A \ltimes N \to \mathfrak{a} \oplus \mathcal{N} \tag{4.36}$$

by the formula

$$c_N(c(a)\exp n) = \mathrm{Ad}\exp n(a) + n \qquad a \in \quad, n \in \mathcal{N}.$$

We refer to operation as the N-Cayley transform.

Partly as a justification for this definition but mainly for later use, we show that the N-Cayley transform is actually the restriction of Cayley transform in an appropriate representation of N.

Lemma 4.10: There exists a symplectic space W and an isotropic subspace X and an embedding

$$\beta : A \ltimes N \to Sp(W_1) \ltimes N_X \tag{4.37}$$

such that

i) $\beta(N) \subseteq N_X$

ii) $\beta(A) \subseteq Sp(W_1)$ where W_1 is an appropriate complement to X X, as in formula (4.30)

iii) $\beta(A \times N)$ is Cayley-stable and the diagram

$$\begin{array}{ccc} A \times N & \xrightarrow{\beta} & Sp(W) \\ c_N \downarrow & & \downarrow c \\ \mathfrak{a} \oplus \mathcal{N} & \xrightarrow{d\beta} & \mathfrak{sp}(W) \end{array}$$

commutes.

Proof: First, let us observe that this is already proved by explicit calculation in §3(cf. formula (3.22)) if N is Heisenberg. (Actually in §3 we do not deal with the entire $A(H_m) \simeq Sp(W)$, but only with the unipotent subgroup U_m; but the extension is straightforward.) Next, observe that if it is true for two (wH)-groups N^1 and N^2, then it is true for $N^1 \times N^2$. The key point here is that $A(N^1 \times N^2) = A(N^1) \times A(N^2)$. To see this, observe that if $\lambda = \lambda^1 + \lambda^2$ is a generic linear functional on $\mathcal{N}_2^1 \oplus \mathcal{N}_2^2$, then the radical of the antisymmetric form

$\lambda^i \circ [\ ,\ ]$ on $N_1^1 \oplus N_1^2$ is N_1^j where $i + j = 3$. Hence both N_1^1 and N_2^1 must be invariant by $A(N^1 \times N^2)$. Therefore given maps $\beta_i : A(N^i) \ltimes N^i \to Sp(W^i)$ satisfying the lemma, we can take

$$\beta = \beta_1 \times \beta_2 : (A(N^1) \ltimes N^1) \times (A(N^2) \ltimes N^2) \to Sp(W^1) \times Sp(W^2) \subsetneq Sp(W^1 W^2).$$

It is easy to check that this β satisfies the conditions of the lemma.

Thus to prove the lemma for general N, we need only show that we can embed N in a direct sum of Heisenberg groups in a way that is consistent with Cayley transform. But this is easy to see. Choose a basis $\{\gamma\}$ of generic elements of N_2^*. Then the quotients $N/\ker \gamma_i$ are Heisenberg and the natural map

$$N \to \sum_i N/\ker \gamma_i$$

is easily seen to extend to a map

$$A(N) \ltimes N \to \prod_i A(N/\exp \ker \gamma_i) \ltimes N/\exp \ker \gamma_i$$

having the desired compatibility with Cayley transform.

<u>Definition 4.11:</u> Let E be an (OKP)-group. Define the associated groups P and M as in formula (4.20). Recall that M is a (wH)-group. Let $\alpha : E \to P \ltimes M$ be the embedding of formula (4.25). Then E is Cayley-stable provided the M-Cayley transform maps $\alpha(E)$ to $d\alpha(\ )$. If E is Cayley-stable then the resulting map

$$c_E : E \to \mathfrak{E} \tag{4.38}$$

given by

$$d\alpha(c_E(e)) = c_M(\alpha(e)) \qquad e \in E$$

is called the Cayley transform on E.

Remarks: a) Since $Ad\alpha(E)$ is a unipotent subgroup of $GL(M_1)$, c_E will be a diffeomorphism between E and $\mathfrak{E}$.

b) Although the requirement for E to be Cayley stable is quite restrictive, one can check that the nilradicals of parabolics described in remark b) following definition 4.4 are all Cayley-stable.

We come finally to the main goal of this section.

Theorem 4.12: Let E be a Cayley-stable (OKP)-group, and let c_E be its Cayley transform. Define a symbol mapping

$$s:\mathcal{S}(E) \to \mathcal{S}(\mathfrak{E}^*)$$

by setting

$$s(e)(\nu) = \chi(-2\nu(c_E(e))) \qquad e \in E, \nu \in \mathfrak{E}^* \tag{4.39}$$

and

$$s(f) = \int_E f(e)s(e)de.$$

Then s is an isomorphism from $\mathcal{S}(E)$ to $\mathcal{S}(\mathfrak{E}^*)$ which extends to a unitary map from $L^2(E)$ to $L^2(\mathfrak{E}^*)$ and such that the analogues of properties (4.9) hold for a generic family of representations of E.

Proof: According to formula (4.11), the theorem will follow if, given a generic irreducible representation σ of E we can find a homomorphism $\phi:E \to U_{m+1}$ such that $\sigma \simeq \omega\circ\phi$, where ω is the canonical (oscillator) representation of U_{m+1} studied in §3, and such that ϕ preserves Cayley transform, i.e., such that the diagram

$$\begin{array}{ccc} E & \xrightarrow{\phi} & U_{m+1} \\ c_E \big\downarrow & & \big\downarrow c \\ \mathfrak{E} & \xrightarrow{d\phi} & \mathfrak{u}_{m+1} \end{array}$$

commutes. Using proposition 4.5 and the definition of Cayley-stability for (OKP)-groups, we see it is enough to construct such a ϕ for the representations ω_λ of the group $P \times M$ of fourmula (4.23).

Consider the semidirect product $Sp(W_1) \times N_X$, with W_1 and X as in equation (4.30). Suppose $X' \subseteq X$ is a subspace. Then X' and $X'^{\perp}$, its orthogonal complement in W, are invariant under $Sp(W_1) \times N_X$, so there is a natural homomorphism

$$r : Sp(W_1) \times N_X \to Sp(W_1) \times N_{X/X'} \subseteq Sp(X'^{\perp}/X'),$$

Since r results from first restricting to $X'^{\perp}$ and then factoring to $X'^{\perp}/X'$, we see that r preserves Cayley transform. If X' has codimension 1 in X, then $N_{X/X'}$ will be a Heisenberg group based on W_1.

Thus to find the homomorphism we want from $P \times M$ to U_{m+1} it suffices to find a homomorphism $\tilde{\phi}$ from $P \times M$ to $Sp(W_1) \times N_X$ for suitable W_1 and X, such that $\tilde{\phi}$ preserves Cayley transform, and such that $M/(\exp \ker \lambda)$ can be realized as a quotient of $\tilde{\phi}(M)$ in the fashion of the previous paragraph. But in fact this is already done in the proof of lemma 4.10. This concludes Theorem 4.12.

§5: To conclude we will mention some things that Theorem 4.12 is good for, and how it might be better.

L^2-estimates: Since the symbol s of formula (4.39) simply transports, in a coherent way, the Weyl symbols of the group elements in appropriate realizations of the representations of E, one has the opportunity to transform the very highly developed theory of estimates in terms of Weyl symbols [Be], [Hr], [F] to get L^2 estimates for convolution operators of estimates for operators in unitary representations. The point about these estimates is they will be localizable around generic orbits-the estimate for a operator $\rho(f)$, for $f \in \mathcal{S}(N)$ in a given representation ρ will depend only on the symbol $S(f)$ restricted to the corresponding orbit. In practical terms, this means the estimates will be couched in terms of derivatives in directions tangent to the generic orbits rather than derivatives in all directions, as for example in [Hw3]. This project has been begun in [R] which transfers the Calderòn-Vaillancourt (0,0) estimate to 3-step nilpotent groups. We hope to treat further cases in subsequent articles.

Diagonalization of $ZU(\mathbb{E})$: Because of property (4.9)d), the symbol map (4.39) exhibits the center of the univeral enveloping algebra of E as diagonal, i.e., diagonal operators. This means that in the coordinates on $\mathbb{E}$ defined by $c_E \circ \exp$, the bi-invariant differential operators on E have constant coefficients. Thus all theorems about constant coefficient operators can be transfered to them.

Matrix Coefficients: As is explained in [Wi], the matrix coefficients of generic representations are just the "inverse Fourier transforms" i.e., pull backs to $\mathcal{S}(E)$ via the symbol map, of measures supported on the appropriate orbits. In particular they can be seen to vanish at ∞ modulo the projective kernel of the representation at rates related to the geometry of the orbits. Compare [C].

Primitive Ideals: It is clear from property (4.9)a) that the primative ideal I_ρ in $L^1(G)$ corresponding to a generic representation ρ is simply the space of $f \in L^1(G)$ such that $s(f)|\mathcal{O}\rho = 0$. This makes it very easy to work with I_ρ. One can see fairly easily for example that it is "infitesimally determined" in the sense of Poguntke [Pg]. (This is of course proved by Poguntke for general representations, but not in a straightforward manner.)

Thus various questions in harmonic analysis are amenable to a reasonable symbolic calculus. However it is clear that the symbolic calculus as formulated in Theorem 4.12 has at least two drawbacks: i) it is constructed only on special groups, and ii) even for those groups it is known to have properties (4.19) only for generic representations. With regard to point i), it can be seen that the methods of [Wi] and this article will allow construction of an effective symbol mapping on classes of groups other than (OKP) groups. For example the 3-step groups of [R] are not in general (OKP)-groups; neither is the final example in [Wi]. Furthermore in those examples the symbol map does extend correctly to all representations. Thus the prospects are quite good that for large classes of interesting groups a fully effective symbol mapping can be devised. On the other hand whether an explicit, flexible symbol with all of the properties of Theorem 4.12 exists in general is unclear at this point.

References

[Bl] R. Beals, A general calculus of pseudo-differential operators, Duke Math. J. 42(1975), 1-42.

[Br] F. Berezin, Wick and anti-Wick symbols, Mat. Sb.(1971), 576-610.

[Ca] P. Cartier, Quantum Mechanical Commutation Relations and Theta Functions, Proc. Symp. Pure Math., v.IX, AMS, Providence, R.I., 1966, 361-383.

[Co] L. Corwin, Matrix coefficients of nilpotent Lie groups, Maryland Special Year in Harmonic Analysis, 1982-1983, Springer Lecture Notes, these proceedings.

[F] C. Fefferman, The Uncertainty Principle, B.A.M.S. (new Series), v.9 (1983), 129-266.

[Ge] S. Gelbart, Examples of Dual Reductive Pairs, Proc. Symp. Pure Math., v.XXXIII, AMS, Providence, R.I., 1979, 287-296.

[GLS] A. Grossman, G. Loupias, and E.M. Stein, An algebra of pseudo-differential operators and quantum mechanics in phase space, Ann. Inst. Fourier (Grenoble), v.18(1969), 343-368.

[Hr] L. Hörmander, The Weyl calculus of pseudo-differential operators, Comm. Pure and App. Math. 32(1979), 359-443.

[Hw1] R. Howe, On a connection between nilpotent groups and oscillatory integrals associated to singularities, Pac. J. Math. 73(1977), 329-364.

[Hw2] R. Howe, Quantum Mechanics and Partial Differential Equations, J. Fun. Anal. 38(1980).

[Hw3] R. Howe, A symbolic calculus for nilpotent groups, Conference on Operator Algebras and Group Representations, Neptun, Romania, Sept. 1980, Pitman Publishers, to appear.

[K] A. Kirillov, Unitary representations of nilpotent Lie groups, Uspehi Mat. Nauk. 17(1972) no. 4(106), 57-110(translated in Russian Math. Surveys 17(1962), no. 4, 53-104).

[KN] J. Kohn and L. Nirenberg, An algebra of pseudo-differential operators, Comm. Pure and App. Math. 18(1965), 443-492.

[L] J. Ludwig, Good ideals in the group algebra of a nilpotent group, Math. Zeits. 16(1978), 195-210.

[Ma] G. Mackey, Unitary representations of group extensions, Acta Math. 99(1958), 265-311.

[Me] G. Metivier, Hypoellipticité analytique sur des groupes nilpotents de rang 2, Duke Math. J., v.47(1980), 195-221.

[Mi] K. Miller, Parametrices for hypoelliptic operators on two-step nilpotent groups, Comm. P.D.E. 5(1980), 1153-1184.

[Pg] D. Poguntke, Über das Synthese-Problem für nilpotente Liesche Gruppen, preprint.

[Pk1] L. Pukanszky, Lecons sur les représentations des groupes, Mon. de la. Soc. Math. de France, no. 2, Dunod, Paris, 1967.

[Pk2] L. Pukansky, On characters and the Plancherel formula of nilpotent groups, J. Fun. Anal. 1(1967), 255-280.

[R] G. Ratcliff, A symbolic calculus for 3-step nilpotent Lie groups, Ph.D. thesis, Yale University, 1983.

[S] R. Strichartz, Invariant pseudo-differential operators on a Lie group, Ann. Sc. Norm. Sup. Pisa, v.26(1972), 587-611.

[We] H. Weyl, The Classical Groups, Princeton University Press, Princeton N.Y., 1939.

[Wi] N. Wildberger, Quantization and harmonic analysis on nilpotent Lie groups, Ph.D. thesis, Yale University, 1983.

Roger Howe
Department of Mathematics
Box 2155 Yale Station
New Haven, CT 06517

Gail Ratcliff
Department of Mathematics
University of Missouri
St. Louis, Missouri 63121

Norman Wildberger
Department of Mathematics
Stanford University
Stanford, California 94305

LEFSCHETZ FORMULAE FOR HECKE OPERATORS*

Henri Moscovici
Department of Mathematics
The Ohio State University
Columbus, Ohio 43210-1174/USA

INTRODUCTION

In this paper we describe an analogue of the Atiyah-Singer Lefschetz fixed point theorem [A-Si] for generalized Hecke operators, acting on a compact locally symmetric space $M_\Gamma = \Gamma\backslash G/K$. Given such a Hecke operator T_α , where α belongs to the commensurator of Γ in G , together with a locally invariant elliptic operator D_Γ on M_Γ , we obtain an explicit expression for the associated Lefschetz number $L(T_\alpha, D_\Gamma)$. When α normalizes Γ and in addition Γ is of finite index in the group generated by α and Γ , this expression can be also obtained from the Atiyah-Singer Lefschetz theorem but, in our context, such a special case is relatively uninteresting.

The formula we prove has roughly the following form:

$$L(T_\alpha, D_\Gamma) = \sum_\xi \ell(\xi, D_\Gamma) \ ;$$

the sum is to be taken over a set of representatives of the finitely many elliptic Γ-conjugacy classes in the double coset $\Gamma\alpha\Gamma$, and $\ell(\xi, D_\Gamma)$ is a certain number attached to the Γ-conjugacy class of ξ . One noteworthy feature here is that, while the right hand side is, a priori, at best an algebraic number, the Lefschetz number is, for certain elliptic operators related to cohomology, an algebraic integer (see (7.4)). Although we shall not pursue this point here, one may hope to draw interesting arithmetic conclusions from this observation, similar in spirit to certain applications of the Atiyah-Bott-Singer Lefschetz theory (cf., for instance, [A-B, §7]).

We also give a cohomological version of the above formula, which is formally similar to the Atiyah-Singer formula, the fixed point set being now replaced by a certain coincidence locus. As an application, we recover a result of Kuga and Sampson concerning the Euler number $E(T_\alpha)$, which does not seem to be readily ob-

*Research partially supported by a National Science Foundation grant

tainable from our explicit formula (5.1) for $E(T_\alpha)$.

A few words about the method being used are now in order. It is based on an idea similar to that suggested in [Mo] and implemented in [B-M] in connection with the index formula, which in essence consists in applying a version of the Selberg trace formula to the operator $T_\alpha R_\Gamma(h_t)$, acting on $L^2(\Gamma\backslash G)$, where h_t is the "difference heat-kernel" attached to D_Γ. While in principle applicable to the case of an arbitrary non-uniform lattice, this approach obviously requires a much better understanding of the general Selberg trace formula than presently available. We hope however to discuss an extension for rank 1 locally symmetric spaces of finite volume, along the lines of [B-M], in a future publication.

§1. HECKE OPERATORS

Let G be a unimodular locally compact group and Γ a discrete subgroup. We fix an invariant measure on $\Gamma\backslash G$ and denote by $L^2(\Gamma\backslash G)$ the corresponding L^2-space, on which G acts unitarily via the right quasi-regular representation R_Γ. The so-called "Hecke operators" form a special class of intertwining operators for R_Γ and this section is devoted to their definition.

Let $\Gamma^\#$ denote the commensurator of Γ, i.e., the subgroup of G consisting of all $\alpha \in G$ such that $\Gamma(\alpha) = \Gamma \cap \alpha\Gamma\alpha^{-1}$ has finite index in both Γ and $\alpha\Gamma\alpha^{-1}$. We fix an element $\alpha \in \Gamma^\#$ and proceed to define the associated Hecke operator. First of all we note that the double coset $\Gamma\alpha\Gamma$ splits as a disjoint union of Γ-cosets. More precisely,

(1.1) there exist $\alpha_1, \dots, \alpha_N \in G$ such that $\Gamma\alpha\Gamma = \bigsqcup_{i=1}^{N} \alpha_i \Gamma$

(disjoint union); in fact, one has $\Gamma\alpha\Gamma = \bigsqcup_{i=1}^{N} \alpha_i \Gamma$ iff

$\Gamma = \bigsqcup_{i=1}^{N} \alpha_i \alpha^{-1} \Gamma(\alpha)$, in particular $N = [\Gamma : \Gamma(\alpha)]$.

Let now $f \in C_c(\Gamma\backslash G)$. Regarding f as a Γ-invariant function on G, let us set

$$(1.2) \qquad f_\alpha(x) = \sum_{i=1}^{N} f(\alpha_i^{-1}x) \quad , \quad x \in G ,$$

where $\alpha_1, \dots, \alpha_N$ are as in (1.1).

Because of the invariance of f under left translations by elements in Γ, the function f_α thus defined is independent of the choice of all the α_i's satisfying (1.1). In particular, since for $\gamma \in \Gamma$ and α_i as above

$$\Gamma\alpha\Gamma = \bigsqcup_{i=1}^{N} \gamma^{-1}\alpha_i\Gamma ,$$

it follows that

$$f_\alpha(\gamma x) = \sum_{i=1}^{N} f(\alpha_i^{-1} \gamma x) = f_\alpha(x) ,$$

i.e., $f_\alpha \in C(\Gamma\backslash G)$. Moreover, one actually has

$$(1.3) \qquad f_\alpha \in C_c(\Gamma\backslash G) .$$

Indeed, let F denote the support of f in G and set

$$F_\alpha = \bigcup_{i=1}^{N} \alpha_i F .$$

There exists a compact Q in G such that

$$F = \Gamma(\alpha^{-1})Q$$

and so

$$F_\alpha = \bigcup_{i=1}^{N} \alpha_i \alpha^{-1} \Gamma(\alpha)\alpha\, Q = \Gamma\alpha\, Q ,$$

the second equality being a consequence of (1.1). Since the support f_α in G is contained in F_α, it is clearly compact modulo Γ.

Concerning the L^2-norm of f_α, one has

$$(1.4) \qquad \|f_\alpha\|_{L^2(\Gamma\backslash G)} < (NN')^{1/2} \|f\|_{L^2(\Gamma\backslash G)} ,$$

where $\qquad N = [\Gamma : \Gamma(\alpha)]$ and $N' = [\Gamma : \Gamma(\alpha^{-1})]$.

To check this we choose, as we may, a function $h > 0$ on G whose support has compact intersection with any set of the form ΓQ , with Q a compact in G , and such that

$$\sum_{\gamma\in\Gamma} h(\gamma x) = 1 \quad , \quad \text{for any } x \in G ,$$

and then remark that

$$\int_{\Gamma\backslash G} |f_\alpha(x)|^2 \, d(\Gamma x) = \int_G h(x)|f_\alpha(x)|^2 \, dx =$$

$$\sum_{i,j=1}^{N} \int_G h(x)|f(\alpha_i^{-1}x)||f(\alpha_j^{-1}x)|dx < \Big(\sum_{i=1}^{N} (\int_G h(x)|f(\alpha_i^{-1}x)|^2 dx)^{1/2} \Big)^2$$

$$< N \sum_{i=1}^{N} \int_G h(x)|f(\alpha_i^{-1}x)|^2 dx = N \sum_{i=1}^{N} \int h(\alpha_i x)|f(x)|^2 dx$$

$$= N \int_{\Gamma(\alpha^{-1})\backslash G} \sum_{\gamma\in\Gamma(\alpha^{-1})} \sum_{i=1}^{N} h(\alpha_i \gamma x)|f(x)|^2 \, d(\Gamma(\alpha^{-1})x) ;$$

here and hereafter dx stands for a fixed Haar measure on G and $d(\Gamma x)$ for the corresponding invariant measure on $\Gamma\backslash G$. Noting now that

$$\sum_{\gamma\in\Gamma(\alpha^{-1})} \sum_{i=1}^{N} h(\alpha_i \gamma x) = \sum_{\delta\in\Gamma(\alpha)} \sum_{i=1}^{N} h(\alpha_i \alpha^{-1} \delta \alpha x)$$

and so, by (1.1), this is also equal to

$$\sum_{\gamma\in\Gamma} h(\gamma\alpha\, x) = 1 ,$$

it follows that

$$\int_{\Gamma\backslash G} |f_\alpha(x)|^2 \, d(\Gamma x) < N \int_{\Gamma(\alpha^{-1})\backslash G} |f(x)|^2 \, d(\Gamma(\alpha^{-1})x) = NN' \int_{\Gamma\backslash G} |f(x)|^2 \, d(\Gamma x) ,$$

as claimed.

In conclusion, the assignment

$$(1.5) \qquad T_\alpha f = f_\alpha$$

defines a bounded linear operator on $L^2(\Gamma\backslash G)$, which only depends on the double coset $\Gamma\alpha\Gamma$. In addition, as formula (1.2) readily implies,

$$T_\alpha R_\Gamma(x) = R_\Gamma(x)\, T_\alpha \quad , \quad x \in G .$$

<u>Definition.</u> The intertwining operator T_α is called the "Hecke operator" associated to $\Gamma\alpha\Gamma$.

§2. INVARIANT AND LOCALLY INVARIANT ELLIPTIC OPERATORS

In this section G denotes a unimodular Lie group, K a compact subgroup and $M = G/K$. For the reader's convenience, we recall here some basic facts concerning the objects mentioned in the title, referring to [Mo] for a more detailed discussion.

Given a unitary representation ε of K on a finite-dimensional complex vector space E, we let $\mathscr{E}$ denote the associated homogeneous vector bundle over M and identify the space $C^\infty(\mathscr{E})$ of C^∞-sections of $\mathscr{E}$ with the subspace $(C^\infty(G)\otimes E)^K$ of all K-invariant elements in $C^\infty(G;E) \cong C^\infty(G)\otimes E$, where K acts on $C^\infty(G)$ via the restriction of the right regular representation R of G. Similarly, $(L^2(G)\otimes E)^K$ will be regarded as the space of L^2-sections of $\mathscr{E}$.

We let $\mathfrak{g}$ denote the Lie algebra of G, $\mathfrak{g}_C$ its complexification and $\mathfrak{A}(\mathfrak{g}_C)$ the universal enveloping algebra of $\mathfrak{g}_C$. If $\varepsilon^\pm$ are finite-dimensional representations of K on $E^\pm$ respectively, we let $(\mathfrak{A}(\mathfrak{g}_C)\otimes \mathrm{Hom}(E^+,E^-))^K$ denote the subspace of all elements fixed by the representation $k \longmapsto \mathrm{Ad}(k)\otimes(\varepsilon^-(k)\circ\cdot\circ\varepsilon^+(k)^{-1})$ of K on $\mathfrak{A}(\mathfrak{g}_C)\otimes \mathrm{Hom}(E^+,E^-)$.

Let now $D_M^+ : C_c^\infty(\mathscr{E}^+) \to C_c^\infty(\mathscr{E}^-)$ be a G-invariant differential operator on M. It is well-known that D_M^+ can be written under the form:

$$(2.1)\qquad D_M^+ = \sum_{i=1}^{n} R(X_i)\otimes C_i \quad , \quad \text{where } D^+ = \sum_{i=1}^{n} X_i\otimes C_i$$

belongs to $(\mathfrak{A}(\mathfrak{g}_C)\otimes \mathrm{Hom}(E^+,E^-))^K$.

Also, the formal adjoint of D_M^+ is the operator:

$$(2.2)\qquad D_M^- = \sum_{i=1}^{n} R(X_i)^*\otimes C_i^* \text{ , with } D^- = \sum_{i=1}^{n} X_i^*\otimes C_i^* \in (\mathfrak{A}(\mathfrak{g}_C)\otimes \mathrm{Hom}(E^-,E^+))^K .$$

Thus, $D_M^\pm$ are obtained from $D^\pm \in (\mathfrak{A}(\mathfrak{g}_C)\otimes \mathrm{Hom}(E^\pm,E^\mp))^K$ via the right regular representation R .

More generally, let π be an arbitrary unitary representation of G on a Hilbert space $\mathcal{H}_\pi$, and let $\mathcal{H}_\pi^\infty$ denote the subspace of C^∞ vectors for π . With $D^\pm$ as above, we set:

$$D_\pi^\pm = \sum_{i=1}^{n} \pi(X_i) \otimes C_i : (\mathcal{H}_\pi^\infty \otimes E^\pm)^K \to (\mathcal{H}_\pi^\infty \otimes E^\mp)^K . \tag{2.3}$$

Then $D_\pi^\pm$ are formally adjoint to each other and thus extend to closed operators $D_\pi^{\pm cl}$ from $(\mathcal{H}_\pi \otimes E^\pm)^K$ to $(\mathcal{H}_\pi \otimes E^\mp)^K$.

Our concern lies in the case when D_M^+ , and hence D_M^- too, is elliptic. Under this hypothesis, it was proved in [Mo] (Corollary 1.2) that

$$D_\pi^{\pm cl} = D_\pi^{\mp\,*} , \tag{2.4}$$

i.e., the closure of $D_\pi^\pm$ coincides with the Hilbert space adjoint of $D_\pi^\mp$. Also (cf. [Mo], Proposition 1.1),

$$\operatorname{Ker} D_\pi^{\pm cl} = \operatorname{Ker} D_\pi^\pm \subset (\mathcal{H}_\pi^\infty \otimes E^\pm)^K . \tag{2.5}$$

Consider now a discrete subgroup Γ of G . Specializing the above construction to the quasi-regular representation R_Γ of G on $L^2(\Gamma\backslash G)$, we let

$$D_\Gamma^\pm = D_{R_\Gamma}^\pm . \tag{2.6}$$

It can be viewed as a differential operator on the V-manifold $M_\Gamma = \Gamma\backslash G/K$. Such an operator, with $D_M^\pm$ elliptic, will be called a "locally invariant elliptic operator" (abbreviated l.i.e. operator) on M_Γ .

§3. LEFSCHETZ NUMBERS

We now assume that G is a connected real semisimple Lie group with finite center and that Γ is a lattice in G , i.e., a discrete subgroup of G such that $\Gamma\backslash G$ has finite invariant volume. We denote by K a maximal compact subgroup of G and set, as before, $M = G/K$ and $M_\Gamma = \Gamma\backslash G/K$.

The quasi-regular representation R_Γ of G on $L^2(\Gamma\backslash G)$ admits, on formal grounds alone, a decomposition of the form

$$L^2(\Gamma\backslash G) = L^2_d(\Gamma\backslash G) \oplus L^2_c(\Gamma\backslash G) \quad , \quad R_\Gamma = R^d_\Gamma \oplus R^c_\Gamma ,$$

with R^d_Γ a direct sum of irreducible representations and R^c_Γ a direct integral with no irreducible subrepresentations. In our context, and in the presence of a certain technical condition on Γ (cf. [La], p. 16 or [O-W], p. 62) which we assume to be satisfied, much more is known about the decomposition. While referring to the above cited sources for exact statements, we content ourselves to mention here that:

(i) the multiplicity of any irreducible representation of G in $L^2_d(\Gamma\backslash G)$ is finite, and

(ii) via the theory of Eisenstein series, $L^2_c(\Gamma\backslash G)$ breaks up into finitely many invariant subspaces, each of which can be written as a direct integral over a certain "analytic" family of induced representations, analogous to the principal series, associated to a Γ-cuspidal parabolic subgroup.

Using these properties we proved in [Mo] that:

(3.1) If D^+_Γ is an l.i.e. operator on M_Γ, then:

a) $\operatorname{Ker} D^\pm_\Gamma = \operatorname{Ker} D^\pm_{\Gamma,d}$, where $D^\pm_{\Gamma,d} = D^\pm_{R^d_\Gamma}$;

b) $\dim \operatorname{Ker} D^\pm_\Gamma < \infty$.

Let now T_α, with $\alpha \in \Gamma^\#$, be a Hecke operator on $L^2(\Gamma\backslash G)$ and let D^+_Γ be an l.i.e. operator on M_Γ. Denote

$$T^\pm_\alpha = T_\alpha \otimes I \,|\, (L^2(\Gamma\backslash G) \otimes E^\pm)^K ,$$

which makes sense because T_α intertwines R_Γ. Furthermore, by the same token,

$$D^\pm_\Gamma T^\pm_\alpha = T^\mp_\alpha D^\pm_\Gamma$$

and so $T^\pm_\alpha$ leaves $\operatorname{Ker} D^\pm_\Gamma$ stable. We can thus define the "Lefschetz number" of

T_α with respect to D_Γ^+ by the formula:

$$L(T_\alpha, D_\Gamma^+) = Tr(T_\alpha^+ | Ker\ D_\Gamma^+) - Tr(T_\alpha^- | Ker\ D_\Gamma^-) .$$

In the remainder of this section, we shall derive a few primary properties of these Lefschetz numbers.

Given $\pi \in \hat{G}$, let $L_\pi^2(\Gamma \backslash G)$ denote the corresponding isotypical component of $L^2(\Gamma \backslash G)$ and let R_Γ^π denote the restriction of R_Γ to $L_\pi^2(\Gamma \backslash G)$. Thus,

$$L_d^2(\Gamma \backslash G) = \sum_{\pi \in \hat{G}}^{\oplus} L_\pi^2(\Gamma \backslash G) \quad \text{and} \quad R_\Gamma^d = \sum_{\pi \in \hat{G}}^{\oplus} R_\Gamma^\pi .$$

Since T_α leaves each isotypical component invariant, it makes sense to set

$$T_{\alpha,d}^\pm = T_\alpha^\pm \otimes I | (L_d^2(\Gamma \backslash G) \otimes E^\pm)^K , \quad T_{\alpha,\pi}^\pm = T_\alpha^\pm \otimes I | (L_\pi^2(\Gamma \backslash G) \otimes E^\pm)^K ,$$

and one has

$$T_{\alpha,d}^\pm = \sum_{\pi \in \hat{G}}^{\oplus} T_{\alpha,\pi}^\pm .$$

On the other hand,

$$Ker\ D_\Gamma^\pm = Ker\ D_{\Gamma,d}^\pm = \sum_{\pi \in \hat{G}}^{\oplus} Ker\ D_{\Gamma,\pi}^\pm ,$$

where $D_{\Gamma,\pi}^\pm = D_{R_\Gamma^\pi}^\pm$. Since $\dim Ker\ D_\Gamma^\pm < \infty$, it follows that

$$Ker\ D_{\Gamma,\pi}^\pm = 0 \quad , \quad \text{for all but finitely many} \quad \pi \in \hat{G} .$$

Furthermore,

$$T_\alpha^\pm | Ker\ D_\Gamma^\pm = \sum_{\pi \in \hat{G}}^{\oplus} T_{\alpha,\pi}^\pm | Ker\ D_{\Gamma,\pi}^\pm ,$$

therefore

$$L(T_\alpha, D_\Gamma^+) = \sum_{\pi \in \hat{G}} (Tr(T_{\alpha,\pi}^+ | Ker\ D_{\Gamma,\pi}^+) - Tr(T_{\alpha,\pi}^- | Ker\ D_{\Gamma,\pi}^-)) ,$$

where the sum involves finitely many non-zero terms.

We now observe that, in view of (i),

$$\dim (L_\pi^2(\Gamma \backslash G) \otimes E^\pm)^K = \dim Hom_K(L_\pi^2(\Gamma \backslash G), E^{\pm *}) < \infty ,$$

and so, by elementary linear algebra,

$$Tr(T_{\alpha,\pi}^+ | Ker\ D_{\Gamma,\pi}^+) - Tr(T_{\alpha,\pi}^- | Ker\ D_{\Gamma,\pi}^-) = Tr T_{\alpha,\pi}^+ - Tr T_{\alpha,\pi}^- .$$

The above discussion is summarized in the following statement.

(3.2) Proposition. Let T_α be a Hecke operator on $L^2(\Gamma\backslash G)$ and let D_Γ^+ be an l.i.e. operator on M_Γ. Then

$$\mathrm{Tr}T^+_{\alpha,\pi} = \mathrm{Tr}T^-_{\alpha,\pi}\ , \text{ for all but finitely many } \pi \subset \hat{G}$$

and

$$L(T_\alpha, D_\Gamma^+) = \sum_{\pi\in\hat{G}} (\mathrm{Tr}T^+_{\alpha,\pi} - \mathrm{Tr}T^-_{\alpha,\pi})\ .$$

As a consequence of this result, the Lefschetz number $L(T_\alpha, D_\Gamma^+)$ is seen to depend not on the operator D_Γ^+ itself but only upon its "homogeneous symbol"

$$\sigma(D^+) = E^+ - E^- \in R(K)\ ,$$

where $R(K)$ is the ring of virtual representations of K. In other words, one has:

(3.3) Let D_Γ^+ and $\tilde{D}_\Gamma^+$ be two l.i.e. on M_Γ such that $\sigma(D^+) = \sigma(\tilde{D}^+)$. Then $L(T_\alpha, D_\Gamma^+) = L(T_\alpha, \tilde{D}_\Gamma^+)$.

Proof. By hypothesis, $E^+ - E^- = \tilde{E}^+ - \tilde{E}^-$ in $R(K)$, i.e.,

$$E^+ \oplus \tilde{E}^- \cong \tilde{E}^+ \oplus E^-\ .$$

It follows that, for any $\pi \in \hat{G}$,

$$\mathrm{Tr}T^+_{\alpha,\pi} + \mathrm{Tr}\tilde{T}^-_{\alpha,\pi} = \mathrm{Tr}\tilde{T}^+_{\alpha,\pi} + \mathrm{Tr}T^-_{\alpha,\pi}$$

(the notation being self-explanatory), hence the statement.

Another noteworthy consequence is the following.

(3.4) Suppose that $\operatorname{rank} K \neq \operatorname{rank} G$. Then

$$L(T_\alpha, D_\Gamma^+) = 0\ ,$$

for any $\alpha \in \Gamma^\#$ and any l.i.e. D_Γ^+ on M_Γ.

<u>Proof.</u> The assumption on the rank together with the ellipicity of D_Γ^+ imply, as in [B-M, Prop. (1.2.5)], that $E^+ \cong E^-$, therefore the claim.

This observation disposes, in an admittedly trivial way, of the case when rank $K \neq$ rank G. So, from now on, we may (and do) assume that:

$$\text{rank } K = \text{rank } G .$$

Before stating the results in this case we must introduce some more notation.

Let $H \subset K$ be a maximal torus, with Lie algebra $\mathfrak{h} \subset \mathfrak{k}$, where $\mathfrak{k}$ denotes the Lie algebra of K, and let Φ and Φ_k, respectively, denote the root systems of $(\mathfrak{g}_C, \mathfrak{h}_C)$ and $(\mathfrak{k}_C, \mathfrak{h}_C)$. We fix compatible systems of positive roots $\Psi_k \subset \Psi$ and denote $\Psi_n = \Psi - \Psi_k$, $\rho_k = \frac{1}{2} \sum_{\beta \in \Psi_k} \beta$, $\rho_n = \rho - \rho_k$.

For $\mu \in \mathfrak{h}_C^*$, Ψ_k-dominant integral, we let (V_μ, τ_μ) denote the irreducible representation of $\mathfrak{k}_C$ with highest weight μ.

Let $S = S^+ \oplus S^-$, $s = s^+ \oplus s^-$ be the spin representation of $\mathfrak{k}_C$. We recall that, letting W denote the Weyl group of Φ, W_k the Weyl group of Φ_k and

$$W^1 = \{w \in W \ ; \ w\Psi \supset \Psi_k\} ,$$

one has:

$$S^\pm = \sum_{w \in W^1, \det(w) = \pm 1} V_{w\rho - \rho_k} .$$

Next, we recall the following result, proved in [Mi]:

(3.5) Let $E^\pm$ be two representations of K such that the corresponding homogeneous bundles $\mathcal{E}^\pm$ afford a G-invariant elliptic differential operator $D_M^+ : C^\infty(\mathcal{E}^+) \to C^\infty(\mathcal{E}^-)$. Then, in the representation ring $R(\mathfrak{k}_C)$, $E^+ - E^-$ is divisible by $S^+ - S^-$, that is

$$E^+ - E^- = \sigma_r(D^+) \otimes (S^+ - S^-) \text{ , with } \sigma_r(D^+) \in R(\mathfrak{k}_C) .$$

Moreover, $\sigma_r(D^+) \otimes S^\pm \in R(K)$.

The element $\sigma_r(D^+) \in R(\mathfrak{k}_C)$ will be referred to as the "reduced homogeneous symbol" of D_M^+ (or of D^+ or yet of D_Γ^+).

Finally let us recall the definition of the $\pm$-Dirac operators with coefficients in an irreducible representation (V_μ, τ_μ) of $\mathfrak{k}_C$. Set $S_\mu^\pm = V_\mu \otimes S^\pm$,

$s^{\pm}_{\mu} = \tau_{\mu} \otimes s^{\pm}$ and assume that $S^{\pm}_{\mu} \in R(K)$. Let $\mathfrak{p}$ denote the orthogonal complement of $\mathfrak{k}$ in $\mathfrak{g}$, relative to the Cartan-Killing form. Choose an orthonormal basis $\{X_1, \ldots, X_p\}$ of $\mathfrak{p}$ and set

$$D^{\pm}_{\mu} = \sum_{i=1}^{p} X_i \otimes I \otimes c(X_i) \in \mathfrak{A}(\mathfrak{g}_C) \otimes \mathrm{Hom}(S^{\pm}_{\mu}, S^{\mp}_{\mu}) ,$$

where $c(X)$ denotes the Clifford multiplication by $X \in \mathfrak{p}$. Then $D^{\pm}_{\mu}$ are independent of the choice of the orthonormal basis of $\mathfrak{p}$ and, in particular, K-invariant. So they give rise to two l.i.e. operators on M_{Γ} , $D^{\pm}_{\mu,\Gamma}$, which are adjoint to each other, called the " ± - Dirac operators with coefficients in V_{μ} ". Let us also note that:

$$\sigma(D^{+}_{\mu}) = V_{\mu} \otimes (S^{+} - S^{-}) \quad , \quad \text{therefore} \quad \sigma_{\Gamma}(D^{+}_{\mu}) = V_{\mu} .$$

Let now T_{α} , with $\alpha \in \Gamma^{\#}$, be a Hecke operator on $L^2(\Gamma\backslash G)$. The Lefschetz number

$$\mathrm{Spin}(T_{\alpha}, V_{\mu}) = L(T_{\alpha}, D^{+}_{\mu,\Gamma})$$

will be called the " V_{μ} - spinor number " of T_{α} . More generally, if

$$\sigma = \textstyle\sum_{\mu} n_{\mu} V_{\mu} \quad , \quad n_{\mu} \in \mathbb{Z} , \quad n_{\mu} \geq 0 ,$$

is an arbitrary element in $R(\mathfrak{k}_C)$ such that

$$S^{\pm}_{\mu} = S^{\pm} \otimes V_{\mu} \in R(K) \quad , \quad \text{for any } \mu \text{ with } n_{\mu} \neq 0 ,$$

we set

$$\mathrm{Spin}(T_{\alpha}, \sigma) = \textstyle\sum_{\mu} n_{\mu} L(T_{\alpha}, D^{+}_{\mu,\Gamma})$$

and call it the " σ-spinor number " of T_{α} .

In this notation, (3.2) and (3.5) yield:

(3.6) THEOREM. Let T_{α} be a Hecke operator on $L^2(\Gamma\backslash G)$ and let D^{+}_{Γ} be an l.i.e. operator on M_{Γ} . Then

$$L(T_{\alpha}, D^{+}_{\Gamma}) = \mathrm{Spin}(T_{\alpha}, \sigma_{\Gamma}(D^{+})) .$$

§4. THE LEFSCHETZ FORMULA

Turning now to the problem of evaluating the spinor numbers of a Hecke operator, we shall follow an approach similar to that used in [B-M] for computing spinor indices. So, let us begin by briefly reviewing its basic ingredients. The notation is the same as in the previous section.

First of all, we recall that the "formal" spinor Laplacians $\Delta_\mu^\pm = D_\mu^\mp D_\mu^\pm$ are given by the following simple formula:

$$\Delta_\mu^\pm = -\Omega \otimes I + c_\mu I \in (\mathfrak{A}(\mathfrak{g}_C) \otimes \mathrm{End}\, S_\mu^\pm)^K ,$$

where

$$c_\mu = \|\mu + \rho_k\|^2 - \|\rho\|^2$$

and $\Omega \in \mathfrak{A}(\mathfrak{g}_C)$ is the Casimir element. Therefore, for any unitary representation π of G, one has:

$$(4.1) \qquad \Delta_{\mu,\pi}^\pm = -\pi(\Omega) \otimes I + c_\mu I .$$

In particular,

$$D_{\mu,R}^\mp D_{\mu,R}^\pm = \Delta_{\mu,R}^\pm = -R(\Omega) \otimes I + c_\mu I .$$

Let us now form the "heat semigroups" of bounded operators on $(L^2(G) \otimes S_\mu^\pm)^K$,

$$H_{\mu,t}^\pm = e^{-t\,\Delta_{\mu,R}^\pm} , \quad t > 0 .$$

Each $H_{\mu,t}^\pm$ is a smoothing pseudo-differential operator which commutes with the action of G, therefore of the form

$$H_{\mu,t}^\pm u(x) = \int_G h_{\mu,t}^\pm (x^{-1}y)\, u(y)\, dy , \quad u \in (L^2(G) \otimes S_\mu^\pm)^K .$$

The kernels $h_{\mu,t}^\pm : G \to \mathrm{End}\, S_\mu^\pm$ are C^∞, L^2 and covariant with respect to the action of $K \times K$, that is

$$h_{\mu,t}^\pm(x) = s_\mu^\pm(k_1)\, h_{\mu,t}^\pm(k_1^{-1} x\, k_2) s_\mu^\pm(k_2)^{-1} , \quad x \in G , \quad k_1 , k_2 \in K .$$

Moreover, by [B-M, Prop. 2.4],

$$h^{\pm}_{\mu,t} \in (\mathscr{C}^p(G) \otimes \mathrm{End}\, S^{\pm}_{\mu})^{K \times K}, \quad \text{for all} \quad p > 0,$$

where $\mathscr{C}^p(G)$ denotes the Harish-Chandra L^p-Schwartz space. ALso (cf. loc. cit., Prop. 2.1), if π is a unitary representation of G, then

$$\pi(h^{\pm}_{\mu,t}) = e^{-t\, \Delta^{+}_{\mu,\pi}}, \quad t > 0. \tag{4.2}$$

In particular, if π is irreducible, and so

$$\pi(\Omega) = X_{\pi}(\Omega) I, \quad \text{with} \quad X_{\pi}(\Omega) \in \mathbb{R},$$

one has:

$$\pi(h^{\pm}_{\mu,t}) = e^{t(X_{\pi}(\Omega) - c_{\mu})} I. \tag{4.2'}$$

The point of departure of our method is provided by the following observation.

(4.3) Suppose that $R^d_{\Gamma}(h^{\pm}_{\mu,t})$ is a trace class operator. Then

$$\mathrm{Spin}(T_{\alpha}, V_{\mu}) = \mathrm{Tr}(T^{+}_{\alpha,d}\, R^d_{\Gamma}(h^{+}_{\mu,t})) - \mathrm{Tr}(T^{-}_{\alpha,d}\, R^d_{\Gamma}(h^{-}_{\mu,t})).$$

Proof. For abstract operator-theoretic reasons, one has

$$\mathrm{Tr}(T^{+}_{\alpha,d} | \mathrm{Ker}\, D^{+}_{\mu,\Gamma,d}) - \mathrm{Tr}(T^{-}_{\alpha,d} | \mathrm{Ker}\, D^{-}_{\mu,\Gamma,d}) = \mathrm{Tr}(T^{+}_{\alpha,d} e^{-t\, \Delta^{+}_{\mu,\Gamma,d}})$$

$$- \mathrm{Tr}(T^{-}_{\alpha,d}\, e^{-t\, \Delta^{-}_{\mu,\Gamma,d}})$$

and so the statement follows from the following two facts:

$$\mathrm{Ker}\, D^{\pm}_{\mu,\Gamma} = \mathrm{Ker}\, D^{\pm}_{\mu,\Gamma,d} \qquad \text{(cf. Proposition 3.1)}$$

and

$$R^d_{\Gamma}(h^{\pm}_{\mu,t}) = e^{-t\, \Delta^{\pm}_{\mu,\Gamma,d}} \qquad \text{(cf. (4.2))}.$$

Another, more direct argument, which takes advantage of the specific form of the spinor Laplacians, runs as follows. Since

$$T^{\pm}_{\alpha,d}\, R^{d}_{\Gamma}(h^{\pm}_{\mu,t}) = \sum^{\oplus}_{\pi\in\hat{G}} T^{\pm}_{\alpha,\pi}\, R^{\pi}_{\Gamma}(h^{\pm}_{\mu,t}) = \sum^{\oplus}_{\pi\in\hat{G}} e^{t(\chi_\pi(\Omega) - c_\mu)}\, T^{\pm}_{\alpha,\pi} ,$$

one has

$$\mathrm{Tr}(T^{+}_{\alpha,d}\, R^{d}_{\Gamma}(h^{+}_{\mu,t})) - \mathrm{Tr}(T^{-}_{\alpha,d}\, R^{d}_{\Gamma}(h^{-}_{\mu,t})) = \sum_{\pi\in\hat{G}} e^{t(\chi_\pi(\Omega) - c_\mu)}\, (\mathrm{Tr}\, T^{+}_{\alpha,\pi} - \mathrm{Tr}\, T^{-}_{\alpha,\pi}) .$$

But

$$\mathrm{Tr}\, T^{+}_{\alpha,\pi} - \mathrm{Tr}\, T^{-}_{\alpha,\pi} = \mathrm{Tr}(T^{+}_{\alpha,\pi}|\mathrm{Ker}\, D^{+}_{\mu,\Gamma,\pi}) - \mathrm{Tr}(T^{-}_{\alpha,\pi}|\mathrm{Ker}\, D^{-}_{\mu,\Gamma,\pi})$$

and, on the other hand, in view of (4.1) ,

$$\mathrm{Ker}\, D^{\pm}_{\mu,\Gamma,\pi} = \mathrm{Ker}\, \Delta^{\pm}_{\mu,\Gamma,\pi} = \mathrm{Ker}((-\chi_\pi(\Omega) + c_\mu)I) ,$$

that is

$$\mathrm{Ker}\, D^{\pm}_{\mu,\Gamma,\pi} = 0 , \text{ if } \chi_\pi(\Omega) \neq c_\mu$$

and

$$\mathrm{Ker}\, D^{\pm}_{\mu,\Gamma,\pi} = (L^2_\pi(\Gamma\backslash G) \otimes S^{\pm}_{\mu})^K , \quad \text{if } \chi_\pi(\Omega) = c_\mu .$$

Therefore

$$\mathrm{Tr}\, (T^{+}_{\alpha,d}\, R^{d}_{\Gamma}(h^{+}_{\mu,t})) - \mathrm{Tr}(T^{-}_{\alpha,d}\, R^{d}_{\Gamma}(h^{-}_{\mu,t})) = \sum_{\pi\in\hat{G},\, \chi_\pi(\Omega) = c_\mu} (\mathrm{Tr}\, T^{+}_{\alpha,\pi} - \mathrm{Tr}\, T^{-}_{\alpha,\pi}) =$$

$$= \sum_{\pi\in\hat{G}} (\mathrm{Tr}\, T^{+}_{\alpha,\pi} - \mathrm{Tr}\, T^{-}_{\alpha,\pi}) = L(T_\alpha, D^{+}_{\mu,\Gamma}) ,$$

by Proposition (3.2). Q.e.d.

At this juncture, in [B-M] we introduced the assumption that G has real-rank 1 , which made possible the application of the Osborne-Warner version of the Selberg Trace Formula. Here, while allowing G to be of arbitrary real-rank, we instead impose the following restriction on Γ :

(4-Γ) $\Gamma\backslash G$ is compact .

Also, for simplicity, we shall assume that G satisfies the following additional hypothesis:

(4-G) G is embedded in the simply connected complex analytic group G_C with Lie algebra $\mathfrak{g}_C$.

It is well-known [G-G-P] that the compactness assumption (4-Γ) implies that $L^2(\Gamma\backslash G) = L^2_d(\Gamma\backslash G)$. It also implies the following "soft" version of the Selberg

Trace Formula (cf. [Se], (2.14)):

(4.4) Let $h \in \mathscr{C}^1(G)$ and $\alpha \in \Gamma^\#$. The operator $T_\alpha R_\Gamma(h)$ is of trace class and one has

$$\mathrm{Tr}(T_\alpha R_\Gamma(h)) = \sum_{[\xi] \in [\Gamma\alpha\Gamma]} \mathrm{vol}(\Gamma_\xi\backslash G_\xi) \int_{G_\xi\backslash G} h(x^{-1}\xi x)\, d(G_\xi x),$$

where $[\Gamma\alpha\Gamma]$ is the set of all Γ-conjugacy classes in $\Gamma\alpha\Gamma$, $[\xi] = \{\gamma\xi\gamma^{-1}\,;\, \gamma \in \Gamma\}$ is such a conjugacy class, G_ξ stands for the centralizer of ξ in G, $\Gamma_\xi = \Gamma \cap G_\xi$ and $\mathrm{vol}(\Gamma_\xi\backslash G_\xi)$ is the volume of $\Gamma_\xi\backslash G_\xi$ relative to some choice of a G_ξ-invariant measure. All the integrals that occur as well as the series, are absolutely convergent.

Proof. This is certainly known, although not exactly stated in this form. For the reader's convenience, we shall sketch below the arguments, which are slight variations of the known ones.

Write $\Gamma\alpha\Gamma = \bigsqcup_{i=1}^{N} \alpha_i\Gamma$ and let $f \in \mathscr{C}^1(G)$ and $u \in C(\Gamma\backslash G)$. Then

$$(T_\alpha R_\Gamma(f)u)(\Gamma x) = \sum_{i=1}^{N} (R_\Gamma(f)u)(\alpha_i^{-1} x) = \sum_{i=1}^{N} \int_G f(y)\, u(\alpha_i^{-1} xy)\, dy =$$

$$= \sum_{i=1}^{N} \int_G f(x^{-1}\alpha_i y)\, u(y)\, dy = \int_{\Gamma\backslash G} \sum_{i=1}^{N} \sum_{\gamma\in\Gamma} f(x^{-1}\alpha_i\gamma y)\, u(y)\, dy$$

$$= \int_{\Gamma\backslash G} \sum_{\xi\in\Gamma\alpha\Gamma} f(x^{-1}\xi y)\, u(y)\, d(\Gamma y)\,.$$

Thus, $T_\alpha R_\Gamma(f)$ is an integral operator on $L^2(\Gamma\backslash G)$ with kernel

$$K_{\alpha,f}(\Gamma x, \Gamma y) = \sum_{\xi \in \Gamma\alpha\Gamma} f(x^{-1}\xi y)\,.$$

The hypothesis $f \in \mathscr{C}^1(G)$ implies (see, for instance, [O-W], p. 352) that f is of regular growth. Reasoning now as in [G-G-P], Ch. I, §2, one sees that the series defining $K_{\alpha,f}$ converges absolutely and uniformly on compacta of $G \times G$. In particular, $K_{\alpha,f}$ is C^∞ and therefore , by standard theory, the associated operator $T_\alpha R_\Gamma(f)$ is Hilbert-Schmidt.

We now invoke the parametrix argument of [O-W, Remark, p. 21] to write our function $h \in \mathscr{C}^1(G)$ as a sum

$$h = f * \mu + h * \nu ,$$

where f is a derivative of h and so $f \in \mathscr{C}^1(G)$, $\mu \in C_c^p(G)$ (with $p > 1$ fixed, but arbitrarily chosen) and $\nu \in C_c^\infty(G)$. As noted before, $R_\Gamma(h)$ and $R_\Gamma(f)$ are Hilbert-Schmidt operators. For similar but more elementary reasons, $R_\Gamma(\mu)$ and $R_\Gamma(\nu)$ are also Hilbert-Schmidt. Therefore $R_\Gamma(h)$ is a trace class operator. This being established, it is just a formality to conclude that

$$\mathrm{Tr}(T_\alpha R_\Gamma(h)) = \int_{\Gamma\backslash G} K_{\alpha,h}(\Gamma x, \Gamma x)d(\Gamma x) = \int_{\Gamma\backslash G} \sum_{\xi\in\Gamma\alpha\Gamma} h(x^{-1}\xi x)\, d(\Gamma x) ,$$

and the formula claimed in the statement follows now by elementary manipulations (cf. [Se] or [G-G-P]).

We are going to apply the above "trace formula" to the function

$$h_{\mu,t}(x) = \mathrm{tr}\, h^+_{\mu,t}(x) - \mathrm{tr}\, h^-_{\mu,t}(x) \quad , \quad x \in G ,$$

where tr denotes the trace function on $\mathrm{End}\, S^\pm_\mu$. The reason is obvious, in view of

(4.5) $$\mathrm{Spin}(T_\alpha, V_\mu) = \mathrm{Tr}(T_\alpha R_\Gamma(h_{\mu,t})) .$$

This, in turn, follows from (4.3) and the following result, which we shall state, for later use, in a slightly more general form than presently needed.

(4.6) Let π be a unitary representation of G and let $T : \mathscr{H}(\pi) \to \mathscr{H}(\pi)$ be an intertwining operator for π . Assuming that

$$\pi(h^\pm_{\mu,t}) : (\mathscr{H}(\pi) \otimes S^\pm_\mu)^K \to (\mathscr{H}(\pi) \otimes S^\pm_\mu)^K$$

are trace class operators, one has

$$\mathrm{Tr}(T\pi(h_{\mu,t})) = \mathrm{Tr}(T^+\pi(h^+_{\mu,t})) - \mathrm{Tr}(T^-\pi(h^-_{\mu,t})) ,$$

where $T^\pm = T \otimes I | (\mathscr{H}(\pi) \otimes S^\pm_\mu)^K$.

<u>Proof.</u> Similar to that of Lemma 3.3 in [B-M] (cf. also Lemma 5.1, loc. cit.).

Combining now (4.4) and (4.5), one obtains:

(4.7) $$\mathrm{Spin}(T_\alpha, V_\mu) = \sum_{[\xi]\in[\Gamma\alpha\Gamma]} \mathrm{vol}(\Gamma_\xi\backslash G_\xi) \int_{G_\xi\backslash G} h_{\mu,t}(x^{-1}\xi x)\, d(G_\xi x)\ .$$

Thus, our problem is to express the orbital integrals

$$\Lambda_\xi(h_{\mu,t}) = \int_{G_\xi\backslash G} h_{\mu,t}(x^{-1}\xi x)\, d(G_\xi x)$$

in elementary terms.

Invariant distributions of the type Λ_ξ have been and still are, a central object of study in the harmonic analysis on semisimple Lie groups. A principal goal of this study is to calculate their Fourier transform in the sense of Harish-Chandra, that is to expand them in terms of the distributional characters associated to the irreducible representations of G. For example, when $\xi = 1$, Λ_1 is the Dirac distribution on G and its Fourier expansion is precisely Harish-Chandra's Plancherel formula. For a general $\xi \in G$, the problem is far from being solved. However, the case when ξ is semisimple is nowadays well-understood (the most general explicit results to date being due to R. Herb [He]) and fortunately, for the purposes of the present paper, this is the only case we have to worry about. Indeed, because of the compactness of $\Gamma\backslash G$, it is true that:

(4.8) All the elements of $\Gamma^\#$, in particular those of $\Gamma\alpha\Gamma$, are semisimple.

This is, of course, similar to (but more elementary than) the well-known "compactness criterion" for quotients by arithmetic sub-groups. For the convenience of the reader, we shall include here a proof. Recall first that an element of G is semisimple if and only if its orbit under inner conjugation in G is closed in G. So, let $\beta \in \Gamma^\#$ and assume that the sequence $g_n^{-1}\beta g_n$, with $g_n \in G$, converges to some $x \in G$. Choose a compact F in G such that $G = \Gamma F$ and then write $g_n = \gamma_n f_n$, with $\gamma_n \in \Gamma$, $f_n \in F$. Passing, if need, to a subsequence, we can assume that f_n converges to an $f \in F$. Then $\gamma_n^{-1}\beta\gamma_n$ converges to fxf^{-1}. But $\{\gamma_n^{-1}\beta\gamma_n\ ;\ n \in \mathbb{N}\}$ is contained in $\Gamma\beta\Gamma$, which, as a finite union of Γ-cosets, is discrete and closed. Therefore $\gamma_n^{-1}\beta\gamma_n = fxf^{-1}$ for all n's sufficiently large. Thus, x is conjugated to β. Q.e.d.

Let us now review, very briefly, the results we need concerning the Fourier expansion of the orbital integrals associated with semisimple orbits. For details and/or appropriate references, the reader is referred to [He].

So, let us fix a semisimple element $\xi \in G$. Then, there exists a Cartan subgroup B of G, containing ξ, and a differential operator Π_ξ on B, of degree $r_\xi = \frac{1}{2} \dim G_\xi/B$, such that, for any $f \in \mathscr{C}^2(G)$,

$$(4.9) \qquad \Lambda_\xi(f) = \lim_{b \to \xi} \Pi_\xi F_f^B(b)$$

where F_f^B is Harish-Chandra's invariant integral relative to B and the limit is taken through regular elements of B.

In order to describe now the Fourier transform of F_f^B, some more notation is needed. Assuming, as we may, that B is stable under the Cartan involution θ determined by the choice of the maximal compact subgroup K, and letting $\mathfrak{b}$ denote its Lie algebra, write

$$\mathfrak{b} = \mathfrak{b}_k + \mathfrak{b}_p \quad , \quad \text{where } \mathfrak{b}_k = \mathfrak{b} \cap \mathfrak{k} \, , \, \mathfrak{b}_p = \mathfrak{b} \cap \mathfrak{p}$$

and $B = B_k B_p$, where $B_k = B \cap K$ and $B_p = \exp \mathfrak{b}_p$. Let $C_G(B_p)$ be the centralizer in G of B_p; then $C_G(B_p) = MB_p$, where M is reductive, with compact Cartan subgroup B_k. Let $\mathfrak{C}_M$ denote a full set of θ-stable representives of M-conjugacy classes of Cartan subgroups of M. We can (and do) choose these representatives $C \in \mathfrak{C}_M$ such that $C_k^o \subset B_k$. Also, we note that, for $C \in \mathfrak{C}_M$, $\tilde{C} = CB_p$ is a Cartan subgroup of G.

Given a θ-stable Cartan subgroup B of G, its unitary character group is parametrized by pairs (b^*, ν) where $b^* \in \hat{B}_k$ and $\nu \in \mathfrak{b}_p^*$, the real dual of $\mathfrak{b}_p$. To each such pair (b^*, ν) there corresponds a certain tempered invariant distribution $\Theta(B, b^*, \nu)$ on G. When b^* is regular, this distribution is, up to a sign, the character of a tempered unitary representation of G induced from a parabolic subgroup of G with split part B_p. Otherwise, $\Theta(B, b^*, \nu)$ is a linear combination of characters which can be embedded in a unitary principal series representation associated to a different class of cuspidal parabolics.

Now the main result in [He] (Theorem 1) says that:

(4.10) If $b = b_k b_p$ is a regular element of B, then

$$F_f^B(b) = \sum_{C\in\mathfrak{C}_M} \sum_{c^*\in\hat{C}_k} \sum_{w\in W(M,B_k)} \int_{\mathfrak{b}_p^*} b_p^{-i\nu'} \int_{\mathfrak{c}_p^*} \det(w)\, \ell(M,C,c^*,\nu,wb_k)\, \Theta(\tilde{C},c^*,\nu'\otimes\nu)(f)\,d\nu\, d\nu'$$

where $W(M,B_k) = N_M(B_k)/B_k$, with $N_M(B_k)$ denoting the normalizer of B_k in M and $\mathfrak{b}_p$ (resp. $\mathfrak{c}_p$) is the Lie algebra of B_p (resp. C_p) ; the coefficients $\ell(M,C,c^*,\nu,b_k)$ are explicitly determined.

The application of the above result to our problem is considerably facilitated by the following essential feature of the "difference heat-kernel" $h_{\mu,t}$.

(4.11) The map sending $\pi \in \hat{G}$ to $\operatorname{Tr} \pi(h_{\mu,t})$ has finite support. More precisely, denoting by $\hat{G}_{(\mu+\rho_k)}$, the set of all $\pi \in \hat{G}$ whose infinitesimal character χ_π coincides with the character $\chi_{\mu+\rho_k}$ associated to $\mu + \rho_k \in \mathfrak{h}_C^*$, one has:

$$\operatorname{Tr} \pi(h_{\mu,t}) = \dim(\mathcal{H}(\pi) \otimes S_\mu^+)^K - \dim(\mathcal{H}(\pi) \otimes S_\mu^-)^K ,$$

if $\bar{\pi}$ = the contragredient of π belongs to $\hat{G}(_{\mu+\rho_k})$, or otherwise

$$\operatorname{Tr} \pi(h_{\mu,t}) = 0 .$$

<u>Proof.</u> By (4.6) and (4.2')

$$\operatorname{Tr} \pi(h_{\mu,t}) = e^{t(\chi_\pi(\Omega) - \chi_{\mu+\rho_k}(\Omega))} (\dim (\mathcal{H}(\pi) \otimes S_\mu^+)^K - \dim(\mathcal{H}(\pi) \otimes S_\mu^-)^K) .$$

On the other hand, according to Proposition (4.20) in [A-Sc],

$$\dim (\mathcal{H}(\pi) \otimes S_\mu^+)^K - \dim(\mathcal{H}(\pi) \otimes S_\mu^-)^K = 0 \quad , \quad \text{unless } \chi_{\bar{\pi}} = \chi_{\mu+\rho_k} .$$

Q.e.d.

We shall now use the preceding results to simplify the expression (4.7) of the spinor number $\operatorname{Spin}(T_\alpha, V_\mu)$.

According to (4.8), the summation in the right hand side of (4.7) ranges over semisimple Γ-conjugacy classes. Furthermore, we claim that only the elliptic Γ-conjugacy classes can make a non-zero contribution to that sum. Indeed, this follows from:

(4.12) If ξ is semisimple and not elliptic then $\Lambda_\xi(h_{\mu,t}) = 0$.

Proof. Let B denote a Cartan subgroup containing ξ . Without loss of generality, we can assume that B is θ-stable. Thus, $B = B_k B_p$ with $\mathfrak{b}_p \neq 0$. Due to the character-theoretic nature of the distributions $\Theta(CB_p, b^*, \nu' \otimes \nu)$ occuring in (4.10), it follows from (4.11) that

$$\Theta(CB_p, c^*, \nu' \otimes \nu)(h_{\mu,t}) = 0 \quad , \quad \text{a.e. relative to } d\nu' \otimes d\nu .$$

Therefore, by (4.10), $F^B_{h_{\mu,t}}$ vanishes identically on the set B' of regular elements of G , and the claim now follows from (4.9).

So, letting $[\Gamma\alpha\Gamma]_e$ denote the set of all Γ-conjugacy classes in $\Gamma\alpha\Gamma$ which consist of elliptic elements, one has:

(4.7') $$\operatorname{Spin}(T_\alpha, V_\mu) = \sum_{[\xi] \in [\Gamma\alpha\Gamma]_e} \operatorname{vol}(\Gamma_\xi \backslash G_\xi)\, \Lambda_\xi(h_{\mu,t}) .$$

Before proceeding any further, let us pause to remark that:

(4.13) The set $[\Gamma\alpha\Gamma]_e$ is finite.

Indeed, if $\xi \in \Gamma\alpha\Gamma$ is elliptic then $g^{-1}\xi g \in K$ for some $g \in G$. Choosing a compact F such that $G = \Gamma F$ and then representing g as a product $g = \gamma f$, with $\gamma \in \Gamma$ and $f \in F$, one sees that $\gamma^{-1}\xi\gamma \in FKF^{-1} \cap \Gamma\alpha\Gamma$. But this latter set is clearly finite. Q.e.d.

Consider now an elliptic element $\xi \in G$ and suppose, for the moment, that $\xi \in H$, the fixed compact Cartan subgroup of G . As usual, we identify $\hat{H}$ with a lattice L_H in $i\mathfrak{h}^*$; for each $\lambda \in L_H$, the corresponding character on H will be denoted e^λ . Let G°_ξ be the connected component of G_ξ , $\Phi_\xi = \{\alpha \in \Phi \, ; \, e^\alpha(\xi) = 1\}$, $\Psi_\xi = \Phi_\xi \cap \Psi$, $q_\xi = \frac{1}{2} \dim G_\xi/K_\xi$ (where $K_\xi = G_\xi \cap K$) and $r_\xi = \frac{1}{2} \dim G_\xi/H$; when $\xi = 1$, we drop the subscript ξ from the notation. The formula

$$\mathrm{spin}_\mu(\xi) = \frac{(-1)^{q+q_\xi} \sum_{w \in W_k} \det w \prod_{\alpha \in \Psi_\xi} (w(\mu+\rho_k),\alpha) e^{w(\mu+\rho_k)}(\xi)}{(2\pi)^{r_\xi}[G_\xi : G_\xi^\circ]\, e^{\rho}(\xi) \prod_{\alpha \in \Psi - \Psi_\xi} (1 - e^{-\alpha}(\xi))}$$

defines a function on H, which is invariant under the action of $W_k = W(G,H)$ and therefore extends to an invariant (under inner automorphisms) function on the set of elliptic elements of G.

In this notation, one has:

(4.14) If $\xi \in G$ is elliptic, then

$$\Lambda_\xi(h_{\mu,t}) = \mathrm{spin}_\mu(\xi) ,$$

provided that the Haar measure on G has been normalized as in [Wa, 8.1.2] and a similar normalization has been chosen for the Haar measure on G_ξ°, by use of the Lebesgue measure on $\mathfrak{g}_\xi$ (the Lie algebra of G_ξ) obtained from the Euclidean structure inherited from the Cartan-Killing form of $\mathfrak{g}$.

<u>Proof.</u> We start by applying Herb's formula (4.10) to the case at hand, when $B = H$ is compact and $f = h_{\mu,t}$. As already remarked in the proof of (4.12), $\Theta(C,c^*,\nu)(h_{\mu,t}) = 0$, a.e. relative to $d\nu$, for any $C \in \mathfrak{C}_G$ with $\mathfrak{c}_p \neq 0$. Therefore, the only Cartan subgroup contributing a non-zero term to $F^H_{h_{\mu,t}}$ is H itself, i.e.,

(4.14a) $$F^H_{h_{\mu,t}}(b) = (-1)^r \sum_{\lambda \in L_H} \Theta(H,\lambda)(h_{\mu,t}) e^{-\lambda}(b) , \quad b \in H' ,$$

where H' is the set of regular elements in H.

At this point, we recall two important properties of the distributions $\Theta(H,\lambda)$, $\lambda \in L_H$; firstly that

(4.14b) $$\Theta(H,\lambda)|H' = \frac{\sum_{w\in W_k} \det(w) e^{w\lambda}}{\prod_{\alpha \in \Psi} (e^{\alpha/2} - e^{-\alpha/2})} ,$$

and secondly that, as already mentioned before,

(4.14c) $\Theta(H,\lambda) = \sum_{\pi \in \hat{G}_{(\mu+\rho_k)}} c_\pi \Theta_\pi$, where c_π are real (in fact, rational) numbers and Θ_π is the (distributional) character of $\pi \in \hat{G}$.

From the latter property and formula (4.11), one gets:

$$\Theta(H,\lambda)(h_{\mu,t}) = \sum_{\pi \in \hat{G}_{(\mu+\rho_k)}} c_\pi(\dim(\mathfrak{h}(\pi) \otimes S^+_\mu)^K - \dim(\mathfrak{h}(\pi) \otimes S^-_\mu)^K) .$$

On the other hand, arguing as in [A-Sc, §4], one sees that

$\dim(\mathfrak{h}(\pi) \otimes S^+_\mu)^K - \dim(\mathfrak{h}(\pi) \otimes S^-_\mu)^K$ = the multiplicity with which the character ch V_μ occurs in the formal series $(-1)^q$ (ch S^+ - ch S^-) $\bar{\Theta}_\pi|K'$ = the coefficient $e^{\mu+\rho_k}$ in the finite Fourier series

$$(-1)^q \prod_{\alpha \in \Psi} (e^{\alpha/2} - e^{-\alpha/2}) \bar{\Theta}_\pi|H' ,$$

and therefore

$\Theta(H,\lambda)(h_{\mu,t})$ = the coefficient of $e^{\mu+\rho_k}$ in the finite Fourier series $(-1)^q \prod_{\alpha \in \Psi} (e^{\alpha/2} - e^{-\alpha/2}) \overline{\Theta(H,\lambda)}|H'$.

This, together with (4.14b), gives:

$$\Theta(H,\lambda)(h_{\mu,t}) = \begin{cases} (-1)^{q+r} \det w \ , & \text{if } \lambda = -w(\mu + \rho_k) \text{ with } w \in W_k \\ 0 \ , & \text{otherwise.} \end{cases}$$

So, (4.14a) becomes

$$F^H_{h_{\mu,t}}|H' = (-1)^q \sum_{w \in W_k} \det w \ e^{w(\mu+\rho_k)}. \tag{4.14d}$$

Let now ξ be an arbitrary element of H . We recall again that

(4.14e)
$$\Lambda_\xi(h_{\mu,t}) = M_\xi \lim_{\substack{b \to \xi \\ b \in H'}} \left(\prod_{\alpha \in \Psi_\xi} H_\alpha \right) F^H_{h_{\mu,t}}(b) .$$

Here M_ξ is a constant which, relative to the normalizations of measures specified above, has the expression (cf. [R-W, p. 297]):-

(4.14f)
$$M_\xi^{-1} = (-1)^{q_\xi}(2\pi)^{r_\xi}[G_\xi : G_\xi^o]\, e^{\rho}(\xi) \prod_{\alpha \in \Psi - \Psi_\xi} (1 - e^{-\alpha}(\xi)) .$$

Combining (4.14e) with (4.14d) and (4.14f) yields the stated result. Q.e.d.

From (4.7') and (4.14), we may conclude that:

(4.15)
$$\mathrm{Spin}(T_\alpha, V_\mu) = \sum_{[\xi] \in [\Gamma\alpha\Gamma]_e} \mathrm{vol}(\Gamma_\xi \backslash G_\xi)\, \mathrm{spin}_\mu(\xi) ,$$

the volume of $\Gamma_\xi \backslash G_\xi$ being taken with respect to the Haar measure on G_ξ normalized as in (4.14).

Together with Theorem (3.6), this result leads to the following general Lefschetz formula for Hecke operators. We recall that G is a connected linear semisimple Lie group, assumed for convenience to satisfy (4 - G), and that Γ is a discrete co-compact subgroup. Also, G is assumed to possess a compact Cartan subgroup.

(4.16) THEOREM. Let $\alpha \in \Gamma^\#$ and let D_Γ^+ be an l.i.e. operator on M_Γ. Then

$$L(T_\alpha, D_\Gamma^+) = \sum_{[\xi] \in [\Gamma\alpha\Gamma]_e} \mathrm{vol}(\Gamma_\xi \backslash G_\xi) \sum_\mu n_\mu\, \mathrm{spin}_\mu(\xi) ,$$

where the integers n_μ are determined by the relation

$$\sum_\mu n_\mu V_\mu = \sigma_r(D^+) .$$

To illustrate the application of this formula, we shall quickly examine in the next section what it reduces to in a few important special cases.

§5. CLASSICAL ELLIPTIC OPERATORS

To begin with, let us consider the de Rham operator. In our context, it can be described as follows. Pick up an orthonormal basis $\{X_1, \ldots, X_p\}$ of $\mathfrak{p}$ $(p = 2q)$ and set

$$d = \sum_{j=1}^{p} X_j \otimes \varepsilon(X_j) \in \mathfrak{A}(\mathfrak{g}_C) \otimes \mathrm{End}(\Lambda\mathfrak{p}_C)$$

where $\varepsilon(X)$ denotes the exterior multiplications by $X \in \mathfrak{p}$. Since the definition is clearly independent of the choice of the orthonormal basis, $d \in (\mathfrak{A}(\mathfrak{g}_C) \otimes \mathrm{End}(\Lambda\mathfrak{p}_C))^K$. The homogeneous vector bundle over M induced by the K-module $\mathfrak{p}_C \cong \mathfrak{p}_C^*$ can be identified with the cotangent bundle T_C^*M. Under this identification, the operator d_M on $C^\infty(\Lambda T_C^*M)$ induced by d is precisely the exterior differentiation. The de Rham operator on M is, by definition, the elliptic differential operator

$$d_M^{ev} = d_M + d_M^* : C^\infty(\Lambda^{ev} T_C^*M) \to C^\infty(\Lambda^{odd}\, T_C^*M) .$$

Similarly,

$$d_\Gamma^{ev} = d_\Gamma + d_\Gamma^* \quad \text{from} \quad (L^2(\Gamma\backslash G) \otimes \Lambda^{ev}\mathfrak{p}_C)^K \quad \text{to} \quad (L^2(\Gamma\backslash G) \otimes \Lambda^{odd}\mathfrak{p}_C)^K$$

is the de Rham operator on M_Γ. Its adjoint is, obviously,

$$d_\Gamma^{odd} = d_\Gamma + d_\Gamma^* \quad \text{from} \quad (L^2(\Gamma\backslash G) \otimes \Lambda^{odd}\mathfrak{p}_C)^K \quad \text{to} \quad (L^2(\Gamma\backslash G) \otimes \Lambda^{ev}\mathfrak{p}_C)^K .$$

Using Hodge theory on V-manifolds, the space of complex-valued harmonic k-forms

$$\mathscr{H}^k(M_\Gamma;\mathbb{C}) = \mathrm{Ker}(d_\Gamma + d_\Gamma^*) \cap (L^2(\Gamma\backslash G) \otimes \Lambda^k\, \mathfrak{p}_C)^K$$

may be identified with the kth cohomology group $H^k(M_\Gamma;\mathbb{C})$ of the complex

$$\cdots \to C^\infty(\Gamma\backslash G) \otimes \Lambda^{k-1}\mathfrak{p}_C)^K \xrightarrow{d_\Gamma} (C^\infty(\Gamma\backslash G) \otimes \Lambda^k\, \mathfrak{p}_C)^K \xrightarrow{d_\Gamma} (C^\infty(\Gamma\backslash G) \otimes \Lambda^{k+1}\mathfrak{p}_C)^K \to \cdots$$

which in turn, essentially by de Rham's theorem, is naturally isomorphic to the ordinary complex k^{th} cohomology group of M_Γ. So, the Lefschetz number $L(T_\alpha, d_\Gamma^{ev})$ coincides with the Euler number of T_α

$$E(T_\alpha) = \sum_{k=0}^{2q} (-1)^k \,\mathrm{Tr}\, T_\alpha^{(k)} ,$$

where $T_\alpha^{(k)}$ is the operator on $H^k(M_\Gamma;\mathbb{C})$ induced by $T_\alpha \otimes I$, acting on $(C^\infty(\Gamma\backslash G) \otimes \Lambda^k \mathfrak{p}_C)^K$.

Now the homogeneous symbol of d_Γ^{ev} is

$$\sigma(d^{ev}) = \Lambda^{ev}\mathfrak{p}_C - \Lambda^{odd}\mathfrak{p}_C$$

and it is well-known that, in the representation ring of Spin(2q) and therefore a fortiori in $R(\mathfrak{k}_C)$, one has

$$\Lambda^{ev}\mathfrak{p}_C - \Lambda^{odd}\mathfrak{p}_C = (-1)^q (S^+ - S^-)^2 .$$

Hence, its reduced symbol is

$$\sigma_r(d^{ev}) = (-1)^q (S^+ - S^-) = (-1)^q \sum_{u \in W^1} \det(u) V_{u\rho-\rho_k} .$$

Now

$$(-1)^q \sum_{u \in W^1} \det(u)\,\mathrm{spin}_{u\rho-\rho_k}(\xi) = \frac{(-1)^{q_\xi} \sum_{u\in W^1} \sum_{v\in W_k} \det(vu) \prod_{\alpha\in\Psi_\xi} (vu\rho,\alpha) e^{vu\rho}(\xi)}{(2\pi)^{r_\xi}[G_\xi:G_\xi^\circ]\, e^{\rho}(\xi) \prod_{\alpha\in\Psi-\Psi_\xi} (1 - e^{-\alpha}(\xi))} =$$

$$= \frac{(-1)^{q_\xi} \sum_{w\in W} \det w \prod_{\alpha\in\Psi_\xi} (w\rho,\alpha) e^{w\rho}(\xi)}{(2\pi)^{r_\xi}[G_\xi:G_\xi^\rho]\, e^{\rho}(\xi) \prod_{\alpha\in\Psi-\Psi_\xi} (1 - e^{-\alpha}(\xi))}$$

and thus Theorem (4.16) gives:

$$(5.1)\quad E(T_\alpha) = \sum_{[\xi]\in[\Gamma\alpha\Gamma]_e} \frac{(-1)^{q_\xi} \sum_{w\in W} \det w \prod_{\alpha\in\Psi_\xi} (w\rho,\alpha) e^{w\rho}(\xi)}{(2\pi)^{r_\xi}[G_\xi:G_\xi^\circ]\, e^{\rho}(\xi) \prod_{\alpha\in\Psi-\Psi_\xi} (1 - e^{-\alpha}(\xi))} \mathrm{vol}(\Gamma_\xi\backslash G_\xi) .$$

Another, and in some ways more interesting, operator which can be fashioned out of the de Rham complex is the signature operator

$$d_\Gamma^+ = d_\Gamma + d_\Gamma^* \quad \text{from} \quad (L^2(\Gamma\backslash G) \otimes \Lambda^+\mathfrak{p}_C)^K \quad \text{to} \quad (L^2(\Gamma\backslash G) \otimes \Lambda^-\mathfrak{p}_C)^K ;$$

where $\Lambda^\pm\mathfrak{p}_C$ is the ±1-eigenspace of the involution τ on $\Lambda\mathfrak{p}_C$ defined as follows:

$$\tau(\omega) = i^{k(k-1)+q} \star \omega \quad , \quad \text{for} \quad \omega \in \Lambda^k \mathfrak{p}_C \,,$$

where $\star$ stands for the "star operator" defined by the inner product on $\mathfrak{p}$. One can easily verify that

$$(d_\Gamma + d_\Gamma^\star)(I \otimes \tau) = -(I \otimes \tau)(d_\Gamma + d_\Gamma^\star) \,,$$

therefore $I \otimes \tau$ induces an involution on

$$\mathscr{H}(M_\Gamma;\mathbb{C}) = \mathrm{Ker}(d_\Gamma + d_\Gamma^\star) \,,$$

whose ± 1-eigenspaces are precisely

$$\mathscr{H}^\pm(M_\Gamma;\mathbb{C}) = \mathrm{Ker}\ d_\Gamma^\pm \,.$$

Furthermore, this involution leaves stable the subspaces

$$S_k = \mathscr{H}^k \oplus \mathscr{H}^{2q-k} \quad , \quad 0 \leq k < q$$

and in fact it switches the two factors in this decomposition. Since, on the other hand, $I \otimes \tau$ commutes with $T_\alpha \otimes I$, it follows that

$$\mathrm{Tr}(T_\alpha \otimes I | S_k^+) = \mathrm{Tr}(T_\alpha \otimes I | S_k^-) \quad , \quad 0 \leq k < q$$

and therefore the only dimension contributing to the Lefschetz number of T_α is q, i.e., $L(T_\alpha, d_\Gamma^+)$ coincides with the "signature number":

$$\mathrm{Sign}(T_\alpha) = \mathrm{Tr}(T_\alpha \otimes I|\mathscr{H}^q(M_\Gamma\ ;\ C)^+) - \mathrm{Tr}(T_\alpha \otimes I|\mathscr{H}^q(M_\Gamma\ ;\ C)^-) \,.$$

Now, in the representation ring of $\mathfrak{k}_C$, one has

$$\sigma(d_\Gamma^+) = \Lambda^+\mathfrak{p}_C - \Lambda^-\mathfrak{p}_C = (-1)^q\ (S^+ + S^-) \otimes (S^+ - S^-)$$

and so

$$\sigma_\Gamma(d_\Gamma^+) = (-1)^q(S^+ + S^-) = (-1)^q \sum_{u \in W^1} V_{u\rho - \rho_k} \,.$$

Thus, applying Theorem (4.16) to d_Γ^+, one obtains:

(5.2)

$$\mathrm{Sign}(T_\alpha) = \sum_{[\xi]\in[\Gamma\alpha\Gamma]_e} \frac{(-1)^{q_\xi} \sum_{w\in W} \det w_k \prod_{\alpha\in\Psi_\xi} (w\rho,\alpha) e^{w\rho}(\xi)}{(2\pi)^{r_\xi} [G_\xi:G_\xi^o]\, e^{\rho}(\xi) \prod_{\alpha\in\Psi-\Psi_\xi} (1 - e^{-\alpha}(\xi))} \mathrm{vol}(\Gamma_\xi\backslash G_\xi) \ .$$

where, for $w \in W$, w_k is the unique element in W_k such that $w_k^{-1}w \in W^1$.

Finally, we present our version of the holomorphic Lefshetz theorem. So, let us assume that the symmetric space $M = G/K$ is Hermitian, i.e., it has a G-invariant complex structure. This amounts to the existence of a splitting

$$\mathfrak{p}_C = \mathfrak{p}_+ \oplus \mathfrak{p}_-$$

such that

$$\mathfrak{p}_- = \overline{\mathfrak{p}}_+ \ , \quad [\mathfrak{k}_C , \mathfrak{p}_+] \subset \mathfrak{p}_+ \quad \text{and} \quad [\mathfrak{p}_+,\mathfrak{p}_+] = 0 \ .$$

Regarding $\mathfrak{p}$ as the tangent space of M at the origin $o = 1K$, $\mathfrak{p}_+$ (resp. $\mathfrak{p}_-$) becomes the holomorphic (resp. anti-holomorphic) tangent space at $o \in M$. We fix a positive root system $\Psi \supset \Psi_k$ such that

$$\mathfrak{p}_\pm = \sum_{\alpha\in\Psi_n}^{\oplus} \mathfrak{g}_C^{\pm\alpha} \ .$$

It is known that the quotient space M_Γ is a normal projective variety. Also, to any irreducible K-module V_λ one can associate in a natural way a sheaf $\mathcal{F}_\lambda$ over M_Γ. The kth cohomology space of this sheaf, $H^k(M_\Gamma ;\mathcal{F}_\lambda)$, can be identified with the space of Γ-invariant, V_λ-valued harmonic (0,k)-forms on M (see [Sc, Lemma 10] for the proof of a similar statement):

$$\mathscr{H}^k(M_\Gamma;V_\lambda) = \mathrm{Ker}(\overline{\partial}_{\lambda,\Gamma} + \overline{\partial}^*_{\lambda,\Gamma}) \cap (L^2(\Gamma\backslash G)\otimes \Lambda^k \mathfrak{p}_+ \otimes V_\lambda)^K \ ,$$

where the operator $\overline{\partial}_{\lambda,\Gamma}$ on $(L^2(\Gamma\backslash G) \otimes \Lambda\mathfrak{p}_+ \otimes V_\lambda)^K$ is induced by the usual Cauchy-Riemann operator $\overline{\partial}_{\lambda,M}$ acting on

$$C^\infty(\Lambda\overline{T}^*M \otimes V_\lambda) \cong (C^\infty(G) \otimes \Lambda\mathfrak{p}_+ \otimes V_\lambda)^K \ .$$

Thus, the $\mathcal{F}_\lambda$-Euler number of T_α, which is initially defined in terms of the action of T_α on the sheaf cohomology $H^*(M_\Gamma;\mathcal{F}_\lambda)$, has the expression

$$\chi(T_\alpha;\mathscr{F}_\lambda) = \sum_{k=0}^{q} (-1)^k \operatorname{Tr}(T_\alpha \otimes I \otimes L\mathscr{H}^k(M_\Gamma;V_\lambda)) ,$$

i.e., it is the Lefschetz number of T_α relative to the l.i.e. operator

$$\partial^{ev}_{\lambda,\Gamma} = \bar\partial_{\lambda,\Gamma} + \bar\partial^{*}_{\lambda,\Gamma} \text{ from } L^2(\Gamma\backslash G) \otimes \Lambda^{ev}\mathfrak{p}_+ \otimes V_\lambda)^K \text{ to } (L^2(\Gamma\backslash G) \otimes \Lambda^{odd}\mathfrak{p}_+ \otimes V_\lambda)^K .$$

In order to compute the reduced homogeneous symbol of $\bar\partial^{ev}_{\lambda,\Gamma}$, we recall from [H-P, §4] that

$$\sum_{k\text{ even}}^{\oplus} \Lambda^{q-k}\mathfrak{p}_+ \cong S^+ \otimes V_{\rho_n} ,$$

$$\sum_{k\text{ odd}}^{\oplus} \Lambda^{q-k}\mathfrak{p}_+ \cong S^- \otimes V_{\rho_n} ,$$

as K-modules, and also that V_{ρ_n} is a 1-dimensional representation (of $\mathfrak{k}$).

So,

$$\sigma_r(\bar\partial^{ev}_\lambda) = (-1)^q V_{\lambda+\rho_n} ,$$

which immediately implies that

$$\chi(T_\alpha;\mathscr{F}_\lambda) = (-1)^q \operatorname{Spin}(T_\alpha, V_{\lambda+\rho_n}) = \tag{5.3}$$

$$= \sum_{[\xi]\in[\Gamma\alpha\Gamma]_e} \frac{(-1)^{q_\xi} \sum_{w\in W_K} \det w \prod_{\alpha\in\Psi_\xi} (w(\lambda+\rho),\alpha)\, e^{w(\lambda+\rho)}}{(2\pi)^{r_\xi}[G_\xi:G_\xi^\circ]\, e^{\rho}(\xi) \prod_{\alpha\in\Psi-\Psi_\xi} (1-e^{-\alpha}(\xi))} \operatorname{vol}(\Gamma_\xi\backslash G_\xi) .$$

§6. COHOMOLOGICAL INTERPRETATION

In addition to the previous hypotheses, we shall assume in this section that Γ has no elliptic elements other than the identity. Thus, Γ acts without fixed points on the symmetric space $M = G/K$ and the quotient space $M_\Gamma = \Gamma\backslash M$ is a smooth manifold.

We begin by introducing a pair of maps closely related to the Hecke operator

T_α. Let $\Gamma' = \Gamma \cap \alpha^{-1}\Gamma\alpha$. Then $M_{\Gamma'} = \Gamma'\backslash M$ is a finite unramified covering of M_Γ. We denote by f_1 the natural projection of $M_{\Gamma'}$ onto M_Γ,

$$f_1(\Gamma' m) = \Gamma m \ , \ m \in M \ ,$$

and by f_α the projection induced by α,

$$f_\alpha(\Gamma' m) = \Gamma\alpha m \ , \ m \in M \ .$$

Let us briefly examine their coincidence locus, i.e. the set

$$F = \{u \in M_{\Gamma'};\ f_1(u) = f_\alpha(u)\} \ .$$

Clearly, $u = \Gamma' m \in F$ if and only if there exists $\gamma \in \Gamma$ so that

$$m \in M_{\gamma\alpha} = \{m' \in M;\ \gamma\alpha m' = m'\} \ .$$

Since Γ acts without fixed points on M, such a $\gamma \in \Gamma$ is unique; also, $\gamma\alpha$ is necessarily elliptic. Thus,

$$F = \bigcup_{\xi \in (\Gamma\alpha)_e} F_\xi \ ,$$

where $(\Gamma\alpha)_e$ stands for the set of elliptic elements in $\Gamma\alpha$ and F_ξ is the image in $M_{\Gamma'}$ of $M_\xi \subset M$. It is easy to see that if $\xi_1, \xi_2 \in (\Gamma\alpha)_e$ and $\gamma \in \Gamma'$ then γM_{ξ_1} and M_{ξ_2} either have no point in common or coincide; the latter happens precisely when $\xi_2 = \gamma\,\xi_1\,\gamma^{-1}$. Therefore, F_ξ depends only on the Γ'-conjugacy class $[\xi]' = \{\gamma\,\xi\,\gamma^{-1};\ \gamma \in \Gamma'\}$, and to emphasize this we shall write $F_{[\xi]'}$ instead of F_ξ; also

$$F_{[\xi_1]'} \cap F_{[\xi_2]'} = \phi \ , \ \text{if} \ [\xi_1]' \neq [\xi_2]' \ ,$$

and moreover

$$F_{[\xi]'} \cong \Gamma'_\xi \backslash M_\xi \ , \ \text{for any} \ \xi \in (\Gamma\alpha)_e \ .$$

We now recall that, according to [H-P, Lemma 1] (cf. also [K-S], §17), for any elliptic $\xi \in G$, M_ξ is a connected symmetric space on which G_ξ and G°_ξ act transitively, i.e.

$$M_\xi \cong G_\xi / K_{(\xi)} \cong G^\circ_\xi / K^\circ_{(\xi)} \ ,$$

where $K_{(\xi)}$ is a maximal compact subgroup of G_ξ and $K^\circ_{(\xi)}$ its identity component; when $\xi \in K$, one can choose $K_{(\xi)} = K_\xi = K \cap G_\xi$.

The above discussion may be summarized as follows (cf. [K-S, §17]):

(6.1.a) $$F = \bigsqcup_{[\xi]' \in [\Gamma\alpha]'_e} F_{[\xi]'} \qquad \text{(disjoint union)},$$

where $[\Gamma\alpha]'_e$ denotes the set of elliptic Γ'-conjugacy classes in in $\Gamma\alpha$;

(6.1.b) $$F_{[\xi]'} \cong \Gamma'_\xi \backslash G_\xi / K_{(\xi)} \,, \text{ for each } \xi \in (\Gamma\alpha)_e \,.$$

Let us remark that

(6.2) $$\Gamma'_\xi = \Gamma_\xi \,, \text{ if } \xi \in \Gamma\alpha$$

and so one can rewrite (6.1.b) as

(6.1.c) $$F_{[\xi]'} \cong \Gamma_\xi \backslash G_\xi / K_{(\xi)} \,, \ \xi \in (\Gamma\alpha)_e \,.$$

To check (6.2), we write $\xi \in \Gamma\alpha$ as a product

$$\xi = \gamma\alpha \,, \text{ with } \gamma \in \Gamma$$

and note that, if $\delta \in \Gamma_\xi$, then

$$\delta = \xi^{-1}\delta\xi = \alpha^{-1} \gamma^{-1} \delta\gamma\alpha \in \alpha^{-1} \Gamma\alpha \,;$$

thus $\delta \in \Gamma \cap \alpha^{-1} \Gamma\alpha = \Gamma'$ and so $\delta \in \Gamma'_\xi$. Since on the other hand $\Gamma'_\xi \subset \Gamma_\xi$, it follows that $\Gamma'_\xi = \Gamma_\xi$, Q.e.d.

A consequence of (6.1.c.), of independent interest, is:

(6.3) If $\varepsilon \in \Gamma^\#$ is elliptic, then $\Gamma_\varepsilon \backslash G_\varepsilon$ is compact.

Indeed, it is enough to apply (6.1.c.) to the situation when $\xi = \alpha = \varepsilon$ and to observe that $F_{[\varepsilon]'}$, as a connected component of the coincidence locus of the pair $f_1, f_\varepsilon : \Gamma \cap \varepsilon^{-1} \Gamma \varepsilon \backslash M \to \Gamma\backslash M$, is compact.

Let us now consider a G-invariant elliptic differential operator $D^+_M : C^\infty_c(\mathscr{E}^+) \to C^\infty_c(\mathscr{E}^-)$, where $\mathscr{E}^\pm = G \times_K E^\pm$, and also an elliptic element $\xi \in \Gamma^\#$.

The symbol of D_M^+ defines a G-complex ΣD_M^+ over T^*M, which is exact outside the zero-section. Let $\Sigma_\xi D_M^+$ denote the restriction of this complex to M_ξ. It is clearly Γ_ξ-invariant and so it drops down to $X_\xi = \Gamma_\xi\backslash M_\xi$ giving rise to a quotient complex $\Gamma_\xi\backslash\Sigma_\xi D_M^+$. By (6.3), the manifold X_ξ is compact, hence $\Gamma_\xi\ \Sigma_\xi D_M^+$ is a complex with compact support. On the other hand, the (finite) cyclic group Z_ξ generated by ξ acts naturally on it. Thus, $\Gamma_\xi\backslash\Sigma_\xi D_M^+$ defines an element in $K_{Z_\xi}(T^*X_\xi)$, the Z_ξ-equivariant K-theory (with compact supports) of T^*X_ξ. Since the action of Z_ξ on $T^*X_\xi \cong \Gamma_\xi\backslash T^*M_\xi$ is trivial, $K_{Z_\xi}(T^*X_\xi) \cong K(T^*X_\xi) \otimes R(Z_\xi)$ and so we can define, as in [A-Si, §3], the cohomology class

$$ch(\Gamma_\xi\backslash\Sigma_\xi D_M^+)(\xi) \in H^*(T^*X_\xi; \mathbb{C})$$

Denote by N^ξ the normal bundle of M_ξ in M. Each fibre N_m^ξ splits under the action of Z_ξ

$$N_m^\xi = N_m^\xi(-1) \oplus \sum\nolimits^{\oplus}_{0 < \theta < \pi} N_m^\xi(\theta) ,$$

as in [A-Si, (3.2)]. This splitting yields a bundle decomposition of the vector bundle $\Gamma_\xi\backslash N^\xi$ over X_ξ

$$\Gamma_\xi\backslash N^\xi = \Gamma_\xi\backslash N^\xi(-1) \oplus \sum\nolimits^{\oplus}_{0 < \theta < \pi} \Gamma_\xi\backslash N^\xi(\theta)$$

and we can then define the characteristic classes

$\mathscr{R}(\Gamma_\xi\backslash N^\xi(-1))$, $\mathscr{S}(\Gamma_\xi\backslash N^\xi(\theta)) \in H^*(X_\xi; \mathbb{C})$, as well as the index class $\mathscr{I}(X_\xi) \in H^*(X_\xi; \mathbb{C})$, as in [A-Si, §3].

The value $\det(1-\xi|N_m^\xi)$ is independent of $m \in M_\xi$ and thus can be regarded as an element of $H^\circ(X_\xi; \mathbb{C})$; we shall write this element as $\det(1-\xi|\ \Gamma_\xi\backslash N^\xi)$.

Recalling now that $H^*(T^*X_\xi; \mathbb{C})$ has a natural $H^*(X_\xi; \mathbb{C})$-module structure let us set

$$L(\xi, D_\Gamma^+) = \left\{ \frac{ch(\Gamma_\xi\backslash\Sigma_\xi(D_M^+))(\xi)\mathscr{R}(\Gamma_\xi\backslash N^\xi(-1))\ \Pi_{0 < \theta < \pi}\ \mathscr{S}(\Gamma_\xi\backslash N^\xi(\theta)).\mathscr{I}(X_\xi)}{\det(1-\xi|\Gamma_\xi\backslash N^\xi)} \right\}[T^*X_\xi]$$

where, on the right, we evaluate the top-dimensional component of the cohomology class enclosed within braces on the fundamental homology class of T^*X_ξ (which is oriented as an almost complex manifold).

Translation by an element $g \in G$, acting on $M = G/K$, has the effect of changing Γ to $g\Gamma g^{-1}$, ξ to $g\,\xi\,g^{-1}$ and so forth, without altering the cohomological invariants introduced in the preceding paragraphs. Thus,

(6.4.a) $L(g\,\xi\,g^{-1}, D^+_{g\Gamma g^{-1}}) = L(\xi, D^+_\Gamma)$, for any elliptic $\xi \in \Gamma^\#$ and any $g \in G$;

in particular,

(6.4.b.) $L(\gamma\,\xi\,\gamma^{-1}, D^+_\Gamma) = L(\xi, D^+_\Gamma)$, for $\gamma \in \Gamma$.

With these notations, the cohomological form of our Lefschetz formula is:

(6.5) THEOREM $$L(T_\alpha, D^+_\Gamma) = \sum_{[\xi] \in [\Gamma\alpha\Gamma]_e} L(\xi, D^+_\Gamma) .$$

Before proving the theorem, we pause to comment on the seemingly formidable expression of $L(\xi, D^+_\Gamma)$. Following [H-P, §2], we shall apply a number of standard operations to rewrite it in a more utilizable form. In view of (6.4.a), there is no loss of generality in assuming, as we do, that $\xi \in H$. Now, by first passing from X_ξ to its orientable covering

$$\tilde{X}_\xi = \Gamma_\xi \backslash G_\xi / K^\circ_\xi \xrightarrow{\;p_\xi\;} \Gamma_\xi \backslash G_\xi / K_\xi \cong X_\xi ,$$

then replacing the evaluation on $T^*\tilde{X}_\xi$ by evaluation on $\tilde{X}_\xi$, via the Thom isomorphism, and finally using the fact that we are in the presence of a K-structure, one obtains, exactly as in [H-P, §2],

(6.6)

$$L(\xi, D^+_\Gamma) = \frac{(-1)^q}{[K_\xi : K^\circ_\xi]} \{ (\frac{\mathrm{ch}\ \sigma_r(D^+)(\xi)}{\prod_{\alpha \in \Psi_n} (e^{\alpha/2}(\xi)e^{\alpha/2} - e^{-\alpha/2}(\xi)e^{-\alpha/2})} \prod_{\alpha \in \Psi_{n,\xi}} \alpha \ )(\Gamma_\xi \backslash G_\xi)\}[\tilde{X}_\xi] .$$

Here we are using the same notational conventions as in [H-P]. Thus, given a K°_ξ-module V , $\mathrm{ch}\ V(\xi)$ denotes the element of $H^*_{K^\circ_\xi}(\mathbb{C}) = H^{**}(BK^\circ_\xi; \mathbb{C})$ defined by

$$\mathrm{ch}\ V(\xi)(P) = \mathrm{ch}(P \times_{K^\circ_\xi} V) \in H^*(X; \mathbb{C}) ,$$

for any principal K°_{ξ}-bundle over a compact manifold X, where $Px_{K^{\circ}_{\xi}} V$ is viewed as an element of $K_{Z_{\xi}}(X) \cong K(X) \otimes R(Z_{\xi})$. Also, we use the natural identification of $H^{*}_{K^{\circ}_{\xi}}(\mathbb{C})$ with $H^{*}_{H}(\mathbb{C})^{W_{k,\xi}}$, where $W_{k,\xi}$ is the Weyl group for (K°_{ξ}, H). In particular,

$$(6.7)\qquad \mathrm{ch}\, V(\xi) = \sum_{\mu} m_{\mu}\, e^{\mu}(\xi) e^{\mu}\ , \text{ if } \mathrm{ch}(V)|H = \sum_{\mu} m_{\mu}\, e^{\mu}\ .$$

<u>Proof of Theorem (6.5)</u>. Let $\sigma_r(D^{+}) = \sum_{\mu} n_{\mu} v_{\mu}$. In view of (4.16), it suffices to show that if $[\xi] \in [\Gamma\alpha\Gamma]_e$,

$$(6.8)\qquad L(\xi, D^{+}_{\Gamma}) = \sum_{\mu} n_{\mu} \operatorname{vol}(\Gamma_{\xi}\backslash G_{\xi}) \operatorname{spin}_{\mu}(\xi) \quad .$$

By (6.4.a) the left hand side of this identity is invariant under conjugation by inner automorphisms, and so is the expression on the right. So we may assume that $\xi \in H$. It follows then from (6.6) that

$$(6.9)\qquad L(\xi, D^{+}_{\Gamma}) = \sum_{\mu} n_{\mu} L(\xi, D^{+}_{\mu,\Gamma}) \quad .$$

On the other hand, modulo obvious adjustments of notation and Haar measure normalizations, Lemma 3 in [H-P] asserts that

$$(6.10)\qquad L(\xi, D^{+}_{\mu,\Gamma}) = \operatorname{vol}(\Gamma_{\xi}\backslash G_{\xi}) \operatorname{spin}_{\mu}(\xi)\ .$$

Putting (6.9) and (6.10) together one obtains (6.8) and hence the theorem.

To illustrate the usefulness of the cohomological version of the Lefschetz formula, we shall apply it to the deRham operator to recover a result of Kuga and Sampson [K-S, §17, Theorem] :

(6.11) COROLLARY. The Euler number of a Hecke operator T_{α} is equal to the Euler characteristic of the coincidence locus of the associated pair of covering maps

$f_1, f_\alpha : \Gamma \cap \alpha^{-1} \Gamma\alpha \backslash M \to \Gamma \backslash M$.

Proof. By (6.5) and using the notation in §5, one has

$$E(T_\alpha) = L(T_\alpha, d_\Gamma^{ev}) = \sum_{[\xi] \in [\Gamma\alpha\Gamma]_e} L(\xi, d_\Gamma^{ev}) .$$

The map sending $[\gamma\alpha] \in [\Gamma\alpha\Gamma]$ to $[\gamma\alpha]' \in [\Gamma\alpha]'$ is a bijection and so we may rewrite the above formula as

$$E(T_\alpha) = \sum_{[\xi]' \in [\Gamma\alpha]'_e} L(\xi, d_\Gamma^{ev})$$

On the other hand, by (6.1.a - c), the Euler characteristic $E(F)$ of the coincidence locus F can be expressed as a sum

$$E(F) = \sum_{[\xi]' \in [\Gamma\alpha]'_e} E(X_\xi) ,$$

where $X_\xi = \Gamma_\xi \backslash G_\xi / K_{(\xi)}$. Thus, it suffices to check that

$$L(\xi, d_\Gamma^{ev}) = E(X_\xi) , \text{ for } \xi \in \Gamma^\# \text{ and elliptic.}$$

Furthermore, due to the invariance of this identity under inner automorphisms, we can assume that $\xi \in H$.

So let $\xi \in H \cap \Gamma^\#$. By (6.6),

$$L(\xi, d_\Gamma^{ev}) = \frac{1}{[K_\xi : K_\xi^\circ]} \{ \frac{\sum_{u \in W^1} \det u \cdot \operatorname{ch} V_{u\rho-\rho_k}(\xi)}{\prod_{\alpha \in \Psi_n} (e^{\alpha/2}(\xi) e^{\alpha/2} - e^{-\alpha/2}(\xi) e^{-\alpha/2})} \prod_{\alpha \in \Psi_{n,\xi}} \alpha)(\Gamma_\xi \backslash G_\xi)\} [\tilde{X}_\xi]$$

and using now (6.7) and Weyl's character formula, we get

$$L(\xi, d_\Gamma^{ev}) = \frac{1}{[K_\xi : K_\xi^\circ]} \{(\frac{\sum_{w \in W} \det w \cdot e^{w\rho}(\xi)\, e^{w\rho}}{\prod_{\alpha \in \Psi} (e^{\alpha/2}(\xi) e^{\alpha/2} - e^{-\alpha/2}(\xi) e^{-\alpha/2})} \prod_{\alpha \in \Psi_{n,\xi}} \alpha)(\Gamma_\xi \backslash G_\xi)\} [\tilde{X}_\xi] =$$

$$= \frac{1}{[K_\xi : K_\xi^\circ]} \{(\prod_{\alpha \in \Psi_{n,\xi}} \alpha) \quad (\Gamma_\xi \backslash G_\xi)\} [\tilde{X}_\xi] .$$

But, assuming that the orientation of $\tilde{X}_\xi$ is appropriately chosen,

$(\prod_{\alpha \in \Psi_{n,\xi}} \alpha)(\Gamma_\xi \backslash G_\xi) \in H^*(\tilde{X}_\xi; \mathbb{C})$ is precisely the Euler class of $\tilde{X}_\xi$; therefore

$$L(\xi, d_\Gamma^{ev}) = \frac{1}{[K_\xi : K_\xi^\circ]} E(\tilde{X}_\xi) = E(X_\xi) ,$$

q.e.d.

7. FURTHER REMARKS

We collect here a few additional comments concerning the Lefschetz formula for Hecke operators. G and Γ are assumed to satisfy the hypotheses (4-G) and (4-Γ) respectively.

(7.1) Let $\mu \in \mathfrak{h}_C^*$ be Ψ_k-dominant integral and such that $\mu - \rho_n \in L_H$. The latter property implies that $S_\mu^\pm = V_\mu \otimes S^\pm \in R(K)$, so we may form the Dirac operators $D_{\mu,\Gamma}^\pm$. Consider the positive root system

$$\Psi^{(\mu)} = \{\alpha \in \Phi;\ (\mu + \rho_k, \alpha) > 0\}$$

and assume that μ satisfies the condition:

(P) $$(\mu - \rho_n^{(\mu)}, \alpha) > 0 \quad \text{for every} \quad \alpha \in \Psi_n^{(\mu)} .$$

According to Parthasarathy's vanishing theorem (see [H-P], Theorem 1) and denoting by $w_\mu \in W$ the unique Weyl group element such that $\Psi^{(\mu)} = w_\mu \Psi$, one has:

$$\operatorname{Ker} D_{\mu,\Gamma}^+ = 0 \quad \text{if} \quad \det(w_\mu) = -1 ,$$

$$\operatorname{Ker} D_{\mu,\Gamma}^- = 0 \quad \text{if} \quad \det(w_\mu) = 1 .$$

It follows that, for any $\alpha \in \Gamma^\#$,

$$\operatorname{Spin}(T_\alpha, V_\mu) = \det(w_\mu) \operatorname{Tr}(T_\alpha \otimes I | \operatorname{Ker} D_{\mu,\Gamma}) ,$$

where $D_{\mu,\Gamma}$ is the full Dirac operator $D_{\mu,\Gamma}^+ \oplus D_{\mu,\Gamma}^-$. Thus, (4.15) becomes:

$$\operatorname{Tr}(T_\alpha \otimes I | \operatorname{Ker} D_{\mu,\Gamma}) = \det(w_\mu) \sum_{[\xi] \in [\Gamma\alpha\Gamma]_e} \operatorname{vol}(\Gamma_\xi \backslash G_\xi) \operatorname{spin}_\mu(\xi) .$$

Actually, as shown in [Wi] ,

$$\operatorname{Ker} D_{\mu,\Gamma} = (L^2_\pi(\Gamma\backslash G) \otimes S_\mu)^K ,$$

where $\pi = \pi_{\mu+\rho_k}$ is the discrete series representation corresponding to $\mu + \rho_k$ and $S_\mu = V_\mu \otimes S = S^+_\mu \oplus S^-_\mu$. Therefore, the above identity can be rewritten as

$$\operatorname{Tr}(T_\alpha \otimes I | (L^2_\pi(\Gamma\backslash G) \otimes S_\mu)^K) = \det(w_\mu) \sum_{[\xi] \in [\Gamma\alpha\Gamma]_e} \operatorname{vol}(\Gamma_\xi\backslash G_\xi) \operatorname{spin}_\mu(\xi) .$$

Let us also remark that this formula holds, in particular, for all integrable discrete series. Indeed, by a well-known criterion of Trombi and Varadarajan, if $\pi_{\mu+\rho_k}$ is integrable then μ necessarily satisfies the condition (P).

(7.2) Suppose now that $M = G/K$ is Hermitian symmetric. Let V_λ be an irreducible K-module and denote by $\mathcal{F}_\lambda$ the sheaf over $M_\Gamma = \Gamma\backslash G/K$ associated to it. We assume that $\mu = \lambda - \rho_n$ satisfies (P). Then, by Proposition 1 in [H-P],

$$H^k(M_\Gamma; \mathcal{F}_\lambda) = 0 \quad \text{for} \quad k \neq k(\lambda) = [\Psi_n \cap \Psi_n^{(\mu)}] .$$

Denoting by $T^{(k)}_{\alpha,\lambda}$ the operator induced by T_α on $H^k(M_\Gamma; \mathcal{F}_\lambda)$, one then has

$$\operatorname{Tr} T^{(k(\lambda))}_{\alpha,\lambda} = (-1)^{k(\lambda)} \chi(T_\alpha; \mathcal{F}_\lambda) ;$$

this, together with (5.3), provides an explicit expression for the trace of the Hecke operator T_α acting on the sheaf cohomology space $H^{k(\lambda)}(M_\Gamma; \mathcal{F}_\lambda)$.

(7.3) The cohomological formula (6.5) for the Lefschetz number is, clearly, similar to Atiyah-Singer's Lefschetz fixed point formula [A-Si, Theorem (3.9)]. There is an obvious dictionary, which can be used to translate the explicit cohomological expressions obtained in [A-Si] and [A-B] in various important special cases, from their context into ours, and thus derive interesting conclusions.

(7.4) The action of the Hecke operator T_α on the complex cohomology $H^*(M_\Gamma; \mathbb{C})$ preserves the integral cohomology classes (see [K, Ch. III, §2]) and can thus be represented by a matrix whose entries are all integers. Hence the eigenvalues of the operators $T_\alpha^{(k)}$, are algebraic integers. In particular, the signature number $\mathrm{Sign}(T_\alpha)$ is an algebraic integer, whereas the right hand side of (5.2) is, a priori, at best an algebraic number. This observation, which can be extended to the case of cohomology with local coefficients, leads to "algebraic-integrality theorems" analogous to the integrality theorems arising from the Atiyah-Bott-Singer theory.

REFERENCES

[A-B] M. F. ATIYAH and R. BOTT, "A Lefschetz fixed point formula for elliptic complexes: II, Applications", Ann. of Math. 88 (1968), 451-491.

[A-Sc] M. F. ATIYAH and W. SCHMID, "A geometric construction of the discrete series for semisimple Lie groups", Inventiones Math. 42 (1977), 1-62.

[A-Si] M. F. ATIYAH and I. M. SINGER, "The index of elliptic operators: III", Ann. of Math. 87 (1968), 546-604.

[B-M] D. BARBASCH and H. MOSCOVICI, "L^2-index and the Selberg trace formula", J. Funct. Anal., 53 (1983), 151-201.

[G-G-P] I. M. GELFAND, M. I. GRAEV and I. I. PIATETSKII - SHAPIRO, "Representation theory and automorphic forms", W. B. Saunders Co., Philadelphia, 1969.

[He] R. HERB, "Discrete series characters and Fourier inversion on semisimple real Lie groups", Trans. Amer. Math. Soc. 277 (1983), 241-262.

[H-P] R. HOTTA and R. PARTHASARATHY, "A geometric meaning of the multiplicity of integrable discrete classes in $L^2(\Gamma\backslash G)$ ", Osaka J. Math 10 (1973), 211-234.

[K] M. KUGA, "Fiber varieties over a symmetric space whose fibers are abelian varieties", II, Lecture Notes, Univ. of Chicago, 1963/64.

[K-S] M. KUGA and J. H. SAMPSON, "A coincidence formula for locally symmetric spaces", Amer. J. Math. 94 (1972), 486-500.

[La] R. P. LANGLANDS, "On the functional equations satisfied by Eisenstein series", Lecture Notes in Math., Vol 544, Springer-Verlag, 1976.

[Mi] R. J. MIATELLO, "Alternating sum formulas for multiplicities in $L^2(\Gamma\backslash G)$, II", Math. Zeitschrift, 182 (1983), 35-44.

[Mo] H. MOSCOVICI, "L^2-index of elliptic operators on locally symmetric spaces of finite volume", Contemporary Math., Vol. 10, A.M.S. 1982, 129-138.

[O-W] M. S. OSBORNE and G. WARNER, "The theory of Eisenstein Systems", Academic Press, 1981.

[R-W] D. L. RAGOZIN and G. WARNER, "On a method for computing multiplicities in $L^2(\Gamma\backslash G)$ ", Symposia Math., Vol. 22, Academic Press, 1977, 291-314.

[Sc] W. SCHMID, "On a conjecture of Langlands", Ann. of Math. 93 (1971), 1-42.

[Se] A. SELBERG, "Harmonic analysis and discontinuous groups in weakly symmetric spaces with applications to Dirichlet series", J. Indian Math. Soc. 20 (1956), 47-87.

[Wa] G. WARNER, "Harmonic analysis on semisimple Lie groups" Vol. II, Springer-Verlag, 1972.

[Wi] F. L. WILLIAMS, "Discrete series multiplicities in $L^2(\Gamma\backslash G)$ ", Amer. J. Math., to appear.

Harmonic Analysis on Unbounded Homogeneous Domains in $\mathbb{C}^n$

By Richard Penney*

Let $\Omega \subset \mathbb{C}^n$ be a domain (open and connected). Ω is said to be homogeneous under a Lie group G if G acts transitively (as a Lie group) on Ω by means of bi-holomorphic mappings. The action is said to be rational if G is a closed subgroup of $Gl(m,\mathbb{R})$ for some m and the mapping

$$\mu : G \times \Omega \to \Omega$$

given by the G action extends to a rational mapping of $Gl(m,\mathbb{R}) \times \mathbb{C}^n \to \mathbb{C}^n$.

In this talk, our goal is to study certain operators defined on the boundary of rational homogeneous domains by means of group representations.

More explicitly, under the appropriate smoothness and connectedness conditions, it will turn out that such Ω have boundaries that are homogeneous spaces of nilpotent Lie groups N. Let $x_o \in \Omega$ and let H_{x_o} be the Levi form at x_o. This is a Hermitian-linear, two-form on the space $\mathcal{T}^{(0,1)}$ of anti-homomorphic vector fields on $\partial\Omega$. H_{x_o} is uniquely defined up to a scalar. We assume that H_{x_o} is non-degenerate, but not necessarily positive. Using the N action on $\partial\Omega$, we may translate H_{x_o} to other points of $\partial\Omega$ and hence obtain a uniquely defined (modulo scalars), N-invariant, Levi form H on all of $\partial\Omega$. (It follows from the nilpotency that H_{x_o} is invariant under the isotropy subgroup of x_o).

Let $\Lambda^{(0,1)}(\partial\Omega)$ be the complex dual of $\mathcal{T}^{(0,1)}$. Let $\Lambda^{(0,p)}(\partial\Omega)$ be the pth wedge product of $\Lambda^{(0,1)}(\Omega)$ with itself. Let $C^{(0,p)}$ be the space of C^∞ sections of $\Lambda^{(0,p)}(\Omega)$. Following standard constructions, there is a boundary operator

$$\bar{\partial}_b^p : C^{(0,p)} \to C^{(0,p+1)} .$$

Using H, we get an identification of $\mathcal{T}^{(0,1)}$ and $\Lambda^{(0,1)}(\partial\Omega)$. Imitating the construction of the Laplace-Beltrami operator on a complex manifold, we obtain a dual mapping

$$\delta_b^p : C^{(0,p+1)} \to C^{(0,p)}$$

We then set

$$\Box_b^p = \bar{\partial}_b^p \delta_b^p + \delta_b^p \bar{\partial}_b^p .$$

If H is positive, this is the usual $\Box_b^p$ operator. However, if H is non-positive then $\Box_b^0$, for example, will in general be expressible in the form

$$\square_b^0 = \sum \varepsilon_i X_i^2 + d_i X_i$$

where $\varepsilon_i = \pm 1$ and X_i are certain vector fields on $\partial\Omega$. Thus, it is to be expected that $\square_b^0$ will exhibit certain "hyperbolic" tendencies.

Granted this, it is to be expected that $\square_b^0$ should be a "bad" operator in general. Our results, however, show that on the contrary, $\square_b^0$ is an extremely nice operator. Under an additional technical assumption on Ω, we are able to explicitly invert $\square_b^0$. We obtain precise conditions under which certain associated operators are invertible and we obtain a curious regularity theorem that says that for these associated operators, we have regularity in the real direction (but not other directions).

In order to obtain our results we first must describe the structure of the domains in question. Our description is motivated by the description of bounded homogeneous domains discovered by Pjateckii-Sapiro [4]. We feel that this structure theory is significant in its own right because it clarifies the role of nilpotent Lie groups in the boundary theory of homogeneous domains. One consequence of this description is a facinating connection between the concept of a totally complex polarization in a nilpotent Lie group and the concept of a smooth, rational domain in $\mathbb{C}^n$ with non-degenerate Levi form.

Section I - Structure Theory.

In 1963 Pjateckii-Sapiro began the work which eventually resulted in a description of the general bounded, homogeneous domain in $\mathbb{C}^n$ in terms of what Pjateckii-Sapiro called Siegel domains of type I or type II. Probably the best known example of the Pjateckii-Sapiro realization is the so called unbounded realization of the unit ball in $\mathbb{C}^2$. Specifically, let $\Omega \subset \mathbb{C}^2$ be

$$\Omega = \{(z,w) \mid \operatorname{Im} w > |z|^2\}.$$

Then Ω is a homogeneous domain which is bi-holomorphically equivalent with the unit ball in $\mathbb{C}^2$ under the mapping

$$(z,w) \to \left(\frac{2z-1}{w+i} , \frac{w-i}{w+i} \right).$$

It is well known that the Heisenberg group acts transitively on $\partial\Omega$. To motivate our general theory, let us show how Ω may be described in terms of the Heisenberg group.

Let $N = \mathbb{R}^3$ with the group structure

$$(x,y,s)(x',y',t) = (x+x',y+y',s+t+(xy'-yx')/2).$$

Let N_c be $\mathbb{C}^3$ with the group structure defined by the same formula. Let $P \subset N_c$ be the subgroup

$$P = \{(z,iz,0)\,|\,z\in\mathbb{C}\}.$$

The space

$$X = N_c/P$$

is a two dimensional complex manifold. In fact, the subgroup

$$Q = \{(w,-iw,s)\,|\,w,s\in\mathbb{C}\}$$

is a cross section for P in N_c and hence defines the isomorphism of X with $\mathbb{C}^2$. Let Π be the projection mapping of N_c to X. Since $N \cap P = \{e\}$, Π is injective on N. $\Pi(N)$ is a three (real) dimensional submanifold of X which splits X into two components X^+ and X^-. Let $V \subset N_c$ be

$$V = \{(0,0,ri)\,|\,r\in\mathbb{R}^+\}.$$

Then $X^+ = \Pi(NV)$ is one such component. It is easily seen that under the identification of X with $\mathbb{C}^2$, X^+ is identified with Ω.

We shall indicate how, in general, any connected, contractible, rational homogeneous domain is bi-holomorphically equivalent with a domain which can be described in a similar manner in terms of some nilpotent Lie group N.

We begin by introducing the model spaces, which we refer to as Siegel domains of type (N,P).

Let N be a connected, simply connected nilpotent Lie group with Lie algebra $\mathcal{N}$. Let $\mathcal{N}_c$ be the complexification of $\mathcal{N}$ and let N_c be the corresponding connected, simply connected Lie group. Let $\mathcal{P}$ be a complex subalgebra of $\mathcal{N}_c$ and let $P \subset N_c$ be the corresponding subgroup. Any automorphism of N extends holomorphically to N_c. Hence it makes sense to consider the group A_P of automorphisms of N which leave P invariant. A_P is a real algebraic group which acts on the double quotient space $N\backslash N_c/P = Y$.

Definition (1.1) <u>We shall say that</u> P <u>is fat if there is an</u> $\mathbb{R}$-<u>split torus</u> $T \subset A_P$ <u>which has an open orbit in</u> Y.

It is easily seen that if $T \subset A_P$ has an open orbit, then any closed subgroup of A_P containing T has an open orbit. Thus, we may replace T by any maximal isotropic torus of A_P which contains T. Since all such tori are conjugate in A_P we may define fatness by saying that <u>every</u> maximal isotropic torus of A_P has an open orbit.

Let T be a maximal isotropic torus of A_P. It is easily seen that there are only a finite number of open orbits for T. Let $v_i \in N_c$ be a complete set of representatives for these orbits, $i = 1\ldots,d$. Let $V_i = T\,v_i$. We refer to each V_i as a "cone". Let $\Pi : N_c \to N_c/P = X$ be the projection mapping and let Ω_i be the image of NV_i under Π. By the open orbit assumption, Ω_i is an open subset of X. It is easily seen that X may be identified with $\mathbb{C}^n$ so the Ω_i may be identified with domains in $\mathbb{C}^n$. Both N and T act holomorphically on each Ω_i and the Ω_i are homogeneous rational domains under this action.

Definition (1.2) The domains Ω_i are referred to as the Siegel domains of type (N,P).

Strictly speaking, our notation should include the dependence of the Ω_i on T. However, since all maximal isotropic tori are conjugate, it is easily seen that the Ω_i for different T are all bi-holomorphically equivalent. Our main result is

Theorem (1.3) Let N and P be given as above, P fat. Then the Siegel domains of type (N,P) are contractible, rational, homogeneous domains. Conversely, any contractible, rational homogeneous domain is bi-holomorphically equivalent to a Siegel domain of type (N,P) for some N and P, P fat.

Example (1.4) The purpose of this example is to justify calling our domains "Siegel domains". Recall that the Siegel generalized upper half plane $\mathcal{H}_n$ is the set of complex $n \times n$ matrices Z such that the Hermitian imaginary part of Z is positive definite. $\mathcal{H}_n$ is homogeneous under the set of transformations

$$Z \to AZA^* + B \tag{1.1}$$

where A ranges over all non-singular, $n \times n$, complex, upper-triangular matricies with positive diagonal entries and B is an arbitrary Hermitian matrix ($B^* = B$). We denote the transformation defined by formula (1.1) by $\tau(A,B)$ and we let S devote the solvable Lie group generated by the $\tau(A,B)$. Then S acts simply transitively on Ω. The Lie algebra $\mathcal{S}$ of S is identified with the set of pairs (A,B) where A is a complex, $n \times n$, upper-triangular matrix with real diagonal and B is a Hermitian matrix. The bracket structure is defined by

$$[(A,B),(A',B')] = ([A,A'],AB' + B'A^* - (A'B + B(A')^*)).$$

We let $\mathcal{N}$ be the set of $(A,B) \in \mathcal{S}$ such that A has zero diagonal. The subalgebra $\mathcal{P} \subset \mathcal{N}_c$ is defined by $\mathcal{P} = \operatorname{span}_{\mathbb{C}} \{(A,B) \mid B = \sqrt{-1}(A+A^*)/2,$ $\operatorname{diag} A = 0\}$. ($\sqrt{-1}$ denotes complex multiplication by i in $\mathcal{N}_c$).

Let $\mathcal{B}_c \subset \mathcal{N}_c$ be the space of all $(0, B_1 + \sqrt{-1}B_2)$ where $B_1^* = B_1$ and $B_2^* = B_2$. Let $B_c \subset N_c$ be the corresponding group. Then $\mathcal{B}_c \oplus \mathcal{P} = \mathcal{N}_c$ so $N_c/P = B_c$. We may identify B_c, and hence N_c/P, with the space of all $n \times n$ complex matrices.

Let D denote the set of $\tau(A,0)$ in S with A diagonal. D acts by conjugation on N and this action leaves $\mathcal{P}$ invariant. The element $v_o = (I, \sqrt{-1}\ I) \in N_c$ represents an open orbit in $N\backslash N_c/P$ and the corresponding Siegel domain is easily identified with $\mathcal{H}_n$.

Example (1.5) Let $\mathcal{N}$ be the Heisenberg Lie algebra with generators X_i, Y_j and Λ, $i = 1,\ldots,n$ subject to the relations

$$[X_i, Y_j] = \delta_{ij}\Lambda.$$

We identify N with $\mathcal{N}$, equiped with the Campbell-Hausdorff product:

$$A\ B = A+B+\frac{1}{2}[A,B].$$

Let $\varepsilon_i \in \{\pm 1\}$, $i=1,\ldots, n$ and let $\mathcal{P}$ be the subalgebra of $\mathcal{N}_c$ spanned by the elements

$$Z_j = X_j + i\varepsilon_j Y_j \ , \ j = 1,\ldots, n.$$

We also identify N_c and $\mathcal{N}_c$. The center Z_c of N_c is then $\mathbb{C}\Lambda$. The equality

$$\mathcal{N}_c = \bar{\mathcal{P}} + \mathcal{Z}_c + \mathcal{P}$$

gives rise to the decomposition

$$N_c = \bar{P}\ Z_c\ P$$

where P is $\mathcal{P}$, thought of as a group. Hence N_c/P is identified with $\bar{P}Z_c$ which may be identified with $\mathbb{C}^{n+1}$ using the $\bar{Z}_j$ bases, $j = 1,\ldots,n$. If we define Ω to be the image of $N(i\mathbb{R}^+\Lambda)$ in N_c/P, then Ω is identified with

$$\{(z,w) \mid z \in \mathbb{C}^n,\ w \in \mathbb{C},\ \operatorname{Im} w > \sum \varepsilon_i |z_i|^2\}\ .$$

This domain is bi-rationally equivalent with the "hyperboloid"

$$\{(z,w) \in \mathbb{C}^{n+1} \mid 1 > |w|^2 - \sum \varepsilon_i |z_i|^2\}\ .$$

Example (1.5) is a specific case of a general class of domains called nil-balls. A nil-ball is a Siegel domain of type N-P for which

(i) N acts simply transitively on $\partial\Omega$.

(ii) the Levi-form is non-degenerate at each point of $\partial\Omega$.

These conditions can be stated in terms of the algebra $\mathcal{P}$. Condition (i) implies (i)' $\mathcal{P} \cap \overline{\mathcal{P}} = \{0\}$ and $\mathcal{P} + \overline{\mathcal{P}}$ has co-dimension one in $\mathcal{N}_c$.

There is, then, a real, non-zero linear functional $\lambda \in \mathcal{N}^*$ which is zero on $(\mathcal{P} + \overline{\mathcal{P}}) \cap \mathcal{N}$. Let H be the Hermitian linear form on $\mathcal{P} \times \mathcal{P}$ defined by

$$H(Z,W) = \frac{1}{2i} \lambda([Z,\overline{W}]). \qquad (1.2)$$

H is the Levi form. Condition (ii) is equivalent with the stipulation

(ii)' H is non-degenerate on $\mathcal{P} \times \mathcal{P}$.

Condition (ii) has another formulation which is directly relevant to representation theory. Let $\mathcal{Z}_c \subset \mathcal{N}_c$ be the center of $\mathcal{N}_c$. The non-degeneracy of H implies that $\mathcal{Z}_c \cap (\mathcal{P} + \overline{\mathcal{P}}) = 0$ so that $\mathcal{Z}_c$ is a **one** complex-dimensional compliment to $\mathcal{P} + \overline{\mathcal{P}}$ in $\mathcal{N}_c$. Let $\mathcal{P}_o = \mathcal{P} + \mathcal{Z}_c$. Then $\mathcal{P}_o$ is a subalgebra of $\mathcal{N}_c$ which satisfies

(a) $\mathcal{P}_o + \overline{\mathcal{P}}_o = \mathcal{N}_c$

(b) $\mathcal{P}_o \cap \overline{\mathcal{P}}_o = \mathcal{Z}_c$ and $\mathcal{Z}_c$ is one dimensional

(c) $W \in \mathcal{P}_o$ if and only if $\lambda([W,X]) = 0$ for all $X \in \mathcal{N}_c$.

$\mathcal{P}_o$ then is what would be referred to as a totally complex (condition (a)), relatively ideal (condition (b)), polarization (condition (c)) for λ. (see [3]).

Such polarizations form the basis of the construction of the harmonically induced representations of H. (See [3]) Conversely, any such polarization gives rise in a conical fashion to a domain in $\mathbb{C}^n$. Thus domains in $\mathbb{C}^n$ and representation theory are intimately tied.

Nil-balls may always be realized as the set of points in $\mathbb{C}^n \times \mathbb{C}$ of the form (z,w) with $\operatorname{Im} w > q(z)$ where q is same real valued polynomial on $\mathbb{C}^n$. This realization is obtained by generalizing the construction described in example (1.5) above. We shall not describe the construction in detail. In this manner large number of "non-classical", homogeneous domains may be produced. One needs only to produce nilpotent Lie algebras with the appropriate polarizations. There is a general theory for doing this which arises out of [2]. Our next example is a consequence of this general construction.

Example (1.6) The domain associated with an associative algebra.

This example is meant to generalize example (1.5). Let $\mathcal{A}$ be a nilpotent associative algebra (possibly abelian). Suppose $\mathcal{A}$ carries a symmetric, bi-linear form B which satisfies

$$B(xy,z) = B(x,yz) \qquad (1.3)$$

for all $x,y,z \in \mathcal{A}$. We define a polynomial function ℓ (the scalar log function) on $\mathcal{A}$ by the formula

$$\ell(x) = \sum_{n=1}^{\infty} (-1)^{n} B(x,x^n)/(n+1). \tag{1.4}$$

Note that this is really a finite sum. Let $\mathcal{A}_c$ be the complexification of $\mathcal{A}$. We extend ℓ to $\mathcal{A}_c$ holomorphically. Let $\Omega \subset \mathcal{A}_c \times \mathbb{C}$ be defined by $\Omega = \{(z,w) | z \in \mathcal{A}_c, w \in \mathbb{C}, \text{Im } w > q(z)\}$ where $q(z) = \ell(\text{Re } z)$. For example, suppose $\mathcal{A} = \mathbb{R}^n$ with the trivial algebra structure (all products equal zero). Let B be any symmetric bilinear form. Then

$$\ell(x) = -B(x,x)/2.$$

The corresponding domain is bi-holomorphically equivalent with the domain of example (1.5) for an appropriate choice of B.

Each point $\tau = \sigma + i\eta$ of $\mathcal{A}_c$ defines a bi-holomorphic mapping A_τ of Ω by means of the formula

$$A_\tau(z,w) = (z+\tau+\sigma z, w+i(\ell(\sigma)-B(\sigma,z))).$$

The crucial property of ℓ that is used in proving that A_τ maps Ω into Ω is the identity

$$\ell(z+w+zw) = \ell(z) + \ell(w) - B(z,w)$$

for all $z,w + \mathcal{A}_c$. We won't go into the proof of this identity here. Translation in the real direction in the w variable also leaves Ω invariant. The $\mathcal{A}_\tau$ together with these translations generate a nilpotent Lie group which acts transitively on the boundary of Ω.

Ω may not be homogeneous. However, if there is a one parameter family of automorphisms δ_t of $\mathcal{A}$ such that

$$B(\delta_t x, \delta_t y) = t^d B(x,y) \tag{1.5}$$

for some d, then $(z,w) \to (\delta_t z, t^d w)$ leaves Ω invariant and Ω is then homogeneous.

Examples of algebras $\mathcal{A}$ possessing forms B which satisfy (1.3) and (1.5) are easily produced. For example let $\mathcal{A}_k$ be the algebra on one generator e subject to the sole relation $e^{k+1} = 0$. Let B be the bi-linear form defined by

$$B(e^i, e^j) = \delta_{i+j,k+1}$$

where $\delta_{k,\ell}$ is the usual Kronecker symbol. More specifically, if $k = 2$ we obtain the domain

$$\Omega = \{(z,w) | z \in \mathbb{C}^2, w \in \mathbb{C}, \text{Im } z > q(z)\}$$

where

$$q(z_1,z_2) = -x_1x_2 + x_1^3/3$$

and $x_i = \text{Re } z_i$.

Now we shall discuss briefly the proof of theorem (1.3). Let $w_o \in \Omega$ be a fixed base point. Let G_c be the algebraic closure of G in $Gl(m,\mathbb{C})$. Since G acts rationally, the G action extends uniquely to a rational mapping of $G_c \times \mathbb{C}^n \to \mathbb{C}^n$. This mapping may of course have singularities, however. Let $P = \{g \in G_c | g\cdot w_o$ is defined and equal to $w_o\}$. P is a subgroup which is open in its closure and hence is a closed subgroup of G_c. Let

$$V_o = \{g \in G_c | g\cdot w_o \text{ is defined and is in } \Omega\}.$$

It is easily seen that V_o is exactly GP. We conclude that GP is open in G_c in the Euclidian topology.

The fact that GP is open is very important. It allows us to assume, first of all, that G is the component of e of the real algebraic group $G_{\mathbb{R}} = G_c \cap Gl(m,\mathbb{R})$.

Next we set $P_G = P \cap G$. The following lemma is due to the contractibility of Ω. The idea of the proof (which we omit) is that if P_G did not contain a maximal compact subgroup of G, then the cohomology groups of Ω would be non-trivial.

Lemma (1.7) P_G contains a maximal compact subgroup of G.

This lemma allows us to assume that G is solvable. In fact, since $G_{\mathbb{R}}$ is algebraic, G may be written as a semi-direct product $G = G_u \times_s Q$ where Q is a connected, reductive group and G_u is the maximal normal unipotent subgroup of $G_{\mathbb{R}}$. We may choose Q so that a maximal compact subgroup of Q is contained in P_G. Let ANK be an Iwasawa decomposition of Q with $K \subset P_G$. The group $G_u X_s AN$ is solvable and acts transitively on Ω. Thus we assume that K is trivial so that G is a solvable Lie group without compact subgroups. In particular, we see that G contains no anistropic tori. Hence G is exponential solvable and its maximal torus is split over $\mathbb{R}$. The next important lemma is the following.

Lemma (1.8) P contains a maximal torus of G_c.

The argument here is roughly as follows. Let N_c be the unipotent radical of G_c. The lemma is equivalent to saying that $PN_c = G_c$. This we prove by considering the quotient space G_c/PN_c. The openness of GP and the connectedness of G_c imply that $G_c = GPN_c$. Hence

$$G_c/PN_c = G/G \cap PN_c.$$

It is not hard to show that the space on the right is contractible (since G is exponential solvable and $PN_{\mathbb{C}}$ is algebraic). Thus, the space on the left is contractible. This implies that $PN_{\mathbb{C}}$ contains a maximal compact subgroup K (a torus) of $G_{\mathbb{C}}$. The algebraic closure of K is a maximal torus of $G_{\mathbb{C}}$ contained in $PN_{\mathbb{C}}$, proving our claim.

Corollary (1.9) Let $P_N = P \cap N_{\mathbb{C}}$. Ω is bi-holomorphically equivalent to an open subset of $N_{\mathbb{C}}/P_N$.

Proof From the lemma $G_{\mathbb{C}} = N_{\mathbb{C}}P$, then $\Omega = GP/P \subset G_{\mathbb{C}}/P = N_{\mathbb{C}}/P_N$.

The above corollary suggests that P_N will be the "fat" subgroup of theorem 1. Actually, this is not quite the case. $G_{\mathbb{C}}$ has a maximal torus $T_{\mathbb{C}}$ which is defined over $\mathbb{R}$. Lemma (1.8) implies that there is a $v \in N_{\mathbb{C}}$ such that

$$vT_{\mathbb{C}}v^{-1} \subset P.$$

The Jordan decomposition of P is then

$$P = vT_{\mathbb{C}}v^{-1}P_N = vT_{\mathbb{C}}\tilde{P}v^{-1}$$

where $\tilde{P} = v^{-1}P_Nv$. $\tilde{P}$ is our "fat" subgroup. Note that $\tilde{P}$ is normalized by $T_{\mathbb{C}}$.

As complex manifolds, $G_{\mathbb{C}}/T_{\mathbb{C}}\tilde{P}$ and $G_{\mathbb{C}}/P$ are isomorphic. We take as our isomorphism the mapping

$$\phi : gP \to gPv = gv\, v^{-1}Pv = gvT_{\mathbb{C}}\tilde{P}.$$

Now, let $T = T_{\mathbb{C}} \cap Gl(m,\mathbb{R})$. G is a semi-direct product

$$G = N \times_S T.$$

The typical element nt of G satisfies

$$\phi(ntP) = ntvT_{\mathbb{C}}\tilde{P} = nv^t tT_{\mathbb{C}}\tilde{P} = nv^t T_{\mathbb{C}}\tilde{P} \qquad (1.6)$$

where $v^t = tvt^{-1}$. As in Corollary (1.9), we may write

$$G_{\mathbb{C}}/T_{\mathbb{C}}\tilde{P} = N_{\mathbb{C}}/\tilde{P}\ .$$

Ω is bi-holomorphically equivalent with the image of G under this isomorphism which, due to formula (1.6), is the image of NV in $N_{\mathbb{C}}/\tilde{P}$ where $V = \{tvt^{-1} | t \in T\}$. It is easily seen that $\tilde{P}$ is fat and the realization of Ω just described is as a Siegel domain of type N-$\tilde{P}$. This finishes the description of the proof of theorem (1.3).

Section II – Applications to complex analysis.

In this section we assume that Ω is a nil-ball. Hence Ω is a Siegel domain of type N-P where the Lie algebra $\mathcal{P}$ meets conditions (i)' and (ii)' of section I above. We shall strengthen our assumption on Ω in the following way. The properties (i)' and (ii)' imply that $\mathcal{P}+\mathcal{N}$ has real co-dimension one in $\mathcal{N}_c$. Let Λ be a generator of the one-dimensional center of $\mathcal{N}$. Then $\mathcal{N}_c = \mathcal{P}\oplus\mathcal{N}\oplus i\mathbb{R}\Lambda$. Hence

$$N\backslash N_c/P = \exp_{N_c} i\mathbb{R}\Lambda \ .$$

The condition that P be fat is equivalent with the statement that A_p (the space of automorphisms of N which leave P invariant) acts non-trivially on $\mathcal{Z}$. We strengthen this condition by requiring that there exist an element of A_p with only positive eigenvalues which acts non-trivially on $\mathcal{Z}$. This is equivalent with requiring that there is a one-parameter group of dilations of $\mathcal{N}$ which leaves $\mathcal{P}$ invariant. We refer to such domains as dilated nil-balls.

Now, let $\lambda\varepsilon\mathcal{N}^*$ and H be as in formula (1.2). H is non-degenerate on $\mathcal{P}$. There is a basis $\{Z_i\}$ of $\mathcal{P}$ such that

$$H(Z_i,Z_j) = \delta_{ij}\varepsilon_i$$

where $\varepsilon_i = \pm 1$ (depending on i).

Let $\Box = \sum\varepsilon_i\bar{Z}_i Z_i$. We may consider $\Box$ as either a left invariant differential operator on $C^\infty(N)$ or as an element of the universal enveloping algebra of $\mathcal{N}$.

By assumption, $\mathcal{N}$ acts simply - transitively on $\partial\Omega$. Identifying $\partial\Omega$ with N, we see that we are interested in studying $\Box$ on $C^\infty(N)$. We shall in fact consider $\Box$ as an unbounded operator on $L^2(N)$.

Let R denote the right regular representation of N in $L^2(N)$. We begin by decomposing R into a direct integral of more simple representations

$$R = \int_X \oplus\, U^\alpha d\mu(\alpha)$$

where X is some topological space. For the sake of exposition, we shall assume that this is the irreducible decomposition, although in practice one uses a somewhat coarser decomposition (the primary decomposition).

If U is any unitary representation of N we let $C^\infty(U)$ denote the usual C^∞ space of U. $\Box$ acts on $C^\infty(R)$. In fact $\Box$ may be identified with $R(\Box)$. As an operator on $C^\infty(R)$, $R(\Box)$ is a direct integral of the operators $U^\alpha(\Box)$ acting on $C^\infty(U^\alpha)$. We propose studying $\Box$ by studying $U^\alpha(\Box)$ for each α. For this to be feasible, there must be some way in which $U^\alpha(\Box)$ is "better" than $R(\Box)$. The way in which $U^\alpha(\Box)$ is better is described in the following definition.

<u>Definition</u> (2.1) <u>Let</u> U <u>be a unitary representation of</u> N. <u>Let</u> $\mathcal{T}_U$

be the Lie algebra of operators on $C^{\infty}(U)$ generated over $\mathbb{R}$ by $U(i\Box)$ and $U(\mathcal{N})$. We say that $\Box$ is finite at U if $\mathcal{T}_U$ is finite dimensional and contains $\mathcal{N}$ in its nil-radical.

Note that under the assumption of finiteness, $\mathcal{T}_U$ will always be solvable. Since the nil-radical contains $\mathcal{N}$, the nil-radical will be at most co-dimension one.

To go along with this definition we have the following theorem, which is the basis of our analysis.

Theorem (2.2) Suppose N is the boundary of a dilated nil-ball. Then $\Box$ is finite at any representation of N which is scalar on the center $\mathcal{Z}$ of N. In particular $\Box$ is finite at any irreducible representation of N.

At this point, an example might help in understanding the general theory. We return to the unit ball in $\mathbb{C}^2$. The Lie algebra $\mathcal{N}$ is the Heisenberg algebra spanned by the elements X, Y and Λ which satisfy the commutation relations

$$[X,Y] = \Lambda .$$

The operator $\Box$ is

$$\Box = (X + iY)(X - iY) = X^2 + Y^2 - i\Lambda.$$

A little experimentation will convince one that the Lie subalgebra of the enveloping algebra of $\mathcal{N}$ generated by $i\Box$ and X, Y, Λ is spanned by the set

$$\{i\Box, \Lambda^p X, \Lambda^q Y, \Lambda^r \mid p,q,r \in \mathbb{Z}\}. \tag{2.1}$$

Thus $i\Box \cup \mathcal{N}$ does not generate a finite dimensional Lie subalgebra of the enveloping algebra. However, the image of the set (2.1) under any representation which is scalar on Λ is finite dimensional and is spanned by the images of $i\Box, X, Y$ and Λ. The new commutation relations are

$$[i\Box', X'] = 2\lambda Y' \quad \text{and}$$

$$[i\Box', Y'] = -2\lambda X'$$

where the prime denotes the image of the corresponding element and where the image of Λ is $i\lambda I$.

To understand how we use theorem (2.2), let U be any irreducible representation of N. It follows from results of E. Nelson and from theorem (2.2) that $U(\Box)$ has a self-adjoint closure so that the operator $\exp U(it\Box)$ exists and is unitary for all t. Let $\tilde{T}_U$ be the group of unitary operators on $\mathcal{H}(U)$ generated by the sets $U(N)$ and $\exp U(it\Box)$, $t \in \mathbb{R}$. It follows (again from results of E. Nelson) that $\tilde{T}_U$ is a finite dimensional solvable Lie group with Lie algebra $\mathcal{T}_U$. What this really says is that the Lie algebra of operators $\mathcal{T}_U$ can be exponentiated to a Lie group of unitary operators.

Our goal is to describe in some explicit way the group $\tilde{T}_U$, for then we would have an explicit description of exp itU(□). This would then yield information concerning spectral properties, inverses, etc.

In principal, we know $\tilde{T}_U$, at least up to local isomorphism for we can construct a connected, simply connected Lie group T_U with Lie algebra $\mathcal{T}_U$. There is then a surjective local isomorphism

$$\rho_U : T_U \to \tilde{T}_U.$$

Since T_U is a group of unitary operators on $\mathcal{H}(U)$, ρ_U is, by definition, a representation. It is in fact an irreducible representation because $U(N)$ acts irreducibly on $\mathcal{H}(U)$ and $U(N) \subset \tilde{T}_U$.

Since T_U is solvable, ρ_U should be, (with any luck), equivalent with an induced representation (or a holomorphically induced representation). Luck is with us and ρ_U is an induced representation.

Let us write

$$\rho_U = \mathrm{ind}(M_U, T_U, \lambda_U)$$

where $\mathcal{M}_U \subset (\mathcal{T}_U)$, and $\lambda_U \in \mathcal{T}_U^*$ are the polarization and Kirillov functional for ρ_U respectively. Let $\Delta \in \mathcal{T}_U$ be $\Delta = U(i\Box)$. Then by definition

$$U(i\Box) = \rho_U(\Delta).$$

But now a remarkable thing has happened. Δ is a first order element in the Lie algebra $\mathcal{T}_U$. Hence $\rho_U(\Delta)$, in the realization of ρ_U as an induced representation, is a <u>first order operator</u>!

More is true, however. If we can explicitly realize U as an induced representation, then we obtain an <u>explicit</u> description of exp itU(□). In fact

$$\rho_U(\exp_{T_U} t\Delta) = e^{it\rho_U(\Delta)} = e^{itU(i\Box)}$$

Thus, our program is the following:

(i) Decompose R into a direct integral of irreducible representations U^α.

(ii) Realize each ρ_{U^α} as an induced representation.

This procedure can be carried out explicitly in a large number of cases. However, even without explicitly describing the irreducible decomposition of R, much can be said.

Since Ω is a nil-ball, the center $\mathcal{Z}$ of $\mathcal{N}$ is one dimensional.

Let Λ be a basis element for $\mathcal{Z}$, which we consider as a left invariant vector field on N. We let

$$\mathcal{L}_\alpha = \Box + i\alpha\Lambda.$$

The operator $\mathcal{L}_\alpha$ becomes important when one considers $\overline{\partial}_b$ harmonic forms on $\partial\Omega$. In the case of the unit ball in $\mathbb{C}^n$, $\mathcal{L}_\alpha$ is the operator considered by Folland, Greiner and Stein. One of the main questions concerning $\mathcal{L}_\alpha$ is for which α (if any) can $\mathcal{L}_\alpha$ have a non-trivial kernel on $\mathcal{S}(N)$ (the Schwartz space). It turns out that this question relates to some rather interesting algebraic properties of $\mathcal{T}_U$.

Theorem (2.3) *$\mathcal{L}_\alpha$ has a non-trivial kernel on $\mathcal{S}(N)$ for some value of α if and only if there exists an irreducible representation U of N which is non-trivial on the center of N such that Δ is a semi-simple element of the Lie algebra $\mathcal{T}_U$. In this case there is a real number α_o such that the set of α such that $\mathcal{L}_\alpha$ has a non-trivial kernel on $\mathcal{S}(N)$ is contained in $\alpha_o + 2\mathbb{Z}$. Furthermore*

$$L^2(N) = \sum_\alpha \oplus \mathcal{H}_\alpha$$

where $\mathcal{H}_\alpha$ is the kernel of $\mathcal{L}_\alpha$.

Corollary (2.4) *If Δ is a semi-simple element in $\mathcal{T}_U$, then $\mathcal{L}_\alpha$ is not locally solvable for some sequence of $\alpha \in \alpha_o + 2\mathbb{Z}$.*

This last corollary follows from some results of Corwin-Rothschild. [1].

We shall sketch the proof of theorem (2.3)

Proof Since U is scalar on Z, $U(i(\Box + i\alpha\Lambda)) = U(i\Box) - i\alpha\lambda I$. Hence for $\Box + i\alpha\Lambda$ to have a non-trivial solution, $U(i\Box)$ must have a non-trivial spectrum for a.e.U. We interpret $U(i\Box)$ as $\rho_U(\Delta)$, $\Delta \in \mathcal{T}_U$.

There is a theorem of C. Moore which tells us precisely when $\rho_U(\Delta)$ has a non-trivial discrete spectrum. To state Moore's result (the Mautner theorem), let G be a Lie group with Lie algebra $\mathcal{G}$. Let V be an irreducible representation of G and let $X \in \mathcal{G}$. We define $\mathcal{I}_X$ to be the smallest $\mathcal{G}$ ideal such that ad X acts semi-simply with purely imaginary eigenvalues on $\mathcal{G}/\mathcal{I}_X$. Moore's theorem states that for V(X) to have a non-trivial discrete spectrum, it is necessary that $V(\mathcal{I}_X) = 0$. In our case, ρ_U is injective on $\mathcal{T}_U$ so for $\rho_U(\Delta)$ to have a discrete spectrum, it is necessary that $\mathcal{I}_\Delta = 0$. Hence it is necessary that ad Δ act semi-simply with purely imaginary eigenvalues on $\mathcal{T}_U$.

Actually, a very close inspection of the Lie algebras involved allows one to show that the eigenvalues of ad Δ on $\mathcal{T}_U$ are always of the form $2ki\lambda, k\epsilon\mathbb{Z}$. Thus, the necessary condition is just that Δ be a semi-simple element of $\mathcal{T}_U$.

We shall omit the proof of the converse as we shall shortly derive a more powerful result.

Let U be an irreducible representation of N which is non-trivial on Z. Let

$$V_U(t) = e^{itU(\Box)} = \rho_U(\exp_{T_U} t\Delta).$$

Assume that Δ is a semi-simple element of $\mathcal{T}_U$ and that $U(\Lambda) = i\lambda$. The following lemma is very important to the analysis of $\Box$.

Lemma (2.5) There is a constant α_o independent of U, such that $e^{-i\alpha_o t} V_U(\lambda t)$ is periodic of period π in t.

Proof Let us first prove the periodicity. As commented above semi-simple implies that its eigenvalues belong to $2i\lambda\mathbb{Z}$. From this it follows that

$$e^{\beta \mathrm{ad}\Delta} = I$$

for all $\beta \in \pi\lambda^{-1}\mathbb{Z}$. Hence $\exp_{T_U}(\pi\lambda^{-1}\mathbb{Z}\Delta) = \Gamma_U$ is a central subgroup in T_U. Since ρ_U is irreducible, ρ_U (and thus V_U) is scalar on Γ_U. Hence

$$e^{-i\alpha_o t} V_U(\lambda t)$$

is periodic of period π for some α_o.

Although not immediately clear, the independence of α_o on λ follows from the fact that our nil-ball is dilated.

The above lemma allows us to invert $\Box + i\alpha\Lambda$, for $\alpha \notin (\alpha_o + 2\mathbb{Z})$. In fact, let $\alpha\epsilon\mathbb{R}$. We set

$$B_U^\alpha = \pi^{-1} \int_0^\pi e^{-i\alpha s} V_U(\lambda s)\, ds. \tag{2.2}$$

B_U^α is a bounded operator on $\mathcal{H}(U)$.

Theorem (2.6)

$$U(\mathcal{L}_\alpha) B_U^\alpha = B_U^\alpha U(\mathcal{L}_\alpha) = \pi^{-1} i(1-e^{-i\pi(\alpha-\alpha_o)})\lambda I. \tag{2.3}$$

Furthermore $\{B_U^\alpha \mid \alpha\epsilon\alpha_o+2\mathbb{Z}\}$ is a commuting family of self-adjoint projections which sums to I on $\mathcal{H}(U)$.

Note that equality (2.3) together with the rest of the theorem implies that B_U^α projects onto the kernel of $U(\mathcal{L}_\alpha)$ for $\alpha\epsilon\alpha_o+2\mathbb{Z}$.

Proof of theorem:

Note that

$$U(i\Box) = \frac{d}{dt}\Big|_{t=o} V_U(t) \quad .$$

Hence

$$U(i\Box)B_U^{\alpha} = \pi^{-1}\int_0^{\pi} \lambda^{-1}e^{-i\alpha s}\frac{d}{ds}V_U(\lambda s)ds.$$

We integrate by parts, using lemma (2.5) to evaluate the boundary terms. One easily rearranges the resulting expression to obtain formula (2.3) Q.E.D.

Next we let B^{α} be the direct integral of the B_U^{α} with respect to U. B^{α} is an operator on $L^2(N)$.

Corollary (2.7) $\mathcal{L}_{\alpha}B^{\alpha} = B^{\alpha}\mathcal{L}_{\alpha} = \pi^{-1}(1-e^{-i\pi(\alpha-\alpha_o)})\Lambda.$

If $\alpha\epsilon\alpha_o+2\mathbb{Z}$, then B^{α} is a projection onto the kernel of $\mathcal{L}_{\alpha}$.

Corollary (2.8) If $\alpha\notin\alpha_o+2\mathbb{Z}$, then the equation

$$\mathcal{L}_{\alpha}u = f$$

is (globally) solvable for f contained in the image of $C^{\infty}(R)$ under Λ. Furthermore we have the estimate

$$\|\Lambda f\| \le \pi\ |\ 1-e^{-i\pi(\alpha-\alpha_o)}|^{-1}\|\mathcal{L}_{\alpha}f\|.$$

for all $f \in C^{\infty}(R)$.

To obtain more specific results, we must make the group more specific. Let $\mathcal{N}$, N and $\mathcal{P}$ be as in example (1.5) above. N is then a 2n+1 dimensional Heisenberg algebra. The operator $\Box$ is then

$$\Box = \sum \varepsilon_j \bar{Z}_j Z_j$$

$$= \sum \varepsilon_j (X_j^2 + Y_j^2) + ni\Lambda$$

In example (1.5) we identified N with $\mathcal{N}$. We carry this identification one step further by identifying $\mathcal{N}$ with $\mathbb{C}^n \times \mathbb{R}$ by means of the mapping

$$(z,s) \to \sum x_i X_i + y_i Y_i + s\Lambda$$

where $z = (z_1 \ldots, z_n)$ and $z_i = x_i + iy_i$.

In this case all of the constructs introduced in the proof of theorem (2.6) may be explicitly described. Without going into the details, we find that for f in the image of $C^{\infty}(R)$ under Λ,

$$\mathcal{L}_{\alpha}^{-1}f(0) = c^{\alpha}\int_{-\infty}^{\infty}\int_{\mathbb{C}^n}(\Lambda)^{n-1}f(z,uQ)\left(\frac{i-u}{i+u}\right)^{\alpha/2}(i-u)^n du/(1+u^2)$$

where $Q = Q(z)$ is the function

$$Q(z) = (1/4)\sum \varepsilon_i |z_i|^2$$

and

$$c^{\alpha} = i(-1)^{n-p}|2\sin\pi(n+2)/2)|^{-1}(2\pi)^{-n}$$

(p is the number of positive ε_i).

The double integral converges in the order stated. However, the convergence is not absolute and the order of the integrals cannot be reversed in general. The significance of this formula is that it converges even if H is nonpositive.

Bibliography

1. Corwin, Rothschild, Necessary conditions for local solvability of homogeneous left invariant differential operators on nilpotent Lie groups, Acta 147(1981) 265-288.

2. Penney, The theory of ad-associative Lie algebras, Pacific J. of Math. 99 (1982) 459-472.

3. Penney, Holomorphically induced representations on nilpotent Lie groups and automorphic forms on nilmanifolds, Trans. Amer. Math. Soc. 260 (1980) 123-145.

4. Pjateckii-Sapiro, Geometry and classification of homogeneous bounded domains, Uspehi Math. Nauk, 20 (1965), 3-51 Russian Math. Surv. 20 (1966), 1-48.

*Purdue University

CHARACTERS AS CONTOUR INTEGRALS

W. Rossmann
University of Ottawa

1. Orbital contour integrals. The starting point is the observation that the orbital integrals which give the tempered irreducible characters of a real reductive group G may be interpreted as contour integrals on a complex orbit as follows. A regular character Θ of this type (with regular infinitesimal character) corresponds to a regular orbit Ω in g^* (closed and of maximal dimension), and the formula in question may be written as (see ROS):

$$<\Theta,\phi> = \int_\Omega (J\phi)\mu_\Omega \ . \qquad (1)$$

Here: ϕ is a C^∞ function on G with compact support in a suitable coordinate domain G_{exp} for exp, which may be taken to be the image of the x in g for which the eigenvalues λ of ad(x) satisfy $|Im(\lambda)| < \pi$; $J\phi$ is the function on g^* defined by

$$J\phi(\xi) = \int_g p(x)\phi(\exp x)\, e^{i<\xi,x>}\, dx$$

with

$$p(x) = \det^{\frac{1}{2}}\left(\frac{\sinh ad(x/2)}{ad(x/2)}\right);$$

and μ_Ω the canonical m-form (m = dim Ω) associated to the symplectic structure. According to Paley-Wiener, $J\phi$ extends under these assumptions to an entire analytic function of exponential type on the complex dual $\mathbf{g}^*$ of the complexified Lie algebra $\mathbf{g}$ of G(*); so ϕ satisfies for each n = 1,2.. an estimate of the type

$$|\phi(\zeta)| < \frac{Ae^{B|Im(\zeta)|}}{1 + |\zeta|^n}$$

where Im is the imaginary part in $\mathbf{g}^* = g^* + ig^*$ and $|\ |$ is some norm on $\mathbf{g}^*$. On the other hand, μ_Ω extends to a holomorphic, complex m-form on the complex orbit $\mathbf{\Omega}$ of the complex group $\mathbf{G}$. (I assume throughout that G is contained in a connected complexification $\mathbf{G}$, al-

(*) Complex Lie algebras, groups, orbits etc. are denoted by 'boldface' letters $\mathbf{g}$, $\mathbf{G}$, $\mathbf{\Omega}$ etc. to avoid proliferation of subscripts $\mathbb{C}$.

although this is not essential.) So $(J\phi)\mu_{\mathfrak{N}}$ is a holomorphic m-form on $\mathfrak{N}$ ($m = \dim_{\mathbb{R}} \Omega = \dim_{\mathbb{C}} \mathfrak{N}$), and therefore closed. The integral (1) can consequently be interpreted as a contour integral

$$\langle \Theta, \phi \rangle = \int_{\Gamma} (J\phi)\mu_{\mathfrak{N}} \tag{2}$$

where Γ is any contour on $\mathfrak{N}$ equivalent to Ω. To make this precise, define an <u>admissible</u> (<u>singular</u>) <u>d-chain</u> on $\mathfrak{N}$ to be a locally finite formal sum $\gamma = \Sigma_k m_k \sigma_k$ of singular C^∞ cubes $\sigma_k: I^d \to \mathfrak{N}$ with the following two properties:

(a) The imaginary part of γ is bounded: $|\mathrm{Im}\, \sigma_k| < C$ for some C independent of k.
(b) γ is tempered: $\sum_k |m_k| \max \dfrac{1}{1 + |\sigma_k|^n} < \infty$ for some n.

The coefficients m_k are understood to be integral. In particular, an admissible m-chain $\gamma = \Sigma_k m_k \sigma_k$ is called an (admissible) <u>contour</u> and we set

$$\int_\gamma \psi = \sum m_k \int_{I^m} \sigma_k^*(\psi\mu_{\mathfrak{N}})$$

for any holomorphic function of exponential type on $\mathfrak{N}$. Using these admissible chains one defines a homology $H_*(\mathfrak{N})$ in the usual way. If γ is a cycle (closed chain), then the contour integral (2) depends evidently only on the homology class of γ. This is in particular the case for the regular real orbit Ω (regarded as an m-chain by some parameterization). Also, contour integrals (2) are invariant under <u>admissible homotopy</u>, this being a family γ_t of admissible contours depending continuously on a real parameter t, having the same boundary for all t, and satisfying (a) and (b) uniformly in t.

<u>Example</u>: <u>finite dimensional representations</u>. A finite dimensional representation π of G may equally well be considered as a holomorphic representation of the complex group $\mathbb{G}$, or as a representation of a compact real form U of $\mathbb{G}$. Viewed as the latter, the heighest weight of π may be written in the form $i\lambda - \rho$ where λ is a regular element in the dual of the Lie algebra of a maximal torus of U and ρ is half the sum of the roots α with $\langle \lambda, h_\alpha \rangle > 0$ (h_α the co-root). Regarding λ as an element of $\mathfrak{g}^*$, the U-orbit $U\cdot\lambda$ of λ is a contour on the complex orbit $\mathbb{G}\cdot\lambda$. Because of the compactness of $U\cdot\lambda$ the formula (2) is in

this case equivalent to

$$\Theta(\exp x) = p(x)^{-1} \int_{U\cdot\lambda} e^{i<\xi,x>} \mu_{\Omega}(d\xi); \qquad (3)$$

and this formula is known to be valid for exp x in U_{exp} (see KIR), hence is valid for exp x in $\mathbb{G}_{exp}$, by analyticity. It has the following curious consequence. For the trival representation we have $\lambda = \rho$, and $\Theta(\exp x) \equiv 1$ in (3) gives

$$p(x) = \int_{U\cdot\rho} e^{i<\xi,x>} \mu_{\Omega}(d\xi).$$

This shows that the function p(x) is a globally defined analytic function on $\mathfrak{g}$, in spite of the root in its definition: a fact which may also be deduced from Chevalley's restriction theorem and which is of significance in some applications of the character formula (1) (see VER).

2. Parabolic induction and standard characters. A representation π of G is said to correspond to a contour Γ on a complex orbit Ω if the character Θ of π is given by formula (2) on G_{exp}.

Lemma. Let P = MN be a parabolic subgroup of G (M reductive, N unipotent). Suppose π_M is a representation of M corresponding to a contour Γ_M on an $\mathbb{M}$-orbit $\mathbb{M}\cdot\lambda$ in $\mathfrak{m}^*$ with λ regular in $\mathfrak{g}^*$. Then the induced representation $\pi_G = \mathrm{Ind}_P^G \pi_P$ correpondss to the contour $\Gamma_G = K\cdot(\Gamma_M + p^{\perp})$ on the $\mathbb{G}$-orbit $\mathbb{G}\cdot\lambda$ in $\mathfrak{g}^*$.

Explanation. $\mathfrak{m}^*$ is regarded as a subspace of $\mathfrak{g}^*$ via $\mathfrak{g} = \mathfrak{m} + [\mathfrak{m},\mathfrak{g}]$. For the induction, π_M is extended to a representation π_P trivial on N; the induction procedure is normalized so that unitary representations of M go to unitary representations of G (although π_M need not be unitary, of course). K is a maximal compact subgroup of G. $p^{\perp}$ is the orthogonal of p in the real dual g^*. If Γ_M is an admissible contour on $\mathbb{M}\cdot\lambda$ then $\Gamma_G = K\cdot(\Gamma_M + p^{\perp})$ is an admissible contour on $\mathbb{G}\cdot\lambda$: the regularity of λ in $\mathfrak{g}^*$ insures that $\lambda + \mathfrak{p}^{\perp} \subseteq \mathbb{P}\cdot\lambda$ (the dot denoting the co-adjoint action in $\mathfrak{g}^*$); and $\mathrm{Im}(\Gamma_G) = K\cdot\mathrm{Im}(\Gamma_M)$ insures that $\mathrm{Im}(\Gamma_G)$ is bounded if $\mathrm{Im}(\Gamma_M)$ is.

Proof. Part of the argument follows a familiar pattern, so I omit some of the details (see DUF).

The induced character $\Theta_G = \mathrm{Ind}_P^G\, \Theta_P$ is given by

$$\langle \Theta_G, \phi \rangle = \int_K \langle \Theta_P, \Delta_P^{-\frac{1}{2}} \phi^k \rangle \, dk \tag{4}$$

where $\phi^k(p) = \phi(kpk^{-1})$ and $\Delta_P(p) = |\det \mathrm{Ad}_{p}(p)|$. The character Θ_P of P is simply the character Θ_M of M extended by right N-invariance:

$$\langle \Theta_P, \phi \rangle = \langle \Theta_M, \phi^N \rangle \tag{5}$$

where

$$\phi^N(m) = \int_N \phi(mn)\, dn \, . \tag{6}$$

By hypothesis,

$$\langle \Theta_M, \phi \rangle = \int_{\Gamma_M} J_M \phi \, .$$

Thus

$$\langle \Theta_P, \phi \rangle = \int_{\Gamma_M} J_M \phi^N$$

$$= \int_{\Gamma_M} \int_{\mathfrak{m}} p_M(x)\phi^N(\exp x)\, e^{i\langle \xi, x\rangle} dx \;\; \mu_{\Omega_M}(d\xi)$$

$$= \int_{\Gamma_M} \int_{\mathfrak{m}} \int_{\mathfrak{n}} p_M(x)\phi(\exp x \exp y)\, e^{i\langle \xi, x\rangle} \, dy dx \mu_{\Omega_M}(d\xi).$$

Changing variables $\exp y \to \exp(-x)\exp(x+y)$ in the inner integral one finds (after some calculations with Jacobians and p-functions):

$$\langle \Theta_P, \phi \rangle = \int_{\Gamma_M} \int_{\mathfrak{p}} p_G(x)\Delta_P^{\frac{1}{2}}(\exp x)\phi(x) e^{i\langle \xi, x\rangle} \, dx \; \mu_{\Omega_M}(d\xi).$$

$\xi \in \mathfrak{m}^*$ is here identified with an element of $\mathfrak{p}^*$ vanishing on $\mathfrak{n}$.

Putting things together:

$$\langle \Theta_G, \phi \rangle = \int_K \langle \Theta_P, \Delta_P^{-\frac{1}{2}} \phi^k \rangle \, dk$$

$$= \int_K \int_{\Gamma_M} \int_{\mathfrak{p}} p_G(x)\phi(\exp k\cdot x) e^{i\langle \xi, x\rangle} \, dx \; \mu_{\Omega_M}(d\xi) \; dk$$

$$= \int_K \int_{\Gamma_M} \int_{\mathfrak{p}^\perp} J_G \phi(k\cdot(\xi+\eta)) \; d\eta \; \mu_{\Omega_M}(d\xi) \; dk$$

$$= \int_{K\cdot(\Gamma_M + \mathfrak{p}^\perp)} J_G \phi(\xi) \; \mu_{\Omega_G} \; (d\xi) \, ,$$

provided we show that $dk \times \mu_{\mathcal{O}_M}(d\xi) \times d\eta$ is the pull-back of $\mu_{\mathcal{O}_G}$ under $K/(K\cap M) \times \Gamma_M \times \mathfrak{p}^{\perp} \to \mathcal{O}_G$, $(k, \xi, \eta) \to k\cdot(\xi + \eta)$. (The map requires a choice of coset representatives for $K/(K\cap M)$.) For this we note that this map is the restriction of the holomorphic map

$$\begin{aligned} \mathbb{G}/\mathbb{P} \times \mathcal{O}_M \times \mathfrak{p}^{\perp} &\to \mathcal{O}_G, \\ (g, \xi, \eta) &\to g\cdot(\xi + \eta), \end{aligned} \tag{7}$$

(whose definition requires a choice of coset representatives for $\mathbb{G}/\mathbb{P}$, which may be taken to be holomorphic, locally).

Fix $\lambda \in \mathcal{O}_M$. The tangent space to $\mathcal{O}_M$ at λ is $\mathfrak{m}\cdot\lambda$; the subspace $\mathfrak{p}^{\perp}$ of $\mathfrak{g}^*$ is $\mathfrak{p}\cdot\lambda$, because of the regularity of λ in $\mathfrak{g}^*$. (The dot indicates the co-adjoint action in $\mathfrak{g}^*$.) The tangent map of (7) at $(1,\xi,\eta)$ becomes

$$\begin{aligned} \mathfrak{g}/\mathfrak{p} \times \mathfrak{m}\cdot\lambda \times \mathfrak{p}\cdot\lambda &\to \mathfrak{g}\cdot\lambda \\ (x, y\cdot\lambda, z\cdot\lambda) &\to (x + y + z)\cdot\lambda\ ; \end{aligned}$$

$\mathfrak{g}/\mathfrak{p}$ is here identified with a subspace of $\mathfrak{g}$ complementary to $\mathfrak{p}$.

Choose a basis x_i for $\mathfrak{g}/\mathfrak{p}$, $y_j\cdot\lambda$ for $\mathfrak{m}\cdot\lambda$, $z_k\cdot\lambda$ for $\mathfrak{p}\cdot\lambda$. Then the $u_r\cdot\lambda = x_i\cdot\lambda, y_j\cdot\lambda, z_k\cdot\lambda$ form a basis for $\mathfrak{g}\cdot\lambda$. The value of $\mu_{\mathcal{O}_G}$ on this basis is the Pfaffian of the skew-symmetric matrix $\langle u_r\cdot\lambda, u_s\rangle = \langle\lambda,[u_s,u_r]\rangle$. Since $\langle z_i\cdot\lambda, z_j\rangle = 0$ and $\langle z_i\cdot\lambda, y_j\rangle = 0$ for all values of the indices, the matrix $\langle u_r\cdot\lambda, u_s\rangle$ is of the form

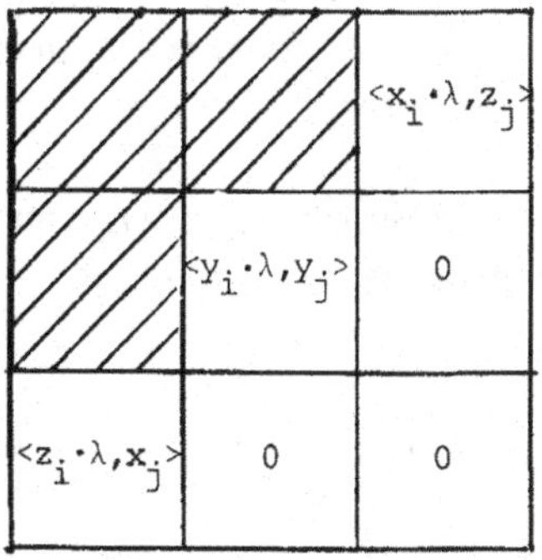

So $\mathrm{Pf}\langle u_r\cdot\lambda, u_s\rangle = \det\langle x_i\cdot\lambda, z_j\rangle \times \mathrm{Pf}\langle y_i\cdot\lambda, y_j\rangle$. The second factor is the value of $\mu_{\mathcal{O}_M}$ on the basis $y_i\cdot\lambda$ of $\mathfrak{m}\cdot\lambda$. The first factor involves the

natural pairing $\langle x, z\cdot\lambda\rangle$ between $\mathfrak{g}/\mathfrak{p}$ and $\mathfrak{p}\cdot\lambda = \mathfrak{p}^{\perp} = (\mathfrak{g}/\mathfrak{p})^*$. The point is, the form $\det\langle x_i\cdot\lambda, z_j\rangle$ on $(\mathfrak{g}/\mathfrak{p}) \times (\mathfrak{g}/\mathfrak{p})^*$ is independent of λ as well as of the choice of the complement of $\mathfrak{p}$ representing $\mathfrak{g}/\mathfrak{p}$; it may equally well be written as a product $\mu_{(\mathfrak{g}/\mathfrak{p})} \times \mu_{(\mathfrak{g}/\mathfrak{p})^*}$ of a form $\mu_{(\mathfrak{g}/\mathfrak{p})}$ on $\mathfrak{g}/\mathfrak{p}$ and a form $\mu_{(\mathfrak{g}/\mathfrak{p})^*}$ on $(\mathfrak{g}/\mathfrak{p})^*$, determined up to mutually reciprocal factors. In particular we may assume that $\mu_{(\mathfrak{g}/\mathfrak{p})}$ and $\mu_{(\mathfrak{g}/\mathfrak{p})^*}$ agree on $k/(k \cap m)$ and on $p^{\perp}$, respectively, with the forms coming from the measures dk on $K/(K \cap M)$ and $d\eta$ on $p^{\perp}$, whose normalization had not been specified so far.

Remarks. (a) In the lemma, K may be replaced by any smooth set of coset representatives for G/P. (b) The lemma generalizes in an obvious way to parabolically induced distributions (not necessarily characters) on $G_{\exp}$, defined as in (4),(5),(6).

In case π_M is square integrable modulo the centre of M (relative discrete series; not necessarily unitary on the centre) the induced representation π_G is called a standard representation of G. Such a π_M is parametrized by a character χ of a Cartan subgroup of M (compact modulo the centre of M), whose differential may be written as $i\lambda - \iota$, with ι half the sum of the roots α in $\mathfrak{m}$ satisfying $\langle\lambda, h_\alpha\rangle > 0$. Write $\pi_M = \pi_M(\chi,\lambda)$ and $\pi_G = \pi_G(\chi,\lambda)$. If λ is regular in $\mathfrak{g}^*$, then the previous lemma, together with the character formula (1) for discrete series, imply that $\pi_G(\chi,\lambda)$ corresponds to the contour $K\cdot(M\cdot\lambda + p^{\perp})$. In general, when λ is possibly singular in $\mathfrak{g}^*$, there is a unique complex orbit of maximal dimension with λ in its closure, denoted $\mathbb{G}\cdot\lambda_+$, $\lambda + p^{\perp}$ has a Zariski-dense intersection with $\mathbb{G}\cdot\lambda_+$, so $K\cdot(M\cdot\lambda + p^{\perp})$ may be regarded as a contour on $\mathbb{G}\cdot\lambda_+$. By the continuity properties of both $\pi_G(\chi,\lambda)$ and $K\cdot(M\cdot\lambda + p^{\perp})$ in λ, we get:

Corollary. The standard representation $\pi_G(\chi,\lambda)$ corresponds to the contour $K\cdot(M\cdot\lambda + p^{\perp})$ on the complex orbit $\mathbb{G}\cdot\lambda_+$ of maximal dimension with λ in its closure.

(See ROS for more details involving orbit structure.)

Recall that a character $\zeta: Z(\mathfrak{g}) \to \mathbb{C}$ of the centre of $U(\mathfrak{g})$ is parametrized by an element λ in the complex dual of a fixed Cartan subalgebra of $\mathfrak{g}$ (Harish-Chandra parameter; $\zeta = \zeta_\lambda$). λ is unique up to conjugation by the Weyl group, so the complex orbit $\Omega = \mathbb{G}\cdot\lambda_+$ of maximal dimension with λ in its closure is uniquely determined by ζ. In this way an irreducible admissible representation of G determines a complex orbit of maximal dimension through its infinitesismal character. ("Admissible" means K-finite, but not necessarily unitary.) Write Rep($\mathcal{R}$) for the additive group generated by the characters of irreducible, admissible representations with infinitesimal character corresponding to the orbit $\mathcal{R}$ (assumed regular). It is known that the standard representations in Rep($\mathcal{R}$) form a $\mathbb{Z}$- basis for Rep($\mathcal{R}$) (see VOG). The correspondence between standard characters and contours therefore extends to all admissible representations with regular infinitesimal character.

Summary: There is a homomorphism Rep($\mathcal{R}$) $\to H_m(\mathcal{R})$, $\Theta \to \Gamma$, given by

$$\langle\Theta,\phi\rangle = \int_\Gamma (J\phi)\mu_{\mathcal{R}}\,. \tag{8}$$

The map Rep($\mathcal{R}$) $\to H_m(\mathcal{R})$ need not be one-to-one, nor onto: characters Θ which agree in a neighborhood of the identity correspond to the same contour Γ; and only contours Γ satisfying an integrality condition (which guarantees that they come from globally defined distributions on G) can arise. It seems of some interest to me to clarify the precise nature of the correspondence Rep($\mathcal{R}$) $\to H_m(\mathcal{R})$, which presupposes determination of the homology group $H_m(\mathcal{R})$. The latter shall be done below.

3. Orbits and the flag manifold. There is a natural map of the set $\mathfrak{g}^*_{reg}$ of regular elements to the flag manifold $\mathcal{B}$ of Borel subalgebras of $\mathfrak{g}$. It is defined as follows. The stabilizer $\mathfrak{h}(\xi)$ of a regular element $\xi \in \mathfrak{g}^*$ is a Cartan subalgebra of $\mathfrak{g}$. Write $\Delta(\xi)$ for the roots α of $\mathfrak{h}(\xi)$ in $\mathfrak{g}$ satisfying

$$\mathrm{Re}\langle\xi,h_\alpha\rangle > 0, \text{ or } \mathrm{Re}\langle\xi,h_\alpha\rangle = 0 \text{ and } \mathrm{Im}\langle\xi,h_\alpha\rangle > 0.$$

$\Delta(\xi)$ is a system of positive roots, and we write $\mathfrak{b}(\xi)$ for the Borel subalgebra of $\mathfrak{g}$ containing $\mathfrak{h}(\xi)$ together with the root spaces of the roots in $\Delta(\xi)$. The resulting map $\mathfrak{g}^*_{reg} \to \mathcal{B}$, $\xi \to \mathfrak{b}(\xi)$, is evidently $\mathbb{G}$-equivariant. It is continuous (even holomorphic) within each chamber of $\mathfrak{g}^*_{reg}$ (by which is meant the $\mathbb{G}$-orbit of the chamber in the dual of a fixed Cartan subalgebra, where $\mathfrak{b}(\xi)$ = constant) but has discontinuities along the 'walls'. (These complex chambers are not open.)

If $\mathcal{R}$ is a regular $\mathbb{G}$-orbit in $\mathfrak{g}^*$, then the fibre of the restriction $\mathcal{R} \to \mathcal{B}$ above $\mathfrak{b}(\xi)$ is exactly $\mathbb{B}(\xi)\cdot\xi = \xi + \mathfrak{b}(\xi)^\perp$, $\mathbb{B}(\xi)$ being the Borel subgroup of $\mathbb{G}$ with Lie algebra $\mathfrak{b}(\xi)$. $\mathfrak{b}(\xi)^{\perp} = (\mathfrak{g}/\mathfrak{b}(\xi))^*$ is just the dual of the tangent space of $\mathcal{B}$ at $\mathfrak{b}(\xi)$. So the fibration $\mathcal{R} \to \mathcal{B}$ is essentially the (complex) cotangent bundle $\mathcal{B}^* \to \mathcal{B}$ over $\mathcal{B}$. To get an explicit isomorphism between these fibrations, fix a compact real form u of $\mathfrak{g}$ so that the conjugation τ of $\mathfrak{g}$ with respect to u leaves g stable. Given $\xi \in \mathcal{R}$, choose $\lambda \in \mathcal{R}$ whose stabilizer $\mathfrak{h}(\lambda)$ is τ-stable. Replacing λ by a conjugate under the compact form U of $\mathbb{G}$, we may arrange further that $\mathfrak{b}(\lambda) = \mathfrak{b}(\xi)$, because U acts transitively on $\mathcal{B}$. This condition now defines λ uniquely. Define

$$\mathcal{R} \to \mathcal{B}^*, \ \xi \to (\mathfrak{b}, \xi - \lambda) \tag{9}$$

by the conditions

$$\text{(a) } \mathfrak{b} = \mathfrak{b}(\xi) = \mathfrak{b}(\lambda)$$
$$\text{(b) } \tau\mathfrak{h}(\lambda) = \mathfrak{h}(\lambda).$$

This is the required isomorphism over $\mathcal{B}$. Note, however, that this map is *not* $\mathbb{G}$-equivariant with respect to the natural action of $\mathbb{G}$ on $\mathcal{B}^*$, but only U-equivariant.

Introduce the real subbundle $\mathcal{B}^{\#}$ of $\mathcal{B}^*$: $\mathcal{B}^{\#}$ consists of those $(\mathfrak{b}, \eta)$ in $\mathcal{B}^*$ for which $\eta \in \mathrm{Re}(\mathfrak{b}^{\perp}) = \mathfrak{b}^{\perp} \cap g^*$. This is a somewhat singular object: the fibre $\mathrm{Re}(\mathfrak{b}^{\perp})$ of $\mathcal{B}^{\#}$ over $\mathfrak{b}$ will generally change dimension as $\mathfrak{b}$ goes from one G-orbit in $\mathcal{B}$ to another; the part of $\mathcal{B}^{\#}$ over a fixed G-orbit $G\cdot\mathfrak{b}$, call it $(G\cdot\mathfrak{b})^{\#}$, is however a perfectly well behaved real vector bundle over $G\cdot\mathfrak{b}$. Its dimension is:

$$\begin{aligned}
\dim_{\mathbb{R}} (G\cdot\mathfrak{b})^{\#} &= \dim_{\mathbb{R}} (G\cdot\mathfrak{b}) + \dim_{\mathbb{R}} (\mathrm{Re}(\mathfrak{b}^{\perp})) \\
&= \dim_{\mathbb{R}} (g/g \cap \mathfrak{b}) + \dim_{\mathbb{R}} (g^* \cap \mathfrak{b}^{\perp}) \\
&= 2 \dim_{\mathbb{C}}(\mathfrak{g}) - \dim_{\mathbb{C}}(\mathfrak{b} \cap \bar{\mathfrak{b}}) - \dim_{\mathbb{C}}(\mathfrak{b} + \bar{\mathfrak{b}}) \\
&= 2 \dim_{\mathbb{C}}(\mathfrak{g}) - 2 \dim_{\mathbb{C}}(\mathfrak{b}) \\
&= \dim_{\mathbb{C}}(\mathcal{R}).
\end{aligned}$$

The decomposition $\mathcal{B}^{\#} = \cup (G\cdot\mathfrak{b}_k)^{\#}$ of $\mathcal{B}^{\#}$ into its parts over the various G-orbits in $\mathcal{B}$ therefore represents $\mathcal{B}^{\#}$ as a union of (not closed, but smooth) real submanifolds of $\mathcal{B}^*$, each of real dimension equal to the complex dimension of the regular orbit $\mathcal{R}$.

In analogy with $\mathcal{R}$ we define admissible singular chains on $\mathcal{B}^*$ and the homology $H_*(\mathcal{B}^*)$. Note that the growth conditions (a) and (b) of section 1 take place fiberwise in $\mathcal{B}^*$: for $(\mathfrak{b}, \eta) \in \mathcal{B}^*$, with $\eta \in \mathfrak{b}^{\perp}$, the norms $|\eta|$, $|\mathrm{Im}(\eta)|$ are taken in $\mathfrak{b}^{\perp} \subset \mathfrak{g}^*$. With this definition the map (9) evidently induces an isomorphism $H_*(\mathcal{R}) \simeq H_*(\mathcal{B}^*)$ for all regular orbits $\mathcal{R}$.

Write $H_*(\mathcal{B}^\#)$ for the subgroup of $H_*(\mathcal{B}^*)$ whose elements are representable by admissible chains which lie in $\mathcal{B}^\#$. (Since $\mathrm{Im}(\eta) \equiv 0$ on $\mathcal{B}^\#$, such chains are just tempered chains which lie in $\mathcal{B}^\#$, and it is clear that $H_*(\mathcal{B}^\#)$ is simply an ordinary singular relative homology of the fiberwise one-point compactification of $\mathcal{B}^\#$, "relative" with respect to the points at infinity. However, I shall not use this observation.) What is of interest here is the following fact.

Theorem. $H_*(\mathcal{B}^*) = H_*(\mathcal{B}^\#)$, i.e. every admissible singular chain is homologous to one in $\mathcal{B}^\#$. In dimension $m = \dim_{\mathbb{R}}(\mathcal{B})$, $H_*(\mathcal{B}^\#)$ is free abelian with $\mathbb{Z}$-dimension equal to the number of orbits of G in $\mathcal{B}$.

Proof. We show that $\mathcal{B}^*$ can be retracted to $\mathcal{B}^\#$ by an admissible homotopy. To see this, let $\mathcal{B}_k$ be the union of the G-orbits in $\mathcal{B}$ of dim $\leq k$, $\mathcal{B}_k^*(\mathcal{B}_k^\#)$ the part of $\mathcal{B}^*(\mathcal{B}^\#)$ over $\mathcal{B}_k$. This gives the filtration

$$\mathcal{B}^* = \mathcal{B}_m^* \supset \mathcal{B}_{m-1}^* \supset \quad \ldots \supset \mathcal{B}_1^* \supset \mathcal{B}_0^*$$

It suffices to show that each $\mathcal{B}_k^*$ can be admissibly retracted to $\mathcal{B}_k^\# \cup \mathcal{B}_{k-1}^*$. $\mathcal{B}_{k-1}^\#$ is a union of a finite number of smooth submanifolds of $\mathcal{B}^*$; it has an admissibly retractible neighborhood in $\mathcal{B}^*$. So it suffices to show that the part of $\mathcal{B}_k^*$ which lies outside of some given neighborhood of $\mathcal{B}_{k-1}^*$ is admissibly retractible to the part of $\mathcal{B}_k^\#$ which lies outside of this neighborhood. Further, since $\mathcal{B}_k - \mathcal{B}_{k-1}$ is a finite union of G-orbits of the same dimension, it suffices to show that the part of $\mathcal{B}^*$ over any given compact subset of a single G-orbit $G\cdot\mathfrak{b}_o$ can be admissibly retracted to $(G\cdot\mathfrak{b}_o)^\#$. That can be done explicitly as follows.

For each point $\mathfrak{b}$ on $G\cdot\mathfrak{b}_o$ choose a splitting of real vector spaces

$$\mathfrak{b}^\perp = \mathrm{Re}(\mathfrak{b}^\perp) \oplus V_{\mathfrak{b}}$$

depending continuously on $\mathfrak{b}$. Choose a real-valued function $a(\mathfrak{b},t)$ defined for $\mathfrak{b}$ in $G\cdot\mathfrak{b}_o$ and $0 \leq t \leq 1$ so that

$$a(\mathfrak{b},0) \equiv 1,$$
$$a(\mathfrak{b},1) \equiv 0, \text{ for } \mathfrak{b} \text{ in the given compact subset of } G\cdot\mathfrak{b}_o.$$

Write, for $(\mathfrak{b},\eta) \in (G\cdot\mathfrak{b}_o)^*$, $\eta = \alpha + \beta$ with $\alpha \in \mathrm{Re}(\mathfrak{b}^\perp)$ and $\beta \in V_{\mathfrak{b}}$.

Set $h_t(\mathfrak{b},\eta) = (\mathfrak{b},\alpha+a(\mathfrak{b},t)\beta)$. This is an admissible homotopy which retracts the part of $(G\cdot\mathfrak{b}_o)^*$ over the $\mathfrak{b}$ with $a(\mathfrak{b},1) = 0$ to $(G\cdot\mathfrak{b}_o)^\#$.

The second assertion follows from a standard argument in singular homology (see MAS, IV.2, for example). An explicit $\mathbb{Z}$-basis for $H_m(\mathcal{B}^\#)$ may be constructed as follows. Select for each G-orbit $G\cdot\mathfrak{b}_k$ in $\mathcal{B}$ an m-cycle γ_k which differs from $(G\cdot\mathfrak{b}_k)^\#$ by an m-chain in $\mathcal{B}^\#$ which lies over orbits of lower dimension. (The m-chain $(G\cdot\mathfrak{b}_k)^\#$ itself need not be closed.) These γ_k represent the required basis for $H_m(\mathcal{B}^\#)$.

4. Contours on $\mathcal{B}^\#$: examples. (a) Finite dimensional representations. Recall that the contour of a finite dimensional representation is an orbit $U\cdot\lambda$ under a compact real form U of $\mathbb{G}$ of an element λ in $\mathfrak{g}^*$ whose stabilizer in U is a maximal torus. The image of $U\cdot\lambda$ in $\mathcal{B}^*$ is exactly the zero section, which may be identified with $\mathcal{B}$ itself. The orientation on $\mathcal{B}$ must be chosen so that the corresponding orientation on $U\cdot\lambda$ becomes positive with respect to the canonical form (which is real valued on $U\cdot\lambda$)

(b) Positive principal series (minimal parabolic). Recall that the contour of a standard representation induced from a parabolic P = MN is $K\cdot(M\cdot\lambda + p^\perp)$ with λ in the dual of a Cartan subalgebra of $\mathfrak{m}$. If P is minimal ($M\cdot\lambda$ compact), this contour becomes $K\cdot(\lambda + p^\perp)$. Assume, in addition, that λ is positive, in the sense that $\mathfrak{b}(\lambda)$ is contained in the complexified Lie algebra $\mathfrak{p}$ of P. (We also assume implicitly that λ is regular in $\mathfrak{g}^*$.) The image of $\lambda + p^\perp$ in $\mathcal{B}^*$ is then exactly the real fibre $\mathrm{Re}(\mathfrak{b}^\perp) = p^\perp$ over $\mathfrak{b} = \mathfrak{b}(\lambda)$, and the image of $K\cdot(\lambda + p^\perp)$ is therefore exactly $(G\cdot\mathfrak{b})^\#$, the part of $\mathcal{B}^\#$ over the closed orbit $K\cdot\mathfrak{b} = G\cdot\mathfrak{b}$. The orientation on $(G\cdot\mathfrak{b})^\#$ must be chosen so that for real λ the corresponding orientation on $K\cdot(\lambda + p^\perp) = G\cdot\lambda$ becomes positive with respect to the canonical form.

(c) $SL(2,\mathbb{R})$. Let PS stand for "principal series" (induced from B_o = upper triangular), $DS_\pm$ for "discrete series" (± = holomorphic and anti-holomorphic), FD for "finite dimensional".

There are three orbits on the flag manifold $\mathcal{B} = \mathbb{P}\mathbb{C}^1$, represented by $\mathfrak{b}_\pm = \mathfrak{b}(\lambda_\pm)$ and $\mathfrak{b}_o = \mathfrak{b}(\lambda_o)$ where

$$\lambda_\pm = \pm\begin{pmatrix}0 & -1\\ 1 & 0\end{pmatrix}, \quad \lambda_o = \begin{pmatrix}1 & 0\\ o & -1\end{pmatrix},$$

regarded as elements of $\mathfrak{g}^* = \mathfrak{g}$.

We already know the contours for PS and FD in $B^{\#}$: they are, respectively, $(G\cdot\lambda_o)^{\#}$ and $B = (G\cdot\lambda_+)^{\#} + (G\cdot\lambda_-)^{\#}$, appropriately oriented. (Positivity is in this case irrelevant for PS, because the two chambers are conjugate under G.) It remains to find the contour for $DS_\pm$ in $B^{\#}$.

Realized in $\mathfrak{g}^*$, the contour of $DS_\pm$ is a real orbit $G\cdot(\ell\lambda_\pm)$ with $\ell = 1,2\ldots$. Its image in B^* lies over the G-orbit $G\mathfrak{b}_\pm$ in B. One sees from this that the contour of $DS_\pm$ in $B^{\#}$ must be of the form $(G\cdot\lambda_\pm)+\gamma_\pm$, where $\gamma_\pm$ lies over the closed orbit $G\cdot\mathfrak{b}_o$. The fibre of $B^{\#}$ over a point in $G\cdot\mathfrak{b}_o$ is $\simeq\mathbb{R}$, so that there are only two possibilities for $\gamma_\pm$: either the part of $(G\cdot\mathfrak{b}_o)^{\#}$ above or below the zero section; call these $(G\cdot\mathfrak{b}_o)^{\#}_\pm$. Which one it is depends on the asymptotic behaviour of $G\cdot(\ell\lambda_\pm)$ along $G\cdot\mathfrak{b}_o$. As a subset of B^*, $G\cdot(\ell\lambda_\pm)$ becomes

$$\{(g\cdot\mathfrak{b}_\pm, \pm i\ell(gc\cdot\lambda_o - u(gc)\cdot\lambda_o) \mid g \in G\}$$

where c is the Cayley transform

$$c = (1/\sqrt{2})\begin{pmatrix} 1 & -i \\ -i & 1 \end{pmatrix}$$

(which satisfies $\lambda_\pm = \pm ic\cdot\lambda_o$) and $u(gc)$ is the U-component of gc in the complex Iwasawa decomposition $\mathbb{G} = UN_oA_o$, $\mathbb{G} = SL(2,\mathbb{C})$, N_o = unit-upper triangular, A_o = positive real diagonal, $U = SU(2)$. We need to find the asymptotic behaviour of $gc\cdot\lambda_o$ as $g \to \infty$ in G. Since $G = KA_o^+K$ with $K = SO(2)$ and

$$A_o^+ = \{\begin{pmatrix} a & 0 \\ 0 & 1/a \end{pmatrix} \mid a > 1\}$$

it suffices to consider $g \in A_o^+$ (because $K = SO(2)$ fixes $c\cdot\lambda_o$). Thus take $g = \begin{pmatrix} a & 0 \\ 0 & 1/a \end{pmatrix}$ and write $g = una$ with $u \in U$, $n \in N_o$, $a \in A_o$. One finds that

$$n = \begin{pmatrix} 1 & b \\ 0 & 1 \end{pmatrix}, \quad b = (1/2i)(a^2 - a^{-2})$$

and $u(gc) \to 1$ as $a \to +\infty$. Thus asymptotically as $a \to +\infty$

$$\pm i\ell(gc\cdot\lambda_o - u(gc)\lambda_o) \sim \pm\ell(a^2 - a^{-2})\begin{pmatrix} 0 & 1 \\ 0 & 0 \end{pmatrix}.$$

It follows that along $G\cdot\mathfrak{b}_o$ the image of $G\cdot(\ell\lambda_\pm)$ in B^* is asymptotic to

$$\{(k\cdot\mathfrak{b}_o, \pm\ell(a^2 - a^{-2})k.\begin{pmatrix} 0 & 1 \\ 0 & 0 \end{pmatrix}) \mid k \in K, a > 1\};$$

this is exactly what we called $(G\cdot\mathfrak{b}_o)^{\#}_{\pm}$.

To summarize, we have the following correspondences :

$$\begin{aligned} PS &= (G\cdot\mathfrak{b}_o)^{\#} \\ FD &= -(G.\mathfrak{b}_+)^{\#} - (G\cdot\mathfrak{b}_-)^{\#} \\ DS_{\pm} &= (G\cdot\mathfrak{b}_{\pm})^{\#} + (G\cdot\mathfrak{b}_o)^{\#}_{\pm}. \end{aligned} \tag{10}$$

(The minus signs for FD have been inserted to suggest the correct orientation in relation to the other contours.) Note that we have the following identity in $H_m(\mathcal{B}^{\#})$:

$$PS = DS_+ + DS_- + FD. \tag{11}$$

This gives the correct decomposition formula for PS for those values of the parameters for which all contours correspond to characters. And that depends on the complex orbit Ω in question, while the contour identity (11) does not. The situation is illustrated schematically in Fig.1.

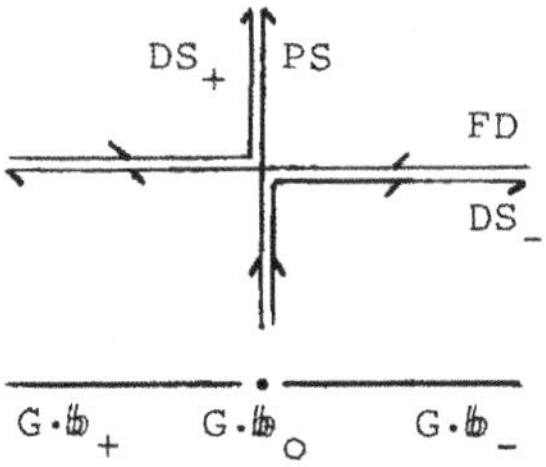

Fig.1

(d) SL(2,$\mathbb{C}$). Let $PS_{\pm}$ stand for "principal series" (induced from B_o = upper triangular in $G = SL(2,\mathbb{C})$) with $\pm$ indicating the two non-G-conjugate chambers in (the dual of) the complexification $\mathfrak{h} = \mathbb{C}\times\mathbb{C}$ of the Cartan subalgebra h of diagonal matrices in g.

The complexification $\mathfrak{g}$ of g is $g\times g$ with conjugation $\overline{(x,y)} = (\bar{y},\bar{x})$, where $\bar{x},\bar{y}$ are the usual conjugates in $s\ell(2,\mathbb{C})$. The flag manifold $\mathcal{B}$ consists of pairs (b_1, b_2) of Borel subalgebras b_1, b_2 of g. There are two orbits of G in $\mathcal{B}$, represented by $\mathfrak{b}_+ = (b_o, b_o)$ and $\mathfrak{b}_- = (wb_o, b_o)$ where b_o = upper triangular and

$$w = \begin{pmatrix} 0 & -1 \\ 1 & 0 \end{pmatrix}.$$

We have $\mathfrak{b}_{\pm} = \mathfrak{b}(\lambda_{\pm})$ where $\lambda_+ = (\lambda_o, \lambda_o)$ and $\lambda_- = (wi\lambda_o, i\lambda_o)$ with

$\lambda_o = \begin{pmatrix} 1 & 0 \\ 0 & -1 \end{pmatrix}$. $PS_\pm$ correspond, respectively, to parameters $\lambda \in \mathfrak{h}^*$ with $\mathfrak{b}(\lambda) = \mathfrak{b}_\pm$. Their contours, when realized in $B^\#$, depend only on the chamber (as is clear from considerations of analyticity in λ); so it suffices to consider $\lambda = \lambda_\pm$.

The contour of the positive principal series PS_+ in $B^\#$ is $(G\cdot\mathfrak{b}_+)^\#$, the part of $B^\#$ over the closed orbit, as we know. The contour of the negative principal series PS_- is the real orbit $G\cdot\lambda_-$ in $\mathfrak{g}^*$ (for the choice $\lambda = \lambda_-$ of parameter). Its image in B^* lies over the open orbit $G\cdot\mathfrak{b}_-$ in B. It follows that the contour of PS_- in $B^\#$ must be of the form $(G\cdot\mathfrak{b}_-)^\# + \gamma$, where γ lies over the closed orbit $G\cdot\mathfrak{b}_+$. The fibre of $B^\#$ over a point in $G\cdot\mathfrak{b}_+$ is $\mathbb{R}^2$, and there are only two possibilities for γ: either γ is all of $(G\cdot\mathfrak{b}_+)^\#$ or it is empty. (In this case $(G\cdot\mathfrak{b}_-)^\#$ *is* a closed chain, as is generally the case when G is complex.) Which one it is depends again on the asymptotics of $G\cdot\lambda_-$ along $G\cdot\mathfrak{b}_+$. As a subset of B^*, $G\cdot\lambda_-$ becomes

$$\{(g\cdot\mathfrak{b}_-,\ i(gw\cdot\lambda_o - k(gw)\cdot\lambda_o,\ \bar{g}\cdot\lambda_o - k(\bar{g})\cdot\lambda_o)) \mid g \in G\}$$

where $k(g)$ is the K-component in the Iwasawa decomposition $G = KN_oA_o$ ($G = SL(2,\mathbb{C})$, $K = SU(2)$, N_o = unit-upper triangular, A_o = positive real diagonal.) We need to find the asymptotics of $gw\cdot\lambda_o - k(gw)\cdot\lambda_o$ as $g\to\infty$ in G. Since $G = KN_oA_o$ and A_o fixes λ_o it suffices to take $g \in N_o$, say $g = \begin{pmatrix} 1 & z \\ 0 & 1 \end{pmatrix}$ with $z \in \mathbb{C}$. Write $gw = kna$ according to $G = KN_oA_o$. From the identity

$$\begin{pmatrix} 1 & b/a \\ 0 & 1 \end{pmatrix}\begin{pmatrix} 0 & -1 \\ 1 & 0 \end{pmatrix} = \begin{pmatrix} b & -\bar{a} \\ a & \bar{b} \end{pmatrix}\begin{pmatrix} 1 & -\bar{b}/a \\ 0 & 1 \end{pmatrix}\begin{pmatrix} 1/a & 0 \\ 0 & a \end{pmatrix},$$

valid for $a\bar{a} + b\bar{b} = 1$, one finds that $n = \begin{pmatrix} 1 & -\bar{z} \\ 0 & 1 \end{pmatrix}$. This gives

$$gw\cdot\lambda_o - k(gw)\cdot\lambda_o = k\cdot\begin{pmatrix} 0 & -2\bar{z} \\ 0 & 0 \end{pmatrix}.$$

It follows that along $G\cdot\mathfrak{b}_+$ the image of $G\cdot\mathfrak{b}_-$ in B^* is asymptotic to

$$\{(k\cdot\mathfrak{b}_+,\ k\cdot\begin{pmatrix} 0 & -2i\bar{z} \\ 0 & 0 \end{pmatrix},\ \bar{k}\cdot\begin{pmatrix} 0 & 2iz \\ 0 & 0 \end{pmatrix}) \mid k \in K\}.$$

This is exactly $(G\cdot\mathfrak{b}_+)^\#$.

To summarize, we have the following correspondences:

$$PS_+ = (G\cdot\mathfrak{b}_+)^\#,\quad PS_- = (G\cdot\mathfrak{b}_-)^\# + (G\cdot\mathfrak{b}_+)^\#,\quad FD = -(G\cdot\mathfrak{b}_-)^\#.$$

This time the identity in $H_m(\mathcal{B}^{\#})$ is:

$$PS_+ = PS_- + FD \ .$$

Its interpretation is analogous to the case G = SL(2,ℝ). The schematic diagram is indicated in Fig.2.

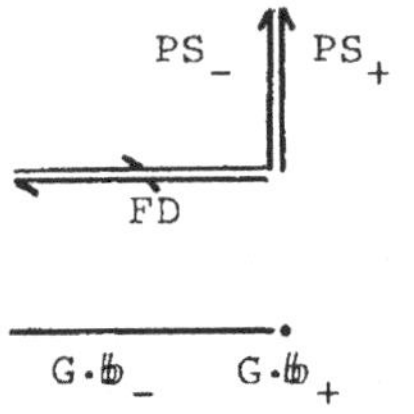

Fig.2

<u>References</u>

(DUF) M.Duflo, Fundamental series representations of a semisimple Lie group, Functional Analysis Appl.4 (1970),122-126.

(KIR) A.A. Kirillov, Characters of unitary representations of Lie groups, Functional Analysis Appl., 2.2 (1967), 40-55.

(ROS) W.Rossmann, Tempered representations and orbits, Duke Math. J. 49 (1982), 215-229.

(VER) M. Vergne,A Poisson-Plancherel formula for semi-simple Lie groups, Annals of Math. 115 (1982), 639-666.

(VOG) D.A.Vogan, Jr., Representations of Real Reductive Lie Groups, Birkhauser, Boston, 1981.

(MAS) W.S.Massey, Singular Homology Theory, Springer Verlag, New York, 1980.

ANALYTICITY OF SOLUTIONS OF PARTIAL DIFFERENTIAL EQUATIONS ON NILPOTENT LIE GROUPS

Linda Preiss Rothschild*
University of Wisconsin, Madison

.. Introduction. A differential operator

$$\text{(1.1)} \qquad P = \sum_{|\alpha| \leq m} a_\alpha(x) D_x^\alpha \qquad D_x^\alpha = \left(\frac{1}{i}\frac{\partial}{\partial x_1}\right)^{\alpha_1}\left(\frac{1}{i}\frac{\partial}{\partial x_2}\right)^{\alpha_2} \cdots \left(\frac{1}{i}\frac{\partial}{\partial x_N}\right)^{\alpha_N}$$

ith $a_\alpha(x)$ real analytic is called analytic hypoelliptic in an open set U if u = f with f analytic in an open subset $V \subset U$ implies u must also be nalytic in V. We survey here some conditions for analytic hypoellipticity when is a left invariant differential operator on a nilpotent Lie group.

For constant coefficient differential operators, analytic hypoellipticity is quivalent to ellipticity (see e.g. [7]). Variable coefficient elliptic differential perators are always analytic hypoelliptic, but the converse is false. Here we will e concerned with nonelliptic variable coefficient operators.

. Homogeneous operators. Now let G be a connected, simply connected nilpotent ie group whose Lie algebra $\mathfrak{g}$ is stratified i.e. $\mathfrak{g} = \mathfrak{g}_1 + \mathfrak{g}_2 + \cdots + \mathfrak{g}_r$, vector pace direct sum with $[\mathfrak{g}_i, \mathfrak{g}_j] \subset \mathfrak{g}_{i+j}$ if $i + j \leq r$, $[\mathfrak{g}_i, \mathfrak{g}_j] = 0$ if $i + j > r$. e assume that $\mathfrak{g}_1$ generates $\mathfrak{g}$. Then $\mathfrak{g}$ carries a natural family of dila- ions which are automorphisms: $\delta_t|\mathfrak{g}_i = t^i$. The dilations may be transferred to via the exponential map, and also extend to the universal enveloping algebra $(\mathfrak{g})$. Thus we may write $\mathfrak{U}(\mathfrak{g}) = \sum_{j=1}^{\infty} \mathfrak{U}_j(\mathfrak{g})$, where each element of $\mathfrak{U}_j$ is homo- eneous of degree j under δ_t.

. Smoothness of solutions. The notion of C^∞ hypoellipticity is defined as for nalyticity, but with real analytic replaced by C^∞. For a homogeneous operator $\in \mathfrak{U}(\mathfrak{g})$ necessary and sufficient conditions for C^∞ hypoellipticity were esta- lished by Helffer and Nourrigat [6], who proved the following conjecture of Rock- and [15]. Let $\hat{G}$ be the set of irreducible unitary representations of G. For

Partially supported by an NSF grant.

$\pi \in \hat{G}$ acting on $L^2(\mathbb{R}^k)$, we denote also by π the corresponding mapping of $\mathfrak{U}(\mathfrak{g})$ into the space of differential operators on $L^2(\mathbb{R}^k)$. Then L is C^∞ hypoelliptic if and only if $\pi(L)$ is injective for all nontrivial $\pi \in \hat{G}$.

4. Nonanalytic hypoelliptic operators. The existence of C^∞ hypoelliptic but not analytic hypoelliptic operators of second order on 2 step groups was suggested by the following example of Baouendi-Goulaouic [1]:

In $\mathbb{R}^{n+2}$ the operator

$$P = \sum_{j=1}^{n} \frac{\partial^2}{\partial x_j^2} + \frac{x_j^2 \partial^2}{\partial y^2} + \frac{\partial^2}{\partial x_{n+1}^2} \tag{4.1}$$

is not analytic hypoelliptic. (It is C^∞ hypoelliptic by a general theorem of Hörmander [8]). P is not a left invariant operator on any group, but it is closely related to

$$L = \sum_{j=1}^{n} \frac{\partial^2}{\partial x_j^2} + \left(\frac{\partial}{\partial t_j} + x_j \frac{\partial}{\partial y}\right)^2 + \frac{\partial^2}{\partial x_{n+1}^2} . \tag{4.2}$$

L is of the form

$$L = \sum_{j=1}^{n} (U_j^2 + V_j^2) + W^2 , \tag{4.3}$$

which is in $\mathfrak{U}^2(\mathfrak{g})$ for $\mathfrak{g} = \mathfrak{h}_{2n+1} \oplus \mathbb{R}$, where $\mathfrak{h}_{2n+1}$ is the $2n + 1$ dimensional Heisenberg algebra. Here $\{U_j, V_j, W, j = 1,2,\ldots,n\}$ is a basis of $\mathfrak{g}_1$, and $\dim \mathfrak{g}_2 = 1$. Now it is easy to see that L cannot be analytic hypoelliptic if P is not. Indeed, if Pu vanishes in an open set in $\mathbb{R}^{n+2}$ then Lu vanishes in an open set in $\mathbb{R}^{2n+2}$. If L were analytic hypoelliptic, then u would have to be analytic. Similar reasoning shows that if $\tilde{\mathfrak{g}}$ is any 2-step nilpotent Lie algebra having a quotient algebra of the form $\mathfrak{h}_{2n+1} \oplus \mathbb{R}$, then the operator L pulls back to $\tilde{L} \in \mathfrak{U}(\tilde{\mathfrak{g}})$, where $\tilde{L}$ is C^∞ but not analytic hypoelliptic.

5. H-groups. One is therefore led to consider 2-step algebras which do not have quotients of the form $\mathfrak{h}_{2n+1} \oplus \mathbb{R}$. These may be characterized as follows. For $\eta \in \mathfrak{g}_2^* \setminus \{0\}$, let $\mathfrak{g}_\eta = \mathfrak{g}/I_\eta$, where $I_\eta = \{Y \in \mathfrak{g}_2 : \eta(Y) = 0\}$. Now let B_η

be the bilinear form on $\mathfrak{g}_1^*$ defined by

$$B_\eta : (X_1, X_2) \longmapsto \eta([X_1, X_2]) .$$

If $[\mathfrak{g}_1, \mathfrak{g}_2] = \mathfrak{g}_2$, then $\det B_\eta \neq 0$, all $\eta \in \mathfrak{g}_2^* - \{0\}$, if and only if none of the quotients $\tilde{\mathfrak{g}}_\eta$ is the Lie algebra direct sum of a Heisenberg algebra with a Euclidean space. In this case the corresponding group G is called a H-group.

6. Analytic regularity of $\Box_b$. Further motivation for positive results on analytic hypoellipticity on H-groups came from the results on analytic hypoellipticity of the boundary Laplacian operator $\Box_b$ on strongly pseudo convex domains. C^∞ regularity for $\Box_b$ had been established much earlier through the work of J. J. Kohn [11], but it was not until the mid '70's that analytic regularity was proved by Trèves [18] and Tartakoff [17]. Their methods were completely different. Tartakoff begins with the well known characterization of analyticity in terms of L^2 norms: a distribution u is analytic near a point x_0 if there is a neighborhood U of x_0 and a constant $C > 0$ such that

$$\| D_x^\alpha u \|_{L^2(U)} \leq C^{|\alpha|+1} |\alpha| ! \tag{6.1}$$

for all multi-indices α. His proof is elementary in the sense that he uses only L^2 estimates with integration by parts.

Trèves methods are microlocal i.e. he works in conic sets in the cotangent space. It is well known that analyticity can be "microlocalized" [9]; a distribution u is analytic near a point $(x_0, \xi_0) \in T^*(U)\setminus 0$ if there is an open cone Γ in $\mathbb{R}^m$ containing ξ_0 and a constant $C > 0$ such that for every integer $N = 0,1,\ldots$ one can find a function $\phi_N \in C_0^\infty(U)$, $\phi_N = 1$ in V, a neighborhood of x_0, $\phi_N = 0$ outside a fixed compact subset K of U such that

$$|(\varphi_N u)\hat{}\,(\xi)| \leq C^{N+1} N! (1+|\xi|)^{-N} \tag{6.2}$$

for all $\xi \in \Gamma$. It can be shown (see e.g. [19, Chapter V]) that u is analytic in a neighborhood of x_0 if and only if it is analytic at (x_0, ξ_0) all $\xi_0 \in \mathbb{R}^n\setminus\{0\}$. The complement in $T^*(U)\setminus 0$ of $\{(x_0, \xi_0) : u$ is analytic at $x_0, \xi_0\}$

is called the <u>wave front set</u> of u, denoted $WF_a u$.

7. <u>Symplectic characteristic varieties</u>. Now recall that the symbol of a differential operator P defined by (1.1) is

$$\sigma(P) = \sum_{|\alpha|=m} a_\alpha(x)\xi^\alpha \qquad \xi^\alpha = \xi_1^{\alpha_1}\xi_2^{\alpha_2}\cdots\xi_n^{\alpha_n},$$

and the characteristic variety Σ of P is $\{(x,\xi) : \sigma(P) = 0,\ \xi \neq 0\}$. Now if $Pu = f$, then $WF_a u \subset \Sigma$ (see e.g. [19]). Hence it suffices to study P microlocally near Σ. For $\Box_b$, the characteristic variety Σ carries a symplectic structure. That is, Σ is given locally by $\Sigma = \{(x,\xi) : v_i(x,\xi) = 0,$ $i = 1,2,\ldots,2k\}$ where the v_i are C^∞ functions such that

$$v_i, v_j \rightarrow \{v_i, v_j\} \quad \text{is a nondegenerate symplectic form,}$$

with

$$\{p,q\} = \sum_{i=1}^{n} \left(\frac{\partial p}{\partial x_i}\frac{\partial q}{\partial \xi_i} - \frac{\partial p}{\partial \xi_i}\frac{\partial q}{\partial x_i}\right).$$

8. <u>Analytic regularity of homogeneous operators on groups</u>. Now let $\mathfrak{g}$ be a stratified Lie algebra as before and $X_1,\ldots,X_n$ a basis of $\mathfrak{g}_1$, and suppose that $L = \sum_{i=1}^{n} X_i^2$. Then the characteristic variety Σ of L is $\{(x,\xi):\sigma(X_i) = 0,\ i = 1,2,\ldots,n,\ \xi \neq 0\}$. Since $\{\sigma(X_j), \sigma(X_k)\} = \sigma([X_j,X_k])$ one easily checks that Σ has a symplectic structure if and only if $\mathfrak{g}$ is of step 2 and G is an H-group.

(8.1) <u>Theorem</u> (Métivier [13]). <u>Suppose</u> G <u>is an H-group and</u> $L \in \mathfrak{U}_m(\mathfrak{g})$. <u>Then</u> L <u>is analytic hypoelliptic if and only if</u> $\pi(L)$ <u>is injective for all nontrivial</u> $\pi \in \hat{G}$ <u>if and only if</u> L <u>is</u> C^∞ <u>hypoelliptic</u>.

Métivier's proof shows that the fundamental solution for L constructed by representations is analytic away from the identity. His methods involve a considerable simplification and extension of those of Trèves [18].

Unfortunately, the question of analytic hypoellipticity is completely resolved only in the case where G is an H-group. It is reasonable to conjecture that if

G is not an H-group then there is no $L \in \mathfrak{U}_m(\mathfrak{g})$ which is analytic hypoelliptic. Partial results on necessary conditions for analytic hypoellipticity have been obtained by Métivier [12] and Helffer [5].

(8.2) Theorem (Helffer [5]). If $\mathfrak{g}$ is a 2-step Lie algebra and $L \in \mathfrak{U}_2(\mathfrak{g})$, then L is not analytic hypoelliptic if G is not an H-group.

9. Non-homogeneous operators. We restrict here to the case where G is a H-group. Our result, which is contained in a recent joint paper with Grigis [3], applies to operators $L \in \mathfrak{U}(\mathfrak{g})$ having the property that $\pi(L) \neq 0$ for all non-trivial one dimensional representations $\pi \in \hat{G}$. Such operators will be called transversally elliptic. Another way of describing these operators is by noting that they are elliptic polynomials in the elements of $\mathfrak{g}_1$.

The elements of $\hat{G}$ are parametrized by $\eta \in \mathfrak{g}_2^*\setminus\{0\}$. We now replace L by $\pi_\eta L$ and study the family of differential operators $\pi_\eta(L)$ as η varies. Now we introduce spherical coordinates $\eta = (\rho,\omega)$ on $\mathfrak{g}_2^* - \{0\}$, and write $L = L_m + L_{m-1} + \ldots + L_0$, with $L_j \in \mathfrak{U}_j(\mathfrak{g})$. Then π_η may be defined so that $\pi_\eta(L_j)$ is homogeneous in η i.e. $\pi_\eta(L_j) = |\eta|^{j/2}\pi_{(1,\omega)}(L_j)$ (see [13]). Then

$$\pi_\eta(L) = |\eta|^{m/2}(\pi_{(1,\omega)}(L_m) + |\eta|^{-1/2}\,\pi_{(1,\omega)}(L_{m-1}) + \ldots + |\eta|^{-m/2}\pi_{(1,\omega)}(L_0)). \tag{9.1}$$

Now let $\lambda = |\eta|^{-1/2}$ and define the operator $A(\lambda,\omega)$ by

$$A(\lambda,\omega) = |\eta|^{-m/2}\pi_\eta(L)\ . \tag{9.2}$$

One can prove that $(\lambda,\omega) \longmapsto A(\lambda,\omega)$ is an analytic family of unbounded operators in the sense of Kato-Rellich [10]. Furthermore, the spectrum of each $\pi_\eta(L)$ is discrete and consists of eigenvalues. For ω_0 fixed, let K_{ω_0} be the multiplicity of 0 as an eigenvalue of $\pi_{(1,\omega_0)}$. Then analytic perturbation theory shows that for $|\lambda|$ small and ω close to ω_0 the product $d(\eta)$ of the K_{ω_0} smallest eigenvalues of $A(\lambda,\omega)$ is analytic and can be expanded

$$d(\eta) = \lambda^{K_{\omega_0}}\,(a_0(\omega) + a_1(\omega_0)\lambda + a_2(\omega)\lambda^2 + \ldots\ )\ . \tag{9.3}$$

In the language of pseudodifferential operators, $d(\eta)$ is a semi-classical analytic symbol on $\mathbb{R}^{n_2}$ which is elliptic near (y_0, η_0) if and only if $a_0(\omega_0) \neq 0$.

Our criterion for analytic hypoellipticity may be stated as follows.

(9.4) Theorem (Grigis-Rothschild [3]). Let G be n H-group, and $L \in U(\mathfrak{g})$ transversally elliptic. Then L is analytic hypoelliptic if and only if for any $\eta \in \mathfrak{g}_2^* - \{0\}$, the product $d(\eta)$ of the small eigenvalues of $\pi_\eta(L^* L)$, given by (9.3), is an elliptic symbol i.e. $a_0(\omega_0) \neq 0$.

In the special case where G is a Heisenberg group, the theorem takes a simpler form.

Corollary. If G is a Heisenberg group, then L is analytic hypoelliptic if and only if $\ker L \cap L^2(G) = \phi$.

To see how the corollary follows from Theorem (8.4), we note that for the Heisenberg group $\mathfrak{g}_2^* - \{0\} = \mathbb{R} - \{0\}$ and hence $d(\eta)$ is elliptic if and only if it is not identically zero. On the other hand, if $d(\eta) \equiv 0$, then one can find a non-zero $f \in L^2(G)$ with $Lf \equiv 0$.

(9.5) Example. Let $\mathfrak{g} = \mathfrak{g}_1 + \mathfrak{g}_2$ be the 3-dimensional Heisenberg algebra with $\{X_1, X_2\}$ a basis of $\mathfrak{g}_1$. Then for any $\alpha, \beta \in \mathbb{C}$ with $\beta \neq 0$, the operator

$$L = X_1^2 + X_2^2 + i\alpha[X_1, X_2] + \beta X_1$$

is analytic hypoelliptic. To prove this, one need only check that $\pi_\eta(L)$ has no zero eigenvalue for $|\eta|$ large.

The proof of Theorem (9.4) borrows heavily from techniques of Sjöstrand [16] and those of Métivier [14]. In [16] the question of C^∞ hypoellipticity for a class of transversally elliptic operators more general than ours is reduced to that of determining the C^∞ hypoellipticity of a pseudodifferential operator in fewer variables. In order to carry out this construction in the analytic category, we use the analytic pseudodifferential operators and approximate inverses constructed by Métivier [14].

References

[1] M. S. Baouendi and C. Goulaouic, "Nonanalytic hypoellipticity for some degenerate elliptic operators," Bull. A.M.S. 78 (1972), 483-486.

[2] L. Boutet de Monvel and P. Kree, "Pseudodifferential operators and Gevrey classes," Ann. Inst. Fourier, Grenoble 17 1 (1967), 295-323.

[3] A. Grigis and L. P. Rothschild, "A criterion for analytic hypoellipticity of of a class of differential operators with polynomial coefficients," Ann. of Math. (to appear).

[4] V. V. Grusin, "On a class of hypoelliptic operators," Mat. Sb. 83 (1970) 456-473 [Math. U.S.S.R. Sb. 12 (1972) 458-476].

[5] B. Helffer, "Conditions nécessaires d'hypoanalyticité pour des opérateurs invariants à gauche homogènes sur un groupe nilpotent gradué," J. Diff. Eq. 44 (1982), 460-581.

[6] ________ and J. Nourrigat, "Caracterisation des operateurs hypoelliptiques homogenes invariants a gauche sur un groupe nilpotent gradue," Comm. P.D.E. 4 (1979), 899-958.

[7] L. Hörmander, Linear partial differential equations, Springer-Verlag, Heidelberg-New York (1969).

[8] ________, "Hypoelliptic second order differential operators", Acta Math. 119 (1967), 147-171.

[9] ________, "Uniqueness theorems and wave front sets," Comm. Pure Appl. Math., 24 (1971), 671-704.

[10] T. Kato, "Perturbation theory of linear operators," 2nd edition, Springer-Verlag, Berlin-Heidelberg-New York (1980).

[11] J. J. Kohn, "Boundaries of complex manifolds," Proc. Conf. on Complex Manifolds, Minneapolis (1964), 81-94.

[12] G. Métivier, "Hypoellipticité analytique sur des groupes nilpotents de rang 2," Duke Math. J. 17 (1980).

[13] ________, "Analytic hypoellipticity for operators with multiple characteristics," Comm. P.D.E. 6 (1) (1981), 1-90.

[14] ________, "Une classe d'opérateurs non-hypoelliptiques analytiques," Seminaire Goulaouic-Schwartz, Ecole Polytechnique (1979).

[15] C. Rockland, "Hypoellipticity on the Heisenberg group," Trans. A. M. S. 240 (1978) no. 517, 1-52.

[16] J. Sjöstrand, "Parametrices for pseudodifferential operators with multiple characteristics," Ark. for Mat. 12 (1974), 85-130.

[17] D. Tartakoff, "The analytic hypoellipticity of $\bar{\partial}_b$ and related operators on nondegenerate C-R manifolds," Acta Math. 145 (3-4) (1980), 177-203.

[18] F. Trèves, "Analytic hypoellipticity of a class of pseudodifferential operators," Comm. P.D.E. 3 (1978), 475-642.

[19] ________, Introduction to pseudodifferential operators and Fourier integral operators, Vol. 1, The University Series in Mathematics, Plenum Press, New York, (1980).

ASYMPTOTIC PROPERTIES OF EIGENVALUES AND EIGENFUNCTIONS OF INVARIANT DIFFERENTIAL OPERATORS ON SYMMETRIC AND LOCALLY SYMMETRIC SPACES

V. S. Varadarajan*
University of California
Los Angeles, California

§0. Introduction

In these lectures I shall be concerned with asymptotic properties of eigenvalues and eigenfunctions of invariant differential operators on symmetric and locally symmetric spaces of negative curvature. The questions that I shall consider are natural and are intimately related to representation theory of and harmonic analysis on semisimple Lie groups; moreover they are a part of the general problem of the spectra of differential operators on manifolds and their relationship to the geometry of the manifold. Due to my own limitations it has not been possible to be really exhaustive in this survey, and I have discussed only those parts of the subject with which I am most familiar. For other results, points of view and the extensive bibliography in this area, the reader should consult [DKV 1,2], [Wa], [S], [Hej], and to the relevant parts of [Y].

The most famous of the asymptotic laws governing such spectra are those of H. Weyl and T. Carleman (see [W]). Here X is a plane domain with piecewise smooth boundary ∂X and $\Delta = -(\partial^2/\partial x^2 + \partial^2/\partial y^2)$. The eigenvalue problem is posed with suitable boundary conditions, for instance, of Dirichlet type:

$$\Delta\varphi = \lambda\varphi, \qquad \varphi = 0 \text{ on } \partial X.$$

Let

---

*Research supported in part by NSF Grant MCS82-00639.

$$0 \leq \lambda_1 \leq \lambda_2 \leq \cdots \leq \lambda_n \leq \cdots$$

be the eigenvalues of this problem, the so-called problem of the vibrating membrane. Put

$$N(\lambda) = \sum_{\lambda_n \leq \lambda} 1.$$

Then Weyl's theorem is

$$N(\lambda) \sim \frac{\text{area }(X)}{4\pi} \lambda. \qquad (\lambda \to \infty).$$

If

$$\varphi_1, \varphi_2, \ldots, \varphi_n, \ldots$$

are the corresponding orthonormal sequence of eigenfunctions, let us define

$$K(x,y:\lambda) = \sum_{\lambda_n \leq \lambda} \varphi_n(x)\overline{\varphi_n(y)}.$$

Then Carleman proved that

$$K(x,y:\lambda) = \begin{cases} \frac{\lambda}{4\pi} + o(\lambda) & x = y \\ o(\lambda) & x \neq y. \end{cases}$$

These problems were reviewed in the late 1940's by Minakshisundaram and Pleijel who considered the case of $\Delta = -$Laplacian on an arbitrary smooth compact Riemannian manifold. Their work was based on a study of the regularity properties of the operator Δ^{-s} and the behaviour, as $t \to 0+$, of the (Schwartz) kernel of the operator $e^{-t\Delta}$; among other things, it led to the estimate, as $\lambda \to \infty$,

$$\sum_{\lambda_n \leq \lambda} \varphi_n(x)\overline{\varphi_n(x)} = (2\sqrt{\pi})^{-n}\Gamma(\tfrac{n}{2}+1)^{-1}\lambda^{n/2} + o(\lambda^{n/2})$$

uniformly for $x \in X$. Integration over X then gives

$$\sum_{\lambda_n \leq \lambda} 1 = (2\sqrt{\pi})^{-n}\Gamma(\tfrac{n}{2}+1)^{-1}\text{vol}(X)\lambda^{n/2} + o(\lambda^{n/2}).$$

Since then there has been a great deal of interest in the eigenvalue problem of the Laplacian on Riemannian manifolds, not only in the spectral asymptotics, but in the question of how much of the geometry of X is determined by the spectrum of the Laplacian (not only on functions as treated above, but also on forms). The reader who wants to know about these developments should start with I. M. Singer's survey [S] in the Vancouver Congress, as well as the references cited there.

The spectral asymptotics of higher order elliptic differential operators were considered by Gårding. Let X be a second countable smooth manifold, P a smooth elliptic differential operator of order m on X which is formally positive, i.e., for some $c > 0$ $(Pu,u) \geq c(u,u)$ for all $u \in C_c^\infty(X)$, the scalar product being with respect to a fixed positive C^∞ density dx. Let $\bar{P}$ be a self-adjoint extension of P with spectrum bounded below by a positive number; such extensions exist by a classical theorem of Friedrichs. Let $(\bar{E}_\lambda)$ be the resolution of the identity of $\bar{P}$. Then $\bar{E}_\lambda$ has a C^∞ kernel $K(x,y:\lambda)$, and Gårding's results give (see [Hö] for references to Gårding's work and other related articles), for $\lambda \to \infty$,

$$K(x,x:\lambda) = (2\pi)^{-n}\left(\int_{p(x,\xi)<1} d\xi_x\right)\cdot\lambda^{n/m} + o(\lambda^{n/m})$$

$$K(x,y:\lambda) = o(\lambda^{n/m}) \qquad (x \neq y).$$

Here $n = \dim(X)$, m = order of P, p is the principal symbol of P, and $d\xi_x$ is the Lebesgue measure on the cotangent space to X at x dual to dx.

A common feature of all of this work is the indirect nature of the study of the spectrum. The basic estimates are obtained first for suitable functions of P: the resolvent $((P-z)^{-1})$, the zeta function (P^{-s}) or the Laplace transform (e^{-sP}). The estimates for the spectrum are then obtained by Tauberian arguments. This seems to be the reason that the error terms showed no improvement at all over those of

Weyl-Carleman. It was Hörmander who took the beautiful and decisive step of considering the Fourier Transform method for this problem; see [Hö] where references are given to Levitan's earlier work on the second order operators based on the cosine transform. The idea of Hörmander was to study the operator e^{-itQ} where $Q = \bar{P}^{-1/m}$ and is defined by the spectral theorem; however Q will now only be a pseudo-differential operator. The mth root is necessary to make $i\frac{\partial}{\partial t} - Q$ hyperbolic. In the special case of compact X, if μ is the spectral measure of Q, i.e., the counting measure on $\mathbb{R}$ defined by the spectrum, then μ is tempered, and its Fourier transform is just the generalized function $t \mapsto tr(e^{-itQ})$. Hörmander's analysis is essentially based on a study of the operator e^{-itQ} as a Fourier integral operator, especially its distribution kernel for $|t| << 1$. The Tauberian arguments are replaced by the simpler and yet more powerful Fourier inversion theorem which gives a direct method of going from $\hat{\mu}$ to μ. Hörmander proved that

$$K(x,x:\lambda) = (2\pi)^{-n}\left(\int_{p(x,\xi)<1} d\xi_x\right)\lambda^{n/m} + O\left(\lambda^{\frac{(n-1)}{m}}\right)$$

as $\lambda \to \infty$, uniformly for x varying in compact subsets of X. The error term cannot be improved for Laplacians on spheres, as was noted by Avakumovič (see [Hö], §6).

The spectral asymptotics arise as a consequence of the analysis of the singularity at $t = 0$ of $\hat{\mu}(t)$, X being compact. The question naturally arises as to the meaning of the other singularities. This was studied by Duistermaat-Guillemin [D-G], and by Chazarain [Ch1] for Laplacians. Let $T^*(X)$ be the cotangent bundle of X and $S^*(X)$ be the subset of $T^*(X)$ defined by $S^*(X) = \{(x,\xi) | q(x,\xi) = 1\}$ where $q = p^{1/m}$; for the Laplacian $S^*(X)$ is the unit sphere bundle in $T^*(X)$. The Hamiltonian flow (ϕ^t) defined by q (with vector field $H_q: (x,\xi) \mapsto \sum_j (\frac{\partial q}{\partial \xi_j}\frac{\partial}{\partial x_j} - \frac{\partial q}{\partial x_j}\frac{\partial}{\partial \xi_j})$) leaves $S^*(X)$ invariant and its trajectories in $S^*(X)$ are the so-called bicharacteristics. We denote by L the set of real numbers which are periods of the periodic bicharacteristics; thus, if $T \in \mathbb{R}$, $T \in L$ if and only if for some $(x,\xi) \in S^*(X)$, $\Phi^T(x,\xi) = (x,\xi)$. For the Laplacian it is clear that the (Φ^t)-flow is the geodesic flow and so L is the set of periods of the periodic geodesics in the unit cosphere bundle. Let us also recall that the singular support of a distribution ψ on $\mathbb{R}$, sing supp(ψ), is the complement of the largest open set on which ψ is a C^∞ function, Then the basic result is that

$$\text{sing supp}(\hat{\mu}) \subset L.$$

For the Laplacian arising from a generic† Riemannian metric we have

$$\text{sing supp}(\hat{\mu}) = L.$$

Thus in the generic case the spectrum of the Laplacian determines the length spectrum of the manifold, a result of Colin de Verdiere (see [Ch. 2]). Furthermore, in the case of the generic Laplacian, the error term for the asymptotic distribution of the eigenvalues can be improved from $O(\lambda^{\frac{n-1}{2}})$ to $o(\lambda^{\frac{n-1}{2}})$, as was proved in [D-G].

It is clearly of interest to supplement this general theory by special examples where one can go beyond these general results and provide greater detail to the picture. The class of examples of the form $X = \Gamma\backslash G/K$ where G is a connected semisimple Lie group with finite center, K is a maximal compact subgroup of G, and Γ is a discrete subgroup of G such that $\Gamma\backslash G$ has finite volume, is especially interesting. We may assume (by passing to a subgroup of finite index in Γ, see [B]) that Γ acts freely on G/K so that X is smooth. Among these are included the classical arithmetically defined manifolds of the quotients of the Poincaré half-plane by the modular group and its congruence subgroups, as well as by the quaternion groups, and the

†This means that for each periodic geodesic γ, the linear Poincaré map does not have 1 as an eigenvalue, and the lengths of any two distinct (up to orientation) periodic geodesics are different; see [Ch 2].

quotient of hyperbolic space by $SL(2,\mathbb{Z}[i])$ and its congruence subgroups. I shall consider only the case when X is compact on this occasion. Of course there are always discrete subgroups $\Gamma \subset G$ such that $\Gamma\backslash G$ is compact and Γ acts freely on G/K; this is also a result of Borel [B]; for instance, when $G = SL(2,\mathbb{R})$, we may take Γ as a quaternion group.

The space $S = G/K$ is a Riemannian symmetric space of noncompact type and G is its group of motions. In this case it is natural to consider not just the Laplacian on $X = \Gamma\backslash S$ but the full algebra $\mathbb{D}(G/K)$ of <u>all</u> G-invariant differential operators on S; these may be viewed as differential operators on X also. Instead of the eigenvalue problem for the Laplacian we now have the <u>simultaneous eigenvalue problem for all the elements of</u> $\mathbb{D}(G/K)$. The eigenvalues are then homomorphisms of $\mathbb{D}(G/K)$ into $\mathbb{C}$ and so are the points of the spectrum of the algebra $\mathbb{D}(G/K)$. This spectrum is in fact an affine algebraic variety V, and the spectrum of X, say $\Lambda(X)$, is a subset of it. The main problem we would be interested is then the determination of the asymptotic nature of $\Lambda(X)$ as a subset of the spectrum of $\mathbb{D}(G/K)$, as well as the extent to which $\Lambda(X)$ determines the geometry of X. The suggestion for studying these multidimensional variants of the classical problems comes from Selberg [Sel] and Gel'fand [G], the latter formulating some conjectures on the asymptotic structure of $\Lambda(X)$. These conjectures were made precise and proved in [DKV1]. It turns out that the essential point is a comparison of the eigenvalue problem and spectral asymptotics on X with those on S. The L^2-eigenvalue problem on S for the algebra $\mathbb{D}(G/K)$ has a continuous spectrum which coincides with the set $V(\mathbb{R})$ of real points of V; and by Harish-Chandra's theory one knows that the corresponding eigenfunction expansion is of multiplicity one and the Plancherel measure is a continuous density of at most polynomial growth on $V(\mathbb{R})$. The fundamental asymptotic result is that $\Lambda(X)\setminus V(\mathbb{R})$ is in a certain sense negligible and that the spectral measure of X on $\Lambda(X) \cap V(\mathbb{R})$ is asymptotic to the Plancherel measure on $V(\mathbb{R})$. This is the generalization of Weyl's asymptotic law to the present case.

The error estimates obtained in [DKV 1] for this asymptotic problem are of the same nature as the $O(\lambda^{\frac{n-1}{m}})$ estimate in Hörmander's theorem. However when $rk(S) = 1$ they can be improved to $O(\lambda^{\frac{n-1}{m}}/\log \lambda)$, and so the question naturally arises whether they can be improved when $rk(S) > 1$ also. This requires new results on the asymptotic properties of the eigenfunctions on S, regarded as functions on $\mathrm{Spec}(\mathbb{D}(G/K)) \times S$. I shall discuss two recently developed methods for this problem. The first is treated in [DKV 2] and its basic idea is to view the Harish-Chandra integral representation of these eigenfunctions as an oscillatory integral on the flag variety of G, and thus to analyse them by a suitable variant of the method of stationary phase. The second, developed by Erik van den Ban [Ba] in his thesis, is a generalization of the classical method of using different contours to study the asymptotics of special functions, and is based on replacing the cycle K that occurs in Harish-Chandra's integral representation by suitable cycles in the complexification of K. Although these questions arise out of spectral asymptotics, the methods and results are very much wider in scope and should be regarded as a part of the general approach to harmonic analysis on Lie groups via methods of complex integration and microlocal analysis.

§1. Preliminaries

We begin with a Gel'fand pair (G,K) where G is a Lie group with biinvariant Haar measure dx, K a compact subgroup such that, for some involutive automorphism θ of G and a closed connected abelian subgroup A we have

$$G = KAK, \qquad a^{\theta} = a^{-1} \quad (a \in A), \qquad k^{\theta} = k \quad (k \in K).$$

Let $\mathbb{D}(G/K)$ be the algebra of G-invariant differential operators on G/K. Then $\mathbb{D}(G/K)$ is abelian; more generally, the convolution algebra of compactly supported K biinvariant distributions on G is also abelian. Let $\mathfrak{g}$, $\mathfrak{k}$, $\mathfrak{a}$ be the Lie algebras of G, K, A respectively, and $U(\mathfrak{g}_C)$ the universal enveloping algebra of the complexification $\mathfrak{g}_C$ of $\mathfrak{g}$. Then there is a natural homomorphism

$$U(\mathfrak{g}_C)^K \to \mathbb{D}(G/K)$$

which is surjective with kernel

$$U(\mathfrak{g}_C)^K \cap (U(\mathfrak{g}_C)\mathfrak{k}) = U(\mathfrak{g}_C)^K \cap (\mathfrak{k}U(\mathfrak{g}_C)).$$

If $\mathfrak{s}$ is any Ad(K)-invariant complementary subspace of $\mathfrak{k}$ in $\mathfrak{g}$ and $\underline{\underline{\lambda}}$ is the symmetrizer map of the symmetric algebra $S(\mathfrak{s}_C)$ into $U(\mathfrak{g}_C)$, the above homomorphism restricts to a linear isomorphism on $\underline{\underline{\lambda}}(S(\mathfrak{s}_C)^K)$:

$$\underline{\underline{\lambda}}(S(\mathfrak{s}_C)^K \simeq \mathbb{D}(G/K).$$

Let $\Gamma \subset G$ be a discrete subgroup such that $\Gamma\backslash G$ is compact and Γ has no torsion, i.e., the identity element e is the only element of Γ of finite order. Then

$$X = \Gamma\backslash G/K$$

is compact and smooth. Elements of $\mathbb{D}(G/K)$ may be regarded as differential operators on X also. Let S = G/K. We define the (unrestricted) spectrum $\Lambda(S)$ of S as the set of all homomorphisms

$$\chi\colon \mathbb{D}(G/K) \to \mathbb{C}$$

having the property that there is a nonzero eigenfunction in $C^\infty(G/\!/K)$ with χ as its eigenhomomorphism. The spectrum $\Lambda(X)$ of X is defined in the same way. Then $\Lambda(X) \subset \Lambda(S)$; $\Lambda(X)$ is nonempty since for example the constant function 1 is an eigenfunction for all members of $\mathbb{D}(G/K)$; and for each $\chi \in \Lambda(X)$ the corresponding eigensubspace has a finite dimension $m(\chi)$. On G/K the situation is simpler; the eigenspaces corresponding to $\chi \in \Lambda(S)$ are one dimensional. For each $\chi \in \Lambda(S)$ we write φ_χ for the unique eigenfunction in $C^\infty(G/\!/K)$ with eigenvalue χ normalized by the condition $\varphi_\chi(e) = 1$.

The self-adjoint elements of $\mathbb{D}(G/K)$ form a commutative algebra (over $\mathbb{R}$) and so have a good L^2-spectral theory. This can be studied via an appropriate transform theory. For any $f \in L^1(G/\!/K)$ we define its transform $\hat{f}$ to be the function on a suitable subset of $\Lambda(S)$ given by $\hat{f}(\chi) = \langle f,\varphi_\chi \rangle = \int_G f(x)\varphi_\chi(x)\,dx$. Now the function φ_χ is not always positive definite and so we introduce the set $\Lambda^+(S)$ of $\chi \in \Lambda(S)$ for which φ_χ is positive definite. From general spectral theory one gets a positive σ-finite measure $d\omega$ on $\Lambda^+(S)$ (which is a standard Borel space) such that for all $g \in C_c(G/K)$

$$\int_G |g(x)|^2 dx = \int_{\Lambda^+(S)} |\hat{g}(\chi)|^2 d\omega(\chi).$$

There is also an accompanying inversion formula, valid for all $g \in C_c(G/\!/K) * C_c(G/\!/K)$ ($*$ = convolution),

$$g(x) = \int_{\Lambda^+(S)} \hat{g}(\chi)\overline{\varphi_\chi(x)}\, d\omega(\chi) \qquad (x \in G).$$

The measure $d\omega$ (the <u>Plancherel measure</u>) is in general supported on a certain naturally defined subset $\Lambda_p(S)$ which we shall now introduce. We note that $L^1(G/\!/K)$ acts on $L^2(G/\!/K)$ via convolution from the left and the image is an abelian algebra of bounded operators of $L^2(G/\!/K)$ closed under taking adjoints; if $\mathfrak{A}$ is the closure of this algebra in the norm topology, there is a natural map of the Gel'fand spectrum of $\mathfrak{A}$ into $\Lambda^+(S)$. This map is a Borel isomorphism onto a Borel subset of $\Lambda^+(S)$ which is $\Lambda_p(S)$ by definition. We write $\Lambda_c(S) = \Lambda^+(S)\setminus\Lambda_p(S)$ and call $\Lambda_p(S)$ (resp. $\Lambda_c(S)$) the <u>principal</u> (resp. <u>complementary</u>) <u>L^2-spectrum</u> of S. Writing $\Lambda_p(X) = \Lambda(X) \cap \Lambda_p(S)$ and $\Lambda_c(X) = \Lambda(X) \cap \Lambda_c(S)$ we call $\Lambda_p(X)$ (resp. $\Lambda_c(X)$) the <u>principal</u> (resp. <u>complementary</u>) <u>spectrum</u> of X (cf. [DKV1]).

It is easy to see that $\mathbb{D}(G/K)$ is finitely generated and so its spectrum (in the sense of commutative algebra) is a complex affine algebraic variety. $\Lambda^{+}(S)$ is imbedded in this variety and the basic problems that arise are (a) to determine at least the "asymptotic" nature of the imbedding $\Lambda_p(X) \hookrightarrow \Lambda_p(S)$; (b) to say something about $\Lambda_c(X)$; (c) to try to determine which aspects of the geometry of X are controlled by $\Lambda(X)$. As I shall explain a little later, these problems can be studied in considerable depth when G is semisimple and K its maximal compact subgroup.

As in all investigations of this kind the starting point is Selberg's trace formula. Let $C(\Gamma)$ be the set of conjugacy classes of the group Γ. For any $\gamma \in \Gamma$ let Γ_γ (resp. G_γ) be its centralizer in Γ (resp. G). Then one knows that the conjugacy class $c_G(\gamma)$ of γ in G is closed and also that $\Gamma_\gamma \backslash G_\gamma$ is compact. The distribution

$$f \mapsto \mathrm{vol}(\Gamma_\gamma \backslash G_\gamma) \int_{G_\gamma \backslash G} f(x^{-1}\gamma x)\, d\bar{x} \qquad (f \in C_c^\infty(G))$$

is then well-defined; if we choose a Haar measure dx_γ in G_γ and take the invariant measure $d\bar{x}$ on $G_\gamma \backslash G$ defined by $d\bar{x}dx_\gamma = dx$, then the above distribution is independent of the choice of dx_γ. It moreover depends on γ only through the Γ-conjugacy class $c = c_\Gamma(\gamma)$ of γ. We denote it by J_c. The trace formula is then the relation

$$\text{(T.F)} \qquad \sum_{\chi \in \Lambda(X)} m(\chi)\hat{f}(\chi) = \mathrm{vol}(\Gamma \backslash G) f(e) + \sum_{\{e\} \neq c \in C(\Gamma)} J_c(f)$$

valid for all $f \in C_c^\infty(G/K)$.

Although a complete analysis of the above formula would require a deep understanding of the distributions J_c, it is interesting to note that in the first approximation they may be neglected. In fact one can easily prove the existence of an open neighborhood U of e in G such that $U = U^{-1} = KUK$ and U does not meet any $c \in C(\Gamma)\backslash\{e\}$. For such a U we have the truncated trace formula:

$$\text{(T.T.F)} \qquad \sum_{\chi \in \Lambda(X)} m(\chi)\hat{f}(\chi) = \mathrm{vol}(\Gamma \backslash G) f(e) \qquad (f \in C_c^\infty(U))$$

One can (roughly) use this truncated version in the following way. If σ is the spectral measure of X (which has the mass $m(\chi)$ at the point χ) and if we write $f(e)$ using the inversion formula as

$$f(e) = \int_{\Lambda^+(S)} \hat{f}(\chi)\,d\omega(\chi)$$

then we have

$$<\sigma - \mathrm{vol}(\Gamma\backslash G)\omega, \hat{f}> = 0$$

for all $f \in C_c^\infty(U)$. If we now suppose that behaviour of a distribution near e is governed by the behaviour at infinity of its Fourier Transform, the above relation suggests the asymptotic relation

$$\sigma \sim \mathrm{vol}(\Gamma\backslash G)\omega.$$

The rigorous proof of such a result would depend on the construction of $f \in C_c^\infty(U)$ such that $\hat{f}$ is a "good" approximation to the characteristic functions of suitable sets $\Omega \subset \Lambda^+(S)$; for such sets Ω, $\sigma(\Omega)$ would be $\approx \mathrm{vol}(\Gamma\backslash G)\int_\Omega d\omega$. It is at this point that the idea of considering the simultaneous eigenvalue problem for the whole of $\mathbb{D}(G/K)$ becomes significant. It forces one to study the transforms $\hat{f}$ for all the functions $f \in C_c^\infty(U)$ rather than work with special classes of functions. As we shall see presently, in the semisimple case one can do this because of the availability of a definitive transform theory for such groups. Moreover, in this case one can even express the $J_c(f)$ in terms of $\hat{f}$, i.e., determine the "Fourier transforms" of the J_c. The trace formula then takes a neater form and may be viewed as a "Poisson formula." I hope that the treatment of these questions in the semisimple case will make it clear to the reader how the transition from single operators to $\mathbb{D}(G/K)$ has introduced considerable simplicity and naturalness.

From now on G will be a connected semisimple Lie group with finite center; we shall choose and fix a maximal compact subgroup K; θ, the Cartan involution that fixes K elementwise, and $G = KAN$ an

Iwasawa decomposition. We write $\mathfrak{g}$, $\mathfrak{k}$, $\mathfrak{a}$, $\mathfrak{n}$ for the Lie algebras of G, K, A, N respectively. We also have $\mathfrak{g} = \mathfrak{k} \oplus \mathfrak{s}$ where $\mathfrak{s}$ is the orthogonal complement of $\mathfrak{k}$ with respect to the Killing form $\langle \cdot , \cdot \rangle$. The map $\exp(\mathfrak{a} \to A)$ is an isomorphism and we write $\log(A \to \mathfrak{a})$ for its inverse. F is the complex dual of $\mathfrak{a}$, F_R the $\mathbb{R}$-linear subspace of F of elements taking only real values on $\mathfrak{a}$, and $F_I = (-1)^{1/2} F_R$. For any $\lambda \in F$ let $\lambda_R \in F_R$, $\lambda_I \in F_I$ be such that $\lambda = \lambda_R + \lambda_I$. W is the Weyl group of $(\mathfrak{g}, \mathfrak{a})$; it acts naturally on F, F_R, F_I. The Killing form on $\mathfrak{a}_c \times \mathfrak{a}_c$ gets transferred to a symmetric nonsingular complex bilinear form $\langle \cdot , \cdot \rangle$ on $F \times F$, and the Hermitian form $(\cdot , \cdot): \lambda, \lambda' \mapsto \langle \lambda, \lambda'^{\mathrm{conj}} \rangle$ is a positive definite scalar product; we shall write $\|\cdot\|$ for the corresponding norm.

In this special case the general discussion of the preceding paragraphs can be enriched with a considerable amount of detail.

<u>$\mathbb{D}(G/K))$, $\Lambda(S)$, $\Lambda^+(S)$</u>: We have the canonical Harish-Chandra isomorphism of $\mathbb{D}(G/K)$ with the algebra of W-invariant polynomials on F (cf. [H 1]). We shall thus view F as the spectrum of $\mathbb{D}(G/K)$, with the proviso that points in the same W-orbit define the same homomorphism $\mathbb{D}(G/K) \to \mathbb{C}$. For $\lambda \in F$ we denote by χ_λ the corresponding homomorphism. Moreover in this case every homomorphism of $\mathbb{D}(G/K)$ into $\mathbb{C}$ admits a corresponding eigenfunction in $C^\infty(G/\!\!/K)$ so that we may take $\Lambda(S) = F$. For χ_λ, this is the elementary spherical function $\varphi_\lambda = \varphi(\lambda:\cdot) \in C^\infty(G/\!\!/K)$ with the integral representation

$$\varphi_\lambda(x) = \varphi(\lambda:x) = \int_K e^{(\lambda-\rho)(H(xk))} dk \qquad (x \in G);$$

Here $\rho \in F_R$ is defined by $\rho(H) = \frac{1}{2} \operatorname{tr}(\operatorname{ad} H)_{\mathfrak{n}}$, $H(x) = \log h(x)$, and $h(x \mapsto h(x))$ is the projection $G \to A$ defined by the decomposition $G=KAN$ ("Iwasawa projection"). We have $\varphi(\lambda:e) = 1$ so that $\varphi_\lambda = \varphi_{\chi_\lambda}$ in the earlier notation. The φ_λ for $\lambda \in F_I$ are the spherical matrix elements of the unitary principal series representations of G and so are positive definite. There are however others. Let F^+ be the set of $\lambda \in F$ for which φ_λ is positive definite. For instance $\varphi_\rho = 1$ is one such. More generally we have

$$\varphi_\lambda \text{ positive definite} \Rightarrow \begin{cases} \text{(i)} & -\lambda^{\text{conj}} \in W\cdot\lambda \\ \text{(ii)} & \|\lambda_R\| \le \|\rho\|. \end{cases}$$

The second is a consequence of the boundedness of φ_λ and follows from the well known result of Helgason-Johnson [H-J], while the first follows from the identity

$$\varphi_{-\lambda^{\text{conj}}} = \varphi_\lambda$$

valid for positive definite φ_λ. For $s \in W$ let

$$F(s) = \{\lambda \in F \mid s\lambda = -\lambda^{\text{conj}}\}.$$

Then the above remarks imply

$$F(1) = F_I \subset F^+ \subset \bigcup_{s\in W} F(s), \qquad \rho \in \Lambda(X).$$

We now define $\Lambda = \Lambda(X)$ to be the set of $\lambda \in F$ such that there is a nonzero eigenfunction on X with eigenhomomorphism χ_λ. Since we are working with F rather than $W\backslash F$ it is natural to define the multiplicity function by

$$m(\lambda) = |W\cdot\lambda|^{-1}\cdot m(\chi_\lambda) \qquad (\lambda \in \Lambda)$$

where $m(\chi_\lambda)$ has the earlier significance. The measure σ (W-invariant) with masses $m(\lambda)$ at the points $\lambda \in \Lambda$ is the <u>spectral measure</u> of X.

<u>L^2-spectral theory of $\mathbb{D}(G/K)$</u> = <u>theory of Harish-Chandra transform. Plancherel measure, inversion formulae, and the Paley-Wiener theorem for $C_c^\infty(G/\!/K)$</u>. We begin with the normalization of Haar measures. For <u>any</u> choice of a Haar measure da on A we define the Haar measure dx on G by the formula $dx = e^{2\rho(\log a)}dkdadn$ where dk is the normalized Haar measure for K and dn is the Haar measure on N that goes over, via the Cartan involution, to the Haar measure $d\bar{n}$ on $\bar{N} = \theta(N)$

normalized by $\int_{\bar{N}} e^{-2\rho(H(\bar{n}))} d\bar{n} = 1$; we shall call dx **standard** if da is the Haar measure corresponding to the Euclidean measure on $\mathfrak{a}$ induced by the Killing form. The basic map is the **Abel transform**

$$A : C_c^\infty(G/K) \to C_c^\infty(A)^W$$

given by

$$(Af)(a) = e^{\rho(\log a)} \int_N f(an) dn \qquad (a \in A).$$

The **Harish-Chandra transform** Hf of f is then the Fourier Transform of Af, and is an entire function on F of the Paley-Wiener class:

$$(Hf)(\lambda) = \int_A (Af)(a) e^{\lambda(\log a)} da = \int_G f(x) \varphi(\lambda:x) dx \qquad (\lambda \in F).$$

Let $\mathbb{P}(F)$ denote the Paley-Wiener space on F, defined as the Fourier Transform of $C_c^\infty(A)$. We then have the commutative diagram

$$
\begin{array}{ccc}
C_c^\infty(G/K) & \xrightarrow{\;\;H\;\;} & \mathbb{P}(F)^W \\
& A \searrow \quad \nearrow & \text{Fourier Transform } (= \wedge) \\
& C_c^\infty(A)^W &
\end{array}
$$

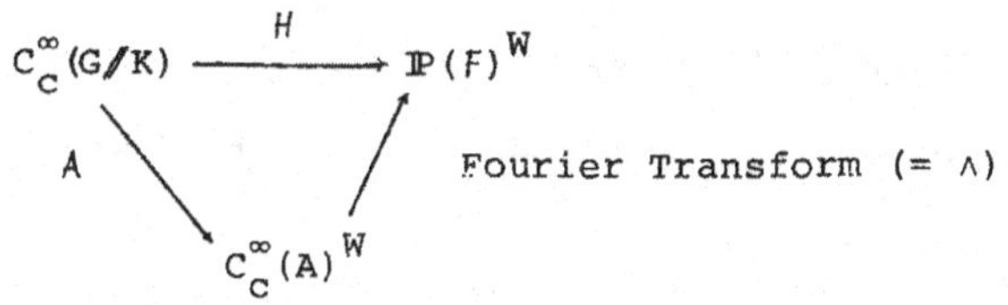

It follows from Harish-Chandra's theory in conjunction with theorems of Gangolli and Helgason ([H1], [H2], [Gal] [He]) that **all maps in** the above diagram are isomorphisms, and in fact that if $b > 0$ and $A(b)$ is the (W-invariant) subset of A of all $a \in A$ with $\|\log a\| \leq b$, than

$$g \in C_c^\infty(A(b))^W \Rightarrow A^{-1}g \in C_c^\infty(KA(b)K).$$

We also need the inversion formula of Harish-Chandra:

$$f(e) = |W|^{-1} \int_{F_I} (Hf)(\nu) \beta(\nu) d\nu = |W|^{-1} < \hat{\beta}, Af >$$

for all $f \in C_c^\infty(G/K)$; here $d\nu$ is the Lebesgue measure on F_I dual to da in the sense of Fourier analysis, and β, the Plancherel density, is W-invariant, smooth, ≥ 0, and is of at most polynomial growth on F_I; and each derivative of β is also of at most polynomial growth on F_I. In particular, we have the Plancherel formula for the eigenfunction expansion with respect to the algebra $\mathbb{D}(G/K)$ in the form

$$\int_G |f(x)|^2 dx = |W|^{-1} \int_{F_I} |Hf(\nu)|^2 \beta(\nu) d\nu$$

for all $f \in C_c^\infty(G/\!/K)$.

The function β can be computed explicitly and so its asymptotic properties can be determined rather quickly. For our purposes we need the following more detailed information regarding β which follows without difficulty from the Gindikin-Karpelevič product formula for β (cf. [DKV 1] for instance):

(i) Let $n = \dim(G/K)$, $r = \mathrm{rk}(G/K)$. Then $\beta(\nu) \leq \mathrm{const.}(1+\|\nu\|)^{n-r}$ $(\nu \in F_I)$

(ii) $\beta(\nu) = 0 \Leftrightarrow \langle\nu,\alpha\rangle = 0$ for some $\alpha \in \Delta^+$

(iii) Let Φ be a nonempty subset of the set of short positive roots. For any short positive root α let $d(\alpha) = \dim(\mathfrak{g}_\alpha) + \dim(\mathfrak{g}_{2\alpha})$ and let $d(\Phi) = \sum_{\alpha\in\Phi} d(\alpha)$. Then, if $T(\Phi)$ is the subspace of all $\nu \in F_I$ such that $\langle\nu,\alpha\rangle = 0$ for all $\alpha \in \Phi$,

$$\beta(\nu) \leq \mathrm{const}(1+\|\nu\|)^{n-r-d(\Phi)} \qquad (\nu \in T(\Phi)).$$

The principal L^2-spectrum for $S = G/K$ is thus F_I whereas Λ is certainly not $\subset F_I$; for instance $\rho \in \Lambda$. It is now natural to define <u>the principal spectrum of</u> X (Λ_p) and the <u>complementary spectrum of</u> X (Λ_p) by

$$\Lambda_p = \Lambda \cap F_I, \qquad \Lambda_c = \Lambda \backslash \Lambda_p.$$

It is clear that Λ_p and Λ_c are both stable under W.

Since

$$\Lambda_c \subset \bigcup_{s\neq 1} F(s)$$

it follows from Chevalley's well known theorem on finite reflexion groups ([V1], Lemma 4.15.15) that for any $\lambda \in \Lambda_c$, its imaginary part λ_I lies on one or more root hyperplanes. Since β is zero precisely on these hyperplanes, the complementary spectrum is related to the degeneracies of the Plancherel measure.

§2. The Poisson Formula for X and its relation to the manifolds of periodic geodesics of X

Since the theorems of Harish-Chandra and Gangolli-Helgason determine the image of $C_c^\infty(G/\!/K)$ under the Harish-Chandra transform, the Selberg trace formula can be refined to yield an analogue for X of the classical Poisson (summation) Formula. Before describing it ([DKV 1], Theorem 5.1) let us first agree on certain normalizations of Haar measures. For G we choose the standard Haar measure dx. Suppose $h \in G$ is a semisimple element. Then there are always elements $y \in G$ conjugate to h and in standard position: this means that the centralizer $\mathfrak{g}_y$ of y in $\mathfrak{g}$ is θ-stable. Then the centralizer G_y of y in G is also θ-stable and we have the notion of the standard Haar measure in G_y. Since y is determined uniquely up to conjugacy by K, this means that we can transfer this Haar measure on G_y to a Haar measure on G_h without any ambiguity. We call this the standard Haar measure on G_h and denote it by dx_h. The standard invariant measure on $G_h\backslash G$ is the measure $d\bar{x}$ such that $d\bar{x}dx_h = dx$. The distribution I_h on G is defined by $I_h(f) = \int_{G_h\backslash G} f(x^{-1}hx)\,d\bar{x}$ ($f \in C_c^\infty(G)$). We now restrict I_h to $C_c^\infty(G/\!/K)$ and use the isomorphism

$$A\colon C_c^\infty(G/\!/K) \cong C_c^\infty(A)^W$$

to define a unique W-invariant distribution T_h on A such that

$$\langle T_h, g\rangle = I_h(A^{-1}g) \qquad (g \in C_c^\infty(A)^W),$$

T_h is the Fourier transform of the Harish-Chandra transform of I_h (or rather the restriction of I_h to the spherical functions).

<u>Poisson Formula</u>. For $c \in C(\Gamma)$ define

$$v_c = \mathrm{vol}(\Gamma_\gamma \backslash G_\gamma), \qquad T_c = T_\gamma \qquad (\gamma \in c)$$

where we use the standard Haar measure on G_γ to calculate v_c. We must also recall that all elements of Γ are semisimple. The Selberg formula then becomes the following identity of distributions on A:

$$\text{(P.F)} \qquad \sum_{\lambda \in \Lambda} m(\lambda) e^{\lambda \circ \log} = \mathrm{vol}(\Gamma \backslash G)|W|^{-1}\hat{\beta} + \sum_{\{e\} \neq c \in C(\Gamma)} v_c T_c .$$

The series $\sum_\lambda m(\lambda)|\hat{g}(\lambda)| < \infty$ for each $g \in C_c^\infty(A)$; moreover, the Paley-Wiener theory mentioned in §1 implies that $\{\mathrm{supp}(T_c)\}$ is a locally finite collection, so that the right side is a finite sum for any $g \in C_c^\infty(A)$. The truncated Poisson Formula is then the following: there is a constant $b > 0$ such that if $A(b) = \{a \in A \mid \|\log a\| < b$, then

$$\text{(T.P.F)} \qquad \sum_{\lambda \in \Lambda} m(\lambda) e^{\lambda \circ \log} = \mathrm{vol}(\Gamma \backslash G)|W|^{-1}\hat{\beta} \quad \text{on } A(b).$$

<u>Properties of T_c: relationship to period geodesics, supports, singularities, and expressions in special cases ([DKV 1], §5)</u>. We regard S as a (complete) Riemannian manifold in the G-invariant metric induced by the Killing form. Let TS (resp. TX) be the tangent bundle of S (resp. X) and $p(S \to X)$ the projection; let $\dot{p}(TS \to TX)$ be its differential. As S is simply connected, the map that assigns to any $\gamma \in \Gamma$ the p-images of curves from y to $\gamma \cdot y$ $(y \in S)$, induces a bijection p from the set $C(\Gamma)$ to the set of <u>free</u> homotopy classes of closed curves in X. Making essential use of the semisimplicity of the elements of Γ one can prove that each $p(c)$ contains periodic geodesics (of period 1, by convention). Regarding the periodic geodesics as curves in TX we obtain in this way a subset $F(c) \subset TX$. It can be proved that $F(c)$ is a compact smooth submanifold of TX for

each c and that they are mutually disjoint as c varies in $C(\Gamma)$. Fix $c \in C(\Gamma)$. Then $F(c)$ is isomorphic to $\Gamma_\gamma \backslash G_\gamma / U_\gamma$ where $\gamma \in c$ and U_γ is a maximal compact subgroup of G_γ; all closed geodesics in $F(c)$ have the same length, say $\ell(c)$; the numbers $\{\ell(c)\}_{c \in C(\Gamma)\backslash\{e\}}$ are bounded away from 0, form a discrete subset of the positive reals, and each value is taken only for finitely many c; moreover, $\ell(c)$ is the minimum of the lengths of the closed curves in $p(c)$. To calculate $\ell(c)$ one can proceed as follows: select an element h in the G-conjugacy class determined by c such that it is in standard position and has the polar decomposition $h = h_K h_R$ where $h_K \in K$, $h_R \in A$; of course h_K and h_R will commute, and the W-orbit $o(c)$ of h_R depends only on c. The number $\ell(c)$ is then $\|\log h_R\|$, the distance of the points of $o(c)$ from e in A.

The distributions T_c seem to be intimately related to $F(c)$. For instance, let c be of regular Iwasawa type, i.e., h_R above is regular in A. Let $\delta_{o(c)}$ be the delta function on $o(c)$. Then

$$T_c = |\det(I-P_c)^{\#}|^{-\frac{1}{2}} |W|^{-1} \delta_{o(c)} ;$$

here P_c may be taken as the linear Poincaré map defined by any point $\gamma \in F(c)$ and $(I-P_c)^{\#}$ as the linear isomorphism $R/N \cong R/N$ where R (resp. N) is the range (resp. null space) of $I-P_\gamma$ (it is true that $N \subset R$).

Fix $c \in C(\Gamma)\backslash\{e\}$. Select h in the G-conjugacy class of c as above and choose a θ-stable Cartan subgroup L containing h such that, if $\mathfrak{L} = \mathrm{Lie}(L)$, $\mathfrak{L}_R = \mathfrak{L} \cap \mathfrak{s}$, $L_R = \exp(\mathfrak{L}_R)$, then $\mathfrak{L}_R \subset \mathfrak{a}$. Let ${}^*\mathfrak{L}$ be the orthogonal complement of $\mathfrak{L}_R$ in $\mathfrak{a}$, ${}^*L = \exp({}^*\mathfrak{L})$, and let $L(c) = \bigcup_{s \in W} s({}^*L \cdot h_R)$. Then $L(c)$ depends only on c and

$$\mathrm{supp}(T_c) \subset L(c).$$

In general, T_c will not live on $L(c)$. If however c is regular, i.e., if its elements are regular in G, T_c lives on $L(c)$ and is a C^∞ function on it.

It is interesting to note that the manifolds X are never generic in the sense mentioned earlier; for instance $I-P_c$ always has a nilpotent component $\neq 0$. I also do not know how from being one-one (up to orientation) the function $c \mapsto \ell(c)$ is. In view of this, the distributions T_c, their singularities, and the contributions they make to the singularities of the Fourier Transforms of the various spectral measures associated with positive operators in $\mathbb{D}(G/K)$ deserve further study beyond what is done in [DKV 1,2].

Suppose now $rk(G/K) = 1$. In this case each $F(c)$ $(c \neq \{e\})$ consists essentially of one periodic geodesic; and given c, we can find c_0 in $C(\Gamma)$ and an integer $m \geq 1$ with the following property: if $\gamma \in c$, $\gamma = \gamma_0^m$ for some γ_0 which lies in c_o and generates Γ_γ. Then c_0 and m are uniquely determined and we call $\ell_0(c) \underset{dfn}{=} \ell(c_0)$ the <u>primitive length</u> corresponding to c; of course $\ell_0(c) = m^{-1}\ell(c)$. As all $c \neq \{e\}$ are now of regular Iwasawa type, the Poisson Formula becomes

$$\text{(P.F. in rank 1)} \quad \sum_{\lambda \in \Lambda} m(\lambda) e^{\lambda \circ \log} = \frac{1}{2} \operatorname{vol}(\Gamma \backslash G)\hat{\beta}$$
$$+ \frac{1}{2} \sum_{c \neq \{e\}} \ell_0(c) \left|\det(I-P_c)^{\#}\right|^{-\frac{1}{2}} (\delta_{\ell(c)} + \delta_{-\ell(c)}) .$$

In this case one can use the above formula to show that if $\hat{\sigma}$ is the Fourier Transform of the spectral measure of $\Delta^{1/2}$, sing supp$(\hat{\sigma})$ = L. It is also worth noting that the right side is a sum of delta functions in the rank one case and singular distributions in the general case, and <u>nothing else</u>, in contrast to what happens in the case of the Fourier Transform of the spectral measure of $\Delta^{1/2}$. This is one of the rewards of passing to the multidimensional spectrum as we have done here

§3. Spectral asymptotics

We have already seen in §1 that one should expect the asymptotic relation

$$\sigma \sim \mathrm{vol}(\Gamma\backslash G)\,|W|^{-1}\beta .$$

However the presence of the complementary spectrum complicates the question and forces one to disentangle the complementary spectrum before dealing with the principal spectrum. Note that this is a nonclassical question and does not arise in the asymptotics of a single operator.

The basic idea (exploited already in [Hö]) is to obtain estimates for

$$\sum_{\lambda\in\Lambda,\,\|\lambda-\mu\|\leq t} m(\lambda)$$

around a variable point $\mu \in F_I$. Here the estimates should be sharp in μ; but as most of the applications of these estimates will be when t is bounded, one can allow the estimate to even grow with large t; this is even necessary, for after all, when $\mu = 0$, the number of spectral points in the complex ball of radius t around the origin grows like t^n. Furthermore, when μ varies in the union of the root hyperplanes and $t = \|\rho\|$ we see that we control the entire complementary spectrum.

The fundamental local spectral estimate is the following:

(L.E) $$\sum_{\lambda\in\Lambda,\,\|\lambda-\mu\|\leq t} m(\lambda) \leq Ct^q\tilde{\beta}(\mu)$$

for all $\mu \in F_I$ and all $t \geq 1$ where $C > 0, q \geq 0$ are constants independent of μ and t, and $\tilde{\beta}$ is a "smoothed out" version of β given by

$$\tilde{\beta}(\mu) = \int_{F_I} (1+\|\nu-\mu\|)^{-m-1}\beta(\nu)\,d\nu \qquad (\mu \in F_I)$$

$(m >> 1)$. The function $\tilde{\beta}$ is of regular growth and so is easier than β to work with when making estimates; nothing is lost however since β and $\tilde{\beta}$ have integrals of the same order of magnitude over bounded sets. The estimates for β lead to the following estimates for $\tilde{\beta}$:

$$(\tilde{\beta}): \begin{cases} \text{(i)} & \tilde{\beta}(\mu) \leq \text{const.}(1+\|\mu\|)^{n-r} \qquad (\mu \in F_I) \\ \text{(ii)} & \text{If } \Phi \text{ is a nonempty subset of the set of short positive roots and } T(\Phi) \text{ is the subspace of } F_I \text{ orthogonal to all elements of } \Phi, \\ & \tilde{\beta}(\mu) \leq \text{const.}(1+\|\mu\|)^{n-r-d(\Phi)} \qquad (\mu \in T(\Phi)). \end{cases}$$

The local estimate (L.E) now gives the following estimate for the part of the complementary spectrum with imaginary parts in $T(\Phi)$:

$$(\text{C.S.}\Phi) \qquad \sum_{\substack{\lambda\in\Lambda_C, \|\lambda\|\leq t \\ \lambda_I \in T(\Phi)}} m(\lambda) \leq \text{const. } t^{n-n(\Phi)}$$

where, writing $\delta(\Phi) = \dim(\text{linear span of } \Phi \text{ in } \mathfrak{a}^*)$

$$n(\Phi) = d(\Phi) + \delta(\Phi)$$

because the set of spectral points in question can be covered by $O(t^{r-\delta(\Phi)})$ balls of radius $\|\rho\|$ with centers located in $T(\Phi)$ at distances $O(t)$ from the origin. The estimates $(\tilde{\beta})$ now yield (C.S.Φ) In particular, if

$$d = \min\{\dim(\mathfrak{g}_\alpha) + \dim(\mathfrak{g}_{2\alpha}) \mid \alpha \text{ a short positive root}\}$$

then $d \geq 1$ and we have the estimate for the entire complementary spectrum given by

$$(\text{C.S}) \qquad \sum_{\lambda\in\Lambda_C, \|\lambda\|\leq t} m(\lambda) \leq \text{const. } t^{n-d-1}.$$

This is the basic estimate for the complementary spectrum. These estimates are sharp for G = a product of rank one groups and Γ = a product of discrete cocompact subgroups of these component groups. Whether these estimates can be improved for simple groups seems to be a difficult question to decide.

For the full spectrum one can prove, using decisively the estimate

(L.E), the following very general theorem: given any $\varepsilon > 0$ we can find a constant $c = c(\varepsilon) > 0$ such that for all bounded Lebesgue measurable sets $\Omega \subset F_I$,

$$\text{(P.S.1)} \qquad \left| \sum_{\lambda \in \Lambda, \lambda_I \in \Omega} m(\lambda) - \mathrm{vol}(\Gamma \backslash G)|W|^{-1} \int_\Omega \beta d\nu \right| \leq c \int_{(\partial\Omega)_\varepsilon} \beta d\nu$$

where $(\partial\Omega)_\varepsilon = \{\nu : \nu \in F_I, \mathrm{dist}(\nu, \partial\Omega) \leq \varepsilon\}$. To get asymptotic results we choose Ω so that the error term is small in comparison with the main term. Thus, take $\Omega = \Omega(t) = \mu + t\Xi$ where $\mu \in F_I$ is fixed, $\Xi \subset F_I$ bounded open and $\partial\Xi$ has finite $(r-1)$-dimensional Hausdorff measure ($r = \dim F_I$). Then, as $t \to +\infty$, the error term is $O(t^{n-1})$, while the complementary spectrum is $O(t^{n-2})$ at most. Hence

$$\text{(P.S.2)} \qquad \sum_{\lambda \in \Lambda_p \cap \Omega(t)} m(\lambda) = \mathrm{vol}(\Gamma \backslash G)|W|^{-1} \int_{\Omega(t)} \beta d\nu + O(t^{n-1}).$$

It may be worthwhile to give a rough indication of how spectral information is obtained from the truncated Poisson Formula, leading to (L.E). As mentioned earlier the key is to construct functions $f \in C_c^\infty(A(b))^W$ whose Fourier transforms are ≥ 1 on suitable parts of Λ. First of all, it is easy to find $h \in C_c^\infty(A(b))$ $(b \ll 1)$ such that

(i) $\hat{h} \geq 0$ on $\bigcup_s F(s)$, in particular on Λ

(ii) $|\hat{h}(\xi)| \geq 1$ if $\xi \in F$ and $\|\xi\| \leq 1$.

If we take $\hat{h}_t = \hat{h}(t^{-1}\xi)$ where $h_t(\exp H) = t^r h(\exp tH)$ then

(i) $\hat{h}_t \geq 0$ on $\bigcup_s F(s)$, in particular on Λ

(ii) $|\hat{h}_t(\xi)| \geq 1$ if $\xi \in F$ and $\|\xi\| \leq t$.

Some crude estimates already follow at this stage. For example,

$$\begin{aligned} \sum_{\lambda \in \Lambda, \|\lambda\| \leq t} m(\lambda) &\leq \sum_{\lambda \in \Lambda} m(\lambda)\hat{h}_t(\lambda) \\ &= v \int_{F_I} \hat{h}_t(\nu)\beta(\nu)d\nu \qquad (v = \mathrm{vol}(\Gamma \backslash G)|W|^{-1}) \\ &\leq \mathrm{const.}\ t^n \qquad (t \geq 1). \end{aligned}$$

To get estimates like this but for λ varying in balls centered at $\mu \in F_I$ we replace h_t by $h_{t,\mu}$ where $h_{t,\mu}(a) = h_t(a)e^{-\mu(\log a)}$ $(a \in A)$; this is permitted since $\mathrm{supp}(h_{t,\mu}) \subset A(b)$ still. Then the truncated Poisson Formula yields

$$(*) \qquad \sum_{\lambda \in \Lambda} m(\lambda)\hat{h}_t(\lambda-\mu) = v \int_{F_I} \hat{h}_t(\nu-\mu)\beta(\nu)d\nu.$$

The term on the right is estimated by

$$ct^m\tilde{\beta}(\mu) \qquad (\mu \in F_I,\ t \geq 1)$$

for some constant $c > 0$, and some integer $m >> 1$ ($\tilde{\beta}$ being defined using this m). We now fix a subspace $T \subset F_I$ of the form $T(\Phi)$, Φ a set of short positive roots, and prove (L.E), by downward induction on $|\Phi|$, for $\mu \in T(\Phi)$. If Φ is the set of all short positive roots, $T(\Phi) = \{0\}$, and (L.E) reduces to the estimate made above for $\sum_{\lambda\in\Lambda, \|\lambda\|\leq t} m(\lambda)$. For a smaller Φ, we begin by writing Λ as a disjoint union $\coprod_s \Lambda(s)$ with $\Lambda(s) \subset F(s)$, so that

$$\sum_{\lambda\in\Lambda, \|\lambda-\mu\|\leq t} m(\lambda) = \sum_{s\in W} \sum_{\lambda\in\Lambda(s), \|\lambda-\mu\|\leq t} m(\lambda).$$

We divide the sum over W into two parts:

$$(**) \qquad \sum_W = \sum_{W_T} + \sum_{W\setminus W_T}$$

where W_T is the centralizer of T in W. If $s \in W_T$ and $\lambda \in \Lambda(s)$, both λ and μ are in $F(s)$, hence $\lambda-\mu \in F(s)$, and

$$\sum_{W_T} \leq ct^m\tilde{\beta}(\mu) + \sum_{s\in W\setminus W_T} \sum_{\lambda\in\Lambda(s)} m(\lambda)\hat{h}_t(\lambda-\mu).$$

The sums over $W\setminus W_T$ in the above estimate and in (**) are now estimated in the same way. If $s \in W\setminus W_T$ and T_1 is the subspace of T fixed by s, $T_1 \subset T(\Phi_1)$ with $|\Phi_1| > |\Phi|$; since for $\lambda \in \Lambda(s)$ one has $\lambda_I \in T_1$, it is not difficult to see that the sums with $\lambda \in \Lambda(s)$, $\mu \in T$ are majorized by the corresponding sums with μ replaced by the orthogonal projection μ_1 of μ on T_1, and are therefore controlled by the induction hypothesis.

If $P \in \mathbb{D}(G/K)$ is a positive elliptic differential operator of order m, we can take Ξ to be the set where $p_m < 1$, p_m being the homogeneous component of degree m of p, p being the polynomial on associated to P by the Harish-Chandra isomorphism. The above result then gives Hörmander's Theorem for such P. We can even take P nonelliptic. Such examples deserve to be worked out.

For any $\lambda \in \Lambda$ the eigenspace in $L^2(X)$ corresponding to χ_λ is finite dimensional and so the projection on it is defined by a smooth kernel. Let $k(x,y:\lambda)$ be $|W\cdot\lambda|^{-1}$ times this kernel. One can then prove the estimates, for $x,y \in X$,

$$\sum_{\lambda\in\Lambda_c\cap\Omega(t)} k(x,y:\lambda) = O(t^{n-2})$$

$$\sum_{\lambda\in\Lambda_p\cap\Omega(t)} k(x,y:\lambda) = |W|^{-1}\delta(x,y)\int_{\Omega(t)} \beta d\nu + O(t^{n-1})$$

the estimate for $x = y$ being uniform for $x \in X$, and $\delta(x,y)$ is the delta function kernel of the identity operator.

It is clearly an interesting problem to examine whether the error estimates $O(t^{n-1})$ above can be improved upon. This can be done for example when $rk(G/K) = 1$ without much difficulty because one can work with the full Poisson Formula instead of the truncated one ([DKV 1], Theorem 9.1); for the error we get the bound $O(t^{n-1}/\log t)$. In the general case the full Poisson Formula is more complicated and the problem lies deeper. It can be shown that nevertheless the $O(t^{n-1})$ can be improved to $o(t^{n-1})$, and even to $O(t^{n-1}/\log t)$, although all the details in the latter improvement have not been fully worked out. If G is complex the $O(t^{n-1}/\log t)$ estimate can be established in full rigour.

Perhaps much better error bounds can be established. To explore this question lower bounds for the difference

$$\sum_{\lambda\in\Lambda_p\cap\Omega(t)} m(\lambda) - \mathrm{vol}(\Gamma\backslash G)|W|^{-1}\int_{\Omega(t)} \beta d\nu = E(t)$$

have to be established. Hejhal [Hej] (cf. Theorem 18.8) considers Selberg's examples for the case $G = SL(2,\mathbb{R})$ and suitable quaternion groups $\Gamma \subset G$ (with compact $\Gamma\backslash G$) for which $E(t)$ is $\geq c\ (t^{1/2}/\log t)$ for infinitely many t tending to $+\infty$, c being a constant >0. It would be interesting to prove such results in greater generality, for instance, when Γ is of the type constructed by Borel [B]. It is perhaps true that when $\mathrm{rk}(G/K) = 1$ and

$$E(x,y:t) = \sum_{\lambda\in\Lambda_p\cap\Omega(t)} k(x,y:\lambda) - \frac{1}{2}\,\delta(x,y)\int_{\Omega(t)} \beta d\nu \qquad (x,y \in X)$$

then for some q, $0 \leq q < n-1$,

$$E(x,y:t) = O(t^q).$$

Randol [R] has proved that for the case $SL(2,\mathbb{R})$,

$$\frac{1}{T}\int_0^T |E(x,y:t)|\,dt = O(T^{\frac{1}{2}+\varepsilon}) \qquad (T \to \infty)$$

for each $\varepsilon > 0$ for almost all (x,y).

Due to lack of time as well as familiarity I have not been able to treat some closely related aspects of these questions. Let me mention the following among others.

a) <u>The Selberg Zeta Function</u>: Defined first by Selberg [Sel] for $G = SL(2,\mathbb{R})$ using length spectrum data, it has since been extensively studied, for instance by [Hej] when $G = SL(2,\mathbb{R})$, and in higher dimensional cases, by Gangolli [Ga 2] and Millson [M]. It is defined by an infinite product converging in a half-plane $\mathrm{Re}(s) >> 1$; it has a meromorphic continuation to the whole s-plane and satisfies a functional equation of the familiar type. Millson studies it when $G = SO(4n-1,1)$ and $X = \Gamma\backslash G/K$ is compact in the context of homogeneous vector bundles, and relates it to the so-called η-invariant of X, while Gangolli studies it for arbitrary G of real rank 1 but using the Laplacian only on functions. Gangolli also shows that it is an integral transform of a "θ-like" function closely related to the fundamental solution of the heat equation on X.

b) Isospectral questions. This is the question whether the spectrum of the Laplacian on X determines the Riemannian structure on it. Milnor's famous example gave a negative answer to this question (see [S]). For an account of more recent work, for instance of Guillemin and Kazhdan, and other related problems, see [Y].

c) Finite volume case. This means that we are allowing X to be noncompact but to have finite volume. The trace formula itself is at a much greater depth, and Warner's lectures will certainly go into this aspect in detail. So far as the spectral asymptotics are concerned, the analogue of Weyl's formula is true when $G = SL(2,\mathbb{R})$ for the discrete spectrum - see for instance [L-P]. For applications of these methods to asymptotics of class numbers of indefinite quadratic forms, see [Sa1]. Sarnak has also studied the arithmetic and geometry of some hyperbolic three manifolds and obtained lower bounds for the first eigenvalue of the Laplace-Beltrami operator on $X = \Gamma\backslash G/K$ where $G = SL(2,\mathbb{C})$ [Sa2]; for $SL(2,\mathbb{R})$, see [Hej] and the references there; see also [Y]. General results involving higher rank symmetric spaces do not seem to be available. This question is closely related to the asymptotics of the length spectrum; see for instance [DeG] in the rank one case.

4. Asymptotic properties of spherical functions

Further progress in some of the questions discussed so far would seem to depend on a deeper study of the distributions T_γ that occur in the Poisson Formula,

$$\langle T_\gamma, f\rangle = \int_{G_\gamma\backslash G} (A^{-1}f)(x^{-1}\gamma x)\,d\bar{x} \qquad (f \in C_c^\infty(A)^W)$$

Since A^{-1} is just the Fourier Transform followed by the inverse of the Harish-Chandra transform, we have, with $\psi \in C_c^\infty(G/\!\!/K)$ being chosen to be 1 on supp $A^{-1}f$,

$$\langle T_\gamma, f\rangle = \int_{G_\gamma\backslash G} \psi(x^{-1}\gamma x)\left(\int_{F_I} \hat{f}(\nu)\varphi(-\nu:x^{-1}\gamma x)\beta(\nu)\,d\nu\right)d\bar{x}$$

$$= \int_{F_I} \hat{f}(-\nu)\beta(\nu)\left(\int_{G_\gamma\backslash G} \varphi(\nu:x^{-1}\gamma x)\psi(x^{-1}\gamma x)\,d\bar{x}\right)d\nu.$$

For a proper study of these integrals it is necessary to have a rather good picture of $\varphi(\nu;y)$ <u>when</u> $\nu \in F_I$ <u>and</u> $y \in G$ <u>are both allowed to go to infinity</u>. This question makes sense for the more general Eisenstein integrals also; however I shall restrict myself to the case of the elementary spherical functions for simplicity.

These functions are far reaching generalizations of the classical "special functions." They have integral representations, obey differential equations with regular singularities, and come up as parametrized families in a natural manner. In analogy with the classical theory of (ordinary) differential equations one may think of the following methods of studying these functions:

(a) The Cauchy-Picard-Lipschitz method of iteration.

(b) The method of oscillatory integrals and stationary phase.

(c) Deforming the "contour" of integration and the paths of steepest descent.

(d) Regular singular differential equations.

These are all closely related; indeed a classical analyst would regard (b)-(d) essentially as parts of a single picture. The method (a) was used in [T-V] to obtain asymptotic expansions for the elementary spherical functions that played a crucial role in an L^1-harmonic analysis of spherical functions. I shall not discuss (a) or (d); the starting point in (d) was Harish-Chandra's [H3,4]; the reader should refer to the recent exposition by [C-M] which also contains many new additions.

I shall restrict myself to (b) and (c).

I. Matrix elements as oscillatory integrals. The method of stationary phase. Let M be a smooth manifold of dimension m; dx, a smooth positive density on M; ψ a smooth real valued function on M, called the phase function. Then an integral of the form

$$I(\psi:a:\tau) = \int_M e^{i\tau\psi(x)} a(x)\,dx \qquad (a \in C_c^\infty(M))$$

is called an oscillatory integral, τ being a real parameter; it is viewed as a distribution (in a) depending on the parameter. Such integrals occur in high frequency optics where one is interested in $\tau \to +\infty$. The principle of stationary phase tells us that the main contributions to the asymptotic behaviour of I come from the points where ψ is stationary; i.e., $d\psi = 0$. The classical case is when the critical points are ψ are non-degenerate, i.e., have nonsingular Hessians at them. The critical points are then isolated; around them ψ can be reduced to a quadratic form in local coordinates (Morse's lemma); and the oscillatory integrals can be explicitly evaluated in these coordinates, allowing one to obtain the asymptotic expansion of I ([G-S],[D]). Such phase functions are called Morse functions, and one has

$$I(\psi:a:\tau) = c(a)\tau^{-\frac{1}{2}\dim(M)} + O(\tau^{-\frac{1}{2}\dim(M)-1})$$

where c is a linear combination of the delta functions at the critical points of ψ (with oscillatory factors involving τ being allowed).

Suppose now, as is often the case, that ψ depends (smoothly) on a parameter θ varying in a manifold Θ. Typically, there will be a subvariety $\Gamma \subset \Theta$ such that if $\theta \in \Theta\setminus\Gamma$, $\psi(\cdot,\theta)$ will be Morse, but that the top coefficient c_θ, which now depends on θ, will be blowing up when $\theta \to \Gamma$. Points of Γ are called caustics (motivated from geometric optics); and for $\theta_0 \in \Gamma$, the decay of $I(\psi:a:\tau)$ in τ will typically be less rapid than for values of θ near θ_0 but off Γ. The basic question is then to write uniform estimates for I when θ

varies around $\theta_0 \in \Gamma$. For generic families (ψ_θ) such estimates can be obtained not in terms of powers of τ but in terms of more complicated functions of the "Airy type" (see [D], [G-S]).

Let us now turn to the framework involving G. The elementary spherical functions can now be represented by the integrals

$$\varphi(\lambda:x) = \int_K e^{(\lambda-\rho)(H(xk))}\,dk \qquad (x \in G).$$

The fundamental idea is to regard this as an oscillatory integral over K for the family $(F_{x,\lambda})$ $(x \in G, \lambda \in F_I)$ of phase functions defined on K by

$$F_{x,\lambda}(k) = \lambda(H(xk)) \qquad (k \in K).$$

An approach to the asymptotics of $\varphi(\lambda:x)$ would then be to use the method of stationary phase. This of course would need a study of the critical data of the $F_{x,\lambda}$. We remark that if M is the centralizer of A in K, the $F_{x,\lambda}$ are actually functions on K/M which is the flag manifold associated to G.

For $G = SL(2,\mathbb{R})$ we have $K = \{u_\theta = \begin{pmatrix} \cos\theta & \sin\theta \\ -\sin\theta, & \cos\theta \end{pmatrix}\}$, $A = \left\{a_t = \begin{pmatrix} e^t & 0 \\ 0 & e^{-t} \end{pmatrix}\right\}$, with θ, t varying in $\mathbb{R}$. We can identify F with $\mathbb{C}$ so that ρ goes to 1. Then, for $\lambda \in \mathbb{C}$,

$$F_{a_t,\lambda}(u_\theta) = \frac{1}{2}\lambda \log(e^{2t}\cos^2\theta + e^{-2t}\sin^2\theta).$$

Fix $\lambda \neq 0$; then, the stationary points (in θ) are the Weyl group points $\theta = 0, \pi/2$, provided $t \neq 0$; we identify θ and $\theta + \pi$ since we work in K/M. Let $\lambda > 0$. Then $\theta = 0$ is a maximum, $\theta = \frac{\pi}{2}$ is a minimum, for $t > 0$; the facts are reversed when $t < 0$. The Hessians and critical values are easily calculated. For $\theta = 0$ the Hessian and critical value are, respectively,

$$-\lambda(1-e^{-4t}), \qquad \lambda t,$$

while for $\theta = \frac{\pi}{2}$, they are, respectively,

$$\lambda(e^{4t} - 1), \qquad -\lambda t.$$

In [DKV 2] (see also [V2]) this method is developed in considerable detail. The critical data of the $F_{a,\lambda}$ ($a \in A$; this is sufficient) are quite remarkable and show that the family $(F_{a,\lambda})$ is extremely non-generic. The critical sets are smooth (not connected in general) and the Hessians transversally non-singular; the latter property means that at any critical point the radical of the Hessian quadratic form is precisely the tangent space at the point to the critical manifold.

The critical sets can be explicitly determined and turn out to be

$$\coprod_{W_a \backslash W / W_\lambda} K_a x_w K_{H_\lambda} \qquad (x_w \text{ a representative for } w \in W)$$

where the suffixes denote the centralizers in the respective groups ((e-g), K_a is the centralizer of a in K) and where H_λ is the image of λ in $\mathfrak{a}$ under the isomorphism of $\mathfrak{a}^*$ with $\mathfrak{a}$ defined by the Killing form. In particular, if a and λ are regular, the critical sets are finite in K/M, coinciding with the Weyl group points, showing that $F_{a,\lambda}$ is a Morse function for regular a and λ. For general a, λ, the Hessians of $F_{a,\lambda}$ at a critical point of it can be determined as follows. Let L_k be the endomorphism of $\mathfrak{k}$ ($\approx$ tangent space to K at k) which is symmetric with respect to $-\langle\cdot,\cdot\rangle$ and whose associated quadratic form is the Hessian at k, k being a critical point; then

$$L_{x_w} = -\frac{1}{2} \sum_{\alpha \in \Delta^+} \langle\alpha,\lambda\rangle (1 - e^{-2w\alpha(\log a)}) F_\alpha$$

where F_α is the orthogonal projection $\mathfrak{k} \to \mathfrak{k}_\alpha = (\mathfrak{g}_\alpha + \mathfrak{g}_{-\alpha}) \cap \mathfrak{k}$. This gives the Hessian on $K_a x_w M$ since $F_{a,\lambda}$ is left K_a-invariant, and even on all of $K_a x_w K_\lambda$ if $\lambda \in Cl(F_R^+)$, because of the remarkable fact that for $a \in A$, $\lambda \in Cl(F_R^+)$, $F_{a,\lambda}$ is right K_λ-invariant. Even, for arbitrary λ, the rank, signature and index are all determined by the above formula since these are constant on $K_a x_w K_\lambda$.

Closely related to the phase functions $F_{a,\lambda}$ are their "infinitesimal (linearised) versions" $f_{X,\lambda}$ $(X = \log a \in \mathfrak{a}, \lambda \in F_R)$ defined by

$$f_{X,\lambda}(k) = \langle H_\lambda, Ad(k^{-1}) \cdot X\rangle \qquad (k \in K).$$

Clearly

$$f_{X,\lambda}(k) = (\frac{d}{dt} F_{\exp tX,\lambda}(k))_{t=0} \qquad (k \in K).$$

The $f_{X,\lambda}$ determine oscillatory integrals which are essentially the matrix elements of the Cartan motion groups (see [DKV2]). Since $K \approx G/AN$, we may regard K as a (left) G-space, with action denoted by $x,k \mapsto \theta_x(k)$; then $F_{a,\lambda}$ is obtained by integrating $f_{X,\lambda}$ along the flow lines of the A-action:

$$F_{\exp X,\lambda}(k) = \int_0^1 f_{X,\lambda}(\theta_{\exp sX}(k))ds \qquad (k \in K).$$

The calculations with $f_{X,\lambda}$ are simpler and the results can be used to determine much of the critical data for F using the above formula For $f_{X,\lambda}$ we have left K_X and right K_λ-invariance for <u>all</u> $X \in \mathfrak{a}$, $\lambda \in F_R$. Although $F_{a,\lambda}$ appears apparently first in [DKV 2], the study of $f_{X,\lambda}$ goes back certainly to Bott and Takeuchi; moreover Takeuchi and Kobayashi use it for a Morse-theoretic study of the flag manifolds (see [DKV 2]).

One also has the formula

$$dF_{\exp X,\lambda}(k,Y) = df_{X,\lambda}(\theta_{\exp X}(k),Z)$$

for $k \in K$, $Z \in \mathfrak{k}$ with $Y = Ad(k^{-1})$ (ad X/sinh ad X) Ad $\theta_{\exp X}(k)(Z)$. Thus $dF_{\exp X,\lambda}$ and $df_{X,\lambda}$ are equivalent, as functions on the tangent bundle of K $(\approx K \times \mathfrak{k})$ under a suitable diffeomorphism of it. One may guess that for suitable diffeomorphisms $g_{X,\lambda}$ of K, $F_{\exp X,\lambda} =$

$f_{X,\lambda} \circ g_{X,\lambda}$ although this seems difficult to decide.

With this preparation let me return to the integrals

$$I(a,\lambda:g) = \int_K e^{i\lambda(H(ak))} g(k)\,dk \qquad (g \in C^\infty(K)).$$

Writing $\lambda = \tau\mu$ where $\tau \geq 0$ and $\|\mu\| = 1$, one may view it as a classical oscillatory integral for the phase function $F_{a,\mu}$, (a,μ) being regarded as parameters. So it is necessary to determine the caustic sets and the way the critical data change when (a,μ) approaches the caustic sets. But the determination of the critical sets reveals a remarkable behaviour of the $F_{a,\mu}$. There are only finitely many possibilities for the critical sets, and when (a,μ) vary equisingularly, the critical sets are rigid; moreover the caustic sets are essentially unions of root hyperplanes. This structure, which is quite different from what happens with generic families of phase functions (see [D]), is at the heart of the possibility of making estimates for $I(a,\lambda:g)$ that are uniform in (a,λ), and expressing them in terms of very simple functions of the product type, in contrast to the Airy-like functions needed in generic situations. As an example of what may be proved by these methods let me mention the following theorm ([DKV 2]; see also [V3]): fix a compact set $\omega \subset A$ and let $\Delta_w^+(\omega)$ for $w \in W$ be the set of all $\alpha \in \Delta^+$ such that $(w\alpha)(\log a) \neq 0 \ \forall\, a \in \omega$; then, for all $\lambda \in F_R$, $a \in \omega$, $g \in C^\infty(K)$, we have,

$$|I(a,\lambda:g)| \leq \nu(g) \sum_{w \in W} \prod_{\alpha \in \Delta_w^+(\omega)} (1+|\langle\alpha,\lambda\rangle|)^{-\frac{1}{2}n(\alpha)}$$

where $n(\alpha)$ = dimension of the root space $\mathfrak{g}_\alpha$ and ν is a continuous seminorm on $C^\infty(K)$. The product structure appearing on the right side in this estimate is in contrast to the situation for generic families. The proof is complicated and I refer the reader to [DVK2]. Let $\{\alpha_1,\ldots,\alpha_r\}$ be a simple system of roots in Δ^+ and $(\alpha_i^\vee)_{1\leq i\leq r}$ the corresponding coroots so that $\langle\alpha_i^\vee,\alpha_j\rangle = \delta_{ij}$. Then we can write

$\lambda = \sum_i \tau_i\alpha_i^\vee$; $(\tau_1,\ldots,\tau_r) \in \mathbb{R}^r$, and $\tau_j \geq 0$ for all j if and only if $\lambda \in Cl(F_R^+)$. Write $\psi_j = F_{a,\alpha_j^\vee}$, $1 \leq j \leq r$. The typical case is to estimate I for g localized around the identity element $e \in K$. The key idea is that the system of phase functions $(\psi_1,\ldots,\psi_r)$ may be trigonalized for a suitable choice of local coordinates at e. For instance, assume that all the τ_j are ≥ 0. If K_m is the centralizer of $\{H_{\alpha_1^\vee},\ldots,H_{\alpha_m^\vee}\}$ in K, we have the filtration

$$K = K_0 \supseteq K_1 \supseteq \cdots \supseteq K_r = \{e\},$$

ψ_j is right K_j-invariant, e is critical for $\psi_j|_{K_{j-1}}$ with explicitly determined critical data; selecting local sections X_j through e for K_{j-1}/K_j gives a diffeomorphism of a neighborhood of e in K with $X_1 \times \cdots \times X_r$, and the system $(\psi_1,\ldots,\psi_r)$ on $X_1 \times \cdots \times X_r$ is trigonal: $\psi_j(x_1,\ldots,x_r)$ depends only on $(x_1,\ldots,x_j)$. If the variables are no longer ≥ 0, i.e., if λ is not in the positive chamber, the trigonalization is more subtle but still available for one can use the relation between $dF_{\exp X,\lambda}$ and $df_{X,\lambda}$ and the right invariance of the $f_{X,\lambda}$ to construct the trigonalizing chart for $(\psi_1,\ldots,\psi_r)$.

I have chosen this example of integrals over the flag variety to illustrate the promise of this method. One can study oscillatory integrals also on conjugacy classes of G in the same way (cf. [DKV2]); such a study gives a much more precise description of the wave front sets of the distributions T_c occuring in the Poisson Formula than given in §2.

One should also try to extend the estimate for $I(a,\lambda:g)$ when a varies on all of A. It is not difficult to predict what the result should be. For the elementary spherical functions $\varphi(\lambda:a)$ with $\lambda \in F_I$ and $a \in \omega$ where ω is now a conical sets in $Cl(A^+)$, the majorant should be essentially the same as in the theorem described above. This is true when G is complex or of real rank 1; it appears that the arguments in [DKV2] extend to give this for general G and the details are being worked out. Such an estimate would give the $O(t^{n-1}/\log t)$ estimate for the error terms in the spectra of compact $\Gamma\backslash G/K$ (cf. §3).

The critical data calculations mentioned above contain a great deal of information which, used properly, imply many results on the structure and geometry of G. For instance Heckman proved the well known convexity theorem of Kostant by this approach: if $a \in A$, the Iwasawa projection in A of the K-conjugacy class of a (namely $\{kak^{-1} | k \in K\}$) is the convex hull of the points a^w, $w \in W$ (see [DKV2]).

These results are only a small part of the application of the methods of the so-called Lagrangian analysis to the theory of analysis on groups and homogeneous spaces. For a general view of some aspects of this approach see [G-S].

<u>II. Matrix elements as integrals over real cycles in G_c. The method of steepest descents</u>. The second method starts with the Harish-Chandra integral representation of the elementary spherical functions

$$\varphi(\lambda:a) = \int_K e^{(\lambda-\rho)(H(ak))}dk \qquad (a \in A,\ \lambda \in F).$$

Its basic idea is to vary the cycle K in its "complexification" K_c. It is developed in detail in the 1982 thesis of Van den Ban [Ba], done in Utrecht under J. J. Duistermaat and J. A. C. Kolk. The point to keep in mind is that when we take the above integral over a smooth cycle $\Gamma: Y \to K_c$ where Y is a smooth real compact oriented manifold of dimension equal to $\dim(K)$ (= smooth $\dim(K)$-cycle), the function $y \mapsto H(a\Gamma(y))$ becomes multivalued and one has to make sure that (a) the cycle avoids the branch locus (b) the branch of $y \mapsto H(a\Gamma(y))$ is actually single valued.

Before giving a description of (some of) the results of van den Ban let us look at the case $G = SL(2,\mathbb{R})$. A simple and standard calculation shows that

$$\varphi(\lambda:a_t) = e^{(\lambda-1)t} \int_{-\infty}^{\infty} e^{\frac{1}{2}(\lambda-1)\log(1+e^{-4t}\tau^2)-\frac{1}{2}(\lambda-1)\log(1+\tau^2)} \frac{d\tau}{\pi}$$

($\lambda \in \mathbb{C}$, $t \in \mathbb{R}$). We now regard τ as a complex variable and note that the integrand has branch points at $\tau = \pm i$, $\tau = \pm ie^{2t}$. The integral over real τ is split as

$$\int_{|\tau|\leq C} + \int_{|\tau|\geq C} .$$

The first of these will now be replaced by the integral over the ("lemniscate") contour L^+

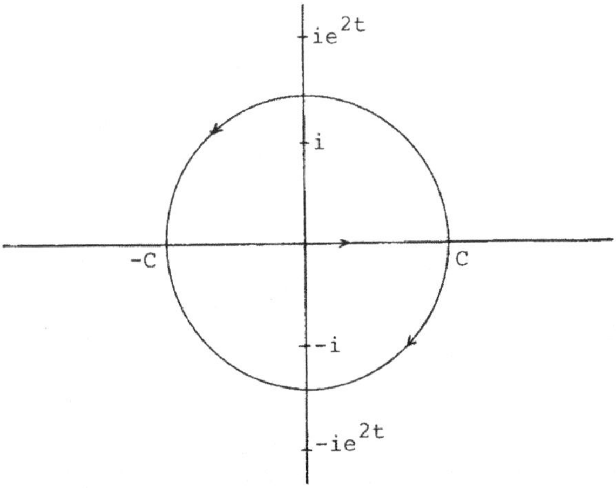

where we go from $-C$ to C on the real axis, then on the upper circular arc from C to $-C$, then again from $-C$ to C on the real axis, followed by the lower arc from C to $-C$. The integral on the real segment is taken twice; but the second time around, the branch of the log is a different one. The "double" circuit guarantees that when we finally return to $-C$, the branch of the log returns to its original choice. The second integral (over $|\tau|\geq C$) is replaced by the integral over the contour L^-:

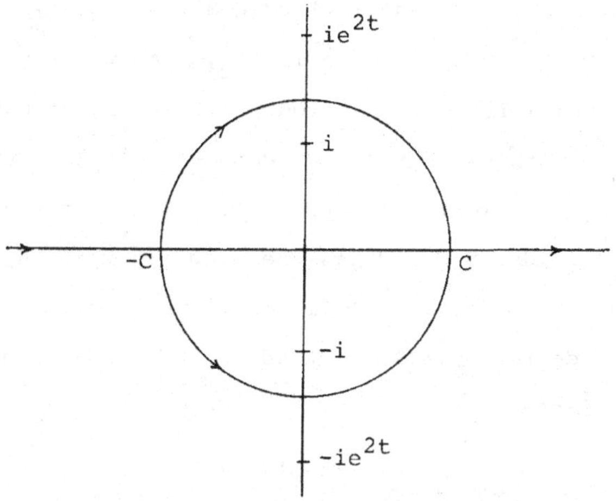

Here we come from $-\infty$ to $-C$ along the real axis, go from $-C$ to C on the lower arc, then go to ∞ from C along real axis, come again from $-\infty$ ($=\infty$ on the sphere) to $-C$ along the real axis, go to C via the upper arc, then from C to ∞ on the real axis followed by $-\infty$ ($=\infty$) to $-C$ on the real axis. Adding the integrals on L^+ and L^- and (a) adjusting for the change of the branch of the log (b) observing that the integrals on the circular arcs cancel, we get

$$(1-e^{-2\pi i\lambda})\varphi(\lambda:a_t) = \int_{L^+} + \int_{L^-}$$

To study the asymptotics of $\varphi(\lambda:a_t)$ is now a question of studying the two integrals $\int_{L^+}$ and $\int_{L^-}$. The integrals $\int_{L^\pm}$ are quite interesting. They satisfy, as functions of t, the <u>same</u> differential equations as $\varphi(\lambda:a_t)$. In this special case (where we are dealing with hypergeometric functions) they go back to Pochhammer.

Let us now go to the general case. I shall assume that G is already the adjoint group; G_c is the complex adjoint group. If $B \subset G$ is a closed subgroup with Lie algebra $\mathfrak{b}$, we write B_c for the complex analytic subgroup of G_c defined by the complexification $\mathfrak{b}_c$ of $\mathfrak{b}$. The projection maps $\kappa(G \to K)$, $h(G \to A)$ associated with the Iwasawa decomposition $G = KAN$ have multivalued analytic extensions κ_c, h_c on $G_c^{\times} = K_c A_c N_c$ which is open dense in G_c; the variety $S_c = G_c \setminus G_c^{\times}$ is the branch locus; moreover, $H = \log h$ $(G \to \mathfrak{a})$ extends to a multivalued analytic map from $G_c^{\times}$ to $\mathfrak{a}_c$, denoted by H_c. Suppose now $a_0 \in A^+$, $\Gamma(Y \to K_c)$ a smooth $\dim(K)$-cycle. Then, if the cycle $a_0\Gamma: y \mapsto a_0\Gamma(y)$ avoids S_c and has the property that H_c has a single valued branch H_Γ over it, and if ω is the holomorphic exterior differential form of degree $\dim(K)$ extending the normalized volume element of K, the integral

$$\phi_\Gamma(a) = \int_\Gamma e^{(\lambda-\rho)(H_\Gamma(ak))} \omega(k)$$

is well-defined for all a in some neighborhood of a_0 in A^+, and is a solution on that neighborhood of the radial component differential equations of Harish-Chandra on A^+ that are satisfied by the $\varphi(\lambda:\cdot)$ (see [H1] for the definition and theory of these radial component differential equations). This theorem is proved roughly the same way as Harish-Chandra's theorem showing that the integrals $\int_K e^{(\lambda-\rho)(H(ak))} dk$ satisfy the appropriate differential equations; the key point is that instead of the bi K-invariance of dk one uses the bi K-invariance of ω together with the invariance of the homotopy class of a smooth cycle under left or right translation by elements of K.

Two rather natural questions come up now. The first is the construction of cycles Γ with the required properties for all a in some neighborhood of infinity on A^+; the second is whether the integrals ϕ_Γ so constructed will form a basis for the space of solutions of the radial component differential equations on A^+. The remarks above for $G = SL(2,\mathbb{R})$ show the answers to be yes in that case for both of these questions. Van den Ban proves that this is so for general G. More precisely, let $C > 1$ and let $A^+(C)$ be the open subset of A^+ of all a with $e^{\alpha(\log a)} > C$ for each $\alpha \in \Delta^+$. Then, if $C >> 1$, there exist smooth oriented compact manifolds Y_w ($w \in W$, $\dim(Y_w) = \dim(K)$), and cycles $\Gamma_w(a)$ ($Y_w \to G_c^\times$) ($a \in A^+(C)$) varying smoothly in a, such that the following are true:

(i) The map H_c has a branch H_w over the map $(a,y) \mapsto a\Gamma_w(a)y$ of $A^+(C) \times Y_w$ into $G_c^\times$

(ii) The functions

$$\phi_{w,\lambda}(a) = \int_{\Gamma_w(a)} e^{(\lambda-\rho)(H_w(ak))} \omega(k) \qquad (a \in A^+(C))$$

form a basis for the space of solutions of the radial component differential equations on $A^+(C)$, provided $\lambda \in F$ is regular and does not lie on any one of a locally finite collection of hyperplanes in F none of which penetrates a "strip" around F_I of the form $\{\lambda \in F \mid \|\lambda_R\| < \varepsilon\}$. The construction of the cycles $\Gamma_w(a)$ is however somewhat complicated. It is first carried out in the real rank one case, more or less imitating the $SL(2,\mathbb{R})$ situation. In the general case it is obtained as a consequence of a generalization and variant of the technique used by Gindikin-Karpelevič in the proof of the product formula for the Harish-Chandra c-function. In addition, these constructions involve a type of "renormalization" of the Bruhat charts around the Weyl group points in K/M that go back to [DKV2].

Because the $\Gamma_w(a)$ are <u>compact</u> cycles no problems of convergence arise and the asymptotic behaviour of the $\phi_{w,\lambda}(a)$ are easily obtained from their integral representation. In fact, for $\lambda \in F_I$ and regular, and $a \in A^+(C)$ with $\alpha(\log a) \to \infty$ for each $\alpha \in \Delta^+$, one has

$$\phi_{w,\lambda}(a) \sim \gamma(\lambda) e^{(w\lambda-\rho)(\log a)}.$$

This relation actually gives the linear independence and the basis theorem. It moreover gives the identification of the $\phi_{w,\lambda}$ with the Harish-Chandra series:

$$\phi_{w,\lambda} = \gamma(\lambda) c(\lambda)^{-1} \Phi(w\lambda : \cdot);$$

here $\Phi(\cdot:\cdot)$ is the Harish-Chandra series solution of the radial component differential equations and γ is a certain meromorphic function which is analytic and nonzero on the regular subset of F_I.

All of this is valid when $C >> 1$ and $a \in A^+(C)$. It is now necessary to consider the case when a goes to infinity on A^+ but may stay close to certain walls. Fix $a_0 \in \mathrm{Cl}(A^+)$, $a_0 \neq 1$; let V be a conical open neighborhood of a_0 in $\mathrm{Cl}(A^+)$; and $V(C)$ the part of V of all $a \in V$ with $e^{\alpha(\log a)} > C$ (>1) for all roots α which do not vanish at $\log a_0$. If we look back at the earlier discussion in the case of $SL(2,\mathbb{R})$ we find that when $t \to 0+$, the lemniscates $L^{\pm}$ get "pinched" by i and ie^{2t} (as well as by $-i$ and $-ie^{2t}$). Somewhat similar but technically more complicated phenomena arise in the general case. These suggest that the cycles $\Gamma_w(a)$ will have to be changed to new ones before the asymptotics of $\varphi(\lambda:a)$ for $a \in V(C)$ can be properly studied. Van den Ban proves that if V is sufficiently small and $C >> 1$, there are smooth $\dim(K)$-cycles $\Gamma_{a_0,w}(a)$ $(Y_{a_0,w} \to G_c^{\times})$ depending smoothly on $a \in V(C)$, and associated branches $H_{a_0,w}$ of H_c over the $\Gamma_{a_0,w}(a)$, such that

$$a \longmapsto \phi_{a_0,w,\lambda}(a) = \int_{\Gamma_{a_0,w}(a)} e^{(\lambda-\rho)(H_{a_0,w}(ak))} \omega(k)$$

are well-defined and satisfy the radial component differential equations on $A^+ \cap V(C)$. These depend on w only through the coset $W_{a_0}w$ where W_{a_0} is the centralizer of a_0 in W; and λ should be restricted as before.

Considerable information on the joint asymptotics of $\varphi(\lambda:a)$ for $\lambda \in F_I$ and $a \in V(C) \cap A^+$ is obtainable from these results and [Ba] carries them out in detail. In particular [Ba] treats a multidimensional variant of the classical method of steepest descent to obtain sharp estimates by deforming the cycles $\Gamma_{a_0,w}(a)$ suitably; however, the results are restricted to the case when $\lambda \in F_I$ stays in a regular cone, and may be obtained also by the arguments in [DKV2] (see also [S-T]. For general variations in λ new difficulties arise which have not been yet overcome (in either of the two methods I and II).

I feel these approaches should be explored further to find out the limits of what can be achieved in these asymptotic questions and also in relating the geometry of the flag varieties to analysis on G.

Acknowledgement. I wish to thank Ron Lipsman and the Department of Mathematics of the University of Maryland for their warm hospitality that made the conference very enjoyable. I am also grateful to Joop Kolk for being very generous with his time, going through most parts of the manuscript and suggesting many improvements.

REFERENCES

[B] A. Borel, Compact Clifford-Klein forms of symmetric spaces. Topology 2 (1963), 111-122.

[Ba] E. P. van den Ban, Asymptotic expansions and integral formulas for eigenfunctions on a semisimple Lie group, Thesis, Utrecht, 1982.

[C-M] W. Casselman and D. Miličic, Asymptotic behaviour of matrix coefficients of admissible representations. Duke Math. J. 49 (1982), 869-930.

[Ch1] J. Chazarain, Formule de Poisson pour les variétés Riemanniennes. Inv. Math. 24 (1974), 65-82.

[Ch2] ______, Spectre des opérateurs elliptiques et flots hamiltoniens. Seminaire Bourbaki Exposé No. 460, 111-123. Springer Lecture Notes in Mathematics No. 514, Springer-Verlag, Berlin, 1976.

[DeG] D. DeGeorge, Length spectrum for compact locally symmetric spaces of strictly negative curvature. Ann. Scient. École Norm. Sup. (4) 10 (1977), 133-152.

[D] J. J. Duistermaat, Oscillatory integrals, Lagrange immersions, and unfolding of singularities. Comm. Pure Appl. Math. 27 (1974), 207-281.

[DKV1] J. J. Duistermaat, J. A. C. Kolk, and V. S. Varadarajan, Spectra of compact locally symmetric manifolds of negative curvature. Inv. Math. 52 (1979), 29-93.

[DKV2] ______, Functions, flows, and oscillatory integrals on flag manifolds and conjugacy classes in real semisimple Lie groups. To appear in Compositio Math.

[D-G] J. J. Duistermaat and V. Guillemin, The spectrum of positive elliptic operators and periodic bicharacteristics. Inv. Math. 29 (1975), 37-79.

[Ga1] R. A. Gangolli, On the Plancherel formula and the Paley-Wiener theorem for spherical functions on semisimple Lie groups. Ann. Math. 93 (1971), 150-165.

[Ga2] ______, Zeta functions of Selberg's type for compact space forms of symmetric spaces of rank one. Ill. J. Math. 21 (1977), 1-41.

[G] I. M. Gel'fand, Automorphic functions and the theory of representations. Proc. Int. Congress of Math. Stockholm 1962, 74-85.

[G-S] V. Guillemin and S. Sternberg, Geometric asymptotics. Mathematical surveys No. 14, American Mathematical Society, Providence, R.I., 1977.

[H1] Harish-Chandra, Spherical functions on a semisimple Lie group, I. Amer. J. Math. 80 (1958), 241-310.

[H2] ______, Spherical functions on a semisimple Lie group, II. Amer. J. Math. 80 (1958), 553-613.

[H3] ______, Some results on differential equations and their applications. Proc. Nat. Acad. Sci. U.S.A. 45 (1959), 1763-1764.

[H4] ______, Some results on differential equations. Preprint, 1960.

[He] S. Helgason, An analogue of the Paley-Wiener theorem for Fourier transform on certain symmetric spaces. Math. Annalen, 165 (1966), 297-308.

[Hej] D. Hejhal, The Selberg trace formula for $PSL(2,\mathbb{R})$, vol. I. Springer Lecture Notes in Math. No. 548, Springer-Verlag, Berlin, 1976.

[H-J] S. Helgason and K. Johnson, The bounded spherical functions on symmetric spaces. Advances in Math. 3 (1969), 586-593.

[Hö] L. Hörmander, The spectral function of an elliptic operator. Acta Math. 121 (1968), 193-218.

[L-P] P. D. Lax and R. S. Phillips, Scattering theory for automorphic functions. Ann. Mathematics. Studies. No. 87, Princeton University Press, Princeton, N.J., 1976.

[M] J. Millson, Closed geodesics and the η-invariant. Ann. Math. 108 (1978), 1-39.

[R] B. Randol, A Dirichlet series of eigenvalue type with applications to asymptotic estimates. Bull. Lond. Math. Soc. 13 (1981), 309-315.

[Sa1] P. Sarnak, Class numbers of indefinite binary quadratic forms. J. Number Theory 15 (1983), 229-247.

[Sa2] ______, The arithmetic and geometry of some hyperbolic three manifolds. Preprint.

[Se1] A. Selberg, Harmonic analysis and discontinuous groups in weakly symmetric Riemannian spaces with applications to Dirichlet series. J. Ind. Math. Soc. 20 (1956), 47-87.

[S] I. M. Singer, Eigenvalues of the Laplacian and invariants of manifolds. Proc. Int. Cong. Math. Vancouver, 1974, 187-200.

[S-T] R. J. Stanton and P. A. Tomas, Expansions for spherical functions on noncompact symmetric spaces. Acta Math. 140 (1978), 251-276.

[T-V] P. C. Trombi and V. S. Varadarajan, Spherical transforms on semisimple Lie groups. Ann. Math. 94 (1971), 246-303.

[V1] V. S. Varadarajan, Lie groups, Lie algebras, and their representations. Prentice-Hall, Englewood Cliffs, N.J., 1974.

[V2] ______, Harmonic analysis on real reductive groups. Springer Lecture Notes in Math. No. 576, Springer-Verlag, Berlin, 1977.

[V3] ______, Oscillatory integrals and their applications to harmonic analysis on real semisimple Lie groups. To appear in the Proceedings of the conference on reductive groups held at Utah, Spring 1982.

[Wa] N. Wallach, On the Selberg trace formula in the case of compact quotient. Bull. A.M.S. 82 (1976), 171-195.

[W] H. Weyl, Ramifications, old and new, of the eigenvalue problem. Bull. A.M.S. 56 (1950), 115-139.

[Y] S. T. Yau, Seminar on differential geometry. Ann. Mathematics studies No. 102, Princeton University Press, Princeton, N.J., 1982.

QUANTUM PHYSICS AND SEMISIMPLE SYMMETRIC SPACES

Gregg J. Zuckerman*
Yale University, New Haven, Connecticut 06520

§1. CONFORMAL GEOMETRY AND MAXWELL'S EQUATIONS

Let M be a four-dimensional Lorentzian manifold with metric g of type $(+,-,-,-)$. Let $|g| = |\det(g_{\mu\nu})|$ and let $*$ be the Hodge $*$-operator on 2-forms, given in local coordinates by

$$(*F)_{\mu\nu} = \sqrt{|g|}\;\varepsilon_{\mu\nu\sigma\tau}\, g^{\sigma\alpha} g^{\tau\beta} F_{\alpha\beta}$$

where we use the Einstein summation convention. If g' is another Lorentz metric, conformally related to g by $g' = \rho^2 g$, then the $*$ operators for g and g' agree on 2-forms, since $\sqrt{\rho^8}\,\rho^{-2}\rho^{-2} = 1$. Thus, Maxwell's equations (in the absence of sources),

1.1 $$dF = 0, \quad d * F = 0,$$

depend only on the conformal structure of M. Let $\mathrm{Maxw}(M)$ be the linear space of all smooth complex-valued solutions of Maxwell's equations above. Let $C(M)$ be the Lie group of conformal automorphisms of M, and $\mathfrak{g}_C(M)$ be the Lie algebra of conformal Killing vector fields on M (X is conformal Killing if the Lie derivative $\mathcal{L}_X g$ equals ωg for some scalar function ω). In general, the Lie algebra of $C(M)$ is strictly contained in $\mathfrak{g}_C(M)$. For instance, if $M = \mathbb{R}^{1,3}$, i.e. Minkowski space with the flat metric of type $(+,-,-,-)$, then $\mathfrak{g}_C(M) \cong so(4,2)$, whereas $C(M)_0 = [\mathbb{R}_0^{\times}\, SO(3,1)_0] \ltimes \mathbb{R}^{1,3}$, i.e. the Poincaré group together with positive scalars. The dimension of $C(M)_0$ is 11, whereas the dimension of $\mathfrak{g}_C(M)$ is 15. Finally, let $I(M)$ be the isometry group of (M,g). $I(\mathbb{R}^{1,3})$ is the Poincaré group.

The solution space $\mathrm{Maxw}(M)$ can be regarded as a module over the Lie algebra $\mathfrak{g}_C(M)$ as well as a module over the Lie groups $C(M)$ and $I(M)$. Of particular interest is the case when M is conformally flat, that is when M is locally conformally equivalent to an open domain in $\mathbb{R}^{1,3}$. If $R(=R_{\mu\nu\sigma\tau})$ is the Riemann curvature tensor for g, one can extract the Weyl tensor, which is the completely traceless part of R. M is conformally flat if and only if the Weyl tensor vanishes.

Example: M = "ESU", the Einstein static universe, $S^3 \times \mathbb{R}$, with the product metric

$$ds^2 = dt^2 - d\sigma^2,$$

where $d\sigma^2$ is the standard metric on S^3. Then,

$$I(M)_0 = SO(4) \times \mathbb{R}.$$

Example: A Robertson-Walker universe, with metric

$$ds^2 = dt^2 - G(t)^2 d\sigma^2$$

on $M = S^3 \times \mathbb{R}$. We introduce the "conformal time"

$$\tau = \int^t \frac{dt}{G(t)} \quad ,$$

and solve (at least locally) for t as a function of τ.

Then, locally,

$$ds^2 = \tilde{G}(\tau)^2 (d\tau^2 - d\sigma^2),$$

so that any Robertson-Walker metric is locally, away from singularities in the map $t \to \tau(t)$, conformally related to ESU.

In fact, ESU, and hence also the Robertson-Walker universes, is conformally flat. One can see this by computing the Weyl tensor, or else by observing that $S^3 \times \mathbb{R}$ is the universal covering of the projectivized null cone in $\mathbb{R}^{4,2}$, relative to which the ESU metric is known to be conformally flat via stereographic projection (see Dirac [7]). $C(\mathrm{ESU})_0$ is $\widetilde{SU}(2,2)$, the infinite sheeted universal covering group of $SU(2,2)$. Thus, $\mathrm{ESU} = \widetilde{SU}(2,2)/P$, where $P = [SL(2,\mathbb{C}) \cdot \mathbb{R}_0^{\times}) \ltimes \mathbb{R}^{1,3}$, where $SL(2,\mathbb{C})$ acts on $\mathbb{R}^{1,3}$ through $SO(1,3)_0$.

Now, Maxw(ESU) is a module for the group $\widetilde{SU}(2,2)$, which is acting via conformal transformations. Define an elementary solution to be a field in Maxw(ESU) that transforms irreducibly under $SU(2) \times SU(2) \times \widetilde{U(1)}$. Let j run over $0, \frac{1}{2}, 1, \ldots$, and let D_j be the $SU(2)$ irreducible module of dimension $2j + 1$. Let $\mathbb{C}_{j+1}$ be the $\widetilde{U(1)}$ module of "frequency" $j + 1$, i.e. $e^{\sqrt{-1}\,t}$ acts via $e^{\sqrt{-1}(j+1)t}$. Then, the elementary solutions classify into $SU(2) \times SU(2) \times \widetilde{U(1)}$ modules of four types:

1.2 i) $D_j \otimes D_{j+1} \otimes \mathbb{C}_{j+1}$

ii) $D_j \otimes D_{j+1} \otimes \mathbb{C}_{-(j+1)}$

iii) $D_{j+1} \otimes D_j \otimes \mathbb{C}_{j+1}$

iv) $D_{j+1} \otimes D_j \otimes \mathbb{C}_{-(j+1)}$,

where again $j = 0, \frac{1}{2}, 1, \ldots$. Each module above occurs exactly once in Maxw(ESU). Let λ index a basis for the algebraic direct sum of the modules of positive frequency type, i.e. i) and iii).

We normalize our basis $F(\lambda)$ as follows: first, let F_1 and F_2 be in Maxw(ESU).

Write $F_i = dA_i$, $i = 1$ or 2, where A_i is a smooth 1-form on ESU. Let

..3 $$B(A_1,A_2) = \frac{1}{2\sqrt{-1}} \int_{S^3} (A_1 \wedge * d\bar{A}_2 - \bar{A}_2 \wedge * dA_1)$$

which depends only on F_1 and F_2, so that we define

..4 $$B(F_1,F_2) = B(A_1,A_2).$$

We now normalize our positive frequency basis so that

$$B(F(\lambda),F(\lambda')) = \delta_{\lambda\lambda'}.$$

hat the inner product is positive definite can be seen by observing that elementary olutions satisfy

.5 $$*F = \pm\sqrt{-1}\, F$$

self-or anti-self-dual). Thus,

.6 $$\begin{aligned} B(F,F) &= \frac{1}{2\sqrt{-1}} \int_{S^3} (A \wedge *\bar{F} - \bar{A} \wedge *F) \\ &= \operatorname{Im} \int_{S^3} A \wedge *\bar{F} \\ &= \operatorname{Im} \int_{S^3} A \wedge \mp\sqrt{-1}\, F \\ &= \operatorname{Im} \int_{S^3} A \wedge \mp\sqrt{-1}\, d\bar{A} \\ &= \operatorname{Re} \int_{S^3} A \wedge d\bar{A} \\ &= \operatorname{Re} \int_{S^3} \langle A, *_3 d\bar{A} \rangle v_3 \end{aligned}$$

here v_3 is the volume form on S^3 and $*_3$ is the Hodge operator on S^3. The ifferential operator $*_3 d$ acting in 1-forms on S^3 is diagonalized by the J(2) × SU(2) decomposition into time-zero elementary solutions. The eigenvalues of $*_3 d$ are negative for the type i) solutions, which are self-dual. The eigenvalues of $*_3 d$ are positive for the type iii) solutions, which are anti-self-dual. Hence, solutions of type i) and iii) have positive norm.

A general real solution to Maxwell's equations can now be written:

1.7 $$F = \sum_\lambda (F(\lambda)a(\lambda) + \bar{F}(\lambda)\bar{a}(\lambda)].$$

As λ runs over the index set, $F(\lambda)$ gives positive frequency "modes" and $\overline{F(\lambda)}$ gives negative frequency modes on ESU. We now quantize the electromagnetic field by the Dirac-Fock prescription: let

1.8 $$\hat{F} = \sum_\lambda [F(\lambda)\hat{A}(\lambda) + \overline{F(\lambda)}\hat{A}^*(\lambda)],$$

where $\hat{A}(\lambda)$ and $\hat{A}^*(\mu)$ are operators satisfying the canonical commutation relations:

1.9 $$[\hat{A}(\lambda),\hat{A}^*(\mu)] = \delta_{\lambda,\mu}$$
$$[\hat{A}(\lambda),\hat{A}(\lambda')] = 0$$
$$[\hat{A}^*(\mu),\hat{A}^*(\mu')] = 0.$$

At a particular point x in ESU we will have

1.10 $$\hat{F}(x) = \sum_\lambda [F(\lambda)(x)\hat{A}(\lambda) + \overline{F(\lambda)}(x)\hat{A}^*(\lambda)].$$

The infinite sum will <u>not</u> converge to an operator, so we regard $\hat{F}(x)$ as a formal symbol. We can for example formally compute the commutator

1.11 $$[\hat{F}_{\mu\nu}(x),F_{\sigma\tau}(y)] = \Delta_{\mu\nu\sigma\tau}(x,y)\cdot\hat{1}$$

and find that $\Delta_{\mu\nu\sigma\tau}$ is a real scalar valued distribution on ESU × ESU.

§2. FIELDS IN GENERAL RELATIVITY: THE SPINOR-TENSOR CALCULUS (See [14])

Once again, let M be a 4-dimensional Lorentzian manifold, which we now assume to be diffeomorphic to $S \times \mathbb{R}$, where S is a 3-manifold. We also assume that for each time $t \in \mathbb{R}$, $S \times \{t\}$ is space-like (i.e. Riemannian), and each line $\{s\} \times \mathbb{R}$ is time-like. Finally, assume S is orientable. It follows that S and $S \times \mathbb{R}$ are parallelizable: there exists on M a smooth $\mathbb{R}^{1,3}$-valued 1-form V such that for any x in M, $V_x : T_x M \to \mathbb{R}^{1,3}$ is invertible and orientation preserving, and if X and Y are vectors in $T_x M$, then $g(X,Y) = \langle V(X),V(Y)\rangle$, where

<,> is the standard bilinear form on Minkowski space, $\mathbb{R}^{1,3}$. The field V isn't unique, since one can alter it by any function ϕ in $G = C^\infty(M, SL(2,\mathbb{C}))$, where $SL(2,\mathbb{C})$ operates on $\mathbb{R}^{1,3}$ through $SO(1,3)_0$. The "gauge group" G operates transitively on the choices of V.

Let W be any finite dimensional real or complex $SL(2,\mathbb{C})$-module. Then, W-fields are elements of $C^\infty(M,W)$. Acting on these fields are <u>two</u> groups: $\mathrm{Diff}(M)$ and G. We think of W-fields as spinor fields. We also have tensor fields, the sections of the tensor algebra on $TM \oplus T^*M$. However, once we choose V, we may convert sections of TM to W-fields; moreover, g identifies TM with T^*M.

A connection 1-form is an $sl(2,\mathbb{C})$-valued 1-form ω on M, where $sl(2,\mathbb{C})$ is the (real) Lie algebra of $SL(2,C)$. We define covariant derivatives as follows:

1) If ψ is a spinor field, X a vector field, write

2.1 $$\nabla_X \psi = X\psi + \omega(X)\psi,$$

where X acts on ψ via differentiation and $\omega(X)$ acts on ψ through a linear transformation of W.

Assume ω and V together satisfy E. Cartan's first structural equation (condition of zero torsion):

2.2 $$dV = \omega \wedge V, \quad \text{or}$$
$$dV(X,Y) = \omega(X)V(Y) - \omega(Y)V(X),$$

for X and Y vector fields. Then, ω is determined <u>uniquely</u>: if X, Y, and Z are vector fields,

2.3 $$\langle V(X), \omega(Y)V(Z)\rangle = \frac{1}{2}\{\langle V(X), dV(Y,Z)\rangle - \langle V(Y), dV(Z,X)\rangle + \langle V(Z), dV(X,Y)\rangle\}.$$

Thus, the space (M,g,V) comes equipped with a unique $sl(2,\mathbb{C})$-valued torsionless connection 1-form and covariant derivative operator ∇.

2) We now think of ∇ as taking W-valued spinor fields to $W \otimes \mathbb{R}^{1,3}$-valued spinor fields.

We call a field a Dirac spinor if it takes values in the $SL(2,\mathbb{C})$ module $\mathbb{C}^2 \oplus \bar{\mathbb{C}}^2$. For each vector v in $\mathbb{R}^{1,3}$ we can associate a linear transformation (v) of $\mathbb{C}^2 \oplus \bar{\mathbb{C}}^2$ to itself in such a way that

$$\gamma(v_1)\gamma(v_2) + \gamma(v_2)\gamma(v_1) = 2\langle v_1, v_2\rangle 1.$$

Thus, there is a map

$$\gamma: (\mathbb{C}^2 \oplus \bar{\mathbb{C}}^2) \otimes \mathbb{R}^{1,3} \to \mathbb{C}^2 \oplus \bar{\mathbb{C}}^2$$

such that if ψ is a Dirac spinor, and v is a vector in $\mathbb{R}^{1,3}$, then

$$\gamma(\psi \otimes v) = \gamma(v)\psi.$$

The map γ will be an $SL(2,\mathbb{C})$-module map.

We now have all the ingredients for the Dirac equation in general relativity: given M, g, V, and the operator ∇ defined above, we call $\gamma\circ\nabla$, restricted to Dirac spinors, the covariant Dirac operator. The Dirac equation reads

2.4 $$\frac{1}{\sqrt{-1}}(\gamma\circ\nabla)\psi = m\psi,$$

where m is a constant (a mass). (When $m = 0$, the equations split into two and become conformally invariant.)

Associated with the Dirac equation is the Dirac inner product

2.5 $$B(\psi_1,\psi_2) = \int_{S=S\times\{0\}} j(\psi_1,\psi_2),$$

where $j(\psi_1,\psi_2)$ is the 3-form on M associated to the $\mathbb{R}^{1,3}$-field

$$\langle\gamma^*\circ\psi_1,\psi_2\rangle,$$

with $\gamma^*: \mathbb{C}^2 \oplus \bar{\mathbb{C}}^2 \to (\mathbb{C}^2 \oplus \bar{\mathbb{C}}^2) \otimes \mathbb{R}^{1,3}$ the adjoint to γ above, and $\langle,\rangle$ the $SL(2,\mathbb{C})$ invariant Hermitian form on $\mathbb{C}^2 \oplus \bar{\mathbb{C}}^2$. Solutions of the Dirac equation are normalized via the Dirac inner product, by analogy with our discussion in Section §1.

§3. COVARIANT LINEAR DIFFERENTIAL EQUATIONS

We now list some general properties of the spinor-tensor calculus on our 4-manifold M (= $S \times \mathbb{R}$, but not in the metrical sense). For any finite dimensional $SL(2,\mathbb{C})$-module W, we write $[W]$ for the space $C^\infty(M,W)$ of W-fields. Let $M_0 = \mathbb{R}^{1,3}$ with the standard action of $SL(2,\mathbb{C})$ (through $SO(1,3)_0$), and let $M = M_0 \otimes_{\mathbb{R}} \mathbb{C}$, complex Minkowski space. Let

$$\nabla: [W] \to [M \otimes_{\mathbb{C}} W]$$

be the covariant derivative as before (W will be complex from now on, unless we state otherwise.) Let $<,>$ be the Hermitian pairing between W and its complex contragredient module W^*. Then,

3.1 1) For $\phi \in [W_1]$, $\psi \in [W_2]$,

$$\nabla(\phi \otimes \psi) = (\nabla\phi) \otimes \psi + \phi \otimes (\nabla\psi).$$

2) For $\phi \in W$, $\psi \in W^*$,

$$\nabla<\phi,\psi> = <\nabla\phi,\psi> + <\phi,\nabla\psi>$$

in $[M]$.

3) For $f \in [\mathbb{C}]$,

$$\alpha\circ\nabla^2 f = 0,$$

where $\alpha: M \otimes M \to \Lambda^2 M$ is the antisymmetrization operator. (Property 3) says ω has zero torsion).

4) If $A^0 \overset{d}{\to} A^1 \to \cdots$ is the de Rham complex of exterior forms on M, one can form an equivalent complex $[\Lambda^0 M] \overset{\delta}{\to} [\Lambda^1 M] \overset{\delta}{\to} \cdots$ with $\delta = \rho\circ\nabla$, where $\rho: M \otimes \Lambda^* M \to \Lambda^{*+1} M$ is again antisymmetrization; we will have $\delta^2 = 0$. (See property 3) for a special case).

5) Write $\nabla = \nabla_V$ to emphasize the dependence on V. Suppose $\psi \in [W]$ and $g \in G$, where again $G = C^\infty(M, SL(2,\mathbb{C}))$. Then,

$$\nabla_{gV}(g\psi) = g\nabla_V\psi.$$

Keeping the above properties in mind, we now consider a general covariant differential operator of the form

$$D = \sum_{i=0}^{m} \varkappa_i \nabla^i : [W_1] \to [W_2],$$

where $\varkappa_i \in \mathrm{Hom}_{SL(2,\mathbb{C})}(M^{\otimes i} \otimes W_1, W_2)$ factors through $\mathrm{Sym}^i M \otimes W_1$. Formally adjoint to ∇ is the operator $\nabla^* = \mathrm{Tr}\circ\nabla$, where $\mathrm{Tr}: [M \otimes M \otimes W^*] \to [W^*]$ is induced by the $SL(2,\mathbb{C})$ map $M \otimes M \to \mathbb{C}$. For D as above, the formal adjoint is $D^* = \Sigma(\nabla^*)^i \varkappa_i^*$, which can in turn be rewritten as $\sum_{i=1}^{m} \lambda_i \nabla^i$. In particular, if $W \cong W^*$, we can consider the class of formally self-adjoint operators D, with $D = D^*$.

In physics, to any such self-adjoint operator D we can associate a formal integral

3.2 $$\int_M \langle D\psi,\psi\rangle v,$$

where $v = \sqrt{|g|}\, dx_1 \wedge dx_2 \wedge dx_3 \wedge dt$ is the Lorentzian volume form on M, and ψ is a W-field. The "action" integral above remains unchanged under the rigid gauge transformation $\psi \to e^{i\alpha}\psi$, $\alpha \in R$. Following E. Noether's method we will be led to an identity of the form

3.3 $$\langle D\phi,\psi\rangle - \langle \phi, D\psi\rangle = \nabla^* J(\phi,\psi)$$

for a suitable [M]-field $J(\phi,\psi)$, which is linear in ϕ and conjugate linear in ψ. Thus, J is a covariant sesquilinear differential operator: $J:[W] \times [W] \to [M]$, and

$$J(\phi,\psi) = \Sigma\varepsilon_{ij}(\nabla^i\phi, \nabla^j\psi)$$

for appropriate $SL(2,C)$ invariant M-valued sesquilinear forms ε_{ij}. We will identify [M] with $[\Lambda^3 M]$ and in turn with A^3, the 3-forms on M. Under this identification, $\nabla^* J$ corresponds to dJ. We have a special case of E. Noether's principle:

3.4 Conservation of Current: If $D\psi = 0$, then $dJ(\psi,\psi) = 0$. More generally, if $D\phi = D\psi = 0$, then $dJ(\phi,\psi) = 0$.

We define the total charge along S of a solution ψ to $D\psi = 0$ to be

$$\int_S J(\psi,\psi) \; ,$$

when that integral converges as an improper integral (a limit of finite volume integrals). If S were already finite volume and without boundary, we could immediately use the total charge to normalize smooth solutions (see Section §1). We would have

3.5 Conservation of Total Charge: For the case S compact without boundary, and for any smooth solution ψ of $D\psi = 0$,

$$\int_{S_t} J(\psi,\psi)$$

is constant in the time t, where $S_t = S \times \{t\}$.

If S is noncompact, we will have conservation of total charge only for solutions ψ which behave well at spatial infinity along all the hypersurfaces S_t,

$t \in \mathbb{R}$. We can't guarantee the existence of such well behaved solutions; yet, such solutions will be a prerequisite for the quantization of our differential equation $D\psi = 0$.

A related problem arises when we consider the effect of an isometry T in $I(M)_0$. By convention, T operates on a spinor field ψ via

$$\psi^T(x) = \psi(T^{-1}x).$$

At the same time, T operates on our field V:

$$V^T = T^*V,$$

where T^*V is the M_0-valued 1-form obtained by pulling back V along the smooth map T. Because T is an orientation-preserving isometry, we will have $T^*V = g_T V$ for some $g_T \in G$, $G = C^\infty(M, SL(2,\mathbb{C}))$.

By the covariance of $\nabla = \nabla_V$ we will have

$$(\nabla_V\psi)^T = \nabla_{V^T}\psi^T.$$

Thus,

$$\begin{aligned}(\nabla_V\psi)^T &= \nabla_{gV}\psi^T \\ &= \nabla_{gV}\, gg^{-1}\psi^T \\ &= g\{\nabla_V g^{-1}\psi^T\},\end{aligned}$$

or

3.6 $$g^{-1}(\nabla_V\psi)^T = \nabla_V g^{-1}\psi^T,$$

where $g = g_T$ above. Hence, if we fix V, we can define a projective representation of $I(M)_0$ on spinor fields by the new rule

3.7 $$\pi(T)\psi = g_T^{-1}\psi^T.$$

We have then, by 3.6,

3.8 $$\pi(T)\nabla_V\psi = \nabla_V\pi(T)\psi.$$

It follows from the form of our operator D that we have

3.9 $$\pi(T)D\psi = D\pi(T)\psi.$$

Hence, if $\mathrm{Sol}(D)$ is the linear space of smooth solutions of $D\psi = 0$, $\pi(T)$ leaves

Sol(D) stable for each $T \in I(M)_0$.

Let us turn now to the "current," $J(\phi,\psi)$. Because of equation 3.9, we will have

$$\begin{aligned} & J(\phi^T,\psi^T) \\ &= J(\pi(T)\phi,\pi(T)\psi) \\ &= T^*J(\phi,\psi), \end{aligned}$$

where T^*J is the pullback along T of the 3-form J. Hence, for well behaved ϕ and ψ, we will have

$$\int_S J(\phi^T,\psi^T) = \int_S T^*J(\phi,\psi) = \int_{T^{-1}S} J(\phi,\psi).$$

Consider again the case S compact without boundary. Then, since T is in $I(M)_0$, $T^{-1}S$ will be homologous to S inside M. Consequently we will have

3.10 Invariance of Total Charge: For the case S compact without boundary, and for a pair of smooth solutions ϕ and ψ of $D(-) = 0$,

$$\int_S J(\pi(T)\phi,\pi(T)\psi) = \int_S J(\phi,\psi)$$

for any isometry $T \in I(M)_0$.

If S is noncompact, we will have invariance of total charge only for solutions ϕ and ψ which behave well at infinity along all the hypersurfaces TS, $T \in I(M)_0$. Again, we can't guarantee the existence of such solutions.

Aside from problems with noncompact space-like S, we encounter an algebra problem: the Hermitian form $\int_S J(\psi,\psi)$ on well behaved solutions may be indefinite; the form may even be degenerate, in that a special solution may be orthogonal to all well behaved solutions. If we regard Maxwell's theory as the physics of the A-field (see §1), then the inner product 1.3 is of the type we are considering in our general theory. If $M = S^3 \times \mathbb{R}$ as in §1, the negative frequency modes will have negative total charge. In addition, there are the "pure gauge" fields $A = d\beta$ which are orthogonal to all of Maxw(ESU).

§4. MASSIVE SCALAR PARTICLES

The "massive" scalar wave equation on a space-time M is

4.1
$$D\phi \stackrel{\text{def.}}{=} (\nabla^*\nabla - \frac{1}{6}R + \mu^2)\phi = 0,$$

where R is the scalar curvature of the metric g, and μ is a constant (the "mass") (assume μ^2 is real). The total charge is

4.2
$$B(\phi,\phi) = -\mathrm{Im}\int_S \phi(*d\bar{\phi}),$$

where $*$ is the Hodge star operator acting on 1-forms on M. (If $D\phi = 0$, $d(\mathrm{Im}\ \phi * d\bar{\phi}) = 0$). Note that $B(\bar{\phi},\bar{\phi}) = -B(\phi,\phi)$, so that B is formally indefinite.

<u>A special case</u>: Take S compact with $ds^2 = dt^2 - d\sigma^2$, $d\sigma^2$ a fixed Riemann metric on S. Then $\nabla^*\nabla = \frac{\partial^2}{\partial t^2} - \Delta$, where Δ is the Laplacian on $(S, d\sigma^2)$. R is time independent and thus gives a smooth function on S. The wave equation becomes

4.3
$$(\Delta + \frac{1}{6}R - \mu^2)\phi = \frac{\partial^2\phi}{\partial t^2}$$

and one can separate variables by finding solutions of the form

$$\phi(s,t) = \phi(s)e^{i\omega t},$$

where $\phi(s)$ satisfies the elliptic partial differential equation

$$-(\Delta + \frac{1}{6}R - \mu^2)\phi = \omega^2\phi.$$

There will be a constant c such that if $\mu^2 > c$, then

$$-\Delta - \frac{1}{6}R + \mu^2$$

is a positive operator in $L^2(S)$, with spectrum $\omega_1^2 < \omega_2^2 < \cdots < \omega_n^2 < \cdots$ with each ω_n a <u>real</u> <u>positive</u> number. The eigenspace S_n attached to ω_n^2 will be finite dimensional for each n; $L^2(S)$ will be an <u>orthogonal</u> direct sum of the spaces S_n.

Suppose we have solutions ϕ and ψ of $D\phi = 0$ such that

$$\phi(s,t) = \phi(s)e^{i\omega_n t} \quad \text{with}$$

$$\phi(s) \ \text{in}\ S_n$$

and

$$\psi(s,t) = \psi(s)e^{i\omega_m t} \quad \text{with}$$

$$\psi(s) \ \text{in}\ S_m.$$

The total charge inner product between ϕ and ψ will be

$$B(\phi,\psi) = \frac{-1}{2\sqrt{-1}} \int_S \phi * d\bar{\psi} - \bar{\psi} * d\phi$$
$$= \frac{1}{2\sqrt{-1}} \int_S (\bar{\psi} \frac{\partial\phi}{\partial t} - \phi\frac{\partial\bar{\psi}}{\partial t})\nu$$

where ν is the volume form on S. We have

$$B(\phi,\psi) = \begin{cases} 0 & \text{if } n \neq m \\ \omega_n \int\bar{\psi}\phi\nu & \text{if } n = m. \end{cases}$$

Hence, we can find a sequence $\phi_1,\phi_2,\ldots,\phi_k,\ldots$ of solutions of $D\phi = 0$ such that:

4.5 a) $\phi_k(s,t) = \phi_k(s)e^{i\omega_{n(k)}t}$

b) $B(\phi_i,\phi_j) = \delta_{ij}$

c) Any smooth solution ϕ of $D\phi = 0$ can be expanded into a series

$$\phi(x) = \sum_{k=1}^{\infty} (\phi_k(x)a_k + \overline{\phi_k(x)}\ \bar{b}_k),$$

where a_k and b_k are independent complex numbers.

d) $L^2(S)$ is spanned by the functions $\phi_1(s),\phi_2(s),\ldots,\phi_k(s),\ldots$.

We now introduce the neutral quantum field,

4.6 $$\hat{\phi}(x) = \sum_{k=1}^{\infty} \phi_k(x)A_k + \overline{\phi_k(x)}A_k^*,$$

where the A_k and A_k^* are the standard creation and annihilation operators in the Fock space $F = \text{Sym}\ \widetilde{Sol(D)}_+$, where $Sol(D)_+$ is the pre-Hilbert space spanned by $\phi_1,\phi_2,\ldots,\phi_k,\ldots$, $\widetilde{Sol(D)}_+$ is the completion in the B-norm, and "Sym" means symmetric Hilbert space on — . We have

4.7 i) $[A_i,A_k] = [A_i^*,A_k^*] = 0$

ii) $[A_i,A_k^*] = \delta_{ik}$

iii) A_k^* is adjoint to A_k

iv) For a unique state $0\rangle$, the "vacuum", we have

$$A_k^*|0\rangle = 0 \quad \text{and} \quad A_k|0\rangle = \phi_k\rangle.$$

Also, $\langle 0|A_i^*A_k|0\rangle = \delta_{ik}$ and $\langle 0|A_iA_k|0\rangle = 0$. We can associate to $\hat{\phi}$:

1) The two-point function $\langle 0|\hat{\phi}(x)\hat{\phi}(y)|0\rangle$

$$= \sum_k \langle 0|\overline{\phi_k(x)}A_k^*\phi_k(y)A_k|0\rangle$$

$$= \sum_k \overline{\phi_k(x)}\phi_k(y),$$

a reproducing kernel for $Sol(D)_+$.

2) The commutator, $[\hat{\phi}(x),\hat{\phi}(y)]$, which turns out to be a scalar-valued function,

$$<0|\hat{\phi}(x)\hat{\phi}(y)|0> - <0|\hat{\phi}(y)\hat{\phi}(x)|0>.$$

3) The propagator

$$S(x,y) = \begin{cases} <0|\hat{\phi}(x)\hat{\phi}(y)|0> & \text{if} \quad t(x) > t(y) \\ <0|\hat{\phi}(y)\hat{\phi}(x)|0> & \text{if} \quad t(x) < t(y) \end{cases}$$

where $t(x)$ is the time-coordinate of the point $x \in M$.

One can ask to what extent the three functions above resemble qualitatively their analogs for flat Minkowski space-time, where a Fourier integral expansion replaces the discrete expansion, 4.6. For example, does the commutator vanish if x and y are not causally related i.e. x and y cannot be connected by a time-like curve in M? If the commutator so vanishes, does the propagator extend to a distribution on all of $M \times M$?

Example 4.8. <u>The deSitter spaces</u>:

a) 4+1 case:

$$SO(4,1)_e/SO(3,1)_e = Sp(1,1)/SL(2,\mathbb{C}).$$

Here $M = S^3 \times \mathbb{R}$, but the metric is <u>not</u> static:

$$ds^2 = dt^2 - (\cosh^2 t)d\sigma^2.$$

The basic difficulty is that we cannot separate solutions into positive and negative frequency components.

b) 3+2 case:

$$SO(3,2)_e/SO(3,1)_e = Sp(4,\mathbb{R})/SL(2,\mathbb{C}).$$

Here $M = H^3 \times \mathbb{R}$, where H^3 is hyperbolic 3-space: however, the static metric is not a product metric:

$$ds^2 = \cos t^2 r\, dt^2 - d\sigma^2$$

where r is the radial distance from the origin in H^3, the origin being the point fixed by the rotation subgroup--SO(3)--of $SO(3,2)_e$ that commutes with the time translation subgroup--SO(2)--in $SO(3,2)_e$.

The 3+2 or "anti-deSitter" space has closed time-like loops. One can pass to the universal covering space,

$$\widetilde{SO}(3,2)_e/SL(2,\mathbb{C})$$

in which there are no time-like loops. However, the noncompactness of $S = H^3$ leads to difficulties in making sensible our formal "total charge" inner product, B(,). Given the form of the metric, we find that

4.9 $$B(\phi,\psi) = \frac{1}{2\sqrt{-1}}\int(\bar{\psi}\frac{\partial\phi}{\partial t} - \phi\frac{\partial\bar{\psi}}{\partial t})\frac{v}{\cosh r}$$

$$= \frac{1}{2\sqrt{-1}}\int\int(\bar{\psi}\frac{\partial\phi}{\partial t} - \phi\frac{\partial\bar{\psi}}{\partial t})\frac{\sinh^2 r}{\cosh r}dr d\Omega$$

where $d\Omega$ is the volume form on the 2-sphere. The explicit form above suggests we look at C^∞ solutions ϕ of the massive wave equation such that for any element u in $\mathcal{U}(\mathfrak{g})$, with $\mathfrak{g} = so(3,2)$, $e^{r/2}(u\phi)(s,0)$ vanishes rapidly as $r \to \infty$. Let us call the space of solutions $Sol(D)_{-1/2}$. Then, if ϕ and ψ are in $Sol(D)_{-1/2}$, $B(\phi,\psi)$ is a convergent integral. In fact, B(,) defines a Hermitian form on $Sol(D)_{-1/2}$ -- however, it is not always the case that B(,) is $\mathfrak{g}$-invariant on $Sol(D)_{-1/2}$. In other words, the local conservation law, $d(\bar{\psi}*d\phi - \phi*d\bar{\psi}) = 0$, for ϕ and ψ in Sol(D), does not necessarily lead to $\mathfrak{g}$-invariance of the global integral $B(\phi,\psi)$. A well-studied case is the so-called massless equation: $\mu = 0$. One can write explicit pairs of solutions of $D\phi = 0$, $\mu = 0$, such that both solutions ϕ and ψ lie in $Sol(D)_{-1/2}$, but $B(\phi,\psi)$ is <u>not</u> conserved under time translation.

4.10. <u>Conservative submodules</u>: We define a $\mathcal{U}(\mathfrak{g})$ submodule $\mathcal{W}$ of $Sol(D)_{-1/2}$ to be conservative if the restriction of B to $\mathcal{W}$ is $\mathfrak{g}$-invariant and positive-definite; moreover, the group $\widetilde{K} = \widetilde{SO}(3) \times \widetilde{SO}(2)$ should act on $\mathcal{W}$ in such a way that each ϕ in $\mathcal{W}$ generates a finite dimensional, completely reducible $\widetilde{K}$-module.

<u>Proposition 4.11</u> (Fronsdal [9]): For each real $\mu \geq 0$, $Sol(D(\mu))_{-1/2}$ contains at least one and possibly two conservative submodules $\mathcal{W}$, each of which is $\mathfrak{g}$-irreducible. Moreover, the $\widetilde{SO}(2)$ spectrum is always positive, i.e. $\frac{1}{\sqrt{-1}}\frac{\partial}{\partial t}$ has positive (discrete) spectrum in the known conservative submodules. Finally, the

restriction map $\phi \to \phi|S$ maps each W injectively to a dense subspace of

$$L^2(S, \frac{1}{\cosh r}\nu).$$

Thus, for each $\mu \geq 0$, we can quantize the massive scalar wave equation in one or possibly two ways: choose an orthonormal basis $\phi_1, \phi_2, \ldots, \phi_k, \ldots$ for a conservative submodule W (obtained from the Proposition); introduce creation and annihilation operators as before and write

4.21 $$\hat{\phi}(x) = \sum_{k=1}^{\infty} (\phi_k(x)A_k + \overline{\phi_k}(x)A_k^*).$$

We may take Fock space, F, i.e. the symmetric Hilbert space on $\widetilde{W}$ (the completion of W) as the Hilbert space on which we represent the commutation relations for the creation and annihilation operators (see 4.7). By a general result of Harish-Chandra's, the Lie algebra representation of g on W will "integrate" to a unitary representation of $G = \widetilde{SO}(3,2)_e$ on $\widetilde{W}$. Because Fronsdal's module W is irreducible as a g-module, $\widetilde{W}$ will be irreducible as a group representation module.

The G action on $\widetilde{W}$ extends functorially to a G action on $F = \mathrm{Sym}\, \widetilde{W}$. Write

$$g \mapsto U(g)$$

for the operator that represents g on F. Then one can show that (formally)

4.13 $$\hat{\phi}(g^{-1}x) = U(g)^{-1}\hat{\phi}(x)U(g)$$

for any $x \in M$ and any $g \in G$. Thus, our quantization is G-equivariant.

It follows immediately that the associate generalized functions $\langle 0|\hat{\phi}(x)\hat{\phi}(y)|0\rangle$, $[\hat{\phi}(x),\hat{\phi}(y)]$, and $S(x,y)$ (see after 4.7), are all examples of G invariant kernels, i.e. generalized functions $K(x,y)$ on $M \times M$ such that for any $g \in G$, $K(g^{-1}x, g^{-1}y) = K(x,y)$. The two-point and commutator functions furthermore satisfy the differential equations

4.14 $$D_x K(x,y) = D_y K(x,y) = 0,$$

where again D is our massive scalar wave operator. If we fix a base point x_0 on M, and let H be the isotropy group at x_0, then we can define three H-invariant distributions on M:

4.15 $$\langle 0|\hat{\phi}(x)\hat{\phi}(x_0)|0\rangle,$$

$$[\hat{\phi}(x),\hat{\phi}(x_0)],$$

and $$S(x,x_0),$$

the first two of which satisfy

4.16 $$D_x K(x) = 0.$$

It would be interesting to check that S really lives on all of $M \times M$, and not just on points (x,y) with $t(x) \neq t(y)$ (see [2]). Then we could ask to determine the distribution

$$D_x S(x,x_0).$$

In the (limiting) case of Minkowski space, $D_x S(x,x_0) = \delta_{x_0}$, the delta-function at x_0.

As a final remark, one can consider $3 = 2 + 1$ dimensional anti-deSitter space:

$$\widetilde{SO}_e(2,2)/\widetilde{SO}_e(2,1),$$

which turns out to be the group manifold $\widetilde{SL}(2,\mathbb{R})$, studied for example in Paul Sally's thesis [13]. If we carry out the same scalar wave analysis as before, we obtain as two-point functions the <u>global characters</u> of a continuous family of unitary irreducible representations of $\widetilde{SL}(2,R)$. We find that the commutator kernel $[\hat{\phi}(x),\hat{\phi}(y)]$ vanishes for pairs of points x and y that can't be connected by a time-like curve.

Returning to $\widetilde{SO}(3,2)_e/\widetilde{SO}(3,1)_e$, we can ask to identify our two-point functions $\langle 0|\hat{\phi}(x)\hat{\phi}(y)|0\rangle$ in terms of the generalized spherical functions studied by several harmonic analysts (see [8]). For the case $\mu = 0$, we expect the following situation: first, we know from Fronsdal's work that there are two conservative submodules W_1 and W_2 in $Sol(D)_{-1/2}$. Let $C_1(x,y)$ and $C_2(x,y)$ be the corresponding commutator kernels. Then we should have

$$C_1(x,y) + C_2(x,y) = 0$$

unless x and y can be connected by a <u>null</u> curve in M. This expected vanishing should be a form of Huyghen's principle (see Prof. Helgason's talk in this proceedings.) The existence of invariant eigendistributions which are not locally L^1

functions is one of the basic difficulties for those who study semisimple symmetric spaces such as anti-deSitter space (see Prof. Flensted-Jensen's talk in this proceedings).

REFERENCES

[1] E. Angelopoulos, M. Flato, C. Fronsdal, and D. Sternheimer. Massless particles, conformal group, and deSitter universe. Physical Review D, Vol. 23, No. 6, (1981), 1278-1289.

[2] S.J. Avis, C.J. Isham, and D. Storey. Quantum field theory in anti-deSitter space-time. Physical Review D, Vol. 18, No. 10, (1978), 3565-3576.

[3] N.D. Birrell and P.C.W. Davies. Quantum fields in curved space. Cambridge Monographs on Mathematical Physics, Vol. 7, Cambridge U. Press, N.Y., (1982).

[4] P. Breitenlohner and D.Z. Freedman. Stability in gauged extended supergravity. Physics Letters 115B, (1982), 197-201.

[5] Davydov, A.S. Quantum Mechanics, especially Chapter XIV. NEU Press Technical Translation Series, (1966).

[6] Dirac, P.A.M. The electron wave equation in deSitter space. Annals of Math., Vol. 36 (1935), 657-669.

[7] Dirac, P.A.M. Wave equations in conformal space. Annals of Math., Vol. 37 (1936), 429-442.

[8] M. Flensted-Jensen. These Proceedings.

[9] Fronsdal, C. Semisimple gauge theories and conformal gravity. Preprint. To appear in "Group Theory and its Applications to Physics and Mathematical Physics," Proceedings of XIV AMS-SIAM Summer Seminar, University of Chicago (1982).

[10] Harish-Chandra, Representations of semi-simple Lie groups I, Trans. Amer. Math. Soc. 75 (1953), 185-243.

[11] Jost, R. The General Theory of Quantized Fields. A.M.S., Providence R.I. (1965).

[12] G. Mack and I. Todorov. Irreducibility of the ladder representations of U(2,2) when restricted to the Poincare subgroup. Journal of Math. Physics, Vol. 10, No. 11 (1969), 2078-2085.

[13] P.J. Sally, Jr. Analytic continuation of the irreducible unitary representations of the universal covering group of SL(2,R). Memoirs of the Amer. Math. Soc. No. 69, A.M.S., Providence, R.I. (1967), 1-94.

[14] J. Weber. General Relativity and Gravitational Waves. Interscience Tracts on Physics and Astronomy, No. 10. London, (1961).

[15] R.O. Wells, Jr. Complex manifolds and mathematical physics. Bull. Amer. Math. Soc. 1 (new series), (1979), 296-336.

[16] G.J. Zuckerman. Induced representations and quantum fields. Preprint. To

appear in "Group Theory and its Applications to Physics and Mathematical physics," Proceedings of XIV AMS-SIAM Summer Seminar, University of Chicago (1982).

[17] G.J. Zuckerman. Non-compact groups and irreducible representations. Proceedings of XI International Colloquium on Group Theoretical Methods in Physics, Istanbul 1982. Edited by M. Serdaroglu. Lecture Notes in Physics, Vol. 180. Springer-Verlag (1983), 1-7.

*The author is supported by NSF grant #MCS80-05151 and by the Alfred P. Sloan Foundation.